Advances in Science, Technology & Innovation

IEREK Interdisciplinary Series for Sustainable Development

Advances in Science, Technology & Innovation (ASTI) is a series of peer-reviewed books based on important emerging research that redefines the current disciplinary boundaries in science, technology, and innovation (ASTI) in order to develop integrated concepts for sustainable development. It not only discusses the progress made towards securing more resources, allocating smarter solutions, and rebalancing the relationship between nature and people, but also provides in-depth insights from comprehensive research that addresses the 17 sustainable development goals (SDGs) as set out by the UN for 2030.

The series promotes the creation and development of viable solutions for a sustainable future and a positive societal transformation with the help of integrated and innovative science-based approaches. Including interdisciplinary contributions, it presents innovative approaches and highlights how they can best support both economic and sustainable development, through better use of data, more effective institutions, and global, local, and individual action, for the welfare of all societies.

The series particularly features conceptual and empirical contributions from various interrelated fields of science, technology, and innovation, with an emphasis on digital transformation, that focus on providing practical solutions to ensure food, water, and energy security to achieve the SDGs. It also presents new case studies offering concrete examples of how to resolve sustainable urbanization and environmental issues in different regions of the world. The series is intended for professionals in research and teaching, consultancies and industry, and government and international organizations. Published in collaboration with IEREK.

The series draws on the best research papers from various IEREK's international conferences, while external conference proceedings proposals are no longer accepted as direct publications. The series does invite significant expansions of selected papers previously submitted to conferences. The Editors' role should be evident in the reviews through an intensive **double-blind peer review** process, helping authors with edits, and collaborating with them to demonstrate **great academic significance** through clear research findings and innovative solutions that go well beyond a typical conference submission.

The ASTI series welcomes external proposals and requests to publish **edited books, contributed volumes, monographs,** and **textbooks**. With the commitment to the aforementioned instructions and policy.

The ASTI series has now been accepted for Scopus (September 2020). All content published in this series will start appearing on the Scopus site in early 2021.

Bertolt Meyer · Ulrike Thomas · Olfa Kanoun
Editors

Hybrid Societies

Humans Interacting with Embodied Technologies, Volume 1

 Springer

Editors
Bertolt Meyer
Department of Psychology
Chemnitz University of Technology
Chemnitz, Germany

Ulrike Thomas
Department of Electrical Engineering
and Information Technology
Chemnitz University of Technology
Chemnitz, Germany

Olfa Kanoun
Chair of Measurement
and Sensor Technology
Chemnitz University of Technology
Chemnitz, Germany

ISSN 2522-8714   ISSN 2522-8722 (electronic)
Advances in Science, Technology & Innovation
ISBN 978-3-032-03487-8   ISBN 978-3-032-03488-5 (eBook)
https://doi.org/10.1007/978-3-032-03488-5

Funded by the Deutsche Forschungsgemeinschaft (DFG, German Research Foundation)—Project-ID 416228727—CRC 1410.

This Springer imprint is published by the registered company Springer Nature Switzerland AG
The registered company address is: Gewerbestrasse 11, 6330 Cham, Switzerland

If disposing of this product, please recycle the paper.

Preface

This book is the result of three years of interdisciplinary research at the Collaborative Research Center "Hybrid Societies—Humans Interacting with Embodied Technology" at Chemnitz University of Technology. In a world increasingly shaped by the technological transformation of our societies, the emergence of Embodied Digital Technologies (EDTs)—such as autonomous service and delivery robots, socially interactive robots, self-driving cars, virtual agents, intelligent wearables, and prostheses—heralds a new era. These technologies, characterized as artificial agents physically present in spaces traditionally occupied by humans, are becoming increasingly integral to our daily lives and are set to proliferate with profound implications for human-technology interaction.

Three years post-establishment, the CRC organized the first International Conference on Hybrid Societies. Our objective was to foster an interdisciplinary research community, extending well beyond the CRC's scope, dedicated to exploring these evolving dynamics. Hybrid societies, as envisioned, are collectives of embodied agents: humans, partly human actors like cyborgs, and non-human EDTs like robots and autonomous vehicles. These agents are capable of engaging in intelligible encounters, of interacting and communicating, and vary in autonomy, agency, and responsibility.

To address the challenges in these hybrid societies—spaces where human and non-human autonomous actors share and navigate in public environments—it is paramount to combine research from diverse fields. Humanities, social sciences, psychology, sociology, cognitive sciences, communication science, alongside engineering sciences like engineering, computer science, and physics, must converge synergically to meet these challenges.

The inaugural conference on Hybrid Societies, held in Chemnitz in March 2023, aimed to lay the groundwork for this necessary research. The interdisciplinary network of researchers that gathered focused on two pivotal research questions:

1. How do humans perceive embodied technologies, how do these perceptions influence their interactions with these technologies, and how do these interactions shape effective human-EDT interactions?
2. What are the design principles and strategies for creating accountable embodied technology, which can effectively signal to others what they can do, want, and how one should interact with them?

The introductory chapter, contributed to by a wide range of experts from the CRC, aims to lay the foundation for addressing these questions. It establishes the concept of hybrid societies and introduced a framework for exploring the key research challenges. The subsequent chapters, authored by researchers from the CRC and researchers from outside the CRC, represent initial steps in tackling these intricate challenges.

We hope that this book not only stimulates further interdisciplinary research into the foundations and boundary conditions for successful coexistence in future hybrid societies but also contributes to the concept and understanding of hybrid societies. Our vision is clear: For the fruitful cohabitation of humans and embodied technologies in society, a new generation of truly interdisciplinary research is imperative.

Chemnitz, Germany
January 2024

Contents

Abbreviations

AGV	Automated Guided Vehicle
AI	Artificial Intelligence
BPMN	Business Process Model and Notation
CDN	Content Delivery Networks
CFA	Confirmatory Factor Analysis
CMM	Capability Maturity Model
COM	Center of Mass
EDT	Embodied Digital Technologies
EPO	European Patent Office
FAIR	Findable, Accessible, Interoperable, Reusable
FIML	Full Information Likelihood Estimation
HMD	Head-Mounted Display
IPA	International Phonetic Alphabet
ISCED	International Standard Classification of Education
ITA	Italian English
LJS	LJ Speech
MaaS	Mobility as a Service
MMDT	Minnesota Manual Dexterity Test
MOS	Mean Opinion Score
PatG	German Patent Act
RWA	Right-Wing Authoritarianism
SEM	Structural Equation Modeling
TAI	Threats of Artificial Intelligence
TCT	Triangle Completion Task
TTS	Text To-Speech

Hybrid Societies: Concepts, Challenges, and Research Agenda

Bertolt Meyer⑩, Olfa Kanoun⑩, Giuseppe Sanseverino⑩,
Michael R. Müller⑩, Christian Pentzold⑩, Andreas Bischof⑩,
and Fred Hamker⑩

Abstract

As part of the technological transformation of societies, embodied digital technologies (EDTs) such as autonomous service and delivery robots, socially interactive robots, self-driving cars, virtual (service) agents, "intelligent" wearables, and protheses (to name only a few) have become a part of our daily lives and will further proliferate in the future, with wide-ranging implications for human-technology interaction. EDTs are artificial agents that are physically represented in areas previously only occupied by humans. We use the term hybrid societies for future societies in which humans, human-technological hybrids, and EDTs interact and share public spaces (e.g., roads, sidewalks, malls, and public buildings). Specifically, we define a hybrid society as a collective of embodied agents (including humans, partly human actors such as cyborgs, and non-human EDTs such as robots) with the capability to engage in intelligible encounters, who interact and communicate, who can meaningfully reference each other as members of society, and who vary in terms of autonomy, agency, and responsibility. We propose that psychological and technological factors governing effective interactions among humans and EDTs in hybrid societies can be subsumed under two research questions: (1) How do humans perceive and interact with embodied technologies? (2) What are design principles and strategies for designing accountable embodied technology, i.e., embodied technologies that effectively signal to others who encounter them what they can do, want, and how one is supposed to interact with them? We lay out a research agenda for addressing these questions and discuss potential implications.

B. Meyer (✉)
Professorship for Work, Organizational, and Economic Psychology,
Chemnitz University of Technology, Chemnitz, Germany
e-mail: bertolt.meyer@psychologie.tu-chemnitz.de

O. Kanoun
Chair of Measurement and Sensor Technology, Chemnitz University of
Technology, Chemnitz, Germany
e-mail: olfa.kanoun@etit.tu-chemnitz.de

G. Sanseverino
Institute for Media Research, Chemnitz University of Technology,
Chemnitz, Germany
e-mail: giuseppe.sanseverino@phil.tu-chemnitz.de

M. R. Müller
Professorship for Visual Communication and Media Sociology,
Chemnitz University of Technology, Chemnitz, Germany
e-mail: michael-rudolf.mueller@phil.tu-chemnitz.de

C. Pentzold
Department of Communication and Media Studies, Leipzig University,
Leipzig, Germany
e-mail: christian.pentzold@uni-leipzig.de

A. Bischof
Junior Professorship of Sociology With Specialization in Technology,
Chemnitz University of Technology, Chemnitz, Germany
e-mail: andreas.bischof@hsw.tu-chemnitz.de

F. Hamker
Professorship Artificial Intelligence, Chemnitz University of
Technology, Chemnitz, Germany
e-mail: fred.hamker@informatik.tu-chemnitz.de

Keywords

Hybrid societies · Public spaces · Technological transformation · Smooth interaction · Embodied digital technologies

1 Introduction

The past decades have witnessed significant technological advancements that continue to transform society and our everyday lives. Driving forces behind these advances include innovation in science and technology, the growth of the global economy, and the proliferation of digital technologies (e.g., [1]). This development has been widely referred to as a digital transformation of society [2].

© The Author(s) 2026
B. Meyer et al. (eds.), *Hybrid Societies*, Advances in Science, Technology & Innovation,
https://doi.org/10.1007/978-3-032-03488-5_1

Embodied digital technologies (EDTs) such as robots, autonomous vehicles, and bionic arms, have a physical body, sensors, and actuators. Specifically, we define them as are artificial agents that are physically represented in areas previously only occupied by humans [3] Contrary to non-embodied systems (e.g., chatbots like ChatGPT®), they literally touch the world, and their actions have real-world consequences. A chatbot produces text, but a driverless vehicle or a drone can injure a human being. Non-embodied technologies (e.g., smartphones) are designed for their users. In contrast, EDTs encounter individuals who are not their intended users (e.g., pedestrians encountering delivery robots on the sidewalk, so-called incidental users).

This advent and spread of embodied digital technologies such as service and delivery robots, socially interactive robots, virtual (AI-based) agents, and self-driving cars (to name only a few), has wide-ranging implications for many aspects of daily life [4], which will reshape society. We refer to societies characterized by a coexistence of humans and EDTs as hybrid societies [3, 5], in which EDTs work alongside humans (and human-technological hybrids) in everyday activities.

We posit that great challenges will accompany the inevitable and rapid transformation of contemporary societies into hybrid societies. Addressing these challenges requires interdisciplinary research among psychology, humanities, social sciences, natural sciences, computer science, and engineering. Only interdisciplinary research can contribute to shaping future hybrid societies such that embodied technologies and humans can coexist and coordinate effectively. This brief introductory chapter first defines and elaborates the related concepts of society and hybrid societies, and seeks to initiate a discourse about why embodied technologies are central to hybrid societies. In so doing, we identify two overarching and intertwined research areas and propose a research agenda to promote effective coexistence and efficient coordination in hybrid societies. These research areas are:

(a) Understanding human–robot perceptions, interactions, and embodied affordances: We require basic interdisciplinary research to understand how humans perceive embodied technologies, how they interact with robots and other non-human agents, and how these interactions impact effective coordination and communication in public spaces.

(b) Making embodied technologies accountable: As we explain below, embodied technologies are not only used by humans, but contrary to other non-embodied technology, are encountered in public space. Therefore, embodied technologies need to signal through their looks and behavior what they can do, want, and how one is supposed to interact with them—they need to be accountable for their actions. This necessitates research to determine

design principles for the accountability of embodied technologies.

Ideally, findings from both areas contribute to enabling effective communication and coordination between humans and EDTs. Achieving this requires novel approaches for efficient communication between humans and embodied technologies (e.g., in space sharing conflicts), as well as understand the interplay of diverse socio-cognitive factors that could impact communication in these contexts.

By addressing these challenges, interdisciplinary research can have a significant impact on future hybrid societies, where human actors and embodied technologies coordinate and interact with one another efficiently, accountably, and successfully. This implies research that benefits from our current understanding of flexible behavior, adaptation, joint action, communication, resilience, and cooperative problem solving, among others.

The rest of this chapter elaborates on the term hybrid society and how it hinges on the central notion of embodied technology. Subsequently, we describe the agents comprising hybrid societies, and detail the two key research areas with a focus on interactions, affordances, social perceptions of embodied technologies, and accountability design. These concepts are elaborated via links to state-of-the-art research in this volume, before we discuss implications for future research on hybrid societies.

2 Hybrid Society as a Metaphor for Interactions Between Humans and Embodied Technology

The term Hybrid Society is a useful *metaphor* for a form of situative encounters and evolving forms of cooperation in which humans frequently engage with embodied digital technologies, as we explain in the following. In contrast to standard definitions of what a society is, our view stresses the micro-level of human–machine interaction in daily life and public spaces that is increasingly happening next to highly structured (work-) environments with clear action protocols and actor roles. Moving outside of these standardized constellations, a Hybrid Society is about to take shape in the mundane and everyday encounters of humans and EDTs. These encounters come with little to no preparation or prior experience where humans and embodied technologies/artificial agents must be aware of others, anticipate aims and trajectories, and manage their movements and actions flexibly along the evolving multi-actor situation. These interaction scenarios imply the ability to meaningfully reference other members of society [6]. While a lot can be learned from research on situations that rely on virtual worlds and augmented reality scenarios [7], engaging with the factors affecting human and non-human encounters in hybrid societies requires addressing the cognitive, spatial, sensory, locomotive, gestural, and communicative dimensions of real-world situations. This begins with acknowledging that humans and EDTs possess a physical body that grounds their motor, practical, articulatory, and sensory abilities. While we acknowledge the relevance of other social levels on the meso and macro level, we focus on actual encounters and embodied interactions that form the building blocks for further organizational and structural conditions.

In this sense, our micro-level interactional usage of the hybrid society metaphor is akin to the psychological concept of society as an experience of being a member of a collective. In the psychological sense, our concept of a hybrid society is in line with a definition of society as "a collection of individuals who are mutually aware of, and at least minimally depend upon, one another. Their mutual awareness includes being cognizant of their interdependence. This, more than anything else, is what makes a group a psychological entity, not merely a physical or statistical one" [8, p. 20]. This does not exclude the possibility of some form of common culture, which we loosely conceptualize as a shared store of knowledge and expectations.

In psychological and legal terms, members of society typically possess some degree of autonomy, agency, and responsibility [3]. Agency, at its core, refers to the "capability to engage in intelligible encounters" [9, p. 1]. Therefore, the growing ability of embodied technology such as robots to engage in intelligible encounters justifies attributing a certain degree or even a certain type of agency toward them [10] and, therefore, justifies conceptualizing them as members of society possessing agency.

In sum, given the increasing proliferation of such embodied technologies that can communicate and interact with humans, we define a hybrid society as *a collective of embodied agents (including human, partly human, and non-human embodied technologies such as robots) with an ability to engage in intelligible encounters, who interact and communicate, who can meaningfully reference each other as interaction partners, and who vary in terms of autonomy, agency, and responsibility* [3].

3 Embodiment as a Precondition for Societal Interactions and Membership

Embodiment enables forms of communication and interaction that are central to partaking in society [11]. In psychology and human–robot interaction, embodiment typically refers to the physical form and appearance of a robot or other non-human agent. This includes the agent's overall appearance, shape, and size, as well as its visual, auditory, and other sensory characteristics (see [4], for a review). Importantly,

"all agents are in some way constrained by their embodiment; they are also highly dependent on affordances, 'the fundamental properties of a device that determine its way of use', which are themselves derived from embodiment [12]" [4, p. 256]. For humans, embodiment implies the possession of capabilities, whereas the embodied design and functionality of machines define their capacities. Affordances accrue in the interplay of human capabilities and machine capacities. More than being static functions of technological bodies, affordances can be understood as enablements that are realized when humans and technology interact. In this sense, affordances have been conceptualized as "collective achievements that emerge within the interplay of humans and machines" [9, p. 1]. In other words, referencing each other and interacting with each other in society are at least partially based on embodied actors' capacities and capabilities, their perception, and corresponding attributional psychological processes.

The above concept of a Hybrid Society hinges on the (social) interactions among agents. Therefore, agents in hybrid societies must be able to engage in effective social interactions. Embodiment enables robots and other embodied technologies to interact with their environment in effective and rich ways that are *unavailable to non-embodied agents* [4], because embodied technologies can use more channels for communicating and interacting, including proxemics [13], gaze [14], and gestures (e.g., [15]). Embodied technologies not only come with affordances, but their embodiment also posits them as potential members of society, as their embodiment sustains rich forms of communication and interaction that constitute a society (among other things) as we argued above. Therefore, a hybrid society entails that non-human (or not-fully human) agents who also (potentially or in the future) belong to society are situated in society, i.e., exhibit some form of embodiment. In sum, this "being in society," this embodiment, is a *conditio sine qua non* for situated interacting, referencing, and communicating that is a constitutive element for the formation of a society in the sociological and psychological sense. Therefore, the concept of a hybrid society and the concept of embodiment necessitate each other. Thus, research on future societies that contain non-human and/or not-fully human (technological) actors is equivalent to research on embodied actors, i.e., embodied technology.

Of note, embodiment is not a binary construct in terms of either an entity exhibits embodiment or not. Instead, embodiment can be conceptualized on an axis from weak to strong [16]. Weak embodiment refers to abstractions of bodies placed in (static) abstractions of the real world, e.g., a virtual robot inside a computer program. Strong embodiment refers to an agent in the real world with the ability to coordinate and explore, who exhibits goal-oriented behavior, who can interact and communicate with the environment in a bi-directional way, and who possesses a level of understanding of the physics of the environment [10].

The above discussion of embodiment builds on the definition of embodiment as physical form and appearance. In contrast to this traditional technology-focused use of the word in the sense of *physical embodiment*, Müller and Sonnenmoser [10] propose a novel, wider-reaching use of the term in the sense of *social embodiment* or sociomorphic display. Social embodiment refers to the design, look, display, or even a behavior of a machine that implies or cues a certain social form or social role in an (upcoming) interaction. This implies signaling social forms or roles such as responsivity, servility, and solipsism through visual displays, sounds, or movements. This social embodiment can be defined as meaning that "the material composition of these machines and the way they react to their surroundings embodies one (or several) social forms, i.e., "the mode or style or formular" (…) illustrating the actual or expected relationship between a person and a machine" [10, p. 5]. For example, two blinking dots on the display on the head of a Sony Aibo® robot dog signal attention, because they embody the social enactment of signaling attention by establishing eye contact. In this way, social embodiment is the form of embodiment that could matter most in hybrid societies, where humans encounter embodied technology in public spaces: Through their looks and behavior, embodied technologies signal the role that they intend to (or are designed to) play as well as their expectations from human actors in an interaction.

4 Actors in Hybrid Societies

Before elaborating on the challenges of hybrid societies, it is necessary to explicate the embodied actors or agents that have the potential to be a member of a (future) hybrid society. Müller and Sonnenmoser [10] define an actor as "the product of a presentation (a designation, design, or ritual action) by which a non-trivially behaving corporeal or intelligible entity (e.g., a machine, a network) is ascribed properties toward which a counterpart can orient itself and which can be used as a standard of social judgment of the actor thus designated" (p. 5). In their review, Meyer et al. [3] argue that, next to humans, other actors could at least potentially achieve the psychological and legal property of *responsibility*, which, in the psychological sense, entails moral agency, explainability, and trust, and which, in the legal sense, entails a legal capacity (to act), autonomy, and liability. For these other actors, Meyer et al. use the two umbrella terms "Machine" and "Autonomous Systems." Next to these broader concepts, they employ the specific terms "Robot," "Hybrid" (i.e., Cyborgs), and "Androids," see Fig. 1.

Note that as of today, only Humans possess full psychological and legal responsibility, and that as of today, so-called hybrids or cyborgs are humans in the legal and psychological sense, because their responsibility does not depend on the fact that some parts of their body are technical in nature [3].

If we combine the potential actors of Hybrid Society from the Meyer et al. review [3] with the preceding embodiment-related arguments that interaction is central to societies and that embodiment affords artificial agents with the richest potentials for interactions, we can conclude that, next to

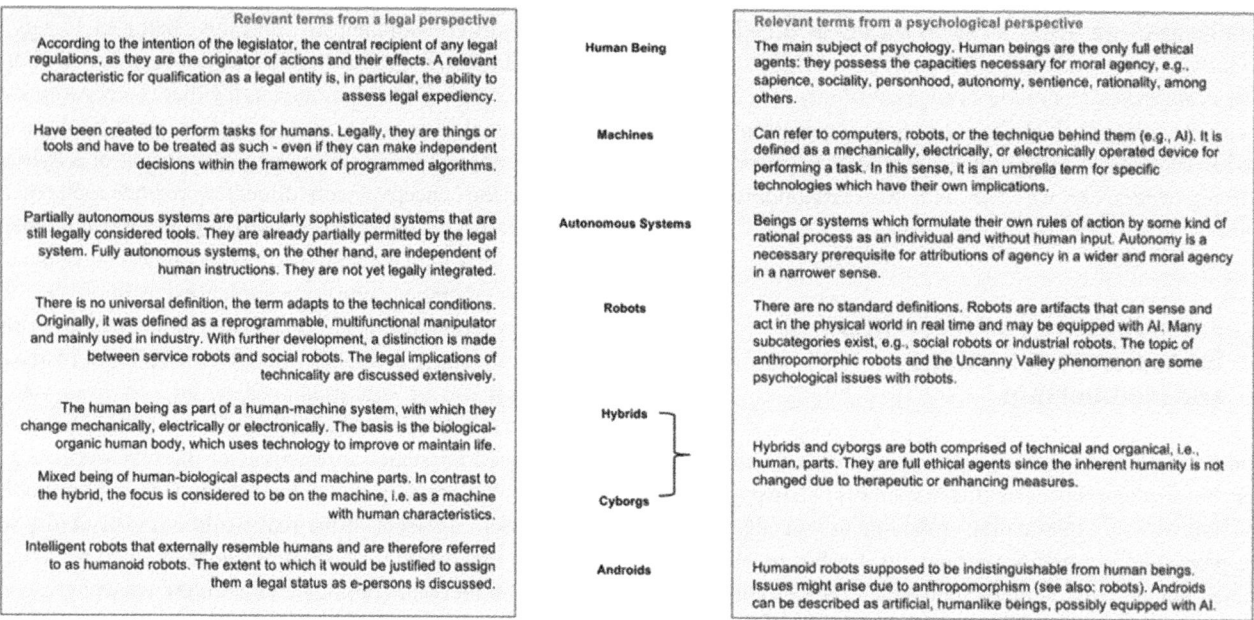

Fig. 1 Agents of hybrid societies according to Meyer et al. [3]

humans and hybrids/cyborgs, embodied actors such as robots constitute potential agents in embodied societies. Therefore, a closer definition of the term "robot" in the context of embodiment appears necessary.

In their review of embodiment in robotics, Deng et al. [4] differentiate between *collaborative* robots and *service* robots. According to these authors, collaborative robots perform dirty, dull, and dangerous work, predominantly in caged manufacturing environments. While Deng et al. acknowledge developments of collaborative robots toward human–robot collaboration at the workplace, they highlight that these robots require "trained professionals to operate them" (p. 9). For the context of our investigations into hybrid societies, industry-based collaborative robots are somewhat less important, because they are typically confined to the workplace and—in contrast to service robots—rarely enter public space and are not designed to encounter less- or non-trained other actors. Thus, while collaborative industry robots undoubtedly have tremendous impacts on value creation and the economy, their effect on society is more distal, because they interact with fewer individuals, are confined to certain places, and are typically designed to interact with certain professionals.

On the other hand, Deng et al. [4] define service robots as robots that can "provide services in everyday life, such as vacuuming and cleaning floors [17–19], folding laundry [20, 21], delivering packages [22, 23], giving museum tours [24], driving autonomously [25], and providing aid to special needs populations in the context of socially assistive robotics [26–28], along with numerous other uses" (p. 10). Therefore, service robots are much more likely to enter public space, encounter other (non-trained) individuals or actors, and are therefore more likely to interact more frequently and directly with other members of society, which justifies focusing research on hybrid societies on service robots, which, based on the above definition, include autonomous vehicles in traffic.

Despite robots being embodied, they generally have no own internal representation of their body. This creates a number of challenges for designing service robots to closely interact with humans. Similarly, while robots could be made responsible for their actions, present robots do not have a representation of agency. Thus, interdisciplinary research needs to better understand the sense of agency and body ownership in humans and implement aspects of those into a robotic minimal self [29]. Forch and Hamker [30] argue that a mechanistic brain-inspired approach to models of a robotic minimal self may have the advantage of avoiding pitfalls similar to present robot models of emotion: Robots, in the eye of the beholder, appear to have emotions, agency, or a sense of their body, but their internal representations are only simulations of those. Embodiment is also an important prerequisite of learning from others. Hybrid societies may fundamentally change, when robots are not just pre-programmed or learn for to optimize a particular task, e.g., reinforcement learning, but are allowed to explore by means of intrinsic motivation and to learn by means of observing humans. Such increasing capabilities of robots may ease human–robot interaction, but can also come with new risks that need to be anticipated.

An important aspect of the development of society toward a hybrid society is the development of the human body toward a human-technologic hybrid body: Just as societies become more hybrid (in the technological sense) and diverse (in the social sense), so are humans. Technology directly interfacing with and attached to the human body is also proliferating (e.g., [31, 32]). This includes modern bionic prostheses such as myoelectric hand prostheses, but also cochlear implants, and body-worn insulin pumps that are permanently connected to the users' circulation. Next to such therapeutic bionic technologies that restore functions that are lost due to disability or illness, bionic devices aimed at enhancing human capabilities, such as exoskeletons, are likely to become more mainstream in the near future. We see the proliferations of what Meyer et al. [3] refer to as *hybrids* (see Fig. 1) as another element of hybrid societies.

5 Challenges in Hybrid Societies: State of Art and Future Research

Society becomes more diverse on a technological level; humans and other agents have to share limited public space, interact, and communicate. The core interactional challenge of a functioning hybrid society, thus, is to ensure effective coordination, communication, and cooperation among societal actors. This implies agents' viable understanding of other agents' intentions and capabilities and corresponding adaptation of behavior in dynamic situations such as space sharing conflicts and road traffic. Ideally, in hybrid societies, the interaction between embodied technologies and humans in such scenarios increases the effectiveness of all actors. For example, in mixed autonomous/human road traffic, Jahn [5] proposes "wizard interfaces" that enable humans to provide additional inputs to automated systems, thereby making them more adaptive and resilient. For conflict situations or problems, he proposes so-called technical first aid by bystanders or passengers who may possess special training or who might be motivated by rewards (so-called Shepards). Wizard interfaces and Shepards are thus two concrete examples of how a reciprocal understanding of situational capabilities and intentionalities can help facilitate cooperation between humans and embodied technologies for efficient resolution of challenging encounters.

These examples highlight that social cognitions/perceptions and attribution processes govern human evaluations of and reactions to others, especially in first encounters. These include, to name only a few, attributions of competence, morality, and sociability, which humans apply to other humans (e.g., [33, 34]) and technology (e.g., [32]). In first encounters, such attributions are made on the basis of superficial visible characteristics of other actors' embodiment: "Their embodiments are used as tools for communication, acceptance, and engagement. These robots primarily interact through their social capabilities in order to achieve their goals. Accordingly, they must be able to both perceive [35–38] and generate communicative signals [39, 40] that their human counterparts are able to intuitively understand, relate to, and accept" [4, p. 10]. In a similar vein, Groom et al. [11, p. 842] highlight: "Bodies are salient indicators of social identity: entities that are embodied, either physically or virtually, are expected to function in the human social context." Therefore, understanding and designing the (psychological) processes surrounding embodied social perception, i.e., how others perceive and interpret embodied communication and affordances, is a central research endeavor in hybrid societies. Attributions, e.g., of properties and intentions, are the key element of communication and interaction [9] and thus constitute a central psychological process for studying hybrid societies.

Second, next to the psychological processes governing human social perception of embodied technology, research needs to better understand optimal ways in which features of the embodied technology (e.g., its looks, movements, and signals) communicate their intentions and abilities to their social environment. For this kind of design-oriented research in hybrid societies, we propose to extend the term usability to accountability (and accordingly, to accountability research or accountability experience, in short AX). In this context, the term accountability [10, 41], in a nutshell, refers to the artificial agents' performance and display of what it is, can do, and want. In public contexts, many humans do not *use* autonomous embodied technologies, but *encounter* them, e.g., on the road. In such spontaneous encounters, for effective coordination, humans must understand what the artificial body is, what it can do, what it wants, and how the human is supposed to behave toward it. For efficient coordination, the machine thus needs to successfully communicate these aspects through its design, appearance, and behavior—i.e., through its performance. If it is able to do so, the machine is *accountable*. The challenge of EDT design therefore "is not to realize *usability*, but to clarify the *accountability* of complex [...] machines to socially predictable role functions and types of behavior" [10, p. 4]. Such an assignment to specific role functions and types of behavior is not least an important source for the normative integration of EDTs into the interaction orders of modern societies.

Designing for accountability implies new research on the addressing of observers by machine designs. Here, it is important to consider both the multi-modal dimensions of technological embodiment (e.g., speech functions, gestures,

behavioral style, overall appearance, use of augmented reality) and the citation and modification of common design languages. For example, many embodied technologies employ anthropomorphic concepts and metaphors to make them accountable to everyday users. However, this strategy is by no means self-evident, nor is it necessary. In fact, a wide variety of anthropomorphic, zoomorphic, fictional, and pictographic forms are currently being used to embody the skills, functions, or states of use of particular machines. According to Müller and Sonnenmoser [10], the embodiment of situative presence, communicative responsivity, subordination or superordination, and other social forms can be accomplished by rather basic, "sociomorphic" presentations instead of sophisticated anthropomorphic designs. Sociomorphic refers to an embodiment of a social function, such as initiating contact, being present, or submission, by simple cues or gestures, such as two dots representing abstract eyes signaling an attempt to establish contact. This argument is mirrored in findings from cognitive science, "that social interactions with robots are not always the result of anthropomorphizing, i.e., the projection of imaginary or fictional human social capacities, but of sociomorphing, i.e., the perception of actual non-human social capacities" [42]. In this way, research on accountability design can—jointly with research on social perceptions of embodied technologies—result in new paths for interacting with and designing embodied technologies.

So far, this section has covered perceptual, attributional, social (i.e., psychological), and sociological research challenges for effective interactions in hybrid societies. However, for accountable EDTs, many technical engineering challenges also require addressing. Specifically, novel technologies still need to be elaborated and advanced for self-organization in hybrid societies, to make the behavior of humans and agents mutually understandable, and to enable a smooth interaction between different human and technical agents. In different contexts, the behavior of humans and agents, including their movements, actions, interactions, and communication, needs to be perceived and tracked. Humans can perceive and understand multimodal actions, predict subsequent actions, and thus recognize higher-level contexts. These abilities are currently lacking in technical agents and robots, resulting in their inability to understand human actions, impeding intuitive interactions between humans and robots.

The tracking of agents and their interaction and communication can be realized with the help of sensors, signal acquisition, and transmission. However, doing the same with humans is particularly challenging. Humans have greater variability in their actions, and the movement of some limbs, especially the fingers, which are engaged in gestures in different patterns, is both important and particularly challenging. Outside of laboratory conditions, the challenges in this regard become especially great, as the necessary technical equipment is not always available, and environmental conditions and lighting vary greatly.

For smooth self-organization in hybrid societies, it is not only necessary to perceive actions, movements, and poses. Building the basis for mutual understanding also requires interpretation, cognition, and prediction. All of these tasks require a multi-modal acquisition of eye gaze, speech, and gestures with sufficient accuracy. Classical methods based on cameras, microphones, and optical methods cannot be adopted in several contexts, especially in public spaces, due to the non-existence of infrastructure, numerous artifacts, and a systematic lack of precision. Novel approaches need to be elaborated based on embodied technologies and the detection of signals directly on humans or agents. By the elaboration of novel sensor technologies, a new generation of smart body-attached wearables can be realized, which are highly sensitive to enable the detection of micro gestures and finger movements independent of specific infrastructure. Gaze-based interaction is well-known as an important means for humans to communicate with each other and express shared attention. This modality could be adapted for intuitive human–robot communication in shared spaces. It is important to identify human gaze direction and to infer the intention underlying gaze allocation from the gaze data. Thereby, it is important to distinguish gaze shifts related to communication with others and to control gaze to express planned movements and or actions in an intuitively understandable manner.

For smooth joint action between agents and humans and agents with each other, although the individual movements are controlled separately, all agents need to coordinate their actions with respect to each other, e.g., meet at a certain position to exchange objects and grab objects with the right force level for realizing smooth transfers. Addressing these challenges requires observing, understanding, and anticipating intentions and actions. For enabling interaction in hybrid societies, agents must be highly adaptive and, most importantly, able to make flexible decisions and realize smooth motor control despite the latencies of sensors, actuators, and cognition processes. For this, novel neuro-computational models must be developed, and the sensory consequences of motor actions must be predicted.

Social interaction in cyberspace requires a high degree of resemblance between the user and their avatar design and movement control so that the avatar movements can be perceived by the observer as natural. Characteristic movement patterns of individuals need to be investigated to enable mimicking manifold individual styles as well as gender-specific and age-specific characteristics.

Sophisticated speech synthesis and recognition systems should have linguistic credibility for a higher acceptance, and are a key factor influencing human learning performance when taught by digital cyber-physical agents. Methods to investigate credibility and how it can be created are therefore central. Design recommendations need to be elaborated for the field of multimedia learning, online education, blended learning concepts, and distance learning.

Robots play an important role in hybrid societies as they interact with humans and enable telepresence and tele-manipulation. It is therefore important that they continuously preserve spatial orientation, track movements, encode locations, and adopt human-like trajectories. Especially for humanoid robots, human-like, predictable motions significantly contribute to increasing their acceptance, and for this, new methods need to be elaborated to achieve functional, collision-free, and human-like motions.

Similar to robots, vehicles need a high level of self-organization, especially in the context of increasingly important autonomous driving. Thereby, the recognition of implicit driving signals onstitutes an important basis for a proactive driving style in highly automated vehicles. For this, vehicle-to-vehicle communication can transmit relevant driving signals, so that they can be characterized in different situations as a basis for automated driving.

6 State of Art in This Volume

This volume assembles pioneering research contributing to the understanding of interactions between humans and embodied technologies and on designing for accountability that is grounded in the hybrid society metaphor to a greater or lesser extent. In the following, we mention some works that fit more narrowly into the above conceptualizations. For example, Siefkes et al. [43] address how novel approaches toward intentionality can increase the effectiveness of communication between humans and embodied technology. With a series of empirical and hypothetical examples of first encounters between humans and embodied technology in public spaces, they demonstrate that an adequate model of intentionality can smooth interaction between humans and embodied technologies and contribute to their social acceptance.

Sanseverino et al. [44] analyze human–machine interaction from a more technical perspective and propose the use of body-attached sensor networks to facilitate the recognition of human gestures by artificial agents. They demonstrate that a wearable smart band is suitable for the recognition of basic hand gestures. The work of Hensch et al. [45] contributes to effective interactions between autonomous vehicles and other road users by investigating the communication behavior of human drivers. Their results show that human road users predict the development of the traffic situation using implicit cues. Therefore, they propose that effective coordination between autonomous vehicles and human road users requires that automated vehicles understand the human divers' implicit cues and adapt their behavior accordingly.

On a micro-level, Gäbert et al. [46] investigate cooperation between humans and embodied technologies in the context of a handover task between a human and a robot. They investigate advancing human–robot interaction by generating a simulated handover training dataset and a novel grasp planner based on a generative neural network and a grasp pose evaluator. This methodology contributes to improving human-to-robot handover scenarios that are common in elderly assistance or agile production scenarios. Kopnarski et al. [47] also contribute to effective human–robot handovers by investigating whether predictions about the weight of one object can be made on the basis of the kinematics of the arm of the person handing over the object. Figuratively speaking, this approach furthers effective coordination between embodied technologies and humans by equipping the embodied technology with a better understanding of human features.

In future hybrid societies, many embodied technologies will act autonomously. However, technological, environmental, or situational constraints will require remote control by humans using teleoperation. Brade et al. [48] explain that smooth coordination of humans and teleoperated EDTs is possible only if teleoperators can orient themselves in remote environments and are able to communicate with on-site humans. They also propose approaches to provide visual support for operating in remote environments.

From the viewpoint of modern machine learning, Teichmann et al. [49] develop a framework for successful spontaneous collaborations between humans and embodied technologies, which they refer to as human–machine teaming. They identify four key aspects for human–machine teaming: (i) shared sub-goals, (ii) establishment of communication between the teaming agents, (iii) learning from the other agent and adapting accordingly, and (iv) ensuring interdependency of the teaming agents. Based on state-of-the-art deep neural networks, they subsequently propose a novel framework for software teaming agents that contributes to understand how communication and conceptual information can be incorporated in the design of future artificial agents. In this way, interdisciplinary research contributes to effective interactions between humans and embodied technologies in first encounters.

Rudolph et al. [50] investigate how the fusion of embodied technology and human bodies into hybrid bodies affects interpersonal social perceptions. For researching perceptions and behaviors in first encounters between humans and hybrids, they introduce a mixed/diminished reality system that can replace a body part of a participant with a bionic prosthesis in real time. This environment allows studying the psychological processes surrounding self-attributions and perceptions of hybrids in novel ways. Their framework overcomes the limited availability of bionic technology and the scarcity

of hybrids for research and contributes to the future design of bionic technologies for effective coordination between users and non-users of these technologies. Tietz [51] also addresses one of the open challenges of social acceptance of embodied technologies with a specific focus on the social-communicative problems surrounding the embodiment of autonomous agents. Tietz argues that it is problematic that embodied technologies appear as if they are operating autonomously, whereas their operation is in fact the result of human and technical actions. To overcome this issue, Tietz proposes accountable display strategies that communicate the level of autonomy or (remote) human control to the environment of the embodied technology. In this way, accountability can contribute to effective interaction between humans and embodied technologies in public spaces.

The concept of a hybrid society hinges on the coexistence of humans and embodied technologies in daily life. This development implies the need to investigate the fit between the concept of a hybrid society and existing legal, psychological, and sociological constructions of societal membership. Accordingly, Meyer et al. [52] conduct an interdisciplinary investigation aiming at understanding whether non-human actors can possess personality, paving the path for what they call hybrid personae. They conclude that from a psychological perspective, only the human being can possess a personality and a sense of agency. On the contrary, sociology and legal science permit the attribution of hybrid personae under certain conditions. In particular, a hybrid persona is conceivable from a sociological perspective if it refers to human parts of an entity. This is theoretically addressable by legal science with the further development of norm addressees and norm recipients, which opens the possibility to establish the concept of a hybrid personae.

7 Outlook and Societal Impact

These are only some examples of how the research assembled in this volume can contribute to understanding and designing interactions between humans and embodied technologies. Despite the advances that this research represents, further challenges for hybrid societies remain. So far, the public and scientific discourses about possible consequences of embodied technologies for societies have primarily centered on questions pertaining to the replacement of human work by robots and on the appropriateness of the (supposed) equation of technical agents and humans. These fundamental questions, of course, remain to be asked and contribute to fundamental understandings about normatively desired or undesired futures with machines. At the same time, research at the intersection of humanities and social sciences and engineering and STEM—some of which are presented in this volume—reveals that the social consequences of the proliferation of embodied technologies hinge on the interplay between humans and technology at different levels that we have only started to systemize. To explore the potential consequences of such distributed agency [53] in hybrid societies, further inter- and transdisciplinary research that combines basic science research and the inclusion of social groups and societal stakeholders appears necessary.

Importantly, successful coordination between humans and embodied technology requires including all members of society in the discourse. Different target groups and stakeholders need to voice their opinions about or demands for participating in a hybrid society. This implies interaction among participants—e.g., in town hall formats—and exchanges between researchers and publics. In order to organize constructive debates, science communication has to offer the necessary resources for people to weigh all the aspects of the discussion. Recommendations, ideas, and criticism generated during deliberative events will be communicated back to the subprojects and included in further participatory efforts. Ultimately, the development of technologies and of society follows paths laid out through research and engineering as well as commercial interests, investments, market research, and usability trials. To step outside the current reality of smart machines and given trajectories thus implies coming up with alternative and hitherto uncharted visions, which ultimately requires the creativity and imaginative force of all societal actors.

Acknowledgements Funded by the Deutsche Forschungsgemeinschaft (DFG, German Research Foundation)—Project-ID 416228727—SFB 1410.

References

1. Brynjolfsson, E., & McAfee, A. (2014). *The second machine age: Work, progress, and prosperity in a time of brilliant technologies.* WW Norton & Company.
2. Bockshecker, A., Hackstein, S., & Baumöl, U. (2018). Systematization of the term digital transformation and its phenomena from a socio-technical perspective—A literature review. *Association for Information Systems ECIS 2018 Proceedings Research Papers.*
3. Meyer, S., Mandl, S., Gesmann-Nuissl, D., & Strobel, A. (2022). Responsibility in hybrid societies: Concepts and terms. *AI and Ethics.* https://doi.org/10.1007/s43681-022-00184-2
4. Deng, E., Mutlu, B., & Mataric, M. J. (2019). Embodiment in socially interactive robots. *Foundations and Trends in Robotics, 7*(4), 251–356. https://doi.org/10.1561/2300000056
5. Jahn, G. (2024). Resilience engineering for highly automated driving, autonomous vehicles, and urban robotics: Wizards and shepherds in hybrid societies. *Theoretical Issues in Ergonomics Science, 25*(6), 680–701. https://doi.org/10.1080/1463922x.2024.2328062
6. Jünger, S. (2002). Gesellschaft, Kultur und Bezugnahme: Ordnung II. In S. Jünger (Ed.), *Kognition, Kommunikation, Kultur* (pp. 104–

138). Deutscher Universitätsverlag. https://doi.org/10.1007/978-3-663-07682-7_8
7. Knorr Cetina, K. (2009). The synthetic situation: Interactionism for a global world. *Symbolic Interaction, 32*(1), 61–87. https://doi.org/10.1525/si.2009.32.1.61
8. Larson, J. R. (2009). *In search of synergy in small group performance.* Psychology Press.
9. Pentzold, C., & Bischof, A. (2019). Making affordances real: Sociomaterial prefiguration, performed agency, and coordinated activities in human–robot communication. *Social Media + Society, 5*(3), 205630511986547. https://doi.org/10.1177/2056305119865472
10. Müller, M., & Sonnenmoser, A. (2026, this volume). Sociomorphic technologies — On the typology of artificial actors. In B. Meyer, O. Kanoun, & U. Thomas (Eds.), *Hybrid Societies: Humans Interacting with Embodied Technologies, Volume 1.* Springer.
11. Groom, V., Nass, C., Chen, T., Nielsen, A., Scarborough, J. K., & Robles, E. (2009). Evaluating the effects of behavioral realism in embodied agents. *International Journal of Human-Computer Studies, 67*(10), 842–849. https://doi.org/10.1016/j.ijhcs.2009.07.001
12. Gibson, E. J. (1982). The concept of affordances in development: The renascence of functionalism. In W. A. Collins (Ed.), *The concept of development: The Minnesota symposia on child psychology* (Vol. 15, pp. 55–81). Lawrence Erlbaum.
13. Mead, R., & Matarić, M. J. (2016). Perceptual models of human-robot proxemics. In M. Hsieh, O. Khatib, & V. Kumar (Eds.), *Experimental robotics. Springer tracts in advanced robotics* (Vol. 109, pp. 261–276). Springer. https://doi.org/10.1007/978-3-319-23778-7_18
14. Drewes, J., Feder, S., & Einhäuser, W. (2021). Gaze during locomotion in virtual reality and the real world. *Frontiers in Neuroscience, 15.* https://doi.org/10.3389/fnins.2021.656913
15. Fricke, E. (2014). 136. Deixis, gesture, and embodiment from a linguistic point of view. In C. Müller, A. Cienki, E. Fricke, S. H. Ladewig, D. McNeill, & J. Bressem (Eds.), *Handbücher zur Sprach- und Kommunikationswissenschaft/Handbooks of Linguistics and Communication Science (HSK) 38/2* (Vol. 2, pp. 1803–1823). De Gruyter. https://doi.org/10.1515/9783110302028.1803
16. Duffy, B., & Joue, G. (2000). Intelligent robots: The question of embodiment. In *Proceedings of the Brain-Machine 2000 Workshop, January 2000* (pp. 20–29).
17. Forlizzi, J., & DiSalvo, C. (2006). Service robots in the domestic environment. In *Proceeding of the 1st ACM SIGCHI/SIGART Conference on Human-Robot Interaction—HRI '06* (pp. 258–265). https://doi.org/10.1145/1121241.1121286
18. Jones, J. L. (2006). Robots at the tipping point: The road to iRobot Roomba. *IEEE Robotics & Automation Magazine, 13*(1), 76–78. https://doi.org/10.1109/MRA.2006.1598056
19. Mutlu, B., & Forlizzi, J. (2008). Robots in organizations. In *Proceedings of the 3rd International Conference on Human Robot Interaction—HRI '08* (p. 287). https://doi.org/10.1145/1349822.1349860
20. Maitin-Shepard, J., Cusumano-Towner, M., Lei, J., & Abbeel, P. (2010). Cloth grasp point detection based on multiple-view geometric cues with application to robotic towel folding. In *IEEE International Conference on Robotics and Automation, 2010* (pp. 2308–2315). https://doi.org/10.1109/ROBOT.2010.5509439
21. Osawa, F., Seki, H., & Kamiya, Y. (2006). Clothes folding task by tool-using robot. *Journal of Robotics and Mechatronics, 18*(5), 618–625. https://doi.org/10.20965/jrm.2006.p0618
22. Coltin, B., & Veloso, M. (2014). Online pickup and delivery planning with transfers for mobile robots. In *IEEE International Conference on Robotics and Automation (ICRA), 2014* (pp. 5786–5791). https://doi.org/10.1109/ICRA.2014.6907709

23. Simmons, R., Goodwin, R., Haigh, K. Z., Koenig, S., & O'Sullivan, J. (1997). A layered architecture for office delivery robots. In *Proceedings of the First International Conference on Autonomous Agents—AGENTS '97* (pp. 245–252). https://doi.org/10.1145/267658.267723

24. Nourbakhsh, I. R., Bobenage, J., Grange, S., Lutz, R., Meyer, R., & Soto, A. (1999). An affective mobile robot educator with a full-time job. *Artificial Intelligence, 114*(1–2), 95–124. https://doi.org/10.1016/S0004-3702(99)00027-2

25. Levinson, J., Askeland, J., Becker, J., Dolson, J., Held, D., Kammel, S., Kolter, J. Z., Langer, D., Pink, O., Pratt, V., Sokolsky, M., Stanek, G., Stavens, D., Teichman, A., Werling, M., & Thrun, S. (2011). Towards fully autonomous driving: Systems and algorithms. In *IEEE Intelligent Vehicles Symposium (IV), 2011* (pp. 163–168). https://doi.org/10.1109/IVS.2011.5940562

26. Bemelmans, R., Gelderblom, G. J., Jonker, P., & de Witte, L. (2012). Socially assistive robots in elderly care: A systematic review into effects and effectiveness. *Journal of the American Medical Directors Association, 13*(2), 114-120.e1. https://doi.org/10.1016/j.jamda.2010.10.002

27. Broekens, J., Heerink, M., & Rosendal, H. (2009). Assistive social robots in elderly care: a review. *Gerontechnology, 8*(2). https://doi.org/10.4017/gt.2009.08.02.002.00

28. Feil-Seifer, D., & Mataric, M. J. (2005). Defining socially assistive robotics. In *9th International Conference on Rehabilitation Robotics, 2005. ICORR 2005* (pp. 465–468). https://doi.org/10.1109/ICORR.2005.1501143

29. Hafner, V., Hommel, B., Kayhan, E., Lee, D., Paulus, M., & Verschoor, S. (2022). Editorial: The mechanisms underlying the human minimal self. *Frontiers in Psychology, 13*, Article 961480. https://doi.org/10.3389/fpsyg.2022.961480

30. Forch, V., & Hamker, F. H. (2021). Building and understanding the minimal self. *Frontiers in Psychology, 12*(716982). https://doi.org/10.3389/fpsyg.2021.716982

31. Meyer, B., & Asbrock, F. (2018). Disabled or cyborg? How bionics affect stereotypes toward people with physical disabilities. *Frontiers in Psychology, 9*(2251), 1–13. https://doi.org/10.3389/fpsyg.2018.02251

32. Mandl, S., Bretschneider, M., Meyer, S., Gesmann-Nuissl, D., Asbrock, F., Meyer, B., & Strobel, A. (2022). Embodied digital technologies: First insights in the social and legal perception of robots and users of prostheses. *Frontiers in Robotics and AI.* Advance Online Publication. https://doi.org/10.3389/frobt.2022.787970

33. Fiske, S. T., Cuddy, A. J., & Glick, P. (2007). Universal dimensions of social cognition: Warmth and competence. *Trends in Cognitive Sciences, 11*, 77–83. https://doi.org/10.1016/j.tics.2006.11.005

34. Fiske, S. T. (2018). Stereotype content: Warmth and competence endure. *Current Directions in Psychological Science, 27*, 67–73. https://doi.org/10.1177/0963721417738825

35. Rani, P., Sarkar, N., Smith, C. A., & Kirby, L. D. (2004). Anxiety detecting robotic system—Towards implicit human-robot collaboration. *Robotica, 22*(1), 85–95. https://doi.org/10.1017/S0263574703005319

36. Kennedy, W. G., Bugajska, M. D., Marge, M., Adams, W., Fransen, B. R., Perzanowski, D., Schultz, A. C., & Gregory Trafton, J. (2007). Spatial representation and reasoning for human-robot collaboration. In *AAAI* (Vol. 7, pp. 1554–1559).

37. Bauer, A., Wollherr, D., & Buss, M. (2008). Human–robot collaboration: A survey. *International Journal of Humanoid Robotics, 5*(1), 47–66. https://doi.org/10.1142/s0219843608001303

38. Scherer, S., Marsella, S., Stratou, G., Xu, Y., Morbini, F., Egan, A., Rizzo, A., & Morency, L.-P. (2012). Perception markup language: Towards a standardized representation of perceived nonverbal behaviors. In Y. Nakano, M. Neff, A. Paiva, & M. Walker (Eds.), *Intelligent virtual agents* (pp. 455–463). Springer. https://doi.org/10.1007/978-3-642-33197-8_47

39. Ono, T., Imai, M., & Ishiguro, H. (2001). A model of embodied communications with gestures between humans and robots. In *Proceedings of 23rd Annual Meeting of the Cognitive Science Society* (pp. 732–737). https://conferences.inf.ed.ac.uk/cogsci2001/pdf-files/0732.pdf

40. Huang, C.-M., & Mutlu, B. (2014). Learning-based modeling of multimodal behaviors for human-like robots. In *Proceedings of the 2014 ACM/IEEE International Conference on Human-Robot Interaction* (pp. 57–64). ACM.

41. Suchman, L. (2007). *Human-machine reconfigurations.* Cambridge University Press.

42. Seibt, J., Vestergaard, C., & Damholdt, M. F. (2020). Sociomorphing, not anthropomorphizing: Towards a typology of experienced sociality. In M. Nørskov, J. Seibt, & O. S. Quick (Eds.), *Frontiers in artificial intelligence and applications, Volume 335: Culturally sustainable social robotics.* IOS Press. https://doi.org/10.3233/FAIA200900

43. Siefkes, M., Fricke, E., Bressem, J., & Charoensit, A. (2026, this volume). Modelling intentional complexity in hybrid interaction scenarios beyond explicit and implicit communication. In B. Meyer, O. Kanoun, & U. Thomas (Eds.), *Hybrid Societies: Humans Interacting with Embodied Technologies, Volume 1.* Springer.

44. Sanseverino, G., Krumm, D., Ramalingame, R., et al. (2026, this volume). Understanding the capabilities of FMG and EMG sensors in recognizing basic gesture components. In B. Meyer, O. Kanoun, & U. Thomas (Eds.), *Hybrid Societies: Humans Interacting with Embodied Technologies, Volume 1.* Springer.

45. Hensch, A.-C., Felbel, K., Beggiato, M., et al. (2026, this volume). Implicit driving cues for coordinating actions when sharing spaces. In B. Meyer, O. Kanoun, & U. Thomas (Eds.), *Hybrid Societies: Humans Interacting with Embodied Technologies, Volume 1.* Springer.

46. Gäbert, C., Bandi, C., & Thomas, U. (2026, this volume). Grasp pose generation for human-to-robot handovers using simulation-to-reality transfer. In B. Meyer, O. Kanoun, & U. Thomas (Eds.), *Hybrid Societies: Humans Interacting with Embodied Technologies, Volume 1.* Springer.

47. Kopnarski, L., Lippert, L., Voelcker-Rehage, C., et al. (2026, this volume). Predicting objects weights from giver's kinematics in handover actions. In B. Meyer, O. Kanoun, & U. Thomas (Eds.), *Hybrid Societies: Humans Interacting with Embodied Technologies, Volume 1.* Springer.

48. Brade, J., Xie, N., Winkler, S., Klimant, P., & Jahn, G. (2026, this volume). Where am I? How to measure and support spatial orientation in teleoperation. In B. Meyer, O. Kanoun, & U. Thomas (Eds.), *Hybrid Societies: Humans Interacting with Embodied Technologies, Volume 1.* Springer.

49. Teichmann, M., Ragni, M., Vitay, J., Gaedke, M., & Hamker, F. (2026, this volume). Human-machine teaming agents: A future perspective. In B. Meyer, O. Kanoun, & U. Thomas (Eds.), *Hybrid Societies: Humans Interacting with Embodied Technologies, Volume 1.* Springer.

50. Rudolph, C., Dadgar, S.-A., Bretschneider, M., et al. (2026, this volume). Towards TechnoSapiens: Experiencing embodied technologies in augmented reality. In B. Meyer, O. Kanoun, & U. Thomas (Eds.), *Hybrid Societies: Humans Interacting with Embodied Technologies, Volume 1.* Springer.

51. Tietz, S. (2026, this volume). What are you: Social displays of 'autonomous' machines. In B. Meyer, O. Kanoun, & U. Thomas (Eds.), *Hybrid Societies: Humans Interacting with Embodied Technologies, Volume 1*. Springer.

52. Meyer, S., Müller, M. R., Sonnenmoser, A., et al. (2026, this volume). Towards hybrid personae? In B. Meyer, O. Kanoun, & U. Thomas (Eds.), *Hybrid Societies: Humans Interacting with Embodied Technologies, Volume 1*. Springer.

53. Rammert, W. (2008). Where the action is. Distributed agency between humans, machines, and programs. In Seifert & Kim & Moore (Eqs.), *Paradoxes of interactivity. Perspectives for media theory, human-computer interaction, and artistic investigations* (pp. 62–91). Transcript. https://doi.org/10.1515/978383940842 1-004

A Mixed Methods Approach for Capturing Interactions of Cyclists with Mobility Space

Martin Loidl⦿, Christian Werner⦿, Elisabeth Füssl⦿, Florian Kratochwil, and Bernd Resch⦿

Abstract

The physiological and cognitive load for cycling is high. A constant interaction with other road users and the environment, consisting of static and dynamic objects, is required while riding. In order to make cycling a pleasant and safe experience, it is of utmost importance to consider these interactions in providing adequate infrastructure and managing traffic accordingly. However, to do so successfully, the multiple layers of interaction, interdependencies, and systemic relations have to be considered. This can hardly be done within a single domain. Thus, we propose an interdisciplinary mixed methods approach that facilitates domain-specific insights as well as an integrated, systemic understanding of cyclists' interaction with mobility space.

Keywords

Cycling · Interaction · Mixed methods · Mobility space · Naturalistic setting · Spatio-temporal context

M. Loidl (✉) · C. Werner
Department of Geoinformatics, University of Salzburg, Salzburg, Austria
e-mail: martin.loidl@plus.ac.at

C. Werner
e-mail: christian.werner@plus.ac.at

E. Füssl
Factum, Vienna, Austria
e-mail: elisabeth.fuessl@factum.at

F. Kratochwil
con.sens Mobility Design, Vienna, Austria
e-mail: kratochwil@cvp.at

B. Resch
Spatial Services GmbH, Salzburg, Austria
e-mail: bernd.resch@spatial-services.com

1 Introduction and Problem Statement

Promoting cycling as a main mode of transport in urban settings is high on the agenda for a number of reasons. Cycling is efficient, flexible, healthy, and economically, ecologically, and socially sustainable [5]. However, most cities in the Global North are planned for car-centered mobility, which is manifested in a road space design that gives priority to the needs of motorized vehicles [2]. In order to make necessary leaps forward in increasing the modal share of cyclists, a re-organization of the road space in favor of cycling (and walking) is fundamental. In this, the preferences and needs of cyclists are to be considered in the provision of adequate infrastructure. Otherwise, the acceptance will be low, and the advantages of cycling mobility cannot be gained.

At a general level, it is well known what makes cycling "irresistible" [14]. Dedicated, preferably physically separated from motorized traffic, well-connected cycling infrastructure is the backbone of any successful cycling promotion. In addition, bicycle parking facilities [6], low speed limits [16], and smooth integration in public transport [13] have effects on cyclists' mode share. Obviously, "soft factors," such as role models, the general perception of cycling, and the prevalent cycling culture, leverage the effects of cycling-friendly road design [1].

There are only rare cases where the provision of adequate cycling infrastructure is systematically planned and implemented completely from scratch. The city of Seville is one of the few examples [11]. Most of the time, cities are implementing infrastructure improvements stepwise, for instance, in the wake of neighborhood development projects, or renovation and maintenance measures. In this context, cities are required to balance out multiple interests and limited financial and personal resources. The challenge can be boiled down to the two questions, "What effects can I expect from certain measures?" and "Where is the money spent most effectively and efficiently?" Often, these questions cannot be answered specifically, for the following reasons:

© The Author(s) 2026
B. Meyer et al. (eds.), *Hybrid Societies*, Advances in Science, Technology & Innovation,
https://doi.org/10.1007/978-3-032-03488-5_2

(a) The effect of interventions is only measured in a single dimension and not investigated in a systemic approach. For instance, traffic counts before and after the intervention are done on the respective spot, but the immediate surroundings, such as parallel roads, is not considered. Consequently, it remains unknown whether the intervention attracted additional users or led only to a redistribution of existing traffic in the network.

(b) The quality of an implemented measure, and with this the acceptance by cyclists, depends on various factors. One and the same type of intervention can be perceived differently in different settings, because of factors that are not considered in the planning phase. For instance, a newly built, separated cycle way might not attract new cyclists, because users are directly exposed to loud noise from the high volume of heavy goods vehicles on the main road next to the cycle way.

(c) Methods for capturing systemic effects of interventions are not established yet. This refers to the spatial and temporal scope as well as to the difficulty of integrating quantitative and qualitative data sources meaningfully. Additional challenges arise from the way data is acquired. Whereas in controlled lab settings the number of co-factors can be reduced and experiments are reproducible under stable conditions, measurements in naturalistic settings are of different characteristics.

(d) Implemented measures are hardly ever evaluated comprehensively. Thus, learning from previous interventions in similar settings within the city and beyond is rarely possible.

Against this backdrop, a multi-disciplinary team of researchers has worked on the development of a mixed methods approach for evaluating the effects of road reorganization in the research project POSITIM. The mixed methods approach itself is described elsewhere in detail [19]. In this paper, we focus primarily on the challenges of integrating data that are generated in naturalistic settings and which represent individual behavior, experience, and perception—the interaction with the mobility space.

2 Measurements in Space–Time Continuum

Investigating the interaction of cyclists with mobility space is challenging not only from a technical point of view, but even more conceptually. All measurements are embedded in a spatio-temporal continuum. Thus, measurements are not independent of each other, but correlated. Consequently, the spatio-temporal context must be considered explicitly, even if the data are sampled or discretized. As we will see in Sect. 3, the spatial and temporal references are key for linking and integrating data. In addition, investigations

in a natural setting require different approaches compared to controlled settings in a laboratory, since the number of uncontrollable co-factors is high, and interactions cannot be reproduced.

2.1 Naturalistic Setting

Naturalistic studies in transport and mobility research primarily focus on road safety [18]. While most of these studies investigate the behavior of car drivers, an increasing number of naturalistic cycling studies have emerged over the past years [4]. This development is closely linked to the miniaturization and price reduction of sensors, as well as to the increasing prevalence of smartphones. Naturalistic cycling studies are well-suited for studying cyclists' behavior in real environments. The study design in a naturalistic setting is ultimately dependent on the research question. To date, no standardized protocol for naturalistic cycling studies exists. Most of the published studies revolve around safety aspects, such as overtaking manoeuvres or near-misses.

The majority of findings on how cyclists interact with the built, the social, and the cultural environment are derived from aggregated data sources, which are then fed into different kinds of models [17]. Few studies have investigated the interaction of individuals with their environment in a naturalistic setting. Studies from Cambridge, MA [20] and Salzburg [15] made use of physiological (stress) in-situ measurements and partly combined these data with geo-located questionnaires and first-person video footage. The authors found negative reactions toward unfavorable road designs and high volumes of motorized traffic. However, the response to different environments varies greatly between cyclists. One of the biggest investigations of cyclists' interaction with the road space might be the cross-sectional pre-post-analysis of the Connect2 program in the UK. The authors conducted two waves of analysis of 84 locations across the UK, using counting data, surveys, and longitudinal cohort data. Overall, an increase in use of newly built and improved infrastructure—particularly if the baseline use was low—and a higher share of people meeting physical activity recommendations was observed [10].

While the first group of naturalistic studies focuses on individual users and their interaction with the environment, Le Gouais et al. [10] conducted a site-specific naturalistic study with a scope on infrastructure use. For the investigation of environmental determinants on mode choice of commuters, Higuera-Mendieta et al. [7] propose an adapted social-ecological model that is well established in public health. This model integrates various layers of influential variables, with individual, socio-demographic factors at the core. The next layer comprises interpersonal factors, such as social norms, and is followed by layers representing the built

Fig. 1 Different types of sensors, which are used to generate a comprehensive data set for analyzing the interaction of cyclists with mobility space in a mixed methods approach

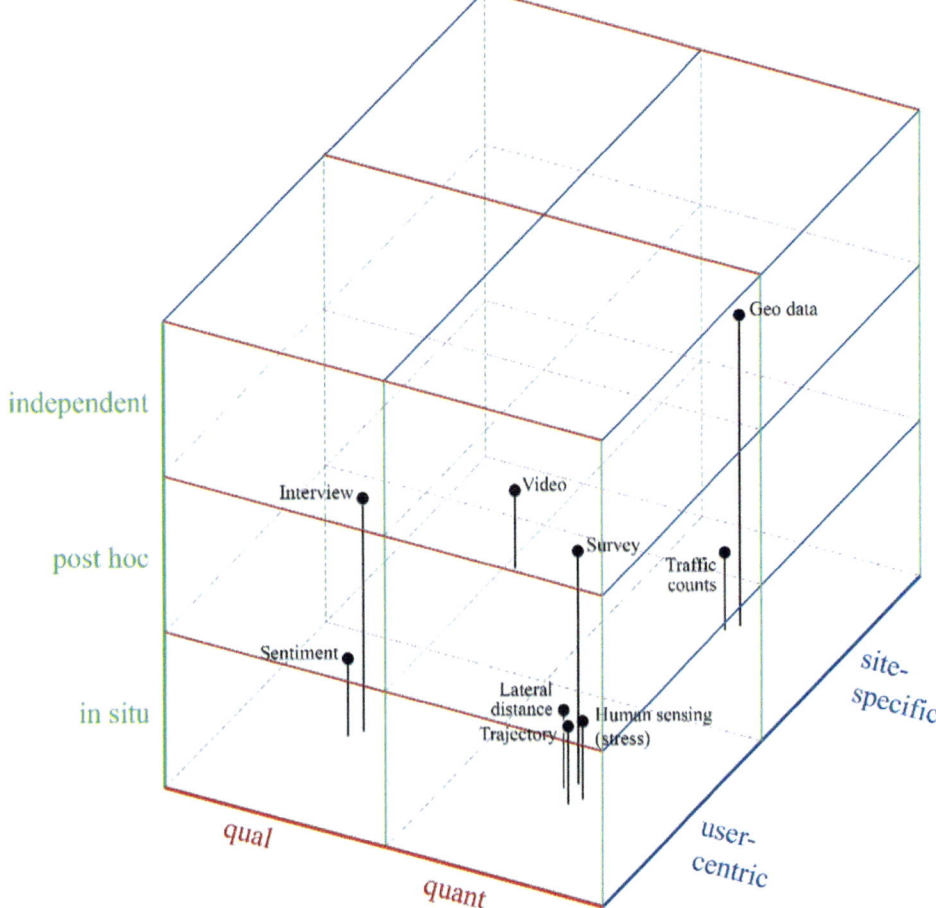

(infrastructure, connectivity, and accessibility, etc.) and the natural (weather, seasonality, terrain, etc.) environment.

To account for the complexity of the interaction of cyclists with their environment, we propose a mixed methods approach, which is applicable in natural settings. In this, we combine site-specific and user-centered, in-situ and post-hoc measurements as well as qualitative methods and quantitative measurements (Fig. 1). With this approach, we anticipate the various layers of the social-ecological model and built the basis for analyzing the relation between these layers.

2.2 Data Sources and Sensors

Data for a comprehensive analysis of cyclists' interaction with mobility space comes from different sensors. We propose using the data sources listed below. This list is neither extensive nor obligatory in its full extent. Sensors can be added for integrating more perspectives, or a subset might be sufficient for specific aspects.

- *Geo data.* For the description of the physical and socio-demographic environment, geo-spatial data are acquired. The road space is commonly represented as an undirected graph with associated attributes that describe the design of the road profile for each segment. The gradient is derived from digital elevation models. Land use data represent the spatial context of the road, with common classes such as green and blue space, building footprints, etc. Data on the natural environment, such as weather, temperature, wind, emissions, etc., can be included retrospectively or linked in real time. Finally, census data provide information on residential population, day population, commuters, demographic structure, income levels, occupancy, etc.
- *Traffic counts.* Stationary traffic counting systems provide information on traffic volume and modal split. These site-specific data sets can be related to massive trajectory data from mobile applications for estimating the distribution of traffic flows in the network [12].
- *Video footage.* Subjects are equipped with first-person cameras. With this, spatio-temporally continuous, audio-visual information on the context is generated.

- *Sentiments.* With a mobile application (eDiary app), subjects can record sentiments during their rides. This geo-located and time-stamped data set is a self-stated, subjective equivalent to objective stress measurements [9].
- *Interviews.* After their rides, subjects are interviewed in a semi-structured way. These interviews serve as a reflection of experiences and perceptions during the ride. Data from the interviews are used for explanatory analysis of quantitative measurements.
- *Survey.* Subjects' demographic characteristics and mobility behavior are surveyed. In addition, perceived stress, safety, and smoothness of the ride are rated.
- *Human sensing.* A chest belt and a wrist band measure physiological parameters during the ride at 4 Hz. From these data, geo-located moments of stress are derived in a post-processing step.
- *Lateral distance.* With ultrasonic sensors, the lateral distance is measured at 10 Hz on both sides. In addition, a button is pressed by subjects to mark overtaking manoeuvres.
- *Trajectory.* Every subject is equipped with a GNSS sensor, which measures the time-stamped geo-location at 1 Hz. In a post-processing step, physical parameters, such as speed, acceleration, number of stops, and duration of stops, are derived.

Data from the different sensors describe different aspects of cyclists' interaction with mobility space. We distinguish between the type of data, the temporal relation of the measurement to a ride, and the focus of the data generation. Qualitative data contain information on how cyclists perceive and experience their environment (recording of sentiments, interview), which is continuously monitored by first-person video footage. Depending on the study design, qualitative data are used for scoping or explanatory purposes. Quantitative data are collected to describe the mobility space (site-specific) and its use on the one side, and how individual users interact with their spatial environment (user-centric). Quantitative data are subject to explorative and descriptive statistical analysis in a first step and serve as input for correlation analysis and regression models consecutively.

Each of the data sources describes a specific perspective of the interaction of cyclists with mobility space, with varying spatial and temporal resolutions. Trajectories, lateral distance measurements, human sensing, video footage, and traffic counts can be regarded as spatio-temporally continuous data, which means that measurements at any given point in time and at any location are available (however, the location does not change for the traffic counts). The rest of the sensed data is discretized, with measurements at specific temporal snapshots and locations. For interview and survey data, the temporal and spatial reference might be reconstructed from mediated information, such as relative time and location infor-

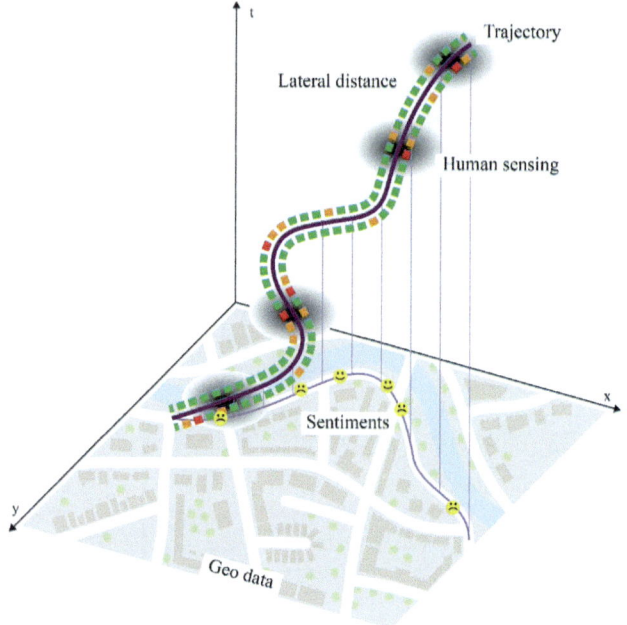

Fig. 2 The spatial (x, y) and temporal (t) reference is used to put data into context

mation ("Approximately three minutes after I started ...," "At the intersection close to the bridge ...").

3 Integration

For investigating the interaction of cyclists with mobility space comprehensively, the different perspectives, represented by the data sources described above, must be integrated. This integration can be done at different levels.

3.1 From Data to Insights

All sensors generate raw data, which are then further processed, either in real time on the device or in a post-processing step. The data-information-knowledge-wisdom (DIKW) pyramid is a well-known metaphor for how value is added by processing data and putting them into context. It is widely anticipated in spatial sciences and applied to data that emerge from transport research [20]. We use the spatial and temporal reference of the data to provide context. Moreover, space and time are used as keys for linking data from different sources (see Fig. 2).

Data that is sensed in-situ, the spatial and temporal references are part of the raw data set. This means that location information (usually coded as a geographic coordinate tuple) and a timestamp are stored with each measurement. This allows for an immediate contextualization. For instance, the device for measuring the lateral distance by ultrasonic sensors is also equipped with a GNSS (Global Navigation Satellite System) unit. Thus, each distance measurement can be directly embedded into a spatial context that is being mapped. This allows for extracting information on overtaking manoeuvres or assessing the sufficiency of dedicated space for cyclists. Based on this information, conclusions for future planning or for adapting the road space, are drawn.

For data that is acquired post-hoc, the spatial and temporal reference has to be either stated (by locating an event on a map or recalling the point of time) or reconstructed from relative descriptions. Obviously, this type of spatial and temporal reference inhibits some vagueness and needs to be treated accordingly. However, conceptually, the contextualized data can be used for further analysis in the same way as in-situ sensed data.

Geo data are used for describing the spatial environment. Static data that describes the physical environment (buildings, road designs, land use, etc.) must be up to date, but the temporal scale is much coarser; for instance, the authoritative road dataset in Austria is updated four times a year. Geo data that describes environmental conditions, such as weather, is commonly provided with a spatial and temporal reference at high resolution.

3.2 Levels of Integration

At all levels of integration, the spatial and temporal references are used as keys. This holds true for data that are generated by various sensors within a single device, for methods—such as human sensing—that rely on more than one device, and for the integration of data and information across sources. The following examples illustrate these levels of integration:

- *Multiple sensors, one device.* Stationary traffic counters have a fixed location. The positioning is either done with a GNSS device or by locating the spot on a map. In the case of a camera-based system, objects are identified from images, and a predefined number of different modes is counted. Each image, and with this each count, is time-stamped. The fixed location, together with the continuous measurement of time and traffic volumes per mode, generates a multidimensional data set that is embedded into a spatio-temporal context.
- *Multiple sensors, multiple devices.* The human sensing method, which is described elsewhere in detail [9], relies on self-reported sentiments, which are entered in a location-based app, video footage, and bio-physical sensors. Data from these three sources is then combined to generate and validate geo-located and time-stamped moments of stress (MOS). This information layer is

subsequently used as one input in a mixed methods approach.

- *Mixed methods approach.* Linking data from different sources and of different types requires semantic and syntactical interoperability. We use the spatial and temporal reference as common ground for integrating the perspectives from quantitative and qualitative data sources. How this is done depends very much on the research design [3] and the hypothesis that needs to be tested in a naturalistic setting. For instance, qualitative results, post-hoc interviews on experience and perception can be related to quantitative data and analysis results representing speed, number of stops, duration of stops, moments of stress, close overtaking manoeuvres, and others. This way, detected events or outliers in the quantitative data, which are geo-located and time-stamped, can be explained by qualitative, self-stated information with any kind of spatial and/or temporal reference.

Evidently, any data set that is acquired in a naturalistic setting is highly autocorrelated in the spatial and temporal dimensions. Hence, the data cannot be treated as independent, but the autocorrelation can be used for detecting clusters of events across layers of data and information. Accounting for autocorrelation in the selection of appropriate sample size and in regression models, as well as by embedding qualitative research into the spatio-temporal continuum, adds benefits to mixed methods research that has remained largely unused [8].

3.3 Benefits of Mixed Methods Approaches

Mixed methods approaches for investigating cyclists' interaction with mobility space are not a fixed set of methods or tools. Depending on the question at hand and the available resources, the research design decides on how to integrate which methods and data, respectively. The terminology for research that combines quantitative and qualitative data is not consolidated either. Drawing from Creswell's categorization [3], we distinguish between exploratory sequential, explanatory sequential, and convergent designs as basic types for mixed methods research. Based on them, he further distinguishes between intervention and multistage evaluation designs. In all cases, mixed methods approaches go beyond data gathering in parallel, but require an integration of methods, data, or information at some point for gaining systemic insights.

Effects of planned and implemented measures for promoting cycling mobility are hard to predict, since many factors influence the attractiveness of the mobility space, and consequently the interaction of cyclists with it. We draw from a rich set of methods that generate data describing the physical environment as well as the use and the perception of mobility space. While each of the data sources represents a certain perspective and is thus not always able to predict effects, the integration of methods, data, and results facilitates comprehensive approaches. The integration of different perspectives can be implemented for individual cyclists, but also for a collective. The latter is particularly relevant for decision makers and planners, since it decides on the acceptance of implemented modifications and new infrastructure.

4 Conclusion

Investigating how cyclists are interacting with mobility space is challenging, yet decisive for multiple aspects in promoting active mobility. In doing so, we need to be aware of the complexity of this interaction, which can hardly ever be captured by single methods, and the particularities of research in naturalistic settings. We argued that the spatial and temporal reference is a powerful key for integrating multiple perspectives at different levels. Moreover, space and time facilitate the embedding of sensed data into a context, a prerequisite for gaining insights from data.

With a better understanding of how mobility space is perceived, experienced, and finally used, decision makers have a comprehensive evidence base at hand that helps in optimizing and prioritizing interventions and allocating resources accordingly. Moreover, an in-depth understanding of how cyclists interact with mobility space supports planners and decision makers in providing adequate infrastructure, which anticipates the heterogeneity among cyclists and serves their different demands, preferences, and habits.

Acknowledgements The Project POSITIM (FFG Nr. 873353) was co-funded by the Austrian Ministry for Climate Action, Environment, Energy, Mobility, Innovation and Technology (BMK) under the program "Mobility of the Future."

References

1. Aldred, R., & Jungnickel, K. (2014). Why culture matters for transport policy: The case of cycling in the UK. *Journal of Transport Geography, 34*, 78–87. https://doi.org/10.1016/j.jtrangeo.2013.11.004
2. Brown, J. R., Morris, E. A., & Taylor, B. D. (2009). Planning for cars in cities: Planners, engineers, and freeways in the 20th century. *Journal of the American Planning Association, 75*(2), 161–177. https://doi.org/10.1080/01944360802640016
3. Creswell, J. W., & Creswell, J. D. (2017). *Research design: Qualitative, quantitative, and mixed methods approaches.* Sage.
4. Dozza, M., & Werneke, J. (2014). Introducing naturalistic cycling data: What factors influence bicyclists' safety in the real world? *Transportation Research Part F: Traffic Psychology and Behaviour, 24*(5), 83–91. https://doi.org/10.1016/j.trf.2014.04.001
5. Fishman, E. (2016). Cycling as transport. *Transport Reviews, 36*(1), 1–8. https://doi.org/10.1080/01441647.2015.1114271
6. Heinen, E., & Buehler, R. (2019). Bicycle parking: A systematic review of scientific literature on parking behaviour, parking preferences, and their influence on cycling and travel behaviour. *Transport Reviews, 39*(5), 630–656. https://doi.org/10.1080/01441647.2019.1590477
7. Higuera-Mendieta, D., Uriza, P. A., Cabrales, S. A., Medaglia, A. L., Guzman, L. A., & Sarmiento, O. L. (2021). Is the built-environment at origin, on route, and at destination associated with bicycle commuting? A gender-informed approach. *Journal of Transport Geography, 94*, 103120. https://doi.org/10.1016/j.jtrangeo.2021.103120
8. Ingram, M. C., & Harbers, I. (2020). Spatial tools for case selection: Using LISA statistics to design mixed-methods research. *Political Science Research and Methods, 8*(4), 747–763. https://doi.org/10.1017/psrm.2019.3
9. Kyriakou, K., Resch, B., Sagl, G., Petutschnig, A., Werner, C., Niederseer, D., Liedlgruber, M., Wilhelm, F. H., Osborne, T., & Pykett, J. (2019). Detecting moments of stress from measurements of wearable physiological sensors. *Sensors, 19*, 17. https://doi.org/10.3390/s19173805
10. Le Gouais, A., Panter, J. R., Cope, A., Powell, J. E., Bird, E. L., Woodcock, J., Ogilvie, D., & Foley, L. (2021). A natural experimental study of new walking and cycling infrastructure across the United Kingdom: The Connect2 programme. *Journal of Transport & Health, 20*, 100968. https://doi.org/10.1016/j.jth.2020.100968
11. Marqués, R., Hernández-Herrador, V., Calvo-Salazar, M., &d García-Cebrián, J. A. (2015). How infrastructure can promote cycling in cities: Lessons from Seville. *Research in Transportation Economics, 53*(11), 31–44. https://doi.org/10.1016/j.retrec.2015.10.017
12. Nelson, T., Ferster, C., Laberee, K., Fuller, D., & Winters, M. (2021). Crowdsourced data for bicycling research and practice. *Transport Reviews, 41*(1), 97–114. https://doi.org/10.1080/01441647.2020.1806943
13. Oeschger, G., Carroll, P., & Caulfield, B. (2020). Micromobility and public transport integration: The current state of knowledge. *Transportation Research Part D: Transport and Environment, 89*, 102628. https://doi.org/10.1016/j.trd.2020.102628
14. Pucher, J., & Buehler, R. (2008). Making cycling irresistible: Lessons from The Netherlands, Denmark and Germany. *Transport Reviews, 28*(4), 495–528. https://doi.org/10.1080/01441640701806612
15. Resch, B., Puetz, I., Bluemke, M., Kyriakou, K., & Miksch, J. (2020). An interdisciplinary mixed-methods approach to analyzing urban spaces: the case of urban walkability and bikeability. *International Journal of Environmental Research and Public Health, 17*, 19. https://doi.org/10.3390/ijerph17196994
16. Schepers, P., Twisk, D., Fishman, E., Fyhri, A., & Jensen, A. (2017). The Dutch road to a high level of cycling safety. *Safety Science, 92*(2), 264–273. https://doi.org/10.1016/j.ssci.2015.06.005
17. Schleinitz, K., Petzoldt, T., Franke-Bartholdt, L., Krems, J. F., & Gehlert, T. (2015). Conflict partners and infrastructure use in safety critical events in cycling—Results from a naturalistic cycling study. *Transportation Research Part F: Traffic Psychology and Behaviour, 31*, 99–111. https://doi.org/10.1016/j.trf.2015.04.002

18. Singh, H., & Kathuria, A. (2021). Analyzing driver behavior under naturalistic driving conditions: A review. *Accident Analysis & Prevention, 150,* 105908. https://doi.org/10.1016/j.aap.2020.105908

19. Werner, C., Füssl, E., Riess, J., Kratochwil, F., Resch, B., & Loidl, M. (in preparation). *A framework to facilitate advanced mixed methods studies for investigating interventions in road space.*

20. Zeile, P., Resch, B., Loidl, M., Petutschnig, A., & Dörrzapf, L. (2016). Urban emotions and cycling experience—Enriching traffic planning for cyclists with human sensor data. *GI_Forum, 1,* 204–216. https://doi.org/10.1553/giscience2016_01_s204

Implicit Driving Cues for Coordinating Actions When Sharing Spaces

Ann-Christin Hensch, Konstantin Felbel, Matthias Beggiato◉,
André Dettmann◉, Josef F. Krems◉,
and Angelika C. Bullinger◉

Abstract

For smooth and efficient vehicle-to-vehicle interactions in mixed traffic, comprising automated vehicles (AVs) and manually driven vehicles, AVs need to be able to anticipate future states of driving scenes and coordinate interactions with surrounding road users. Thus, the communication behavior of manual drivers needs to be investigated and could serve as a basis for AVs' driving functions. Due to the high demand for coordinating and anticipating driving maneuvers in high-speed lane change scenarios and in low-speed shared space interaction scenarios, the behavior of human road users in these scenarios should be further analyzed. Therefore, driving cues that are applied for anticipating lane change maneuvers and manual drivers' gap acceptance for initiating turning actions were investigated in a combined driving simulator study, in which $N = 29$ participants contributed. The results revealed that participants applied driving cues such as speed and position of surrounding vehicles (i.e., implicit driving cues) as a source of information to predict upcoming lane change maneuvers of other road users. Moreover, participants anticipated the development of driving scenes by considering the speed of encountering vehicles when selecting gaps for initiating turning maneuvers. This human driver behavior should therefore be considered for integration into AVs, as this enables an early anticipation of future states of driving scenes and adapting driving actions accordingly. Thus, such established interaction capabilities, based on implicit cues, can allow for intuitive encounters with surrounding road users in mixed traffic.

Keywords

Automated vehicles · Implicit driving cues · Lane change · Gap acceptance · Driving simulator study

1 Introduction

Manually driven and automated vehicles (AV; SAE Level 3 or higher) will share the road in the near future [1]. In the driving context, communicating and anticipating intentions is essential for safe, efficient, and smooth interactions (for an overview, see [2]). Besides explicit communication cues, such as the turn signal, mainly implicit driving cues, as vehicle movements (acceleration, deceleration, swerving) or wheel positions and angles, have an important role when coordinating actions in traffic [3]. For intuitive encounters in mixed traffic, concerning AVs and manually driven vehicles, AVs need to be able to anticipate future states of driving scenes and coordinate interactions with surrounding road users [4]. Therefore, knowledge about established interaction behavior of human drivers, such as applied driving cues and specific maneuver parameters, needs to be determined and could serve as a basis for AV's driving functions [5]. The present research in the project D02 "Implicit Driving Cues" of the CRC "Hybrid Societies" aimed primarily at investigating driving cues in different road traffic scenarios that highly require coordinating interactions when sharing spaces,

A.-C. Hensch (✉) · M. Beggiato · J. F. Krems
Department of Psychology, Cognitive and Engineering Psychology, Chemnitz Technical University, Chemnitz, Germany
e-mail: ann-christin.hensch@psychologie.tu-chemnitz.de

M. Beggiato
e-mail: matthias.beggiato@psychologie.tu-chemnitz.de

J. F. Krems
e-mail: josef.krems@psychologie.tu-chemnitz.de

K. Felbel · A. Dettmann · A. C. Bullinger
Department of Mechanical Engineering, Ergonomics and Innovation, Chemnitz Technical University, Chemnitz, Germany
e-mail: konstantin.felbel@mb.tu-chemnitz.de

A. Dettmann
e-mail: andre.dettmann@mb.tu-chemnitz.de

A. C. Bullinger
e-mail: angelika.bullinger-hoffmann@mb.tu-chemnitz.de

B. Meyer et al. (eds.), *Hybrid Societies*, Advances in Science, Technology & Innovation,
https://doi.org/10.1007/978-3-032-03488-5_3

such as high-speed lane change (LC) scenarios [6, 6] and low-speed intersection scenarios [8]. The main research questions of the project focused on identifying relevant implicit cues for anticipating driving maneuvers and signaling intentions to perform a certain action, including time of occurrence, intensity, and influencing factors such as driver characteristics. Previous results of the project are summarized in the next paragraphs.

The early announcement and anticipation of LCs as complex driving maneuvers is highly important to support traffic safety and flow. Therefore, a study by [9] aimed at identifying typical communication cues and prototypical sequence patterns for announcing prospective LCs in manual driving. In the study, 298 LC maneuvers were annotated in video recordings of a real-world driving data set. The analysis revealed the turn indicator and the vehicles' lateral movement toward lane markings of the target lane as the most frequently and initially applied communication cues to announce LCs. Since the cues were used with a mean time offset of about 1 s, the two cues might prospectively serve as a mutual fallback solution if one of the two communication cues is not detected by the sensors of an AV (e.g., due to occlusion) to predict LCs of surrounding vehicles. An additional study of the authors [10] focused on the identification of implicit and explicit cues during LCs and contextual information in high-speed scenarios (i.e., on highways). Over 1.000 km of 360° video material was recorded on German highways. The data analysis revealed that contextual information (e.g., infrastructure) was applied to anticipate the prospective development of driving scenes. Moreover, specific implicit and explicit communication cues were identified to be particularly relevant in this context for coordinating actions. For instance, small longitudinal distances between two vehicles and lateral lane offsets within the current lane were identified as important cues to anticipate upcoming LCs. Whereas, a steady traffic flow or an increase in the longitudinal distance between two vehicles due to deceleration maneuvers was associated with an absence of LCs. To gain further insight into implicit and explicit communication and the importance of contextual information during LCs, a naturalistic driving study was conducted [11]. Over a 6-month period, 30 participants drove approximately 30.000 km in total and recorded over 900 relevant situations with voice commentary using a dedicated smartphone app. Results showed that participants mainly used implicit cues to predict the driving behavior of surrounding road users. Specifically, participants mainly referred to speed differences, longitudinal distances to other road users, or lateral positioning within the lane. In addition, contextual information (i.e., traffic density or traffic close to on-ramps) in combination with the implicit cues influenced the predicted outcome of LCs. For example, when traffic density was high, the distance between two vehicles ahead was less important than when traffic density was low. These find-

ings were evaluated in a driving simulator study that focused on highway on-ramp scenarios [12]. Results showed that an early, cooperative LC was the preferred decision, resulting in higher trust, comfort, and acceptance ratings. Such LCs are also perceived as pleasant, cooperative, and predictable. Data from speech protocols and a subjective situation assessment indicated that AVs should react as early as they recognize a merging vehicle on the on-ramp.

Since interactions in traffic are frequently coordinated by accepted time gaps (gap acceptance, GA), GA parameters could be derived from manual driving and might serve as a basis in AVs' driving functions for intuitive and transparent interaction behavior [5]. An online study by the authors [13] aimed at investigating manual drivers' GA for initiating left-turn maneuvers at an intersection using real-world video material. In particular, the effect of different vehicle sizes, encountering speeds, and participants' age on GA was investigated. The results showed that participants selected larger gaps when interacting with larger, more threatening vehicles and smaller gaps during higher speeds of the approaching vehicles. In addition, older participants selected larger gaps for initiating turning actions than younger participants. Comparable results were also obtained in a lab study focusing on GA in lower speed levels in a shared space scenario [14]. Since the size and potential threat of interaction partners in traffic are strongly related, [15] investigated the effect of different interaction partners on GA by varying the interaction partners' threat and keeping their size approximately constant. In particular, participants selected the largest gaps when encountering the most threatening interaction partner. The smallest gaps were selected when encountering the least threatening vehicle, which indicated that participants anticipated the potential threat of the interaction partner for their GA when initiating driving maneuvers. A relation of driver personality traits and GA as a specific driving parameter was revealed in a more detailed analysis of driver characteristics in [16]. For instance, the results showed that participants scoring higher in sensation seeking accepted smaller time gaps, resulting in riskier decisions for initiating turning maneuvers than participants scoring lower in sensation seeking.

Due to the high demand for coordinating and anticipating driving maneuvers to achieve smooth vehicle-to-vehicle interactions in LC scenarios [7] and at intersections [8], the behavior of human road users in these scenarios needs to be further analyzed to derive required interaction capabilities that might serve as a basis for AVs' driving functions [5]. To enable AVs for transparently coordinating encounters in mixed traffic, these vehicles need to be aware of *which* driving cues are applied by human road users. Moreover, the impact of influencing factors on *specific* driving parameter values needs to be investigated. Therefore, a combined driving simulator

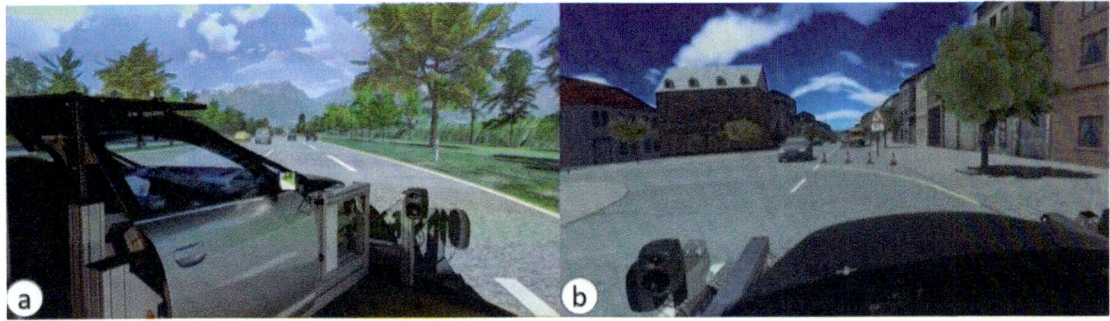

Fig. 1 **a** The fixed-based driving simulator with a 180° field of view, **b** intersection scenario of a simulator drive to collect participants' GA

study was conducted to gain a broader insight into drivers' capabilities for coordinating actions and anticipating future states of driving scenes during LC maneuvers and when coordinating actions at intersections. Besides, the approach allows for a comparison of previous research results, collected by on-road studies [10, 10] and monitor-based studies that applied video material [13, 13], with a standardized but also empirical, realistic implementation. The current driving simulator study aimed at investigating (a) implicit driving cues for anticipating LC maneuvers of surrounding vehicles and (b) manual drivers' GA for initiating left-turn actions as a specific driving parameter for coordinating interactions.

2 Method

2.1 Research Design

The study applied a within-subject design and consisted of three parts: Two simulator drives that focused on (i) applied driving cues to anticipate prospective LC maneuvers of surrounding traffic participants, (ii) participants' GA in traffic flow to initiate left-turn maneuvers at intersections. In addition, (iii) a monitor-based approach was used to collect participants' GA for left-turn actions at intersections in front of oncoming vehicles. The study parts are described below in more detail.

2.2 Apparatus and Material

A fixed-based driving simulator was used for the simulation, which was programmed in SILAB 7.0 [17]. The simulator comprised a mock-up with adjustable seats, steering wheel, speedometer, brake, accelerator pedal for vehicle control, and rear and side mirrors. The driving scenery was projected with three projectors at 180° of field view (resolution of 7680 × 1600 pixels; Fig. 1a). Participants' driving data were logged with a sampling frequency of 60 Hz.

In the drive focusing on LC maneuvers of surrounding road users, participants experienced $N = 13$ highway scenarios where surrounding traffic participants conducted lane changes or not (e.g., a fast car overtakes a slow truck). In each scenario a combination of implicit cues (e.g., longitudinal distances between two vehicles, lateral offset within the current lane, movement toward lane markings), explicit cues (e.g., turn indicator activated/ not activated) and contextual information (e.g., surrounding vehicle on highway on-ramp) could be observed, based on the results of the conducted on-road study [10].

Participants' GA was assessed in a second simulator drive (Fig. 1b). Urban intersections were presented in a balanced order. Straight roads were included between the intersections to reduce the risk of simulator sickness. The ego-vehicle had to give priority to the oncoming vehicle flow that either approached with 30 km/h or 50 km/h, which was varied per intersection. The gaps between the approaching vehicles increased ascendingly (i.e., one second per gap). When the oncoming traffic encountered with 30 km/h, gaps up to 10 s were presented, and gaps up to 12 s were presented at 50 km/h approach speed. The maximum gaps were deduced from previous research results [13, 13].

GA parameters were additionally collected by a monitor-based laptop simulation that allowed a subsequent comparison of the obtained GA results with the values of the driving simulation. The laptop simulation environment (jspsych 6.1.0, [18]) allowed for a precise collection of GA parameters. Real-world videos recorded with a Garmin Virb Ultra 30 (1920 × 1080 pixel, 100 fps) were applied as study material. The videos displayed intersection scenarios from a driver's perspective, including approaching traffic [13, 13], and were comparable with the scenarios of the driving simulation. The video perspective was intended to display potential left-turn maneuvers in front of the approaching vehicles, including hypothetically overlapping trajectories that required a coordination of actions. The oncoming vehicles approached again with either 30 km/h or 50 km/h per trial, which were presented in a randomized order.

2.3 Procedure

Holding a valid driver's license was a precondition for contributing to the study. At the beginning, participants were welcomed, general information about the study were given, and informed consent was obtained. Afterward, sociodemographic information were collected. The simulator drive on the anticipation of LC maneuvers was conducted as the first part of the study. Participants were instructed to express their thoughts about the future outcome of possible LCs aloud in speech protocols that ideally included the following information: Brief description of the observed situation; description of the observed driving behavior of surrounding road users; reference to applied cues that were used to predict prospective driving behavior of surrounding vehicles; information about the confidence of the prediction in percentage (regardless whether the prediction was correct). The participants were instructed with the task by video examples and a test drive. After the experimental drive, a semi-standardized interview was conducted. In the subsequent part of the study, participants' GA was collected by the monitor-based laptop simulation using video material. The participants were instructed to indicate the last moment for initiating a left-turn maneuver at the presented intersections in front of an approaching vehicle by pressing the enter key [13]. Before the 16 trials of data collection, four test trials were conducted. In the last part of the study, an additional simulator drive collected participants GA in traffic flow. The participants were instructed to turn left at the smallest gap in traffic flow they still perceived as comfortable. Again, a test drive was conducted to familiarize the participants with the task. Afterward, data were collected in a drive that included eight intersection scenarios. At the end, participants received 25€ for contributing to the study that lasted about 2.5 h in total.

2.4 Sample

In sum, $N = 41$ participants contributed to the study. However, $n = 12$ participants dropped out due to simulator sickness and were excluded from further data analysis, resulting in a final sample of $n = 29$ participants ($M = 34$ years, $SD = 15.27$). All participants hold a valid driver's license ($M = 16.67$ years, $SD = 13.78$) and reported an annual mileage of $M = 13,017$ km ($SD = 13,605$).

2.5 Data Preparation and Analysis

The named implicit driving cues to anticipate prospective LC maneuvers were exploratively categorized by a qualitative content analysis. The mean GA in traffic flow for conducting left-turn actions during the simulator drive was calculated

according to [19, 19], who applied a similar approach for pedestrians' crossing decisions. Thus, to consider for variations in participants' decisions to accept or reject gaps in traffic flow, logistic regressions with binary outcomes (i.e., yes/no response) were calculated. The decisions were applied to calculate individual regression models for each participant, predicting the probability that a specific gap in traffic flow was accepted to initiate a turning action (i.e., a yes response). The respective transition point of the individual regression models (i.e., 50% probability of accepting a gap in traffic flow) was defined as the mean GA for each participant and was used for further analysis [19, 19]. In the data of the monitor-based laptop simulation, single extreme outliers (i.e., ≥ three interquartile ranges under the first or over the third quartile) were excluded from further analysis. The mean GA values of the respective repetitions per condition were calculated and used for further analysis.

3 Results

Implicit driving cues to anticipate LCs of surrounding road users were investigated exploratively. An essential cue for predicting LCs of other vehicles was revealed in the speed differences between road users. Moreover, participants mentioned the longitudinal distance between surrounding vehicles as an important cue. The higher the speed difference and the smaller the longitudinal distance between surrounding vehicles, the more likely a prospective LC was assumed to be initiated. Moreover, the lateral movement toward lane markings was also noted as a relevant implicit driving cue for prospective LCs of surrounding vehicles. In addition, participants referred to contextual information that strengthened or weakened the influence of implicit driving cues for predicting the outcome of LCs (e.g., the longitudinal distance between two vehicles ahead was less important when traffic density was high compared to a low traffic density). Noticeably, when considering the logged driving data, participants showed an enhanced cooperative driving behavior toward vehicles when they anticipated an LC to be more likely, as they slowed down and gave way to other road users.

In addition, GA for coordinating interactions was investigated by a driving simulation and a monitor-based video simulation. For the driving simulator data where participants accepted or rejected gaps in traffic flow, individual regression models of each participant for the investigated speed levels of 30 km/h (mean Nagelkerke's $R^2 = 0.85$; Fig. 2a) and 50 km/h (mean Nagelkerke's $R^2 = 0.92$; Fig. 2b) were calculated. Smaller time gaps were accepted when turning maneuvers at higher speeds of the encountering vehicles were initiated, which is displayed by the individual regression lines that are distributed closer to the y-axis when the vehicles approached with 50 km/h compared to 30 km/h (Fig. 2). This difference

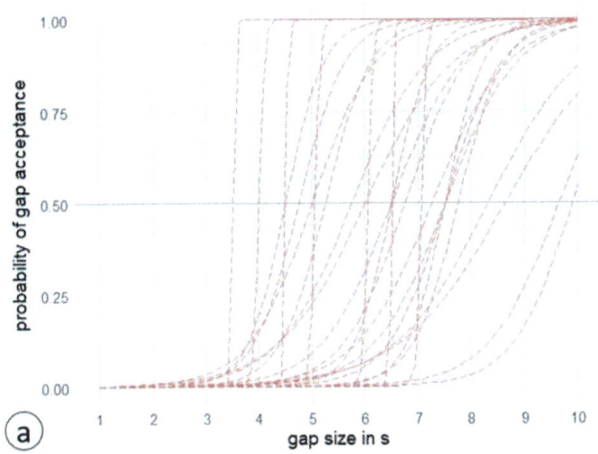
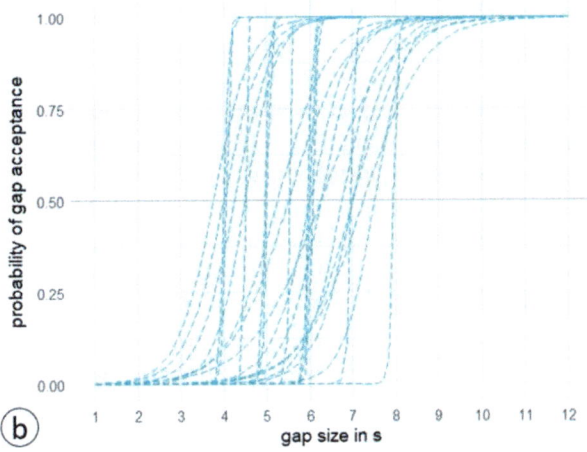

Fig. 2 Individual regression models for participants' gap acceptance behavior at intersections in the driving simulation with approach speeds of **a)** 30 km/h and **b)** 50 km/h of the encountering vehicles. Models might overlap due to participants' identical decision patterns for accepting/ rejecting gaps in traffic flow. Transition points (i.e., 50% probability of accepting a gap in traffic flow) are highlighted with a horizontal line in the graphs.

is also shown when comparing the mean accepted time gaps (M30km/h = 6.25 s, SD30km/h = 1.70; M50km/h = 5.69 s, SD50km/h = 1.22).

Comparable results were revealed for accepted time gaps that were collected monitor-based in a laptop simulation with real-world videos. In detail, participants also indicated smaller GA for an approach speed of 50 km/h (M = 4.51s, SD = 1.22), whereas larger time gaps were indicated for 30 km/h (M = 6.47s, SD = 1.67) of the encountering vehicles. According to the conducted repeated measures ANOVA, the difference between the investigated speed levels was obtained to be significant ($F(1,28) = 94.00$, $p < 0.001$, $\eta^2 p = 0.770$). There was no difference in GA between the applied methods (i.e., driving simulator vs. monitor-based laptop simulation; $F(1,28) = 2.41$, $p = 0.132$, $\eta^2 p = 0.079$). However, a significant interaction effect between the investigated speed levels and the applied methods was revealed ($F(1,28) = 66.38$, $p < 0.001$, $\eta^2 p = 0.703$; Fig. 3), i.e., higher differences between the two methods at 50 km/h.

4 Discussion

For intuitive encounters in mixed traffic, AVs need to be able to anticipate future states of driving scenes and coordinate interactions with surrounding road users [4]. Therefore, knowledge about established interaction behavior of human drivers, such as specific parameters and applied driving cues, needs to be determined and could serve as a basis for AVs' driving functions [5]. To gain a broader understanding of human drivers' interaction behavior, a combined driving simulator study that investigated (a) implicit driving cues for announcing and anticipating LC maneuvers and

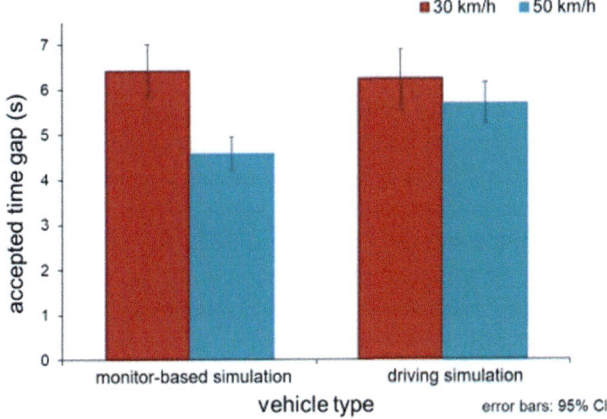

Fig. 3 Mean accepted time gaps in seconds by applied methods and the approach speed of encountering vehicles

(b) drivers' GA for initiating turning actions as a specific driving parameter was conducted, applying a within-subject approach. The results indicate that drivers use various implicit driving cues, such as speed differences or lateral movements toward lane markings of surrounding road users, to anticipate prospective LC actions. Moreover, participants adapted their accepted gaps in traffic flow when initiating turning maneuvers according to the approach speed of oncoming vehicles. The results might serve as a basis for specific driving cues and parameters in AVs' driving functions when anticipating the development of driving scenes during LC maneuvers and at intersections, and coordinating upcoming interactions in mixed traffic intuitively.

Since the coordination of actions is highly required during LC maneuvers [7], communication cues that allow for anticipating prospective LC actions of surrounding road users were investigated within the study. In line with results of previously conducted field studies of the authors (e.g., [10]), the current findings imply that implicit driving cues represent an important source of information to predict the driving behavior of surrounding road users. For instance, speed and distance differences between vehicles or lateral positioning within the lane were used when predicting LC maneuvers. Moreover, contextual information was applied as an additional source of information. The comparable findings indicate that a standardized driving simulation seems a promising approach for further investigating implicit driving cues announcing LC maneuvers. When considering the driving data in closer detail, participants were increasingly cooperative toward vehicles when they assumed LCs to be more likely. This suggests that perceiving implicit driving cues influenced the willingness to cooperate during LC maneuvers, which could support smooth encounters in traffic. However, more detail analysis of the logging data in a larger context of the LC scenarios are necessary. In general, the findings imply that knowledge about implicit driving cues should be implemented in AVs' driving functions to support smooth encounters in mixed traffic. By using such communication cues, AVs would be enabled to predict prospective driving actions of surrounding road users faster and react accordingly to experienced human drivers [21]. Moreover, AVs should apply implicit driving cues by themselves to announce upcoming driving actions transparently for surrounding traffic participants [4]. The results of GA for initiating turning actions indicated that the speed of encountering vehicles influenced participants' accepted gaps for both implementations, the driving simulation (i.e., implementation with higher external validity) and the monitor-based laptop simulation (i.e., cost-efficient implementation). Specifically, smaller time gaps were accepted at higher speeds of approaching vehicles, compared to lower approach speeds in both implementations. The results corroborate previous monitor-based studies of the authors that applied video material aiming at deviating specific GA parameters as a basis for AVs' driving functions. The findings indicate that there is not one single value for adequate time gaps in traffic flow [13, 13]. Rather, manual drivers anticipate future states of driving scenes and consider influencing factors, such as the speed of encountering vehicles, when selecting gaps for initiating turning maneuvers, as it could be shown for both implementations in the current study. These parameters and influencing factors might be used as a basis for AVs' driving functions, to provide intuitive interaction capabilities in these vehicles [13]. However, to gain a broader understanding of GA parameters, additional influencing factors should be considered in prospective driving simulator studies (e.g., potential threat of approaching interaction partner [15] or driver characteristics [16] as shown in previous online studies conducted within the project). The comparison of the applied implementations (i.e., driving simulation vs. monitor-based laptop simulation) did not reveal a significant difference in mean accepted time gaps. However, there was a significant interaction effect between approach speed and applied methods. Moreover, it needs to be mentioned that the operationalization and participants' instructions differed between the implementations. Therefore, the results are potentially limitedly comparable. Thus, a comparison of implementations and resulting differences in GA should be considered in more detail in further studies (i.e., by a more similar operationalization).

5 Conclusion

To support interactions in mixed traffic and user acceptance, AVs should apply established interaction capabilities [4]. Therefore, human road users' behavior needs to be analyzed, and thus derived driving parameters could provide a basis for intuitive AV driving functions [5]. Coordinating actions of multiple agents is, for instance, highly required when conducting LC maneuvers [7] and during turning actions at intersections [8]. The results revealed that implicit driving cues, such as speed differences or the lateral position within a lane, are applied to predict upcoming LC maneuvers of surrounding traffic participants. Moreover, gaps in traffic flow for initiating turning maneuvers at intersections are selected depending on the speed of approaching vehicles. Corresponding to the results, human road users anticipate the prospective development of driving scenes given specific cues and adapt their behavior accordingly. To provide intuitive and transparent encounters in mixed traffic, AVs should also be enabled to anticipate future states of driving scenes due to provided cues and thus adapt their own driving actions accordingly [4].

Acknowledgements The research was funded by the Deutsche Forschungsgemeinschaft (DFG, German Research Foundation—[Project- ID 416228727—SFB 1410]). We would like to thank our student assistants, particularly Maximilian Hentschel, for supporting the programming of the simulator drives and the data acquisition.

References

1. SAE International's new standard J3016. *Levels of driving automation.* SAE International (2018). https://doi.org/P141661
2. Rasouli, A., & Tsotsos, J. K. (2020). Autonomous vehicles that interact with pedestrians: a survey of theory and practice. *IEEE Transactions on Intelligent Transportation Systems, 21*(3), 900–918.
3. Dey, D., & Terken, J. (2017). Pedestrian interaction with vehicles: Roles of explicit and implicit communication. In *AutomotiveUI '17: Proceedings of the 9th International Conference on Automotive User Interfaces and Interactive Vehicular Applications,* pp. 109–113. https://doi.org/10.1145/3122986.3123009

4. Schieben, A., Wilbrink, M., Kettwich, C., Madigan, R., Louw, T., & Merat, N. (2019). Designing the interaction of automated vehicles with other traffic participants: Design considerations based on human needs and expectations. *Cognition, Technology and Work, 21*(1), 69–85. https://doi.org/10.1007/s10111-018-0521-z

5. Beggiato, M., Witzlack, C., Springer, S., & Krems, J. F. (2018). The right moment for braking as informal communication signal between automated vehicles and pedestrians in crossing situations. In N. Stanton (Ed.), *Advances in Human Aspects of Transportation. AHFE 2017*. Advances in Intelligent Systems and Computing (Vol. 597, pp. 1072–1081). Springer. https://doi.org/10.1007/978-3-319-93885-1

6. Bevly, D., Cao, X., Gordon, M., Ozbilgin, G., Kari, D., Nelson, B., Woodruff, J., Barth, M., Murray, C., Kurt, A., Redmill, K., & Ozguner, U. (2016). Lane change and merge maneuvers for connected and automated vehicles: A survey. *IEEE Transactions on Intelligent Vehicles, 1*(1), 105–120. https://doi.org/10.1109/tiv.2015.2503342

7. Malik, S., Khan, M. A., & El-Sayed, H. (2021). Collaborative autonomous driving—A survey of solution approaches and future challenges. *Sensors, 21*(11), 3783. https://doi.org/10.3390/s21113783

8. Hancock, P. A., Caird, J. K., & Shekhar, S. (1991). Factors influencing drivers' left turn decisions. *Proceedings of the Human Factors 35' Annual Meeting, 35*(15), pp. 1139–1143. https://doi.org/10.1177/154193129103501525

9. Hensch, A.-C., Beggiato, M., & Krems, J. F. (2021). Predicting lane changes by identifying sequence patterns of implicit communication cues. In N. Stanton (Ed.), *Advances in Human Aspects of Transportation. AHFE 2021. LNNS 270.* (pp. 3–10). Springer. https://doi.org/10.1007/978-3-030-80012-3_1

10. Felbel, K., Dettmann, A., Lindner, M., & Bullinger, A. C. (2021). Communication of intentions in automated—The importance of implicit cues and contextual information on freeway situations. In H. Kromker (Ed.), *HCI in Mobility, Transport, and Automotive Systems. HCII 2021.* Lecture Notes in Computer Science (Vol. 12791, pp. 252–261), Springer. https://doi.org/10.1007/978-3-030-78358-7_17

11. Felbel, K., Dettmann, A., & Bullinger, A. C., *Identification of implicit driving cues in lane change situations on German highway. A Naturalistic Driving Study* [Manuscript in preparation]. Chair for Ergonomics and Innovation. Chemnitz University of Technology.

12. Felbel, K., Dettmann, A., & Bullinger, A. C. (2022). Analysis of eye gaze given different automated driving styles in an urban environment. *IEEE International Conference on Computational Intelligence and Virtual Environments for Measurement Systems and Applications (IEEE CIVEMSA 2022)*

13. Hensch, A.-C., Beggiato, M., & Krems, J. F. (2022). Should I wait or should I go? - Deciphering implicit communication cues for cooperative interactions in left-turn scenarios. *IEEE International Conference on Computational Intelligence and Virtual Environments for Measurement Systems and Applications* (IEEE CIVEMSA 2022)

14. Hensch, A.-C., Beggiato, M., & Krems, J. F. (2023). Drivers' gap acceptance during parking maneuvers as a basis for initiating driving actions in automated vehicles. *Transportation Research Part F: Traffic Psychology and Behaviour, 92*(3), 133–142. https://doi.org/10.1016/j.trf.2022.11.008

15. Hensch, A.-C., Beggiato, M., Schomann, M. X., & Krems, J. F. (2021). Different types, different speeds—The effect of interaction partners and encountering speeds at intersections on drivers' gap scceptance as an implicit communication signal in automated driving. In H. Kromker (Ed.), *HCI in Mobility, Transport, and Automotive Systems.* HCII 2021. Lecture Notes in Computer Science (Vol. 12791, pp. 517–528). Springer. https://doi.org/10.1007/978-3-030-78358-7_36

16. Hensch, A.-C., Beggiato, M., Mandl, S., Strobel, A., & Krems, J. F. (2022). The interplay of personality traits with drivers' gap acceptance. In K. Plant & G. Praetorius (Eds.), *Human Factors in Transportation. AHFE (2022) International Conference. AHFE Open Access* (Vol. 60, pp. 329–337). AHFE International. https://doi.org/10.54941/ahfe10024641

17. Würzburger Institut für Verkehrswissenschaften GmbH, https://www.wivw.de

18. de Leeuw J. R. (2015). jsPsych: A JavaScript library for creating behavioral experiments in a web browser. *Behavior Research Methods, 47*(1), 1–12. https://doi.org/10.3758/s13428-014-0458-y

19. Petzoldt, T. (2014). On the relationship between pedestrian gap acceptance and time to arrival estimates. *Accident Analysis & Prevention, 72*, 127–133. https://doi.org/10.1016/j.aap.2014.06.019

20. Lobjois, R., Benguigui, N., & Cavallo, V. (2013). The effects of age and traffic densityon street-crossing behavior. *Accident Analysis & Prevention, 53*, 166–175. https://doi.org/10.1016/j.aap.2012.12.028

21. Krems, J. F., & Baumann, M. R. (2009). Driving and situation awareness: A cognitive model of memory-update processes. In M. Kurosu (Ed.), *Lecture Notes in Computer Science, LNCS,* (Vol. 5619, pp. 986–994). Springer. https://doi.org/10.1007/978-3-642-02806-5249_113

Experiencing Automated Vehicles in Real-Life Affects Central Aspects of Drivers' User Experience

Stefan Brandenburg[ID] and Manfred Thüring[ID]

Abstract

Drivers' willingness to use vehicle automation depends on their evaluation of its instrumental and non-instrumental properties. Instrumental properties refer to the usability, utility, etc., of the automation. Non-instrumental properties include emotions, visual aesthetics, etc. This paper presents a study in which we investigated whether drivers' use of vehicle automation in real traffic changes their evaluation of the automation's non-/instrumental properties. In a field study, thirty-eight participants completed a one-hour drive including rural roads and highways. Their user experience evaluations concerning vehicle automation were assessed before and after the drive. The results revealed that driver ratings of the instrumental qualities of the vehicle were higher after using it in real traffic compared to their expectations. No effects were found for non-instrumental qualities. However, driver ratings of usability, status, and positive emotions constantly predicted their intention to use the automated vehicle. Implications for automation design are discussed.

Keywords

Automated vehicles · User experience · Intention to use · Pre-post comparison · Field test

S. Brandenburg (✉)
Cognitive Psychology and Human Factors, Technische Universität Chemnitz, Chemnitz, Germany
e-mail: stefan.brandenburg@psychologie.tuchemnitz.de

M. Thüring
Cognitive Psychology and Ergonomics, Technische Universität Berlin, Berlin, Germany
e-mail: manfred.thuering@tu-berlin.de

1 Introduction

Automated driving fundamentally changes the driving task. Level 3 (L3) automated vehicles control their lateral and longitudinal trajectory for a certain period and specific driving scenarios. During this period, drivers are allowed to disengage from the driving task and to direct their attention toward other activities [1]. Automated vehicles potentially provide significant economic, environmental, and social benefits [2]. They may offer greater mobility to more people [3] and are expected to improve road safety, to reduce emissions [4], and to minimize congestion [5]. However, automated driving systems can only satisfy these expectations when they are accepted and applied in traffic. On-road studies indicate that the driver's experience of vehicle automation may change their adoption of this technology [6].

Factors that determine the acceptance of a technology in general have been proposed in the technology acceptance model (TAM) [7]. According to this approach, perceived usefulness and perceived ease of use greatly affect users' attitude toward a system, which in turn determines its employment. Perceived ease of use represents the degree to which the system is usable and reduces effort, while perceived usefulness represents the degree to which the system is considered as helpful and supports performance. Since the models were developed for the acceptance of technology, it is very likely that subjective usefulness and subjective usability also play a major role for the willingness to use a highly automated vehicle.

This assumption was tested in a simulator study by [8], who not only investigated the influence of perceived usability and usefulness on the intention to use an automated vehicle, but also included trust and factors from the Theory of Planned Behavior (TPB; [9]) as predictors. Participants performed a simulated drive, which consisted of periods of both automated and manual driving, followed by a survey. Data were analyzed with a linear regression model, which showed that three factors from the TBP were significant predictors: attitude

© The Author(s) 2026
B. Meyer et al. (eds.), *Hybrid Societies*, Advances in Science, Technology & Innovation,
https://doi.org/10.1007/978-3-032-03488-5_4

toward a behavior, subjective norms, and perceived behavioral control. In this context, attitude toward a behavior represents favorable or unfavorable expectations using the automation, such as safety gain or energy savings. Subjective norms reflect the belief that relevant others might approve or disapprove this behavior, and perceived behavioral control stands for the expectation of how easy or difficult it is to use the system. As a result, 49% of the variance in usage intentions was explained by the final TPB model and 44% by TAM constructs. With respect to the Technology Acceptance Model, it must be noted that only perceived usability, but not perceived usefulness, proved to be a relevant factor of impact. In addition to these findings, the present study examines whether the drivers' anticipated user experience evaluation of an automated vehicle predicts their intention to use this technology.

While a person's perceived usefulness and usability of technology require its actual employment, anticipated experience may exert an earlier influence. As described in the User Experience Lifecycle Model ContinUE by [10], 'before even interacting with a product or system in a physical sense, the user-to-be has certain expectations. These can be positive (e.g., hopes) as well as negative (e.g., fears). If negative expectations outweigh the positive ones, it might never come to an actual try out of the system.' (p. 311). Following that model, positive and negative expectations are likely to influence the intention to use a system. With respect to highly automated vehicles, this raises the question of how the experience of driving such a car influences the attitude toward the automation. This question was investigated in a simulator study by [11], who assessed trust and attitudes before and after three transitions from highly automated to manual driving. Participants' trust ratings increased from pre- to post-condition, while other measures, such as safety gain and the scanning rate of the driving scene, decreased. Our study extends this research by examining how experiencing an automated vehicle in real traffic might change the drivers' user experience judgments of automation from an a priori expectation to a post-experience impression.

2 Methods

2.1 Participants

Thirty-eight persons (15 females) with ages ranging from 24 to 54 years (M = 32 years, SD = 6 years) took part in the study. They had been holding a valid driving license for an average of 13 years (SD = 5 years). Participants had a normal or corrected-to-normal vision. Thirty-five of them (92%) reported having some experience with driver assistance systems, like adaptive cruise control or parking assistance. None of them had yet used an L2-automated vehicle or Tesla's

autopilot. Three participants (8%) reported to drive less than 8,000 km per year, 29 (76%) between 8,000 and 20,000 km per year, and 5 (16%) more than 20,000 km per year. Participants received student credits (participant hours) for their participation. This research complied with the tenets of the Declaration of Helsinki and was approved by the Institutional Review Board at Technische Universität Berlin. Informed consent was obtained from each participant.

2.2 Automated Vehicle

The automated vehicle was a 2016 Tesla Model S P75D with autopilot 2.0 functionality in standard factory settings. It offered the automation as soon as the driving situation permitted it. Whether the automation of longitudinal and/or lateral control was available was indicated by a speed sign for longitudinal control and a wheel icon for lateral control in the dashboard of the vehicle. The automation controlled the vehicle longitudinally and/or laterally once it was activated.

Longitudinal control was realized by an adaptive ACC. It was activated by pulling an automation lever on the left-hand side of the steering wheel once. It regulated the speed of the vehicle from zero to a maximum of 150 km/h and kept a 3-s time headway to any leading vehicle.

Lateral control was realized by steering assistance. It was activated by pulling the automation lever twice. The steering assistance was combined with the adaptive ACC. Thus, activating and deactivating the steering assistance equalled the activation and deactivation of the complete autopilot. Auditory feedback (two single tones following each other) and a change in color of the respective symbols from gray to blue indicated the activation of the automation. The same two single tones were presented in reversed order, and the icons changed their color from blue to gray when drivers deactivated the automation.

Additionally, several other assistance systems were always activated independently of the autopilot. These systems included a lane change assistance, a collision-avoidance system, and speed assistance. The lane change assistance showed whether the vehicle recognized the lane markings in a small picture in the dashboard. It issued a warning (i.e., vibrating steering wheel) when it detected an unwanted lane change (i.e., lane change without the use of the turn signal) between 70 and 140 km/h. The collision-avoidance system provided a visual and auditory warning in case of possible collisions. It also initiated an automatic emergency braking if necessary. Finally, the speed assistance system recognized speed limits and visualized them in the dashboard. It also provided a visual and an auditory warning (i.e., single tone) if drivers exceeded the actual speed limit.

Tesla's autopilot managed to accomplish a dynamic driving task on its own. Drivers had to remain attentive and

Table 1 Effects of the drivers' experience on the evaluation of the automated vehicle

	Measure	Before usage M (SD)	After usage M (SD)	t(36)	p	d
Instrumental quality	Usefulness	4.30 (.90)	4.63 (1.18)	2.71	0.01 *	0.49
	Usability	5.17 (.99)	5.56 (.99)	2.55	0.01 *	0.42
Non-instrumental quality	Visual Aesthetics	5.52 (.71)	5.49 (.85)	−0.32	0.74	0.05
	Status	4.31 (1.11)	4.22 (1.39)	−0.37	0.71	0.09
	Commitment	1.95 (.86)	1.89 (1.20)	0.41	0.68	0.06
Emotions	Positive emotions	4.16 (.95)	4.23 (.88)	−.39	0.69	0.08
	Negative emotions	2.44 (.79)	2.64 (.73)	−1.49	0.14	0.26
Consequences	Intention to use	4.05 (1.34)	4.45 (1.38)	−2.27	0.02 *	0.36
	Product loyalty	3.43 (1.36)	3.46 (1.42)	0.18	0.85	0.02
	Overall evaluation	3.71 (.97)	4.35 (1.15)	−3.08	0.003 *	0.51

a. *Note* * Denote significant effects

ready to resume controls of the vehicle in case it reached its system limits. The autopilot, therefore, fulfilled the criteria of an L2-automated driving functionality [1].

2.3 Questionnaires

Drivers completed the meCUE questionnaire [12, 13] to capture their expectations and experiences with the automated vehicle. This questionnaire assesses the instrumental (usefulness, usability) and non-instrumental qualities (visual aesthetics, status, commitment) of a technical device as well as users' positive and negative emotions on a 7-point Likert-type scale ranging from (1) strongly disagree, (2) disagree, (3) somewhat disagree, (4) neither agree or nor disagree, (5) somewhat agree, (6) agree, to (7) strongly agree (see also [14]). In addition, it measures the participants' loyalty to a product and their intention to use it, as well as their overall judgment on a 10-point Likert-type scale ranging from -5 to 5.

2.4 Procedure

The study comprised an introduction phase, a test phase, and an interview phase. The *introduction phase* started with the instructions of the participants, including the goals of the study, its procedure, and the capabilities and constraints of the automated vehicle. Then participants answered questionnaires assessing their expectations concerning the vehicles fulfillment of the instrumental and non-instrumental qualities. In the subsequent *test phase,* they drove a 24-km round course in the north of Berlin consisting of 11 km of rural road and 13 km of highway. Here, drivers were free to use vehicle automation whenever they wanted. They were instructed that they were responsible for the safety of the vehicle and the passengers. Drivers had to always comply

with the traffic rules. The *interview phase* started with the drivers filling in the meCUE questionnaire. A short interview assessing the drivers' experiences closed this last phase of the study. The complete procedure lasted for about 2,5 h per participant.

3 Results

A series of t-tests for dependent measures was calculated to test the effects of the automation experience of drivers on their subjective evaluation of the system. Table 1 summarizes means, standard deviations, and t-test results. It shows that the drivers' automation experience increased their ratings of most of the instrumental qualities, their intention to use the system, and their overall evaluation. However, no significant results for the non-instrumental qualities, visual aesthetics, status, and commitment, as well as positive emotions, negative emotions, and product loyalty, were obtained, all $p > .14$.

3.1 Predictive Quality of the a priori User Experience Evaluation Concerning the Intention to Use an Automated Vehicle

We computed three linear models to examine whether the participants' expectations concerning the non-/instrumental qualities of the automated vehicle and their emotions influenced their intention to use it. Table 2 summarizes the results. It shows that the expected usefulness, status, and positive emotions predicted the participants intention to use the automated vehicle before they experienced it. The three linear models quantifying the influence of instrumental and non-instrumental vehicle properties as well as emotions, account for 15% (instrumental qualities) to 26% (non-instrumental

Drivers showed reactions of surprise and interest when activating the automation (positive emotions), but also experienced automation failures (negative emotions). Again, these results were in line with the findings of [18], showing that Tesla owners are quite positive and satisfied with their vehicles despite the experience of automation errors.

Concerning the prediction of drivers' intention to use the automated vehicle, we found slightly different effects. Here, the drivers' intention to use the vehicle was predicted by their ratings of usefulness, status, and positive emotions, regardless of whether they had rated their a priori expectations or their interactions with the vehicle. This finding is not in line with previous studies. For example, [11] found that experiencing vehicle automation led to lower ratings of relief of the driving task and safety gain than expected before. Driver ratings of trust increased because of the driver's experience with the vehicle automation. Other studies assessed the development of drivers' subjective assessments over repeated automated driving experiences and found differences in driver ratings. For example, [19] showed that driver ratings of trust and acceptance stabilized at high levels after five trips with an L2-automated vehicle. [20] found that driver ratings of criticality and their ratings of perceived effort decreased when drivers repeatedly experienced the same takeover situation in a driving simulator study. However, their ratings increased again when the characteristics of the takeover situation changed.

Surprisingly, usability and negative emotions did not turn out to be significant predictors of drivers' intention to use ratings. Anecdotal evidence that was gathered during the drives suggested that participants did occasionally experience automation failures, like sudden and strong automated braking without reason. Also, they were surprised by usability issues like levers that were placed behind the steering wheel and, therefore, were hard to see. Still usability was not a predictor for their intention to use the automated vehicle.

4.1 Limitations, Conclusions, and Future Work

There are some limitations to the present study. First, the automated vehicle included L2-automated driving functionality. It performed the complete driving task and accomplished a permanent object and event detection, including a response (i.e., breaking), but drivers had to survey the actions of the automation closely and were not permitted to engage in secondary tasks. Therefore, it remains unclear whether the results of the study can be transferred to L3-vehicles or higher levels of automation.

Second, a one-hour drive might have been too short for drivers to sufficiently familiarize themselves with vehicle automation. A longer driving period could have led to a more realistic utilization of vehicle automation and in-situ evaluations of drivers [21].

Third, the study setting was still quite artificial, although drivers drove in real traffic. Ethical and safety concerns demanded very clear instructions about the capabilities and constraints of the vehicle automation and close monitoring of the drivers by two experimenters. This setting might have increased drivers' caution and attention, and their automation usage. Upcoming studies should apply field operational tests to assess drivers' behavior when they are alone in their vehicle.

The results of the present study suggest that drivers may apply automated driving functions when they believe that they are useful, go along with an increase in prestige, and evoke positive emotions. These influences remained stable from expectance to experience, which is a new insight and contrasts with the findings of previous studies. The present study is only one step in understanding the User Experience of drivers, how it changes over time, and how it affects their intention to use vehicle automation in real traffic. Future studies should replicate these findings and validate them in naturalistic driving studies. This research can facilitate the understanding of the mid- and long-term implications of automated driving concerning traffic safety.

Acknowledgements We thank Frederice Kuhn, Franziska Telle, and Ronja Schott for their support with the data assessment.

References

1. SAE International, 'Taxonomy and Definitions for Terms Related to Driving Automation Systems for On-Road Motor Vehicles (J3016)'. 2021. Accessed: Nov. 17, 2021. [Online]. Available: https://www.sae.org/standards/content/j3016_202104/
2. Department for Transport. (2015). *The pathway to Driverless Cars.* Department for Transport.
3. Casner, S. M., Hutchins, E. L., & Norman, D. (2016). The challenges of partially automated driving. *Communications of the ACM, 59*(5), 70–77. https://doi.org/10.1145/2830565
4. Tsugawa, S., Kato, S., & Aoki, K. (2011). An automated truck platoon for energy saving. In *2011 IEEE/RSJ International Conference on Intelligent Robots and Systems,* San Francisco, CA, pp. 4109–4114. https://doi.org/10.1109/IROS.2011.6094549.
5. Ertrac, E., & Snet, E. (2017). Automated driving roadmap. *ERTRAC Working Group, 7.*
6. Cooley, E. H., Sanbonmatsu, D. M., Strayer, D. L., White, P. H., & Cooper, J. M. (2022). On-Road vehicle study of the experience of automated driving. *Transportation Research Part F: Traffic Psychology and Behaviour, 87,* 444–453, https://doi.org/10.1016/j.trf.2022.04.014.
7. Davis, F. D.,. Bagozzi, R. P, & Warshaw, P. R. (1989). User acceptance of computer technology: A comparison of two theoretical models. *Management Science, 35*(8), 982–1003, https://doi.org/10.1287/mnsc.35.8.982.
8. Buckley, L., Kaye, S.-A., & Pradhan, A. K. (2018). Psychosocial factors associated with intended use of automated vehicles: A simu-

lated driving study. *Accident Analysis & Prevention, 115,* 202–208, https://doi.org/10.1016/j.aap.2018.03.021.

9. Ajzen, I. (1991). The theory of planned behavior. *Organizational Behavior and Human Decision Processes, 50*(2), 179–211. https://doi.org/10.1016/0749-5978(91)90020-T

10. Pohlmeyer, A. E., Hecht, M., & Blessing, L (2009). User Experience Lifecycle Model ContinUE [Continuous User Experience] .In *Der Mensch im Mittepunkt technischer Systeme. Fortschritt-Berichte VDI Reihe 22 Nr. 29,* A. Lichtenstein, C. Stößel, and C. Clemens, Eds. Düsseldorf, Germany: VDI-Verlag, pp. 314–317.

11. Gold, C., Körber, M., Hohenberger, C., Lechner, D., & Bengler, K. (2015). Trust in automation—before and after the experience of take-over scenarios in a highly automated vehicle. *Procedia Manufacturing, 3,* 3025–3032. https://doi.org/10.1016/j.promfg.2015.07.847

12. Minge, M., Riedel, L., & Thüring, M. (2013). Modulare Evaluation interaktiver Technik. Entwicklung und Validierung des meCUE Fragebogens zur Messung der User Experience. In *Grundlagen und Anwendungen der Mensch-Maschine-Interaktion,* pp. 28–36.

13. Minge, M., Thüring, M., Wagner, I., & Kuhr, C. V. (2016). The meCUE questionnaire: A modular tool for measuring user experience. In *Advances in Ergonomics Modeling, Usability & Special Populations,* vol. 486, M. Soares, C. Falcão, and T. Z. Ahram, Eds. Springer International Publishing, 2016, pp. 115–128. Accessed: Aug. 22, 2016. [Online]. Availablehttps://doi.org/10.1007/978-3-319-41685-4_11

14. Thüring, M., & Mahlke, S. (2007). Usability, aesthetics and emotions in human–technology interaction. *International Journal of Psychology, 42*(4), 253–264. https://doi.org/10.1080/002075907 01396674

15. Minge, M., & Thüring, M. (2018). Hedonic and pragmatic halo effects at early stages of user experience. *International Journal of Human-Computer Studies, 109,* 13–25. https://doi.org/10.1016/j.ijhcs.2017.07.007

16. Carsten, O., & Martens, M. H. (2018). 'How can humans understand their automated cars? HMI principles, problems and solutions', *Cognition. Technology & Work.* https://doi.org/10.1007/s10 111-018-0484-0

17. Thüring, M.,& Minge, M. (2014). Nutzererleben messen—geht das überhaupt? *Mittelstand-Digital: Wissenschaft trifft Praxis, 1*(1), 45–53.

18. Dikmen, M., & Burns, C. M. (2016). Autonomous driving in the real world: Experiences with tesla autopilot and summon, pp. 225–228. https://doi.org/10.1145/3003715.3005465.

19. Beggiato, M., & Krems, J. F. (2013). The evolution of mental model, trust and acceptance of adaptive cruise control in relation to initial information. *Transportation research part F: traffic psychology and behaviour, 18,* 47–57.

20. Brandenburg, S., & Roche, F. (2020). Behavioral changes to repeated takeovers in automated driving: The drivers' ability to transfer knowledge and the effects of takeover request process. *Transportation Research Part F: Traffic Psychology and Behaviour, 73,* 15–28. https://doi.org/10.1016/j.trf.2020.06.002

21. Beggiato, M., Pereira, M., Petzoldt, T., & Krems, J. (2015). Learning and development of trust, acceptance and the mental model of ACC. A longitudinal on-road study. *Transportation Research Part F: Traffic Psychology and Behaviour, 35,* 75–84. https://doi.org/10.1016/j.trf.2015.10.005

What Are You: Social Displays of "Autonomous" Machines

Sabrina Tietz

Abstract

The article addresses the social-communicative problem of the embodiment of so-called "autonomous" machines. Such machines appear as they operate autonomously, whereas their operation modes are in fact fragmented in human and technical actions. The machines' affiliations and role functions are therefore not transparent. This leads to the question of whom or with what one interacts in everyday life. Drawing on three case examples from everyday life, the *accountable* display strategies—the embodiment practices and the understanding of "autonomous" machines—are outlined and discussed alongside the social-communicative problem that arises from this.

Keywords

Embodiment · Agency · Human–machine interaction · Social displays · Autonomy · Design strategies

1 Introduction

In everyday life, for interactions with so-called "autonomous" machines, an understanding of their embodiment is required.[1]

---

[1]I refer to the collectivity of representations, capabilities and functions of machines as "embodiment"; therefore, I follow Krippendorff's perspective on semantics of used objects by the perception of their symbolic characteristics under the premise *form follows meaning* [1, 2]. For an insight on the technical and human embodiments of the social, discussing technologies as independent agencies and constitutive elements of social dynamics and structures, thus, a specific embodiment of the social, see [3–5]. For a discussion on various works relating to the concept of embodiment in the field of socially interactive robots and the characterization of a respective design space, see [6].

S. Tietz (✉)
Visual Communication and Media Sociology Institute for Media Research, Chemnitz University of Technology, Chemnitz, Germany
e-mail: sabrina.tietz@phil.tu-chemnitz.de

These are meta-communicative endeavors that include the machines' operation modes, affiliations, and role functions to be displayed toward human counterparts [4, 7]. However, the interplay of technical and human actions of such machines is often not clearly recognizable from the outside. From this arises the social-communicative problem that human counterparts cannot solely understand the multitude of agencies of an "autonomous" appearing machine through the perceivable embodiment alone. Besides the machines' *pragmatic* understanding—their operating companies and their institutional functionalities, which are usually represented by labels and inscriptions on the machines' *material* and *pictorial* displays (their appearances)—an "autonomous" machine would ideally display its operation mode, thus agency, for a communicative understanding to human counterparts in situ [8–11]. Visible and understandable machine practices of embodiment—in short: "social displays"—are therefore required so that humans can assess the machines' operation mode during everyday encounters [10–12]. Yet, "autonomous" machine embodiments vary in their agency between the operation modes of human (remote) controls, technical-autonomous functions, and shared autonomy. Technical-autonomous functions mean that machines execute tasks and functions independent of time and location of human control, which is commonly achieved by algorithmic processes, whereas shared autonomy consists of action elements that are partially automated as well as controlled by humans. These manifold agency types of different operation modes give rise to the question of whether and when someone interacts with the machine's algorithmic-autonomous program or another human, and, more importantly, how these agencies and operation modes are recognizable.

Following L. Suchman's understanding of agency—that is, the enactment of a reconfiguration of a network of humans and artefacts—the additional question arises of whether and to what extent these multitudes of "autonomous" machine agencies are *accountable* for the machines' embodiment practices [13]. The network of human and technical actions in a machine—as opposed to a holistic technical machine compilation—distributes not only the machine's responsibility, but also its *accountability*. Dating back to H. Garfinkel's conceptualization of accountability, agents permanently produce a commentability and reparability of their actions by organizing them in such a way that their actions' meaning become visible, recognizable and understandable [14, 15]. In terms of an "autonomous" machine, such *accounts* are the explanation of situational understandings embodied in practices—that is, according to the above argumentation, the *pragmatic* understanding of the machines' operation modes and affiliations as well as the attributed role functions within the social situation that let the machines become *accountable*.

In the following, three distinct case examples from everyday life are discussed regarding their ascribed autonomy. In this argumentation, attention is drawn to the machines' embodiments as "autonomous" machines based on empirical field observations. Their embodiment design would ideally make them accessible in regard to their interaction-relevant and cooperative properties, skills or limitations. However, the illustration of the machines' technical interchanges of actions between algorithmic-autonomous program actions and (remotely) controlled human actions depicts that their embodiments are opaque in regard to their *accountable* display strategies.

2 "Autonomous" Machines

Machine operation modes frequently imply the ascription of an automatic or even "autonomous" status. The conceptual idea of autonomy indicates that the machines' functions are fully automated and/or algorithmically controlled. But what actually is understood by "autonomous" machines is not fully clarified in the field of social robotics. G. Lindemann elaborates on "autonomy" in regard to human–machine interaction by explaining that machines possess "behavioral autonomy" when they can automatically process circles of input–output information without human assistance [16]. In concurrence with this reasoning of "autonomy," it needs to be emphasized that even though machines are frequently embodied as "autonomous," they are also partially automated and (remotely) controlled by humans. At this point, it should be further noted that fully "autonomous" vehicles are not yet legally allowed to be operated in German road traffics and they are also not functioning smoothly enough with manifold road traffic scenarios. This gives additional reason for the capability of manifold agency types that are yet not explicitly displayed, but exist in the "autonomous" embodiment during the everyday interactions with such machines.

The below illustrated empirical case examples are (a) the delivery robot *Starship*, (b) the zoomorphic robot *Spot* and (c) the "autonomous" bus *HEAT*. By placing these machines

into the everyday world, their capabilities are tested under criterions of the routine processes of everyday life pragmatics. This is already of relevance as L. Suchman demanded the testing of "developing objects [so that] their appropriability into those [real world] environments become a central criterion of adequacy for their design" [17]. The case examples' *pragmatic* understandings (operation modes and affiliations) and role functions are depicted in regard to their embodied "autonomy" during everyday encounters.[2]

2.1 Starship

The robot *Starship* is a vehicle delivery machine with a height of 55 cm that drives at a speed of 6 km/h on pedestrian sidewalks.[3] The machine is operated by "Starship Technologies" and can be rented by stores for their deliveries of parcels, food, or meals. A browser-based platform or smartphone app—a *digital* display quality—enables an instant delivery, where customers can order items from the distinctive stores that temporarily operate the machine. The machine has inscribed "I am a delivery robot" on its back side, stating the company's name "Starship," their phone and email contact information as well as their website. Imprinted labels such as "Hungry Hamburg? Get the app." and a QR code on the machine's sides advertise for the usage of the delivery robot and its operation via smartphone app (see Fig. 1). In terms of the machine's *pragmatic* understandings, the operating company and its intended use are clearly recognizable for third parties by its labels and vehicle shape. Yet, the actual store that rents the machine and delivers its goods is not perceivable from the machine's embodiment.

The embodied role function of the robot *Starship* is one of a "delivery agent and customer," which is similar to that of a human carrier service. However, the deliveries of meals by car or bicycles from such services usually display their company affiliations, perceivable imprinted on their vehicles, clothes and/or packages. As stores only rent the *Starship* robots, they do not have any visual affiliations from the operating stores, which thus remain anonymous to third parties during the machine's operation. The delivery process is also only known by the person who orders from the offering

---

[2] I refer to role models and the forms of interactions between machines and humans as *role functions*. These include an understanding of the represented machines' embodiments and the humans' function in an interactional situation. For an overview of the interactional role of humans involved in machines' embodiment, such as supervisor, operator, cooperator, or non-participant, see [18]. A typology of user-machine interactions, such as *assisting* or *collaborating* in regard to online platforms, can be found in [19]. Examples of agent role models, as well as explanations on how role modeling can be used to facilitate agent systems analysis and design, can be found in [20].

[3] The empirical fieldwork conducted on the *Starship* delivery robot took place for five days in Hamburg in June 2021.

Fig. 1 Illustration of the delivery robot *Starship's* shape with its app advertisement.[4]

store via the accompanying smartphone app and the store's staff, who send the delivery on its way. Both involved parties can track the machine's delivery in a map-like overview, and they receive location information, such as notifications of the machine's arrival or departure. Additionally, the machine's loading space is mechanically locked and can only be opened by the customer in accordance with the app's functions.

These circumstances lead to the question about the machine's operation mode and its altering store affiliations: The robot *Starship* operates when it is put in place by customers' orders via the accompanying app. This means that the customer telenavigates the machine through the app's order in an interplay with the store's staff, who receive the order and send the machine via their own accompanying app. The driving route of the delivery process is operated autonomously by a mapping technique: The route departure is manually recorded, and the map material is processed so that the machine is able to autonomously navigate in the mapped area.[5] The machine can thus drive from the respective stores to the customer's fixed location using feature detection and

the mapping technique to navigate through the terrain.[6] An algorithmic route planning determines the route's selection by the variables of speed, distance, and safety that enable the machine to operate in an algorithmic-automated mode. During these drives, the machine displays a light rotation signal on an orange flagpole as well as a signal sound for passing crossroads or bicycle lanes (for this, see also children's bicycles and/or flashing lights as conventionalized signs in the sense of A. Schütz). Next driving maneuvers, such as turning or stopping, are not indicated to other road users, neither by features of light nor by sound. Hence, the machine's limited driving skills are anticipated by the orange flagpole that equivales a degraded vehicle status for the machine.

If the machine's "autonomous" operation fails or it gets stuck by an insurmountable obstacle, humans, who monitor all the *Starship* robots that are in operation at a time, can take the machine over by remote control from a control centre.[7] These teleoperated actions must be distinguished from the steering functions by the customer through telenavigated actions: The takeovers by remote control are defined as teleoperated actions. These actions are fully authorized by the human at the control centre, whereas the telenavigated actions are only usability functions offered by the accompanying smartphone app, which allow the customer to steer the machine in terms of the delivery time and location, and to open or close the lid or send the machine away. Besides these telenavigated actions, the smartphone app is also an interface for the first contact with the machine. The store's usage of the machine and the machine's communicative responsiveness are displayed to the customer by means of the smartphone app, which thereby obtains a navigating-instrument character. The involved human instances (manufacturing company, offering stores, and customers) that take part in the machine's operation converge in the smartphone app and make the machine's embodied network of a (supervised) shared autonomy recognizable.

In sum, distinct operation modes of technical as well as human actions are merged within the delivery robot *Starship*. The potential changes of operation modes are not displayed on the machine's *material* and *pictorial* display level. It was not possible to observe whether the takeover by remote human control is displayed in the accompanying smartphone

---

[4] In all figures, the logos and inscriptions are blurred due to image rights protection.

[5] The mapped area comprises a radius of 6km.

[6] The machine's feature detection includes technical equipment with ultrasonic sensors, distance sensors, 10 cameras, radar, and GPS.

[7] The control centre is not necessarily placed in near distance. During the empirical field observations, it was learnt that another human is yet available on site, who can fill in if the operation in remote control also fails. Navigating the machine out of a stuck situation can be difficult to execute by teleoperated actions due to orientation problems of the human. In such cases, the person on site is instructed to release the machine from these stuck situations. Another approach for these problematic situations is the machine's *linguistical* display: When the machine is stuck on the roadside, it executes speech acts that ask for help so that bypassing pedestrians potentially help to put the machine back on the sidewalk.

app. Especially pedestrians, who encounter the machine in everyday road traffic, perceive the machine as partially anonymous, because they are not informed about its operation modes and store affiliations. The agency of the robot *Starship* is therefore three-folded by technical and human operation strategies of (1) algorithmic-automated, (2) telenavigated, and (3) teleoperated actions. The telenavigated and teleoperated human actions, as well as the degraded vehicle status of limited interaction capabilities, are technical interchanges that lead to an unclosed machine unit. Human actions are necessary and involved in the machine's operation, which also becomes evident by the embodied role function of a "delivery agent and customer," but the actual human actions are not indicated or recognizable as such. The distinctive lines of the machine's accountability—its operation modes, affiliations, and role function—blur within its embodiment due to the lack of their (visual) indications.

2.2 Spot

The robot *Spot* is a zoomorphic and mobile machine that was built for navigation in uncertain terrain.[8] The machine is frequently used for routine inspections and security tasks, such as surveying work or property monitoring. *Spot* was developed by the robotic company "Boston Dynamics" that primarily researches and develops in the field of autonomous walking robots. *Spot* runs on an electrical drive and was launched with a charging station on the market in 2021. The machine is controlled via a control unit, but running inspection routes or locating the charging station can be carried out automatically.[9] This allows the machine to conduct repetitive "autonomous" operations by recharging without human assistance. In terms of the machines' *pragmatic* understandings, it is not clearly recognizable at first sight what the machine exactly is or can do: The machine's functions of inspection operations and security patrols are merely indicated by its zoomorphic, though ambiguous shape. The shape with a four-legged locomotion enables the machine for agile movements in uneven terrains by a motion apparatus—that is, the interface between different device segments—that represents technic-mechanical joint mechanisms. The machine's square head segment consists of light and camera bars and interchangeable accessories—such as a camera attachment or a grappler for the top unit—that equip the machine with additional capabilities depending on the operations' tasks (see Fig. 2).

Originally, the machine is colored in yellow with black features and it is labeled with an inscription of the manufac-

Fig. 2 Illustration of the robot *Spot* with the manufacturing company's inscription (top left) and the operating company's color scheme, logo, and inscriptions (bottom right)

turing company "Boston Dynamics." The introduction of the machine on the market resulted into a redesign of various color schemes depending on the respective advertisement colors of the purchasing companies. For instance, the operation of *Spot* by a German police department emerged into a blue and silver color scheme with bright yellow warning stripes, which resembles the color design of police vehicles. Mobile phone providers used pink-white or red–black color schemes, which are their respective advertisement colors. In these cases, the manufacturing company's name was also complemented and/or substituted with their own company's names. Thus, the machines' embodiment in different color schemes respective to the operating company allows the affiliation of the machine to its institutional function, which is in the latter cases more specifically an advertising function for mobile phone providers.

The observed empirical case of *Spot* was operated at a security company, which recolored the machine in their company's colors of red, white, and black. In addition to the inscription of the manufacturing company's name, the operating security company added their own logo with their name "Security Robotics" and their website with an advertising slogan as a label on the machine. The inscription with two company affiliations allows an attribution of the machine to the manufacturer, the operator as well as its current operation as a security patrol machine that is indicated through the additional labels of the security company.

The robot *Spot* embodies the role function of a "dog and its owner," which is primarily achieved by its zoomorphic shape and movements. This form of embodiment anticipates that

---

[8] The empirical fieldwork conducted on the zoomorphic robot *Spot* took place during a demonstration in Leipzig in October 2021 and a promotional tour in Berlin in November 2021.

[9] The machine's control unit is a controller of a game console that consists of a screen display and control buttons.

human instances are in reach during the machine's operation as well, which is, however, only the case when the machine is directly controlled via its control unit (see below). During empirical field observations, it was observed that the machine navigates autonomously in terrain that has been prerecorded in its track for security tasks by placing QR codes as route orientation points. Thereby, the machine maneuvers the route through scanning one after the other QR code along the route. This navigation mode with static orientation points is suitable for security tasks with repetitive routes, because the machine can adjust itself again in case it deviates from the route or it needs to locate its charging station. Such seemingly "autonomous" operation is yet only an automated and tracked route navigation, which is unsuitable as an operation mode for the machine in the everyday world.

In regard to the mentioned machine's embodiment of matching the role function of a dog and its owner, it is implied that a human instance is in visible proximity to control and navigate the machine. Thereby, a human operator remotely navigates *Spot* via a control unit that is based on the operation software "Scout." Due to the human's presence—regardless of whether they are recognizable to others or not—the interactional situation that arises lies between teleoperated and telenavigated actions. The machine's "autonomous" operation is actually achieved by full remote control through teleoperated human actions from a control centre or by steered prompts through telenavigated human actions on site. Currently, the machine is operated by a person in charge on site, because the machine's operation areas are yet not well tested and established for remote control from a distant control centre. The person on site can then steer the machine by prompts only, such as location indications that will instruct the machine where to navigate to. These navigational instructions allow the machine to route its way to the determined location and to cope with emerging obstacles, such as steps, autonomously. The machine's agency is therefore recognizably and visibly linked to human's actions by the control unit on site. This counters not only the machine's anonymous agency control, but it also mitigates for an attributed "autonomy" during everyday encounters. Hence, the machine's operation modes predominantly blend teleoperated and telenavigated actions together. Algorithmic-autonomous program actions are potentially possible in terms of the machine's technical functionalities, but they are not yet fully in place due to the machine's undefined operation areas.

2.3 Heat

The "Hamburg Electric Autonomous Transportation" is an overall technical system infrastructure referred to as the "autonomous" bus *HEAT*.[10] *HEAT* was a research and development project from 2018 to 2021, investigating how fully automated and autonomous driving can successfully be integrated into road traffic. On its *material* and *pictorial* display levels, the vehicle is similarly designed to the regular public service buses of the operating company, the Hamburg HOCHBAHN. Yet, *HEAT* is an electric minibus that only carries a small number of passengers. It has a digital screen on its outside displaying the welcome message "Moin Zukunft—Hello Future," the network of bus stops, and its current location and time information. The overall perceivable system of *HEAT* consists of three subsystems: First, the vehicle's perception by cameras, radar and laser measurements; second, a roadside infrastructure of poles with sensors along the driving route so that the vehicle's field of vision is extended and an anticipatorily drive is enabled and third, an accurate HD road map and the monitoring by humans at a control centre.[11] The *pragmatic* understandings of the *HEAT* bus and its use as a public transportation vehicle are recognizable by its shape and color design of the operating company (see Fig. 3). Further project partners are clearly identifiable by labels and inscriptions on the vehicle's *material* and *pictorial* display level. Digital screens and information boards on and inside the vehicle as well as at the bus stops, give further details about the project.[12] All this information attributes an "autonomous" vehicle status to the bus. However, the bus operates automatically in terms of its communication with, e.g., traffic light systems, and fully autonomous driving maneuvers are still tested and recorded during the vehicle's testing phases. Therefore, vehicle attendants are also on board during the drives so that the vehicle can be taken over manually at any time. A corresponding vehicle design with a windshield that allows the presence of human drivers is therefore also necessary in the embodiment of the bus.

The embodied role function of *HEAT* is one of an "autonomous driver and passengers." The bus transports passengers between specific bus stops on the vehicle's test route. Due to the development project status and the

---

[10] The empirical fieldwork conducted on the "autonomous" bus *HEAT* took place for four days in Hamburg in September 2021.

[11] The video "Autonomous driving: How HEAT works in Hamburg" gives a detailed overview on the *HEAT*-system's interplay. It is accessible under https://www.hochbahn.de/de/projekte/das-projekt-heat, last accessed on 22.08.2022.

[12] A further *digital* display level of the *HEAT* bus is its accompanying smartphone application: The app is an overall information platform that states the vehicle's operating times, its current operational status ("HEAT is on") and its real-time location. It informs passengers about the vehicle's legal and safety regulations and it facilitates them to access a ticket, where they have to accept the vehicle's conditions of carriage and legal protection regulations with which the app then takes on an institutional referral entity.

Fig. 3 Illustration of the "autonomous" bus *HEAT* with the operating company's inscription (right side) and the inscriptions of further project partners (left side)

legally allowed operation regulations, the following scenario occurred during the testing phase in 2021: The vehicle, the accompanying app, the vehicle's attendants, and the humans at the control centre create a *work-sharing* vehicle status. Two attendants—one who will drive the vehicle when a manual maneuver is necessary and one who will attend to passengers' questions—are on board. These attendants are in a constant interchange with the person in the control centre for the documentation and recording of driving maneuvers that are yet not manageable for the vehicle's autonomous operation mode. Such driving maneuvers require that the vehicle be steered manually, and the person at the control centre will track and protocol the exact driving situation so that the vehicle's system can be improved with new information later on.[13] The manual vehicle takeover requires that the bus switches back in its autonomous operation mode at one of its starting positions throughout the test route. These points are at a few selected locations, which means that the bus drives autonomously in certain road sections and manually in others, because the testing conditions still required manual control for certain road traffic situations (e.g., passing of other parked vehicles on the driving lane). Thus, the interchanging operation modes of manual and autonomous agencies are switching back and forth during the vehicle's operation. *HEAT* displays next driving maneuvers, such as turning or pulling into a bus stop by flashing light, but its operation modes are not comprehensively embodied, nor recognizable for third parties.

The embodied role function of an "autonomous driver and passengers" is thus contradicted by the vehicle's attendants, because if human actions were not required for the operation of the bus, they would also be absent on site. Hence, the vehicle's *work-sharing* status constitutes a "performance team" [21], consisting of individual units that cooperate in staging and "fostering a given definition of the situation" [21]. Metaphorically speaking, the performed situation of "autonomy" of the bus *HEAT* is achieved by the overall system infrastructure of human and technical actions.

In sum, the bus *HEAT* embodies an "autonomous" machine unit that is a technical vehicle consisting of human remote and human on-site actions. Digital and analog information displays (screens and boards) further construct the "autonomous" embodiment, and the vehicle's app functions as an embodiment of the vehicle's legal responsibilities. The tasks of the vehicle's attendants on site will prospectively be removed and replaced by the operation and control through humans at the distant control centre. The vehicle's-controlled character by human remote control will nonetheless stay part of the vehicle's features, because the translation of human actions, such as overtaking maneuvers, into technical task areas is difficult to conduct.[14] The advertisements and displays of *HEAT* as an "autonomous" vehicle make the involvement and number of human actions within the technical system appear anonymously, and they blur the vehicle's actual operation modes to third parties on the outside.

3 Conclusion

The addressed social-communicative problem of the embodiment of so-called "autonomous" machines leads to the question of whom or with what one interacts with in everyday life. Especially, the machines' unclear operation modes make the assessment of their agency difficult, leading to potential problems and misunderstandings during the encounter and interaction with these machines.[15] L. Suchman already refers to the ability of artifacts "to understand the actions of the user and to provide for the rationality of its own" as they *explain themselves* [13]. In order to handle "autonomous" machines adequately and to understand these multiple-mode situations, their operation modes need to be transparently displayed to human counterparts in the everyday world. Besides the display of different agencies, the actual operation mode and

---

[13] The intervention into the vehicle's autonomous driving mode during critical driving situations is referred to as "incident management." The control centre and the vehicle's technology will ideally substitute for the vehicle's attendants in the future, cf. video "{ Autonomous driving: How HEAT works in Hamburg".

[14] Based on the current legal situation, the overtaking process in streets with only one lane must be carried out by human authorities in case of oncoming traffic. This raises the question if a vehicle will drive fully autonomously at all when such clearance functions for overtaking processes—or other functionalities such as reversing or honking—are legally not allowed to be translated into technical mechanisms.

[15] For further insights on the construction of mental models for the understanding and prediction of machines' behaviors, see [22].

possible changes must be displayed. The display of various operation modes would thereby facilitate to reduce these multiple-mode situations in their contingency.

In order to fully encompass more transparent—that includes *accountable* display strategies of "autonomous" machines—beyond the outlined empirical field observations, discussions on human perception of machines, machines' autonomy and their explainability are of importance: Beer et al. investigated the autonomy of robots in the context of HRI and they provided a framework for examining levels of robot autonomy as well as the effects on human–robot interaction [23]. Their conceptualization of autonomy as a continuum from teleoperation to full autonomy aligns with the outlined empirical field observations in regard to the involvement and the interchange of human actions in the machines' operation as well as the distribution of certain tasks between human and technical actions (e.g., smartphone app for steering functions).[16] The work of Papagni and Koeszegi [24] proposes a framework to model explanatory interactions with social robots, which emphasizes the role of explanation of machines and the context setting. In respect to the empirical field observations, no clear role explanation is given by the machines themselves was observed in any of the three case examples. Kendall [20] states how role modeling in the field of social robotics can be used to facilitate agent system analysis and design. Facets of agent roles, such as responsibilities (e.g., services or tasks) or collaborators (other roles to interact with) were identified. The dimensions of operation modes and potential multiple-mode situations are yet missing from the agent role models. In regard to the selected case examples, their partial operation in the everyday world is decisive for their current lack of formed role functions. It can be further noted that all case examples identify with machine representations beyond human likeness in respect to their embodied role functions [25]. This shows how different role functions are used for the attempted integration of such "autonomous" machines in the everyday world. However, clearly identifiable role functions as well as recognizable operation modes would facilitate to position these machines to their attributed and used functions. Humans could further perceive the machines' multitude of agencies and act accordingly, which would be of particular interest for bystanders, who are not informed about the machines' operation modes.

Hence, this contribution draws attention to the conceptual gap of *accountable* display strategies of "autonomous" machines. All three case examples demonstrate that these machines are not autonomous in a strict sense. This is mainly because the machines' role functions along their operation modes are only recently forming and emerging in the everyday world. The embodiment of agencies of the delivery robot *Starship*, the zoomorphic robot *Spot,* and the "autonomous" bus *HEAT* demonstrate that human actions are not clearly recognizable, but yet always present next to technical actions in the machines' operations. These non-transparent agencies include algorithmic-automated or autonomous actions, remotely teleoperated human actions from a control centre, and proximally telenavigated human actions on site.[17] These multiple operation modes alternate and also intermingle depending on the respective machine and its operation purpose. The embodied autonomy of the fragmented operation modes within a unified machine representation thus creates an anonymity in regard to by what or whom and at what time the operation is carried out. The "autonomous" machines' aid network, consisting of human actions as well as technical means (e.g., QR codes or light and sound beacons), comprises unidentifiable affiliations (operating stores by *Starship*), double-marked affiliations (manufacturing and operating company by *Spot*), and a multitude of affiliations and components (*HEAT*).

Accountable display strategies of "autonomous" machines thus rely on diverse up to anonymous attributions. The problem that arises from this lack of transparency is that certain interactions are not manageable for these machines, and due to a lack of situational understanding of the human counterparts, smooth interactions are interrupted. From this follows that the "autonomous" machines' usability and functionality are understandable, but their accountability is less recognizable. The involved, yet not perceivable actions, contribute to the machines' practices and humans *behind the scene*—monitoring, controlling and being responsible—are actually part of the machines' *accountable* displays.[18] Barad has aptly summarized that "agency is not an attribute but the ongoing reconfigurings of the world" [28]. In regard to the selected case examples, the fabrication and process of agencies encompass human actions that need to recognizably displayed besides their technical actions. An understanding of the operation modes, affiliations and role functions of "autonomous" machines are necessary so that humans can clearly perceive potential handling and joint cooperations up to necessary interventions for successful and smooth interactions in everyday life.

---

[16] For further information on the mentioned framework by Beer et al. and how it can serve as a taxonomy for categorizing autonomy, see Fig. 5 in [23].

[17] For a study on physical embodiments, including the physical and simulated differences of robots' embodiments and the effects of a co-located or a remote tele-present robot, see [26].

[18] Suchman & Weber point out that „discrete units of analysis [are not] separate from the more extended networks of which [they are] part" [27]. They further state that "this work takes the form of modes of representation that systematically foreground certain sites, bodies and agencies while placing others offstage" [27], which closely links to this analysis of "autonomous" machines as embodied networks.

Acknowledgements This work was funded by the Deutsche Forschungsgemeinschaft (DFG, German Research Foundation) — Project-ID 416228727—CRC 1410 "Hybrid Societies."

References

1. Krippendorff, K. (1988). *Design muß Sinn machen: Zu einer neuen Design Theorie*. Presentation at the IFK Ulm Congress, 03. September 1988. Printed version in DVA. Deutschen Verlagsgesellschaft, Stuttgart.
2. Krippendorff, K. (2006). *The semantic turn: A new foundation for design.* Taylor & Francis.
3. Rammert, W., & Schubert, C. (2017). *Technische und menschliche Verkörperungen des Sozialen.* (TUTS Working Papers, 4–2017). Berlin: Technische Universität Berlin, https://nbn-resolving.org/urn:nbn:de:0168-ssoar-56630-2.
4. Müller, M., & Sonnenmoser, A. (2023/in press). Sociomorphic technologies. On the typology of artificial actors. In B. Meyer, U. Thomas, & O. Kanoun (Eds).*Hybrid Societies—Humans Interacting with Embodied Technologies*, Vol. 1. Springer.
5. Tietz, S., & Sommer, M. (2023/accepted). Accounting AI: (Ro-) bots in algorithmic situations. HIIG Conference 2022, *AI & the Human—Cross-Cultural Perspectives on Science and Fiction.*
6. Deng, E., Mutlu, B., & Mataric, M. J. (2019). Embodiment in socially interactive robots. *Foundations and Trends in Robotics, 7*(4), 251–356, https://doi.org/10.1561/2300000056.
7. Meister, M., & Schulz-Schaeffer, I. (2016). Investigating and designing social robots from a role-theoretical perspective: Response to "Social interaction with robots—three questions". *Gesa Lindemann in AI and Society, 31*(4), https://doi.org/10.1007/s00146-015-0635-2.
8. Goffman, E. (1976). Gender displays. In: *Gender Advertisement. The Society for the Anthropology of Visual Communication*, pp. 69–77.
9. Müller, M. (2022b). Visuelle soziologie. In H. Robert & L. Karl (Eds.), *Goffman-Handbuch. Leben—Werk—Wirkung*, Frankfurt am Main 2020: J. B. Metzler, pp. 471–480.
10. Müller, M. (2023a). Social displays. Creating accountability in robotics. *Austrian Journal of Sociology* 49 (2), https://doi.org/10.1007/s11614-023-00534-2.
11. Tietz, S., & Müller, M. (2023/manuscript under review). *Alice—The Social Displays of a Robot.*
12. Müller, M. (2020). *Social displays. On the Accountability of Embodied Digital Technologies in Everyday Life.* CRC 1410 'Hybrid Societies', subproject D04, https://hybrid-societies.org/research/d04/.
13. Suchman, L. (2007). *Human-machine reconfigurations: Plans and situated actions.* Cambridge University Press.
14. Garfinkel, H. (1967). *Studies in ethnomethodology.* Blackwell.
15. Bergmann, J. R., & Meyer, C. (2021). Ethnomethodology reloaded. New Interpretations of Works and Theoretical Contributions to Harold Garfinkel's Program. Bielefeld: transcript.
16. Lindemann, G. (2016). Social interaction with robots: Three questions. *AI & Society, 31*, 573–575. https://doi.org/10.1007/s00146-015-0633-4
17. Suchman, L. (2000). *Located Accountability in Technology Production*, published by the Centre for Science Studies, Lancaster University, Lancaster LA1 4YN, UK, at http://www.comp.lancs.ac.uk/sociology/papers/Suchman-Located-Accountabilities.pdf.
18. Onnasch, L., Maier, X. & Jürgensohn, T. (2016). *Mensch-Roboter-Interaktion: Eine Taxonomie für alle Anwendungsfälle.* Fokus, https://doi.org/10.21934/baua:fokus20160630.
19. Airoldi, M. (2022). *Machine habitus: Toward a sociology of algorithms.* Polity Press.
20. Kendall, E. A. (1999). Role modelling for agent system analysis, design, and implementation. In *Proceedings. First and Third International Symposium on Agent Systems Applications, and Mobile Agents*, pp. 204–218, https://doi.org/10.1109/ASAMA.1999.805405.
21. Goffman, E. (1959). *The Presentation of self in everyday life.* Penguin Classics.
22. Wortham, R. H., Theodorou, A., & Bryson, J. J. (2016). What does the robot think? Transparency as a fundamental design requirement for intelligent systems. In *Proceedings of the IJCAI Workshop on Ethics for Artificial Intelligence: International Joint Conference on Artificial Intelligence. IJCAI 2016, Ethics for AI Workshop*, New York, USA United States, 9/07/16.
23. Beer, J. M., Fisk, A. D., & Rogers, W. A. (2014). Toward a framework for levels of robot autonomy in human-robot interaction. *Journal of Human- Robot Interaction, 2*, 74–99. https://doi.org/10.5898/JHRI.3.2
24. Papagni, G., & Koeszegi, S. (2021). Understandable and trustworthy explainable robots: A sensemaking perspective. *Paladyn, Journal of Behavioral Robotics, 12*(1), 13–30. https://doi.org/10.1515/pjbr-2021-0002.
25. Müller, M. (2023/in press). Soziomorphe Maschinen. Ein Versuch über die soziale Metaphorik der Robotik. In: E. Fricke & M. Meiler (Eds.), *Transformations—Signs and their Objects in Transition. Abstract Book of the 16th International Congress of the German Society for Semiotics.* Chemnitz: University Press.
26. Wainer, F., Feil-Seifer, D. J., Shell, D. A., & Mataric, M. J. (2006). The role of physical embodiment in human-robot interaction. In *ROMAN 2006—The 15th IEEE International Symposium on Robot and Human Interactive Communication*, pp. 117–122, https://doi.org/10.1109/ROMAN.2006.314404.
27. Suchman, L., & Weber, J. (2016). Human-machine autonomies. In N. Bhuta, S. Beck, R. Geis, H.-Y. Liu, & C. Kreis (Eds.), *Autonomous Weapons Systems* (pp. 75–102). Cambridge University Press.
28. Barad, K. (2007). *Meeting the universe halfway.* Duke University Press.

Influence of an Innovative HMI for Highly Automated Driving on Trust

Nadine Rauh, Tina Günther-Gommlich, Kira-Alyssa Maas, Cornelia Hollander, and Matthias Beggiato

Abstract

For smooth and successful interactions between human drivers and (partly) automated vehicles (AVs), drivers need to be aware of the current status of the AV, know the limitations of the technology, as well as have a correct understanding of possible actions (e.g., handover of the driving task to the automated system). Therefore, new Human Machine Interfaces (HMIs) are needed that support drivers in the respective tasks and foster awareness of the AVs' status and possibilities. To reach the full potential of such supporting HMIs, drivers' trust in the HMI is important, as it plays an important role in the extent to which a system is used. This paper addresses the question of how trust in a newly developed system that mediates between the driver and a (partly) AV (*Mediator* system) developed over time. The *Mediator* system was developed in the European project MEDIATOR. It is designed to coordinate between human drivers and autonomous vehicles based on respective fitness levels to perform current and upcoming driving tasks, including environmental factors. To examine drivers' trust, a driving simulator study was designed. Participants experienced *Mediator* in several short test drives under different conditions. Results show that drivers have initial trust in the system after receiving detailed information about *Mediator* and its functionalities. After experiencing reliable and expected system behavior during two trips, drivers trust increased constantly. In the last drive, participants experienced unexpected system behavior (a close approach to a traffic jam with a late and abrupt braking maneuver). The experience caused a momentary decrease in trust, although the overall trust rating toward *Mediator* after all drives was quite high and even higher as the expected trust. Results indicate that trust in technical systems can be quite robust even after experiencing unexpected system behavior. Adequate trust in the system and its abilities is essential to foster optimal system usage.

Keywords

Automated vehicles · Trust · HMI for (partly) automated driving · Driving simulator study

1 Introduction

Automated vehicles (AVs; SAE Level 3 or higher [1]) will become an integral part of road traffic in the near future [2]. Driving an AV requires new system knowledge and system comprehension from drivers [3]. Monitoring and clear communication of the AVs status (e.g., current driving mode) and (intended) actions (e.g., change of driving mode, need for take-over by the human driver) are essential for efficient, smooth, and safe driving [3]. Hence, new Human Machine Interfaces (HMIs) are needed to enable successful and effective coordination between driver and AV. In the European project MEDIATOR (https://mediatorproject.eu/), a new HMI called *Mediator* was developed. *Mediator* is designed to coordinate between AVs and human drivers based on respective "fitness levels" to perform the current and upcoming driving tasks. *Mediator*, for instance, is able to detect a distracted or drowsy driver, an incoming message, or an upcoming traffic jam and proposes a switch from manual

N. Rauh (✉) · T. Günther-Gommlich · K.-A. Maas · C. Hollander · M. Beggiato
Department of Psychology, Cognitive and Engineering Psychology, Chemnitz Technical University, Chemnitz, Germany
e-mail: nadine.rauh@psychologie.tu-chemnitz.de

T. Günther-Gommlich
e-mail: tina.guenther-gommlich@s2020.tu-chemnitz.de

K.-A. Maas
e-mail: kira-alyssa.maas@s2019.tu-chemnitz.de

C. Hollander
e-mail: cornelia.hollander@psychologie.tu-chemnitz.de

M. Beggiato
e-mail: matthias.beggiato@psychologie.tu-chemnitz.de

B. Meyer et al. (eds.), *Hybrid Societies*, Advances in Science, Technology & Innovation,
https://doi.org/10.1007/978-3-032-03488-5_6

to automated driving if possible. *Mediator* is expected to increase drivers' safety (e.g., prevent distracted or drowsy drivers from driving, perform a smooth approach to the rear end of a traffic jam) and comfort (e.g., allow for reading and answering messages, prevent drivers from driving through a traffic jam). For systems like *Mediator*, which are intended to reduce risks accompanying the transition from manual to automated driving, increase road safety, and improve overall driving experience, it is important that:

(a) People will buy the system.
(b) Drivers will activate the system.
(c) Drivers will follow the system's recommendations.
(d) Drivers will not interfere with the system's performance in inappropriate moments (e.g., take back manual control, although being drowsy).

One important factor for the successful implementation of AVs and the newly developed HMI is drivers' trust in the system [4, 5]. Trust plays an important role in the extent to which a system is used [6, 7] and, hence, for the potential of AVs to maintain and improve road safety [8].

1.1 Trust in the Context of Autonomous Driving

The concept of trust has many different definitions. For the current paper, trust in the newly developed HMI for automated driving (i.e., *Mediator*) was defined, according to Lee and See [5], as the driver's belief that *Mediator* will support drivers in reaching their goals in situations characterized by uncertainty and vulnerability. In complex situations like driving with an (partly) automated vehicle, trust can support decision-making processes when (a) monitoring of all influencing factors is hardly possible (e.g., current and expected drivers' fitness, state of the automated driving system, environmental factors), and/or (b) drivers don't have a complete system understanding (e.g., detailed knowledge about the used algorithms and weighting of influencing factors) [5].

Trust in technical systems is often characterized by initial positive expectations [9]. With further usage, the system's predictability and reliability become increasingly more important factors influencing trust [10, 11]. Users tend to have more trust in a system that supports them in reaching their goals reliably [12]. Reliability is an integral part of a system's performance and is based on (a) the stability of its functions [13] and (b) the frequency of failures and error messages [14]. Repeated, positive experiences with the system can increase trust [15]. However, system errors or unpredicted system behavior can decrease trust [16].

Previous studies revealed that trust is more robust if the system has high reliability at the beginning, even if later on errors occur [17]. Nevertheless, negative experiences with a

technical system have a stronger influence on users' interaction with the system compared to positive experiences [18]. This effect is quite specific, which means that an occurring error in a specific function of the system does not necessarily lead to an overall decrease in trust [19, 20]. For instance, when driving with *Mediator*, errors in the distraction detection system can lead to reduced trust in signals indicating the detection of distracted drivers, but will not necessarily have an influence on drivers' trust in the system's suggestion to change the driving mode.

Trust in a technical system can be adequate (i.e., trust matches the abilities of a system to handle the situation) but also inadequate (e.g., over- or under-trust) [20]. Inadequate trust can lead to usage errors. Over-trust can lead to system usage even in situations that the system cannot handle, and under-trust can lead to non-usage of a reliable system [21].

User characteristics can influence trust in a technical system as well [6]. The concept *affinity for technology* [22] describes a person's tendency to engage and interact with technology. Previous research in the context of autonomous driving revealed that drivers who scored high on the *ATI scale* had higher trust toward technical systems in general [23]. Further, users' age seem to have an influence on trust. Several studies revealed that older people have more trust in automated systems providing support in decision-making compared to younger people [e.g., 24]. Additionally, younger people were shown to be more skeptical toward new technology compared to older people [25].

Adequate trust in a technical system can increase the perceived usefulness and acceptance of a system, which are prerequisites for actual usage of systems like *Mediator* and autonomous vehicles in general [26].

1.2 Research Questions and Hypotheses

Two research questions (Q) were addressed with the current study. Based on existing literature, several hypotheses (H) were formulated.

Q1 How is drivers' trust in *Mediator* developing during system usage?

Previous studies revealed that trust in technical systems is often characterized by initial positive expectations [9].

Hence, the following hypothesis was formulated:

H1.1 Drivers have trust in *Mediator* from the start (i.e., $M > 4$ on the 7-point Likert scale of the *trust in automation questionnaire* from Jian et. al [27] before the first test drive with *Mediator*).

Several authors postulated that repeated, positive experiences with a technical system could increase trust [15]. Hence, the following hypothesis was formulated:

H1.2 Trust in *Mediator* will increase with positive experience after driving two and three.

Previous research could show that system errors or unpredicted system behavior can decrease trust [16]. Further, negative experiences with a technical system can influence users' interaction with the system [18]. Hence, the following hypothesis was formulated:

H1.3 Trust in *Mediator* will decrease after experiencing unpredicted system behavior in drive four.

Existing literature supports the assumption that an occurring error in a specific function of a technical system does not necessarily lead to an overall decrease in trust [19, 20]. Further, previous studies revealed that trust is more robust, if the system has high reliability at the beginning, even if later on errors occur [17]. Hence, the following hypothesis was formulated:

H1.4 Overall trust in *Mediator* will be higher than trust after drive four (i.e., $M > 4$ on the 7-point Likert scale of the *trust in automation questionnaire* from Jian et al. [27] after all four test drives).

Q2 How do driver characteristics influence trust in *Mediator*?

It could be shown that drivers who scored high on the *ATI scale* had higher trust toward technical systems [23].

Hence, the following hypothesis was formulated:

H2.1 Drivers with higher affinity for technology have more trust in *Mediator* compared to drivers with lower affinity for technology.

Previous studies revealed that older people tend to have more trust in automated systems providing support in decision-making compared to younger people [e.g., 24]. Additionally, younger people were shown to be more skeptical toward new technology compared to older people [25]. Hence, the following hypothesis was formulated:

H2.2 Older drivers have more trust in *Mediator* compared to younger drivers.

2 Method

2.1 Research Design

The study consisted of four test drives in a driving simulator equipped with the *Mediator* system. The test drives had a duration of approximately ten minutes each. The driving scenario was the same during each test drive (highway scenario with traffic jam; more details below), but the driving conditions varied in each test drive. In *Condition 1*, drivers completed the whole trip in manual mode. In *Condition 2*, drivers experienced *Mediator* in assisted driving mode (SAE level 2, [1]) in a foggy environment and were instructed to monitor the system. In *Condition 3*, drivers experienced *Mediator* in autopilot mode (SAE level 3) and were instructed to work on a secondary task presented on a laptop (i.e., reading and answering an email). In *Condition 4*, drivers experienced

Mediator again in autopilot mode and continued with the secondary task. In this condition, the approach to the rear end of the traffic jam in automated driving mode was less smooth than in the previous conditions and can be interpreted as unpredicted system behavior causing discomfort. The vehicle approached the traffic jam quite fast, and the braking maneuver was initialized quite late ("close approach"). The study had a within-subject design. All participants experienced all four conditions in a fixed order. The fixed order was chosen to enable participants to experience the advantages of automated driving and of a system like *Mediator* gradually: From manual to assisted driving and, finally, the autopilot with secondary task engagement. The system configuration with the close approach to the traffic jam was presented during the last trip to investigate its influence on drivers' trust after a positive experience with a reliable system.

2.2 Apparatus and Material

A fixed-based driving simulator with three projectors enabling a 180° field of view was used for the test drives. The driving scenarios were programmed with the simulation software SILAB 7.0 [28], presenting highly immersive representations of real driving situations. The simulator was equipped with adjustable seats, pedals for vehicle control, rear and side mirrors, the *Mediator* system, and a laptop for secondary task engagement. Driving data was constantly logged with a sampling frequency of 60 Hz.

Mediator consisted of several components. A display was mounted behind the steering wheel presenting information about the drive (e.g., current driving speed), about the automated driving system (e.g., current and available driving modes, time until driving mode change is possible or necessary), surrounding traffic (e.g., upcoming traffic jam) and *Mediator's* suggestions (e.g., change of driving mode). The display is shown in Fig. 1. The street image on the left side indicates the current and available driving mode. Gray color indicates manual driving mode, orange color indicates assisted driving mode, and purple color indicates autopilot mode. The street image is static (i.e., the image does not display the curvature of the road ahead), but the coloring adapts to current and available driving modes. Additionally, a time budget is shown, which indicates the duration of driving mode availability. In Fig. 1, the display shows driving in manual mode (the lower part of the street image where the steering wheel icon is located is gray colored). A switch to assisted driving is possible (upper part of the street image is orange). Assisted driving will be available for 3 minutes and 40 seconds. Further, information in the upper right part of the display shows that assisted driving is available. Additionally, instructions on how to activate the driving mode are given with the help of the steering wheel icon. In the middle of the display,

Fig. 1 *Mediator* display. The language was changed to English. The original display showed information and instructions in the German language

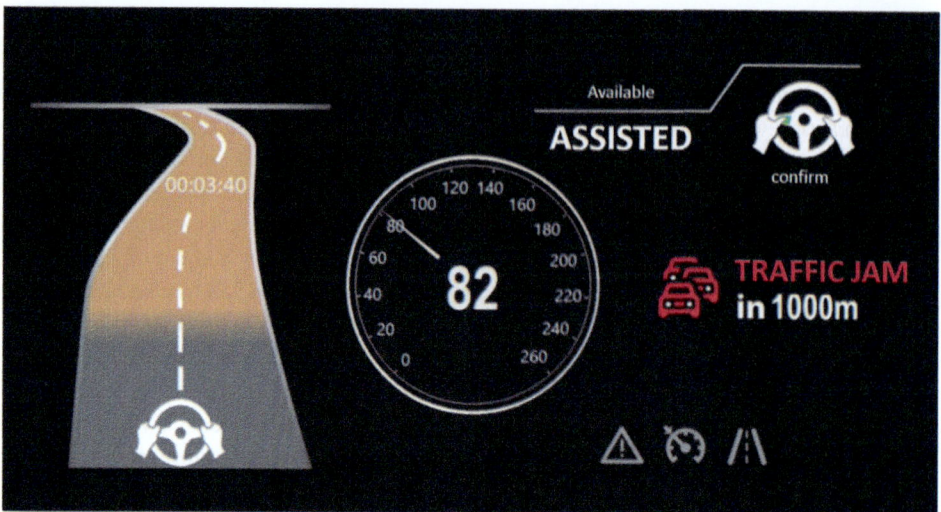

Fig. 2 Cockpit of the driving simulator showing the *Mediator* display behind the steering wheel and LED bars when autopilot mode was active (purple). A laptop was installed for the secondary task

the speedometer is presented. The color of the speedometer changes according to the current driving mode: White for manual driving, orange for assisted driving, and purple for autopilot. Next to the speedometer, additional information provided by *Mediator* can be displayed (e.g., information about an upcoming traffic jam).

In addition to the display, *Mediator* provides spoken messages informing drivers about upcoming traffic jams or incoming messages combined with the availability of assisted driving or autopilot mode. Further, *Mediator* informs about the necessity to take over vehicle control when assisted driving or autopilot mode is no longer available. LED bars on the dashboard and the steering wheel, as well as ambient lighting in the footwell of the vehicle, represent the current driving mode with respective colors. Figure 2 shows the cockpit of the driving simulator when autopilot mode was active.

2.3 Scenarios and Conditions

The driving scenario was the same in all four test drives and was characterized by the following features:

– highway with two lanes in each direction,
– average driving speed of 80 km/h,
– moderate surrounding traffic,
– good weather (clear visibility, blue sky, no rain), except for condition 2
– participants started at a rest stop and entered the highway in manual driving mode
– after several kilometers a traffic jam occurred that dissolved after some minutes

Participants experienced different conditions in each test drive, differing mainly in the available driving modes.

Condition 1 (manual driving). Participants drove the entire trip in manual driving mode including the drive through the traffic jam. *Mediator* did not offer any other driving modes, and driving mode change was not possible.

Condition 2 (assisted driving). This condition is the only one with foggy weather. A few kilometers before the traffic jam, *Mediator* informed the participants that a traffic jam was detected ahead and that assisted driving mode is available. After a successful driving mode change, participants were driven through the traffic jam in assisted driving mode. Participants were instructed to monitor the system. Secondary task engagement was not possible. After a few kilometers, the driving mode suddenly switched back to manual, and *Mediator* informed the driver of the necessity of an immediate takeover.

Condition 3 (autopilot). A few kilometers before the traffic jam, *Mediator* informed the participants about an incoming message and that autopilot mode is available. After a successful driving mode change, participants worked on the secondary task and were driven through the traffic jam in autopilot mode. When approaching the highway exit, *Mediator* informed the drivers of the necessity to take over vehicle control within the next minutes. Participants had enough time to stop working on the secondary task and follow *Mediator's* instructions to take back vehicle control.

Condition 4 (autopilot with close approach). At the beginning, the condition was the same as in *Condition 3*. When reaching the traffic jam, the vehicle approached the rear end of the traffic jam faster and initialized the braking maneuver later compared to the other conditions, leading to a close approach situation. The vehicle stopped right in time without crashing into the vehicle in front. The participants, in case they recognized the situation, had the opportunity to intervene (e.g., brake and switch back to manual driving). Participants were driven through the traffic jam in autopilot mode, or drove in manual mode if they intervened and took back vehicle control.

2.4   Procedure

Participants with a valid driver's license were invited to take part in the study. After participants were welcomed, information about the study's content and procedure was given, and informed consent was obtained. Afterward, sociodemographic information were collected and participants' affinity for technology was examined with the *ATI scale* [22].

Participants completed the first drive in manual mode (*Condition 1*) followed by a short standardized interview and a questionnaire. Afterward, *Mediator* and its functionalities were introduced, and participants filled in a questionnaire regarding their expectations of the system including the

expected trust in *Mediator*. Trust was examined with the *trust in automation questionnaire* by Jian et al. [27] throughout the whole study.

Participants then completed the second drive (*Condition 2*), the third drive (*Condition 3*), and the fourth drive (*Condition 4*), all followed by a short standardized interview and a questionnaire that included questions about participants' experienced trust. Randomly, one of the interviews after drive two, three, or four was extended to get more insights into participants' experiences with *Mediator*.

Afterward, participants filled in a final questionnaire including questions about their overall trust in *Mediator*.

2.5   Sample

In sum, $N = 81$ participants took part in the study. However, $n = 7$ participants dropped out due to simulator sickness resulting in a total of $n = 74$ complete data sets that were included in the analysis. Forty-nine percent of the participants identified themselves as female and 51% as male. Participants' age ranged from 19 to 75 years ($M = 40.5$ years, $SD = 16.57$).

3   Results

3.1   How is Drivers' Trust in Mediator Developing During System Usage?

To answer the first research question, data from the *trust in automation questionnaire* [27] were analyzed for five data collection points (see Fig. 3):

(a) After receiving a detailed explanation about *Mediator* and its functionalities to assess expected trust in *Mediator* (i.e., before drive two).

(b) After each of the three test drives with *Mediator* active (i.e., after drive two, three, and four).

(c) After all drives to assess overall trust in *Mediator*.

The first hypothesis H1.1 stated: Drivers have trust in *Mediator* from the start (i.e., $M > 4$ on the 7-point Likert scale of the *trust in automation questionnaire* from Jian et. al. [27] before the first test drive with *Mediator*). Results show that drivers' expected trust after receiving detailed information about *Mediator* but before actual usage of the system was quite high ($M = 4.73$, $SD = .70$). The results support H1.1.

The second hypothesis H1.2 stated: Trust in *Mediator* will increase with positive experience after drive two and three. Results show that driver's trust in *Mediator* increases after drive two and drive three. An Analysis of Variance (ANOVA) with repeated measures and Greenhaus-Geisser correction revealed a large effect of the data collection points: $F(2.48,$

Fig. 3 Trust in Mediator. Scale ranges from "1—totally disagree" to "7—totally agree". Error bars: 95th-CI

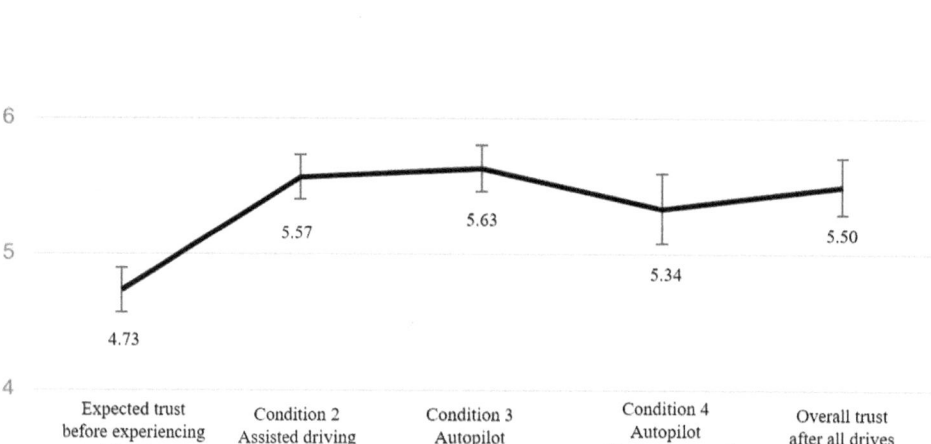

180.94) $= 30.32$, $p < .001$, $\eta p^2 = 0.29$. Post-hoc tests with Bonferroni correction revealed a significant increase from expected trust (after drive two) and actual trust after system usage (after drive two): $MDiff = 0.837$, 95%-CI[0.611, 1.062], $p < .001$. Drivers' trust in *Mediator* increased slightly from drive two to drive three, but the effect was not significant. The results support H1.2.

The third hypothesis H1.3 stated: Trust in *Mediator* will decrease after experiencing unpredicted system behavior in drive four. Results show that drivers trust in *Mediator* decreased after drive four when experiencing the close approach to the rear end of the traffic jam. Post-hoc tests with Bonferroni correction revealed a significant decrease in trust from drive three to drive four: $MDiff = -0.295$, 95%-CI[-0.590, 0.000], $p < .001$. However, the trust rating after drive four ($M = 5.34$, $SD = 1.11$) is still higher than the expected trust ($M = 4.73$, $SD = 0.70$). The results support H 1.3.

The fourth hypothesis H1.4 stated: Overall trust in *Mediator* will be higher than trust after drive four (i.e., $M > 4$ on the 7-point Likert scale of the *trust in automation questionnaire* from Jian et. al [27] after all four test drives). Results show that drivers' overall trust rating after all drives was higher ($M = 5.50$, $SD = 1.89$) compared to the trust rating after drive four ($M = 5.34$, $SD = 1.11$) and even higher than the expected trust ($M = 4.73$, $SD = 0.70$). However, the trust level after all drives did not reach the trust levels after drive two ($M = 5.57$, $SD = 1.70$) or three ($M = 5.63$, $SD = 7.40$). The results support H 1.4.

Participants' statements from the interviews underline the questionnaire results. Eighty-seven percent expressed full trust in *Mediator*, 8% stated that they trusted *Mediator* partly and another 8% expressed only low trust in *Mediator*. Sixty percent of the drivers stated positive system experience during the first drives as important factor for building up trust in

Mediator. A failure of *Mediator* (e.g., wrong or too late reactions, software errors or even accidents) were assumed to reduce trust in *Mediator* by 65% of the drivers. Further, 15% stated that unpredictable or unexpected reactions from *Mediator* might lead to a reduction in trust. Regarding trust development, participants stated:

> Until the third trip, yes. During the fourth trip with the abrupt braking at the end of the traffic jam… it has shaken my trust. (participant ID 36)

> The trust increased from trip to trip. Unfortunately, the last trip put a bit of a stop to it. Basically, when drawing the overall conclusion: Basically, you can trust it [Mediator] (participant ID 70).

Fifty-five percent of the drivers mentioned that more experience with Mediator, for instance, in real traffic, would help them to build up more trust in the system.

3.2 How Do Driver Characteristics Influence Trust in Mediator?

To answer the second research question, data from the *trust in automation questionnaire* [27] were analyzed with regard to drivers' affinity for technology ratings and age.

The first hypothesis H2.1 stated: Drivers with higher affinity for technology have more trust in *Mediator* compared to drivers with lower affinity for technology. The sample scored quite high on affinity for technology ($M = 4.45$, $SD = 0.86$). Results show a positive correlation between drivers' ATI scores and (a) expected trust ($r = 0.32$, $p = 0.005$, medium effect size according to Cohen [29]), (b) trust after drive two ($r = 0.37$, $p = 0.001$, medium effect, c) trust after drive three ($r = 0.36$, $p = 0.002$, medium effect, d) trust after drive four ($r = 0.18$, $p = 0.322$, small effect, and e) overall trust ratings ($r = 0.12$, $p = 0.341$, small effect). Hence, the higher drivers' affinity for technology, the higher their trust in *Mediator*. Results support H2.1.

The second hypothesis H2.2 stated: Older drivers have more trust in *Mediator* compared to younger drivers. The sample was divided into four age groups according to the National Highway Traffic Safety Administration (NHTSA, [30]) with similar sample sizes and gender distribution:

(a) 18–24 years ($n = 20$)
(b) 25–39 years ($n = 20$)
(c) 40–54 years ($n = 16$)
(d) 55 and older ($n = 18$)

A repeated measures ANOVA was conducted for the different data collection points including drivers' age. No significant main effects or interaction effects could be revealed. The results did not support H2.2.

4 Discussion

With increasing numbers of AVs on public roads, new HMIs like *Mediator* are essential to support monitoring and clear communication of the AVs status and (intended) actions to enable efficient, smooth, and safe coordination between AVs and human drivers [2, 3]. One important factor for the successful implementation of AVs and new developed HMIs like *Mediator* is drivers' trust in the technical systems [4, 5]. Trust plays an important role for the extent to which technical systems are used [6, 7, 26] and, hence, for the potential of systems like *Mediator* and AVs in general to maintain and improve road safety and increase driving comfort [8].

In the current study, trust in the new developed HMI *Mediator* was examined over four approx. ten-minute-long trips in a driving simulator under different conditions. $N = 74$ complete data sets were analyzed.

Drivers reported high trust toward *Mediator* after receiving detailed information about the system and its functionalities already before actual experience with *Mediator*. The finding is in line with previous research, which showed that users often trust technical systems in advance [8]. Nevertheless, the selectivity of the sample should be taken into account. Even though the sample was balanced in terms of age and gender, it can be assumed that participants, who volunteered in taking part in a study involving testing a new technology in a driving simulator observed by experimenters, were interested in technology in general and were confident to handle the situation. Actually, the affinity for technology (ATI) score was higher with a lower variance for the sample in the current study (M = 4.45, SD = 0.86) compared to a German public sample with similar age and gender distribution ($M = 3.58$, $SD = 1.09$, [22]). According to Schlüter et al. [23], high affinity for technology leads to a high and robust fundamental trust in technology. Further, drivers expe-

rienced *Mediator* and (partly) automated driving in a safe and controlled driving simulator environment. It can be assumed that drivers' trust in *Mediator* was a bit higher compared to a test drive in a real driving environment, where system failures might have more severe consequences.

After the first test drive with *Mediator* active and operating as expected, drivers' trust in the system increased significantly. Also, after the second drive with *Mediator* without unexpected system behavior or errors, trust increased slightly. Results support the assumption that repeated, positive experiences with a technical system increase trust [9–11, 15]. Nevertheless, it is important that drivers have correct system knowledge and system comprehension. Drivers need to be informed about the abilities and limits of a new system to foster realistic expectations, appropriate trust and ideal system usage (e.g., not using the system in situations it cannot handle due to over-trust, [21]). Hence, the right calibration of drivers' trust is important, and should be taken into account when designing and introducing the system, especially to inexperienced drivers. The driving simulator environment might have an influence on drivers' experienced trust in *Mediator*. The traffic environment were programmed and, hence, no unexpected situations occurred. Drivers experienced an optimal working system in a simulated environment where system errors have no severe consequences. Hence, their trust in *Mediator* might have been higher compared to a test situation in more unpredictable real traffic.

After experiencing the unpredicted system behavior (i.e., close approach to the traffic jam), drivers' trust in *Mediator* decreased supporting the assumption that trust in technical systems is negatively influenced when users' expectations and system behavior differ, even when the system performs the task without failures [16]. However, the trust rating after drive four was higher than the expected trust. Results support the assumption that drivers' trust in *Mediator* is more robust if the system has high reliability at the beginning, even if later on errors or unpredicted behaviors occur [17].

Drivers' overall trust in *Mediator* was higher compared to the trust rating after drive four indicating that unpredicted behavior or an occurring error in a specific function of a technical system does not necessarily lead to an overall decrease in trust [19, 20]. It should be taken into account, that drivers experienced only one situation with unexpected system behavior. Further research is needed to examine the influence of repeatedly unexpected system behavior in the same domain (i.e., approaching other standing or slow driving cars) as well as in different domains. Further, the late and abrupt braking maneuver was unpredictable, but is was no system failure. The vehicle came to a safe stop and no crash was caused. Further research on the influence of actual system failures on drivers' trust is needed to understand drivers' tolerance for system errors with regard to their trust in the overall system as well as in specific functionalities of the system.

Additionally, drivers' trust development after experiencing unexpected driving behavior or system failures in real traffic where consequences might be more severe should be examined. For further development and introduction of system's like *Mediator*, it is important to support drivers in gaining an adequate system understanding, including the limits of the system as well as possible margins for specific driving maneuvers (e.g., braking at the end of a traffic jam). For drivers, it is important to know if the system will handle the situation or if an actual system failure is about to happen. Transparency is important to avoid under-trust in a reliable system. Under-trust might cause drivers to intervene, although it is not necessary, possibly leading to a safety critical situation (e.g., hasty take-over action without complete understanding of the driving situation and the surroundings), or to non-usage of the system wasting the potential the system has to increase driving safety or driving comfort [21].

Nevertheless, the trust level after all drives did not reach the trust level after drive two supporting the assumption that negative experiences with a technical system have a stronger influence on users' trust compared to positive experiences [18]. Further research should examine how trust will develop after further drives with a reliable system (i.e., if trust will reach the initial level drivers had after the first drive with *Mediator*).

Regarding drivers affinity for technology, results show that the current sample has higher scores with lower variance than a comparison sample from German public [22]. Nevertheless, correlations indicate that the higher drivers' affinity for technology the higher their trust in *Mediator* supporting the assumptions by Schlüter et al. [23]. Drivers' age had no influence on trust ratings in the current study, although previous studies could show that older people tend to have more trust in automated systems compared to younger people [e.g., 24]. However, the selectivity of the sample should be taken into account. As stated above, the current sample scored high on affinity for technology. Previous research revealed that high values in affinity for technology are connected to a high and robust fundamental trust in technology [23]. It is possible that the influence of drivers' affinity for technology is stronger than possible age effects. A more diverse group regarding affinity for technology might be more promising to investigate possible age effects. Hence, future studies should focus on attracting a more diverse sample with regard to affinity for technology. One possibility might be a switch to an online study, where participants watch videos presenting *Mediator* and its functionalities. However, the reduced external validity should be taken into account.

In order to better assess drivers' trust in *Mediator* and the system's potential with respect to autonomous driving, it is important to conduct further studies. It would be particularly interesting to create different and more complex driving situations, such as driving on roads with heavier traffic or on roads with traffic lights, intersections, and other obstacles. Further, *Mediator* should be tested in real road traffic with unpredictable situations and real consequences of possible system failures. Hence, participants will have a more realistic risk perception while driving, which could also influence trust.

5 Conclusion

Mediator and autonomous driving systems are intended to increase driving safety and driving comfort, but they must be used appropriately. One important factor influencing the interaction with a technical system is trust. Trust can influence the extent to which a technical system is used. For adequate system usage, appropriate trust is necessary. Over-trust might lead to system usage in situations that the system cannot handle, and under-trust might lead to unnecessary, maybe risky interventions by the human driver, or to non-usage of the system. Hence, optimal system design and introduction supporting drivers understanding of the system including its functionalities and limits are necessary. With optimal calibrated trust and adequate usage behavior, systems like *Mediator* and autonomous driving in general might help increasing driving safety and comfort.

Acknowledgements This research was funded by the European Union's Horizon 2020 research and innovation programme under grant agreement No. 814735 (Project MEDIATOR, *mediatorproject.eu*) and by the Deutsche Forschungsgemeinschaft (DFG, German Research Foundation—[Project-ID 416228727—SFB 1410]).

References

1. SAE International's new standard J3016 (2018) Levels of driving automation. *SAE International.* P141661
2. Maurer, M., Gerdes, J. C., Lenz, B., & Winner, H. (Hrsg.). (2016). *Autonomous driving.* Springer. https://doi.org/10.1007/978-3-662-48847-8
3. Manzey, D. (2012). Systemgestaltung und automatisierung. In *Human Factors* (S. 333–352). Springer.
4. Manchon, J. B., Bueno, M., & Navarro, J. (2021). From manual to automated driving: How does trust evolve? *Theoretical Issues in Ergonomics Science, 22*(5), 528–554. https://doi.org/10.1080/1463922X.2020.1830450
5. Lee, J., & See, K. (2004). Trust in automation: Designing for appropriate reliance. *Human Factors. The Journal of the Human Factors and Ergonomics Society, 46*(1), 50–80. https://doi.org/10.1518/hfes.46.1.50_30392
6. Körber, M., Baseler, E., & Bengler, K. (2018). Introduction matters: Manipulating trust in automation and reliance in automated driving. *Applied Ergonomics, 66*, 18–31. https://doi.org/10.1016/j.apergo.2017.07.006
7. Wintersberger, P., & Riener, A. (2016). Trust in technology as a safety aspect in highly automated driving. *i-com, 15*(3).
8. Zhang, T., Tao, D., Qu, X., Zhang, X., Lin, R., & Zhang, W. (2019). The roles of initial trust and perceived risk in public's acceptance

of automated vehicles. *Transportation Research Part C: Emerging Technologies, 98*, 207–220.

9. Madhavan, P., & Wiegmann, D. A. (2007). Similarities and differences between human–human and human–automation trust: An integrative review. *Theoretical Issues in Ergonomics Science, 8*(4), 277–301.

10. Hoc, J.-M. (2000). From human—machine interaction to human—machine cooperation. *Ergonomics, 43*(7), 833–843. https://doi.org/10.1080/001401300409044

11. Sheridan, T. B. (1992). *Telerobotics, automation, and human supervisory control.* MIT Press.

12. Dzindolet, M. T., Pierce, L. G., Beck, H. P., Dawe, L. A., & Anderson, B. W. (2001). Predicting misuse and disuse of combat identification systems. *Military Psychology, 13*(3), 147–164. https://doi.org/10.1207/S15327876MP1303_2

13. Hoff, K. A., & Bashir, M. (2015). Trust in automation: integrating empirical evidence on factors that influence trust. *Human Factors: The Journal of the Human Factors and Ergonomics Society, 57*(3), 407–434. https://doi.org/10.1177/0018720814547570

14. Merritt, S. M., & Ilgen, D. R. (2008). Not all trust is created equal: Dispositional and history-based trust in human-automation interactions. *Human Factors, 50*(2), 194–210.

15. Mcknight, D. H., Carter, M., Thatcher, J. B., & Clay, P. F. (2011). Trust in a specific technology: An investigation of its components and measures. *ACM Transactions on Management Information Systems (TMIS), 2*(2), 1–25.

16. Rasmussen, J., Pejtersen, A. M. & Goodstein, L. P. (1994). *Cognitive systems engineering.*

17. Fox, J. E., & Boehm-Davis, D. A. (1998). Effects of age and congestion information accuracy of advanced traveler information systems on user trust and compliance. *Transportation Research Record, 1621*(1), 43–49.

18. Kramer, R. M. (1999). Trust and distrust in organizations: Emerging perspectives, enduring questions. *Annual Review of Psychology, 50*, 569.

19. Lee, J. D., & Moray, N. (1994). Trust, self-confidence, and operators' adaptation to automation. *International Journal of Human-Computer Studies, 40*(1), 153–184.

20. Muir, B. M., & Moray, N. (1996). Trust in automation. Part II. Experimental studies of trust and human intervention in a process control simulation. *Ergonomics, 39*(3), 429–460.

21. Parasuraman, R., Mouloua, M., & Molloy, R. (1996). Effects of adaptive task allocation on monitoring of automated systems. *Human Factors, 38*(4), 665–679.

22. Franke, T., Attig, C., & Wessel, D. (2019). A personal resource for technology interaction: Development and validation of the affinity for technology interaction (ATI) scale. *International Journal of Human-Computer Interaction, 35*(6), 456–467. https://doi.org/10.1080/10447318.2018.1456150

23. Schlüter, J., Hellmann, M., & Weyer, J. (2021). Identifikation von Fahrertypen im Kontext des automatisierten Fahrens. *Forschung im Ingenieurwesen, 85*(4), 945–955.

24. Ho, G., Wheatley, D., & Scialfa, C. T. (2005). Age differences in trust and reliance of a medication management system. *Interacting with Computers, 17*(6), 690–710.

25. Renn, O. (1986). Akzeptanzforschung: Technik in der gesellschaftlichen Auseinandersetzung. *Chemie in unserer Zeit, 20*(2), 44–52.

26. Davis, F. D. (1989). Perceived usefulness, perceived ease of use, and user acceptance of information technology. *MIS quarterly*, 319–340.

27. Jian, J.-Y., Bisantz, A. M., & Drury, C. G. (2000). Foundations for an empirically determined scale of trust in automated systems. *International journal of cognitive ergonomics, 4*(1), 53–71.

28. Würzburger Institut für Verkehrswissenschaften GmbH, https://www.wivw.de

29. Cohen, J. (2013). *Statistical power analysis for the behavioral sciences.* Academic Press.

30. NHTSA. (2016). Visual-Manual NHTSA Driver Distraction Guidelines for In-Vehicle Electronic Devices. National Highway Traffic Safety Administration. Washington, DC, USA.

Challenges in Modeling Mental Simulations of Human Drivers

Ragni Marco and Jahn Georg

Abstract

Humans apply many cognitive processes to coordinate smoothly in complex traffic scenarios. While many engineering tasks from perception to the automation of driving have been successfully studied, the core cognitive task, however, remains to be tackled: How can a formally grounded and cognitively inspired representation of the way humans mentally simulate possibilities in specific traffic situations be performed? Based on insights from cognitive science and formal knowledge representation and reasoning, we outline some constraints and formal foundations for such a framework. Limitations are discussed.

Keywords

Mental simulation · Cognitive modeling · Virtual bargaining

1 Introduction

Keeping road traffic and other domains of sharing space in public environments smooth entails avoiding and quickly resolving space-sharing conflicts among agents. Space-sharing conflicts are not exceptional events but denote all occasions, on which agents in pursuing their individual goals need to pass through or (temporarily) occupy the same space. For road traffic, traffic rules with infrastructure such as traffic lights and road signs often determine who goes first. However, there are multiple instances, even in normal, undisturbed traffic, during which agents need to quickly resolve space-sharing conflicts by coordinating their movement patterns and the timing of their movements. For example, merging into, crossing, or otherwise briefly moving into a busy lane (e.g., for overtaking or to move around an obstacle) without having the right of way would be impossible if no agent on the busy lane would step back in this space-sharing conflict.

Coordination among agents in avoiding and resolving space-sharing conflicts can involve explicit communication, but usually, implicit communication is sufficient, which may even be unilateral, in the sense that one agent remains unaware of the other responding to transmitted information for coordinating. In anticipating and predicting others' movements, humans use inferred intentions based on an interpretation of past behavior and the whole situation, including physical constraints, regulations, and social interaction. Most unregulated space-sharing conflicts are not resolved by explicit communication. Humans anticipate and predict assertive or granting behavior in unregulated space-sharing conflicts and choose their behavior complementarily based on shared norms and beliefs about what their community would consider a sensible solution for the respective case. Thus, virtual bargaining embedded in shared intentionality enables smooth coordination in space-sharing among human agents. And even in cases in which space-sharing conflicts are resolved by communication, the common understanding of the physical, regulatory, and social situation is the shared background for explicitly negotiating.

If artificial agents are to coordinate with humans in space-sharing, better than by following rules and avoiding crashes, they need to become capable of at least rudimentary anticipation and prediction of assertiveness (going first) and granting (stepping back) in space-sharing conflicts. This seems not only necessary for smooth mixed traffic, but also

1st International Conference Hybrid Societies, 2023.

R. Marco (✉)
Predictive Analytics TU Chemnitz, Chemnitz, Germany
e-mail: marco.ragni@hsw.tu-chemnitz.de

J. Georg
Applied Geropsychology and Cognition TU Chemnitz, Chemnitz, Germany
e-mail: georg.jahn@hsw.tu-chemnitz.de

B. Meyer et al. (eds.), *Hybrid Societies*, Advances in Science, Technology & Innovation,
https://doi.org/10.1007/978-3-032-03488-5_7

for preventing that defensive artificial agents are taken advantage of, get stuck, and become obstacles. They need a dynamic representation of the situation, with probability predictions for possible further movements based on inferred intentions and attentional and knowledge states of other agents. Intentions expressed by explicit communication need to be identified together with the respective addressees. In anticipating and predicting assertiveness and granting it seems worthwhile to consider variables that humans figure into virtual bargaining, such as the cost for involved agents in terms of how much and for how long an agent would be slowed down or stopped, as well as the apparent willingness to cooperate (grant, step back) also resulting from the time already spent waiting or apparent urgency of the agent.

This paper is structured as follows: In the next section, we will highlight some relevant aspects of driving situations, the role of anticipation, and the virtual bargaining processes as they have been presented in the literature. Section 2 will summarize the processes and what has been already approached by methods from AI and cognitive science. In Sect. 3, we will outline relevant features of cognitive models that can anticipate cognitive states, provide forms of we-reasoning and virtual bargaining, and solve some relevant features. A discussion of the limitations of the approach concludes the paper.

2 The Driving Situation

In all three scenarios in Fig. 1 the anticipation of the other agents' respective action or non-action is relevant for smoothly resolving space-sharing conflicts. Figure 2 illustrates some examples of low-bandwidth signals [5, 19] that agents can use to communicate possible actions. While many cognitive, affective, and social processes can have an impact, in the following, we will shed light from a cognitive perspective on the focus proposed by Chater and colleagues [5]: "we propose that each agent simulates the outcome of a hypo-

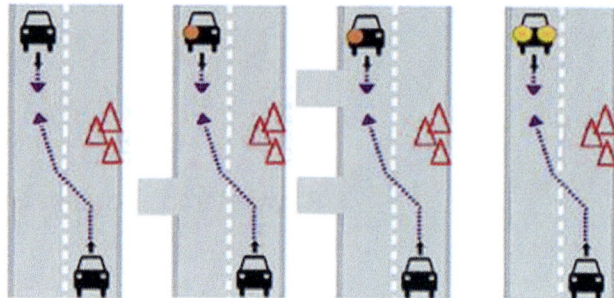

Fig. 2 A space-conflicting situation and the impact of different signals on the respective decision process that requires to take another agent's situation into account

thetical bargaining process, based on the common knowledge among agents of their beliefs and goals" (p. 94). Humans anticipate and predict assertive or granting behavior in unregulated space-sharing conflicts and choose their behavior complementarily based on shared norms and beliefs about what their community would consider a sensible solution for the respective case. For example, one agent slows down and lets a second agent merge into its lane because this is the resolution of the space-sharing conflict that both as members of the community of traffic participants consider sensible to keep traffic flowing. Agents infer what they would agree, were they able to communicate. Thus, virtual bargaining embedded in shared intentionality enables smooth coordination in space-sharing among human agents. And even in cases in which space-sharing conflicts are resolved by communication, the common understanding of the physical, regulatory, and social situation is the shared background for explicitly negotiating. 'If we could negotiate, what agreement would we reach about who does what and when?' [5]. Taking the perspective of algorithmic cognitive modeling, we need to ask and answer a number of questions before we can develop a system that is able to mimic the human decision processes that can be used in turn to inform autonomous systems in such situations to foster smooth coordination (e.g., [17]).

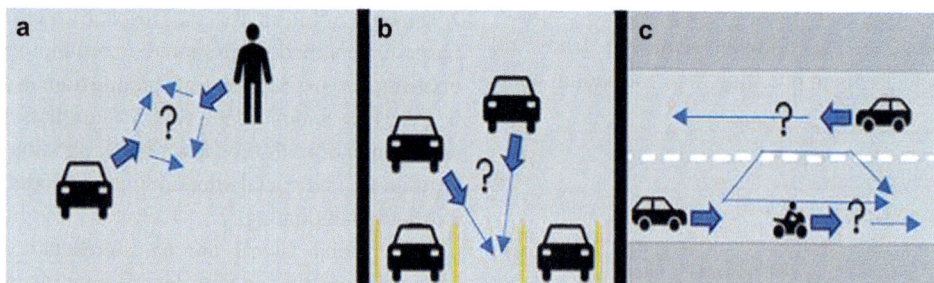

Fig. 1 Three scenarios in space-sharing conflicts. In scenario A, a human and a car need to anticipate the decision of the other partner; the next best coordination would be a quick cooperative response to the first observable action. In scenario B, one driver needs to step back. In scenario C, three vehicles are involved—two cars and a bicycle. The violet dots describe the possible trajectories (the three scenarios have been taken from [16])

1. How do human agents mentally simulate such space-sharing conflict situations? How difficult is this mental simulation computationally?
2. When and how do they identify some situations as more critical than others?
3. When and how do general knowledge (in the sense of experience) and specific individual factors such as impatience have an impact on the decision process?

Before we can analyze these questions, we need a language to formally describe the spatial situations in traffic scenarios. This will allow us in the following to estimate the non-negligible computational effort that is involved in such situations. The scenarios above can all be described by an initial scenario (the current situations the agents are in), a set of possible goal scenarios (a future situation that can be described by some constraints, e.g., circumventing a pedestrian), and a set of actions (including steering, etc.), that each consist of a precondition and an effect, respectively. The set of actions can have a subset of allowed actions (these are actions that are in accordance with traffic regulations), but even non-allowed actions, such as crossing the middle line. A plan is a sequence of actions that transforms the initial scenario into one of the goal scenarios. As the driving scenario is a multi-agent scenario, the execution of an action can depend on the actions of other agents, and the actions need to be coordinated in space and time. Non-actions can lead to a deadlock situation, and if both agents move toward the same location at the same time, an accident can ensue. Hence, the task of the agents is to find a synchronized plan. Ideally, this happens quickly without explicit communication such as giving light signals, honking, or gesturing. Implicit communication is usually involved, and this happens via trajectories and acceleration patterns, for pedestrians and cyclists, also via body posture and gaze. But humans can also anticipate and plan for traffic situations presented as static images. Predicting and planning by virtual bargaining are possible even without implicit communication, and this is the cognitive capability that we focus on here.

Finding a synchronized plan is part of epistemic planning in AI and in general the problem of finding an epistemic plan is even undecidable; however, decidable subclasses exist [2]. Given this pessimistic result, how can humans still find plans in minimal time that are satisfactory, i.e., resulting in successful coordination? Is the driving subdomain potentially easier to deal with? A previous analysis shows that restricting the scenario to a street with cars (that can be represented by intervals that have fixed size [22]), is in general intractable (more specific it is NP-hard; cf. [18, 21, 22]), a result that is demonstrated for all temporalized spatial scenarios (e.g., temporalized cardinal direction scenarios [9, 18]). A consequence of showing that a general problem is intractable is that finding optimal solutions becomes in general too time consuming. Both AI and cognitive science have demonstrated

for decades that computationally intractable problems can either be approached by heuristics or by a different form of representation that allows for the problem to be solved more efficiently. A recent analysis showed that in more than 90% of the problems, epistemic plans can be found by simple human-inspired heuristics [1]. In the following, we will focus on the specifics of the mental simulation that humans may employ in such situations. Cognitive science can already elucidate some factors of the underlying cognitive representations and processes (cf. [5]). On a first look, one might assume that the mental simulation is just of a film-like nature in the mind of the driving agent in Fig. 2. In contrast, mental simulations adhere to at least three basic principles—the iconic nature of the mental models in simulations, the temporal unfolding of the model in time, and that models are schematic and more parsimonious than visual images [11]. The kinematic simulations investigated by Khemlani et al. have dealt with single-person algorithmic representations. Given was an initial scenario of wagons that had to be rearranged into a specific order, and to do so, humans had to mentally envisage the respective steps—to find a plan that can transform the initial scenario into a goal scenario. The case in Fig. 2 above is different, as the task is to transform the initial scenario together with another agent into the final scenario. Hence, if we keep the three basic principles above, how do we need to extend the simulations to take that into account? With another agent, the situation changes, and there are more possibilities available, and possibilities can even change while an agent deliberates (it is a so-called exploration scenario in AI, [23]). How can we represent that? Reference [12] introduced a possible world semantics. It formalizes that we are in a specific world w0 and there are other worlds w1, w2,... that are within a reachable relation of the initial world. This general idea has led to the field of modal logic and is used in many formal domains to describe the different processes.

Each of the worlds in Fig. 3 above represents a possible situation that can be described by propositions that are true. This includes the snapshot of the world state and the actions that another person can take as well as hypotheses about the other agents (we will return later to how we can model another agent's knowledge). The semantics of the possible worlds is investigated in modal logic, but there are some limitations of possible worlds that have been overcome in a mental model-like representation of possibilities, demonstrating the same descriptive power, but in a more cognitively plausible way [10]. How can these be applied to model the situation in Fig. 2? First, propositions that can be evaluated as true or false alone are not sufficient enough to describe the relations between the cars, the street, obstacles, and so on. Humans use spatial qualitative relations in language to describe the spatial scenarios above (see, e.g., [24]), and many cognitive computational models use qualitative relations as mental representations too (see [20]). Qualitative

Fig. 3 A representation of the possible world semantics representing the different possibilities. This reflects state-space representation, that is in terms of time and action the respective world can change. Please note that it is not assumed that any agent builds up a full world model. The respective representation depends on the situation, possible agents' actions, and other aspects. An agent forms this about its knowledge but can form one for other agents too. Each level can be blended into a general representation (by following conceptual blending processes, e.g., [8])

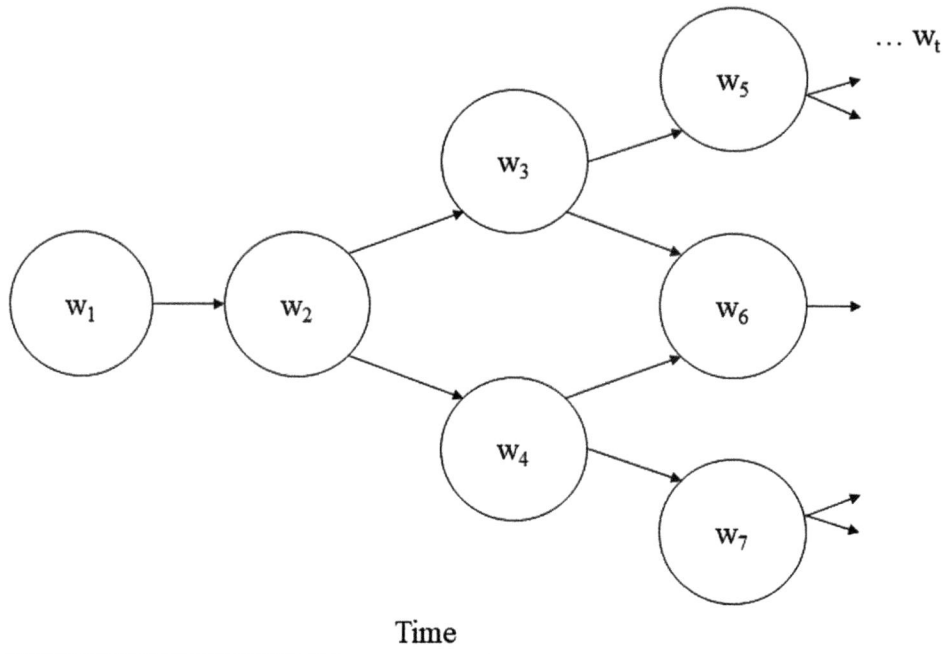

Time

spatial relations abstract from metrical details and summarize similar states into equivalence classes ([18]). AI has identified many relations that can be part of allocentric and egocentric calculi and can describe many spatial situations adequately (for an overview see [6]) and to some extent, the cognitive adequacy of such spatial calculi has been previously demonstrated. Hence, each world is a snapshot of the spatial relations between agents in a respective world at a time point t. Relevant and critical situations emerge when an agent's simulation contains a world wt where the two cars share the same space. The mental simulation process identifies such potentially critical situations, and an agent formulates hypotheses about the possible outcomes of its actions. Probabilistic outcomes of actions in a respective state can be modeled in the classical sense by the fraction of all states that have the preferred outcome divided by all states that are reachable from the respective world. This allows for the description of high- or low-probability outcomes from actions. The agent attempts to avoid potentially critical situations in plans, but since critical situations depend on the synchronization process, they still cannot be excluded with certainty. This happens especially when the projected timelines of both agents do not synchronize. While we may implicitly have assumed a discrete time representation, in principle, the time representation in the mental simulation could be continuous too. It is still an open research question whether mental kinematic simulations with respect to time may assume a discrete timeline (even with very small steps) or require a continuous representation.

What we have not yet represented in our modeling framework are the intentions of other agents and how they can

affect the simulation. The proposed modeling mechanism can be easily extended to the spatial representation of the first agent, the second agent, and any other agent; this follows straightforwardly. Beyond providing a framework for mental simulation, it is, however, important to model perspective-taking, i.e., that agent 1 can model what agent 2 may know and how agent 2 perceives actions (see the major point [5]). How can this be realized within the modeling framework? So far, a world in the framework contains the spatial relational representation and actions that one agent may perform, but to model the knowledge of another agent requires the representation of another agent's epistemic relations and actions. While there is the danger of an infinite regression, as [5] pointed out, it is obvious that an agent does not represent the knowledge of others to an infinite degree. In our modeling framework, this is reflected by a restricted amount of the other's epistemic knowledge, i.e., just the situationally relevant information for the respective situation is represented. Consider again Fig. 2. What agent 1 (the driver that has the obstacle on its path) needs to simulate is the trajectory of agent 2, the possible actions of agent 2, and the possible mental simulation that agent 2 performs on the actions of agent 1. Hence, agent 1 needs to have a mental simulation of the other agent's epistemic knowledge, too. We can assume that each agent formulates hypotheses about the other's planned action. Such formulated hypotheses are a realization of prediction theories (cp. [5]). Now with each action an agent perceives, it updates its simulation of the other agent's model. In this sense, each hypothesis is tested by each ongoing observation, and so the world model is respectively refined.

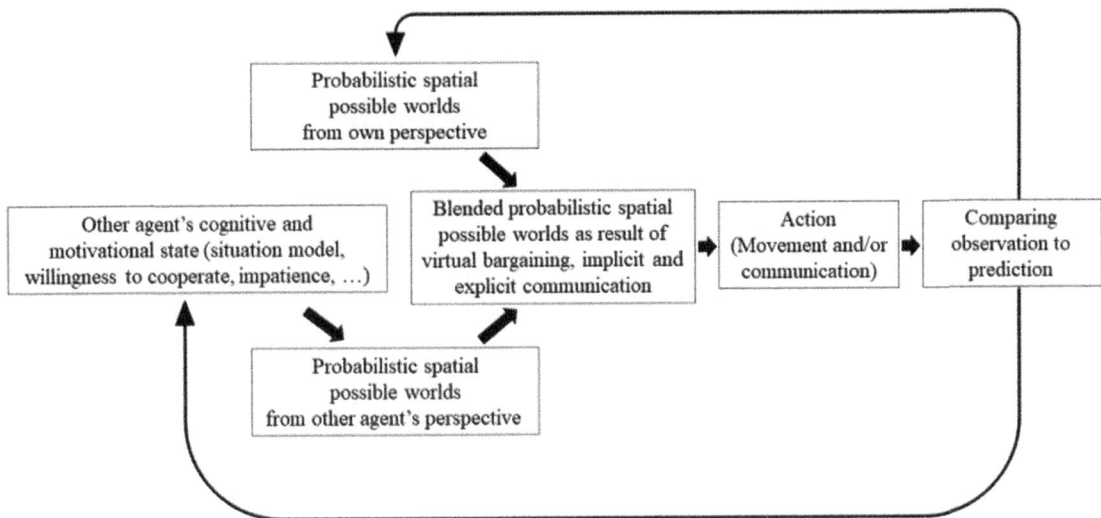

Fig. 4 An abstract representation of the proposed virtual bargaining processes and plan alignment

It becomes necessary to embed the different representations into one general mental model representation (see Fig. 4). References [7] and [8] have outlined a process of how conceptual blending can merge separate connected representations into a.general combined mental representation. This approach has already been applied to many different domains, especially in creative problem-solving. Reference [6] has proposed a general and algorithmic approach based on answer set programming, a nonmonotonic logical framework. The underlying approach is that commonalities are represented and mapped into the same mental space. Within this (technically sophisticated) process an integration of agent's 1 model and its simulation of agent's 2 process is leading to a broader general structure that not only contains one's own simulation, but also the simulation of the other agent. We have now developed an intuition based on algorithmic processes for how finding a synchronized plan may proceed by mental simulation, performed by individual agents who take other agents into account (spatial aspects of virtual bargaining). We have not yet covered a number of additional factors that are relevant in such scenarios, such as implicit communication, an individual's preference, the states and traits of agents (own and those perceived in others) that can have an impact on the selection of actions or the respective mental representation. For instance, the impatience of a driver that can lead to the selection of different actions—even in the same traffic scenario. We all know that time pressure may lead to risky behavior. The intuition is clear—all of them pose constraints on the possible worlds that are represented. These constraints eliminate possible worlds and paths of the simulation of possible worlds in the framework. The semantics of the virtual bargaining process within the framework is then the selection of those actions with which the integration of the different levels can take place.

For example, human impatience can be associated with a mathematical function that maps time to "a temperature": The higher the temperature the less actions that maintain the current state, causing impatience are maintained. How can such models be combined with the existing models? A decision for a plan is deciding on a path through the possible worlds. With an increasing temperature, the worlds that do not change the respective agent's state would be eliminated. This needs to be further formally specified, but it gives an intuition of how behavioral tendencies can be formalized and represented within the framework.

3 Discussion

Autonomous and natural agents should smoothly coordinate their actions. They may synchronize their possible actions by mentally simulating spatial trajectories, forming hypotheses, and going through a process of virtual bargaining. Using a possible world semantics allowed us to introduce a framework to represent possibilities and explain how the spatial trajectories and the other agent's possible and most likely actions can be modeled. It provides a computational and cognitive framework to integrate the simulation theories and prediction-based theories (cp. e.g., Table 2 in [5], p. 422). Both types of theories assume the simulation of possible situations and are based on hypotheses and predictions that are formed, similarly to the mental model inspection process [12]. Through the mental simulation of both—an agent's own world model and the world model of other agents, a virtual bargaining process takes place. Based on these cognitive processes, computationally complex problems can be solved without using too much computational resources. This approach integrates and realizes some theoretically outlined concepts, but it needs to be tested on a still-to-be identified set of benchmark problems (that can be formed based on [14] and relevant scenarios that Chater and colleagues proposed).

The here proposed modeling approach does not yet assume any specific architecture of cognition such as ACT-R and has respectively not implemented any working memory limitations, cognitive bottlenecks, etc. Such an integration becomes especially fruitful if individual differences need to be modeled, or if it is of interest how an agent's attention process can modulate the processing of information in some specific situations.

Machine learning has shown impressive results especially in identifying and classifying general patterns. While a broad variety of test and training materials exists (see [25]) for forming a specific hypothesis based on the other agent in a specific situation, each situation is unique, and each agent is unique, so a situation cannot be regarded as a general inductive problem. However, the respective dynamicity that emerges out of the concert of actions, non-actions, and the passing time requires an updating process on a mental simulation. It requires more of the common-sense models (see [15]) that allow a smooth coordination. Kinematic simulations proposed by [13], prediction, and virtual bargaining are necessary. Social signals, impatience, and other factors can eliminate possible paths. While humans can easily employ common-sense reasoning in many situations, finding algorithmic models or even crisp definitions is difficult and still a grand challenge for AI [3, 4]. Achieving human-like reasoning always relates back to the core quest for finding appropriate models in cognitive architectures ([3], p. 12,247) and poses still hard challenges for AI and machine learning.

Acknowledgements This work was funded by the Deutsche Forschungsgemeinschaft (DFG, German Research Foundation)—Project-ID 416228727—CRC 1410. The authors would like to thank Joshua Blickle for redesigning the pictures and for proof reading.

References

1. Berger, L., Nebel, B., & Ragni, M. (2020). A heuristic agent in multi-agent path finding under destination uncertainty. *In German Conference on Artificial Intelligence (Künstliche Intelligenz)* (pp. 259–266). Springer, Cham.
2. Bolander, T., Jensen, M. H., & Schwarzentruber, F. (2015). Complexity results in epistemic planning. *In Twenty-Fourth International Joint Conference on Artificial Intelligence.*
3. Brachman, R. J., & Levesque, H. J. (2022). Toward a New Science of Common Sense. *Proceedings of the AAAI Conference on Artificial Intelligence, 36*(11), 12245–12249. https://doi.org/10.1609/aaai.v36i11.21485
4. Brachman, R. J., & Levesque, H. J. (2022b). Machines like us: toward ai with common sense. MIT Press.
5. Chater, N., Misyak, J., Watson, D., Griffiths, N., & Mouzakitis, A. (2018). Negotiating the traffic: can cognitive science help make autonomous vehicles a reality? *Trends in Cognitive Sciences, 22*(2), 93–95.https://doi.org/10.1016/j.tics.2017.11.008.
6. Dylla, F., Lee, J. H., Mossakowski, T., Schneider, T., Delden, A. V., Ven, J. V. D., & Wolter, D. (2017). A survey of qualitative spatial and temporal calculi: Algebraic and computational properties. *ACM Computing Surveys (CSUR), 50*(1), 1–39.
7. Eppe, M., Maclean, E., Confalonieri, R., Kutz, O., Schorlemmer, M., Plaza, E., & Kühnberger, K.-U. (2018). A computational framework for conceptual blending. *Artificial Intelligence, 256*, 105–129.
8. Fauconnier, G., & Turner, M. (1998). Conceptual integration networks. *Cognitive Science., 22*(2), 133–187. https://doi.org/10.1207/s15516709cog2202_1
9. Fauconnier, Gilles, Turner Mark (2003). Conceptual blending, form and meaning. recherches en communication. *19*. https://doi.org/10.14428/rec.v19i19.48413
10. Gerevini, A., & Nebel, B. (2002). Qualitative spatio-temporal reasoning with RCC-8 and Allen's interval calculus: Computational complexity. *In ECAI 2*, pp. 312–316.
11. Johnson-Laird, P. N., & Ragni, M. (2019). Possibilities as the foundation of reasoning. *Cognition, 193*, Article 103950.
12. Johnson-Laird, P. (2008). *How we reason.* Oxford University Press.
13. Khemlani, S. S., Mackiewicz, R., Bucciarelli, M., & Johnson-Laird, P. N. (2013). Kinematic mental simulations in abduction and deduction. *In: Proceedings of the National academy of Sciences, 110*(42), 16766–16771.
14. Kripke, S. (1963). Semantical considerations on modal logic. *Acta Philosophica Fennica, 16*, 83–94.
15. Levesque, H. J. (2017). Common sense, the Turing test, and the quest for real AI. Mit Press.
16. Markkula, G., Madigan, R., Nathanael, D., Portouli, E., Lee, Y. M., Dietrich, A., Billington, J., Schieben, A., & Merat, N. (2020). Defining interactions: A conceptual framework for understanding interactive behaviour in human and automated road traffic. *Theoretical Issues in Ergonomics Science, 21*(6), 728–752. https://doi.org/10.1080/1463922X.2020.1736686
17. Millard-Ball, A. (2018). Pedestrians, autonomous vehicles, and cities. *Journal of Planning Education and Research, 38*(1), 6–12. https://doi.org/10.1177/0739456x16675674
18. Moratz, R., & Ragni, M. (2008). Qualitative spatial reasoning about relative point position. *Journal of Visual Languages & Computing, 19*(1), 75–98.
19. Portouli, E., Nathanael, D., & Marmaras, N. (2014). Drivers' communicative interactions: On-road observations and modelling for integration in future automation systems. *Ergonomics, 57*(12), 1795–1805. https://doi.org/10.1080/00140139.2014.952349
20. Ragni, M., Brand, D., & Riesterer, N. (2021). The predictive power of spatial relational reasoning models: a new evaluation approach. frontiers in psychology, 12.
21. Ragni, M., & Wölfl, S. (2006). Temporalizing cardinal directions: from constraint satisfaction to planning. KR, 6, 472–480.
22. Ragni, M. (2006). Reasoning in dynamic environments: temporalized intervals. in workshop at KI 2006 (p. 15).
23. Russel, S. J., & Norvig, P. (2013). Artifical Intelligence—A modern approach. Prentice.
24. Tenbrink, T. (2008). Space, time, and the use of language. In Space, Time, and the Use of Language. De Gruyter Mouton. *J. Clerk Maxwell, A Treatise on Electricity and Magnetism*, 3rd ed., *2*. Oxford: Clarendon, 1892, pp.68–73.
25. https://dataoceanai.com/dataset-category/asr/jsf/jet-engine:all-datasets/tax/languages:170/

A Preliminary Evaluation of a Body-Attached Multisensor Measurement Framework for Hand Gesture Recognition

Rajarajan Ramalingame⊙, Bilel Ben Atitallah⊙, and Olfa Kanoun⊙

Abstract

This paper presents a framework for hand gesture recognition based on the information fusion from a body-attached multisensor technology comprising a smart glove, a smart band, and an inertial measurement unit (IMU). The smart glove is integrated with nanocomposite filament strain sensors developed from carbon nanotubes (CNT) dispersed in thermoplastic polyurethane (TPU) and extruded as filaments using a micro-compounder. The smart band is integrated with nanocomposite pressure sensors developed from CNTs dispersed in a silicone polymer, polydimethylsiloxane (PDMS), and deposited as a thin sheet that is cut as circular disks and coupled with underlying interdigital electrodes. The (IMU) comprising of a three-axis accelerometer, a three-axis gyroscope, and a three-axis magnetometer and is fabricated along with the sensor interface and single processing unit. The paper elaborates on the development of the nanocomposite sensors and the performance of these three sensor technologies by studying ten American Sign Language (ASL) gestures representing numbers 1 to 10 without the involvement of a sophisticated machine learning algorithm for gesture classification. The outcome of this investigation provides valuable insights into the performance of the individual sensor technology in comparison to their counterparts and sets guidelines for further development of such body-attached systems in terms of the type and selection of sensor technology and the required number of sensors. The hardware architecture of the developed framework can be directly implemented as a potential tool for human–machine interface-related activities.

Keywords

Gesture recognition · Filament strain sensors · Nanocomposite pressure sensors · IMU · Body attached sensor network

1 Introduction

Nowadays, human activity monitoring become a more and more interesting and important aspect, especially for the hybrid societies and human–machine interaction field. Human activity recognition techniques are elaborated in many applications such as robotic control [1], gait analysis [2], rehabilitation and healthcare [3], tele-manipulation [4], etc. One of the most interesting and challenging facets of the human activities monitoring field is hand gesture recognition as its demand is increasing due to the huge number of applications [5]. On the other hand, Gesture recognition is a highly challenging research field to the complexity of the hand's muscular structure, the relatively small size, and the similarities of finger movements. Therefore, mainly two well-known measurement approaches are adapted, investigated, and enhanced for recognizing hand signs and gestures, which are vision base and sensor-based respectively. The developed studies based on camera systems show high accuracy in recognizing the hand postures under specific and several conditions, e.g., number of cameras, number of markers, luminosity, angle of the vision, etc. On the other hand, those systems face a lot of challenging aspects in recognizing the hand signs and finger movements or even relatively small movements as the vision-based systems are classified as non-direct measurement systems and highly sensitive to the environmental factor. Therefore, recognizing gestures and signs will be more challenging and requires a very complex hardware system and image processing [6]. To

R. Ramalingame (✉) · B. B. Atitallah · O. Kanoun
Professorship of Electrical Measurements and Sensor Technology,
Chemnitz University of Technology, Chemnitz, Germany
e-mail: rajarajan.ramalingame@etit.tu-chemnitz.de

B. B. Atitallah
e-mail: Bilel.ben-atitallah@etit.tu-chemnitz.de

O. Kanoun
e-mail: olfa.kanoun@etit.tu-chemnitz.de

B. Meyer et al. (eds.), *Hybrid Societies*, Advances in Science, Technology & Innovation,
https://doi.org/10.1007/978-3-032-03488-5_8

overcome these limitations, sensor-based systems are developed and studied, where various measurement techniques are investigated and integrated for the recognition and identification of hand gestures, e.g., electromyography (EMG) for the detection of muscle action potential during performing actions but the output signals are noisy and require complex signal processing for the analysis [7], bioimpedance-based measurements electrodes such as electrical impedance myography (EIM) and electrical impedance tomography (EIT) but they are influenced by the electrodes and skin wet [8]. Furthermore, force myography (FMG) like strain and pressure sensors are directly reflecting the applied forces on the fingers level and forearm respectively, which are also robust to environmental and physiological factors [9]. Similarly, inertial measurement units (IMU) including accelerometer, gyroscope, and magnetometer is a robust solution again all influencing factor, which reflects the information of the hand position in the space. Nevertheless, for detection of individual finger movements the IMU sensors must be installed on each finger which could lead to bulky glove designs.

Each of the techniques has its advantages and limitations, especially in dealing with complex gestures. In this direction, several studies focus on developing and evaluating sophisticated systems. To overcome the inaccuracy aspects, hybrid measurement systems are proposed that combine two or more techniques to reduce the ambiguity between gestures and to have better recognition results. Nowak et al. [10] used the simultaneous measurements of EMG and FMG for hand prosthetic control. In another study, Jiang et al. [11] developed collocated (exact location) EMG and FMG sensing units. A forearm band composed of eight units was used to detect ASL 0 to 9 digits, captured from five subjects. The classification accuracy of the combined data was 91.6% compared to the separate models of FMG, and EMG with 80.6%, and 81.5% values, respectively. Jiang et al. [12] used a wristband composed of four EMG sensors and an inertial measurement unit (IMU) to detect twelve hand gestures (four surface gestures and eight air gestures). The experiments were done on ten subjects and were repeated three times, where the IMU was significant for space movements. From another perspective, Wajahat et al. [13] developed a flexible strain sensor for Human–machine interface and health monitoring requirements. For testing under both tensile and compressive strain, a printed sensor was fixed onto the forefinger of the gloves. During the resistance measurement, only four ranges were selected for human motion detection from 0 to 25% with 6% steps. In the same direction, Jaehwan Ko et al. developed a conductive textile, which is attached to the top of each glove finger for the detection of dynamic movements [14]. From the state of the art, the advantage and limitations of each measurement technique are highlighted. Then, hybrid systems are presenting a potential solution to improve the accuracy of gesture recognition. However, a system that combines both finger movement detection and corresponding muscle movement detection has not been studied. Furthermore, the integration of measurement methods, simultaneous acquisition, and real-time synchronization of signals are important aspects, which are difficult to full fill with multiple discrete systems.

In this paper, a hybrid and compact system is presented including an IMU unit, a pressure band based on nanocomposite pressure sensors, and a smart glove based on a nanocomposite strain sensor. The system is used to evaluate and investigate ten hand gestures belonging to American Sign Language (ASL). A preliminary evaluation with one test subject and a comprehensive study of the performance and potential of each measurement technique (IMU, pressure band, smart glove) in correlation to the selected gestures is carried out.

2 Multisensor Measurement Framework

A. *Nanocomposite Filament Strain Sensors Based on Smart Glove*

The fabrication procedure of the nanocomposite filament strain sensors involves three distinct steps: The first step involves the preparation of the thermoplastic polyurethane (TPU) solution. This is achieved by selecting the right shore harness of the TPU (in this research TPU-Soft565 pellets have been used with a shore hardness of 65) and the suitable organic solvent (in this research tetrahydrofuran (THF) has been used) to completely and homogeneously dissolve the TPU pellets. The procedure involves heating the mixture of the TPU-THF at 70 °C (temperature slightly higher than the boiling point of THF) while constantly agitating the mixture using a magnetic stirrer operated at its maximum capacity of 1500 rpm for 2 h. The time taken is decisively based on the fully dissolved TPU pellets in THF under visual inspection. The second step involves the dispersion of multi-walled carbon nanotubes (CNT) in the TPU solution. To achieve this, the CNTs are pre-dispersed in the same organic solvent (THF) used to dissolve TPUs to maintain compatibility and ease the synthesis procedure. The pre- dispersion involves homogeneous dispersion of 2 wt.% CNTs in THF using ultrasonication at an amplitude of 30% for 15 min and magnetic stirring at a rate of 1500 rpm for 15 min. The prepared pre-dispersion is then gradually introduced into the TPU solution and subjected to the second stage of ultrasonication at 45% amplitude for 30 min and then under a magnetic stirrer at a rate of 1500 rpm for 30 min with a temperature of 70 °C to prevent hardening of TPU. The last step of the fabrication procedure involves the extrusion of the nanocomposite filament strain sensors. To achieve this, the prepared TPU-CNT nanocomposite solution is spread over a Teflon mold (100 mm × 100 mm × 1 mm) and allowed to dry under an extraction hood to facilitate the evaporation of THF and simultaneous hardening of TPU. Roughly after 2 h the complete mixture is cured and peeled off the mold. The nanocomposite sheet is then shredded into small pieces to resemble the original TPU pellet size before being introduced into the micro-compounder. A twin screw co-rotating micro-compounder serves the purpose of extruding the prepared TPU-CNT pellets. The micro-compounder is set at a temperature of 180 °C to melt and extrude the TPU-CNT pellets at a constant force of 5 mN through an exit nozzle with a diameter of 1 mm, resulting in a filament with an average diameter of 1 mm. The extruded filament is cut into small pieces of 5 cm in length and contact wires are attached to either end using the crimp connectors. A nitrile-coated working glove was selected for this research and the prepared filament strain sensors are fixed to the glove using instant glue. One filament strain sensor is attached to each finger with the center of the filament strain sensor corresponding to the middle knuckle of each finger. Figure 1 shows the steps involved in fabricating the glove with filament sensors.

B. *Nanocomposite Pressure Sensors Based Smart Band*

The fabrication process of the nanocomposite pressure sensors involves two distinct steps: The first step involves the pre- dispersion of the CNTs in the organic solvent THF. The aim is to achieve a homogeneous dispersion and de-bundling of individual CNTs from the bundles (agglomerates). As CNTs are held together by Vander walls forces of attraction, they exhibit a higher tendency to form bundles and remain as bundles. Hence, significant energy is required to separate the individual CNTs from the bundles, which is achieved by subjecting the CNT bundles to a combination of shear forces injected by ultrasonication at an amplitude of 15% for 15 min and magnetic stirring at a rate of 1500 rpm for 15 min. The ultrasonication serves the purpose of separating the CNTs from their bundles, the magnetic stirring serves the purpose of uniformly distributing the separated CNTs in the solvent and

the solvent serves the purpose of maintaining the suspended CNTs in a stable state for a defined period. In the second step, the prepared pre-dispersion is then introduced into the soft thermosetting polymer, polydimethylsiloxane (PDMS), and the procedure of dispersion is repeated. At this stage, different dispersion parameters have been used on account of the high viscosity of the PDMS. Thus, the ultrasonication was performed at an amplitude of 30% for 15 min and the magnetic stirring was performed at a rate of 1500 rpm for 30 min. As PDMS is a thermoset polymer, the crosslinking agent must be added to complete the curing process and as per the manufacturer guidelines, the cross-linker was added to the prepared nanocomposite material in the ratio of 10:1. After a subtle mixing to ensure the uniform distribution of the cross-linker the nanocomposite material is drop-casted in a Teflon mold with the dimensions 100 mm × 100 mm × 1 mm. the mold is placed under an extraction hood to allow the evaporation of the solvent and then stored in the oven for over 2 h at a temperature of 120 °C. after the curing process, the nanocomposite sheet is peeled off the Teflon mold and the individual sensor elements are cut out as circular disks with a diameter of 2 cm using a laser. The sensor is combined with an underlying interdigital electrode structure fabricated by screen printing the structure with conductive silver ink on a Kapton substrate of 125 μm thick. Eight such prepared sensors are fixed to a wrist support band from "Actesso Medical Supports" using Alpidex kinesiology tapes. Figure 2 shows the steps involved in fabricating the band with pressure sensors.

C. *Inertial Measurement Unit*

The IMU comprises an LSM6DSO, a combined accelerometer and gyroscope, and a magnetometer LIS2MDL by ST Microelectronics. Both the chips are interfaced with the ESP32 microcontroller over the I2C serial bus interface. The LSM6DSO is a system-in-package featuring a high-performance 3-axis digital accelerometer and 3-axis digital

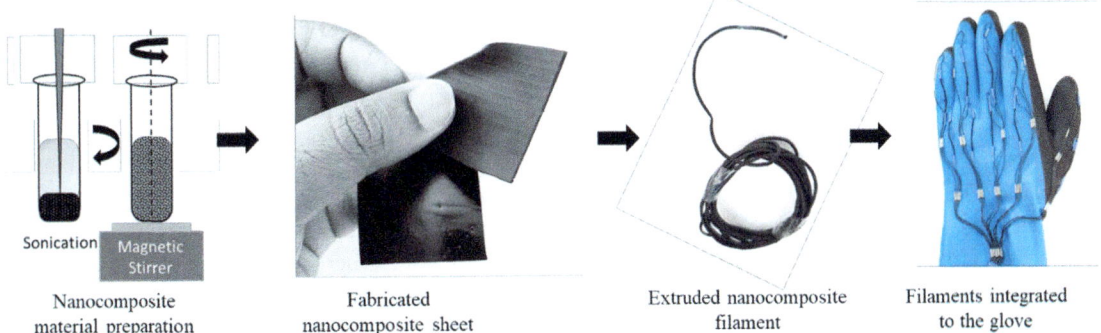

| Nanocomposite material preparation | Fabricated nanocomposite sheet | Extruded nanocomposite filament | Filaments integrated to the glove |

Fig. 1 From left to right: simplified illustration of the nanocomposite preparation, fabricated nanocomposite material as a thin sheet, micro-compounder used to extrude the filaments, picture of the extruded nanocomposite filament, and picture of the glove integrated with the developed nanocomposite filament sensors

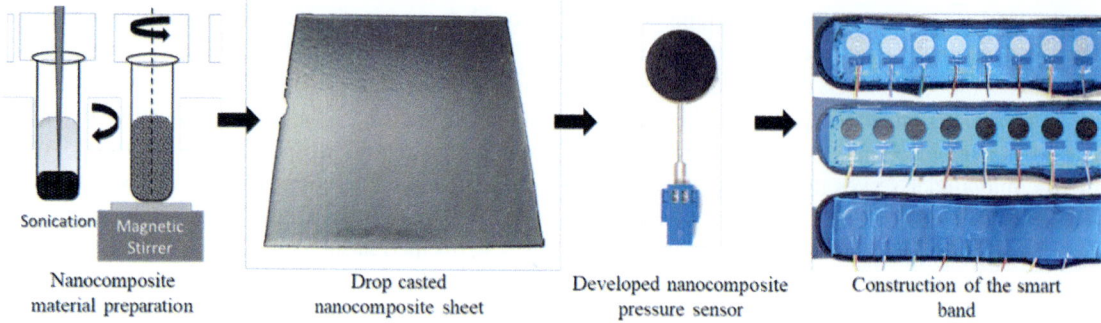

Sonication   Magnetic
              Stirrer

Nanocomposite   Drop casted   Developed nanocomposite   Construction of the smart
material preparation   nanocomposite sheet   pressure sensor   band

Fig. 2 From left to right: simplified illustration of the nanocomposite preparation, fabricated nanocomposite material as a thin sheet, picture of the developed pressure sensor attached to the underlying electrode, pictures of the construction of the smart band, and picture of a user wearing the smart band in the forearm

gyroscope. The LSM6DSO delivers best-in-class motion sensing that can detect orientation and gestures. The event-detection interrupts enable efficient and reliable motion tracking and contextual awareness. It excels in stationary/motion detection, tilt, pedometer functions, timestamping, and supporting the data acquisition of an external magnetometer. The LIS2MDL is an ultra-low-power, high-performance 3-axis digital magnetic sensor with a magnetic field dynamic range of $\pm$ 50 gauss. It supports fast mode, fast mode plus, and high-speed (100 kHz, 400 kHz, 1 MHz, and 3.4 MHz) necessary for static and dynamic gesture/motion tracking. The device can be configured to generate an interrupt signal for magnetic field detection and has a wide operating temperature range from -40 °C to + 85 °C, suitable for wearables.

D. *Sensor Fusion Module*

The IMU discussed in the previous section is a part of the sensor fusion module. In addition to that, the module serves as an interface to the analog sensors namely the filaments sensors in the smart glove and the pressure sensors in the smart band. The module is equipped with 10 × resistance measurement inputs, 10 × impedance measurement inputs, and inertial measurements with the above-mentioned 9-axes. The measurement system is powered via USB and has the provision of a battery supply and is compact at a size of 38 × 48 mm for ease of body attachment as shown in Fig. 3. The test units present accurate resistance measurements and a high acquisition speed of 20 samples per second. To synchronize and visualize the data, a LabVIEW interface is designed. The module communicates via the serial bus with the interface where the synchronization is ensured by the clock of the system. For visualization, three graphs are designed for each hand system, that contains the smart glove, IMU, and smart band sensor signals respectively. In addition, a recording button is created to enable the recording of the data during performing measurements. All data from both systems attached to both

hands at the same measurement trial are saved in one text file to be treated in a further step for classification and recognition.

E. *Basic Gesture Components and Gesture Collection Protocol*

In this study ASL numerical gestures 1 to 10 have been used to evaluate the performance of the developed multisensor system. These are static gestures and are classified as hand shapes with combination of fingers according to the guidelines of the notation system proposed in Bressem [15]. Investigation of these gestures could provide valuable information on sensor-related parameters that could aid in the further development of complex gestures. As a preliminary study that does not involve and machine learning algorithm, the study was conducted with only one healthy male volunteer, 30 years of age. The multisensor system was attached to the forearm of the user as shown in Fig. 3. The total data acquisition time for each gesture is 3 s. The user starts from a rest position where the hand is open and the arm is rested on the tabletop. In total five repetitions for each gesture have been performed and the data is organized according to the acquisition time.

3   Results and Discussion

A. *Analysis of the Raw Data from Both the Gloves*

As this paper aims to understand the performance of each sensor technology used in this multisensor framework, the acquired raw data for the performed set of gestures is analyzed using Microsoft Excel and OrginPro 2022b software. Figure 4a shows sample data recorded from each filament sensor of the smart glove for gesture G1. For the Gesture G1, except for the sensor on the index finger and other sensors are under stress as the fingers are closed. A similar partner can also be observed in Fig. 4a where a significant change in the

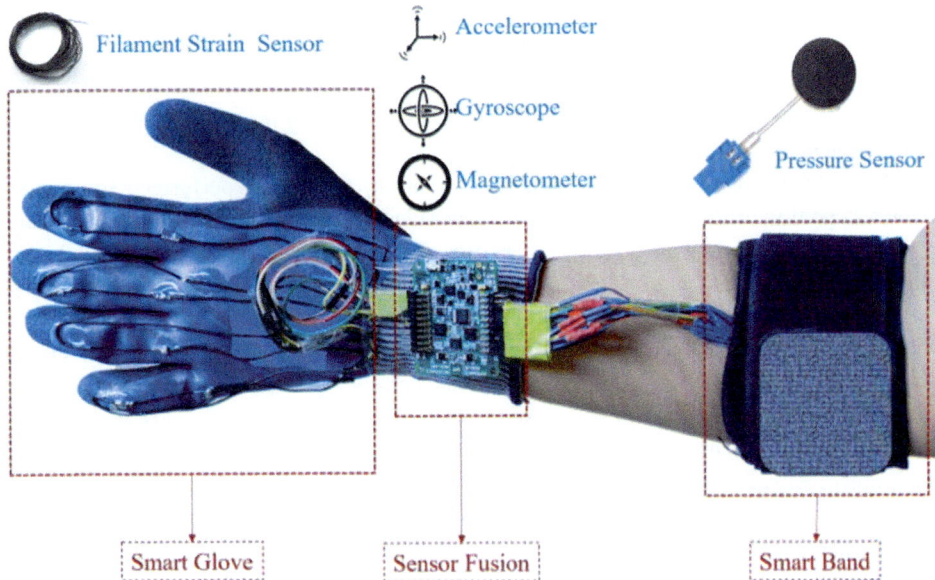

Fig. 3 Picture of a user wearing the complete multisensor system

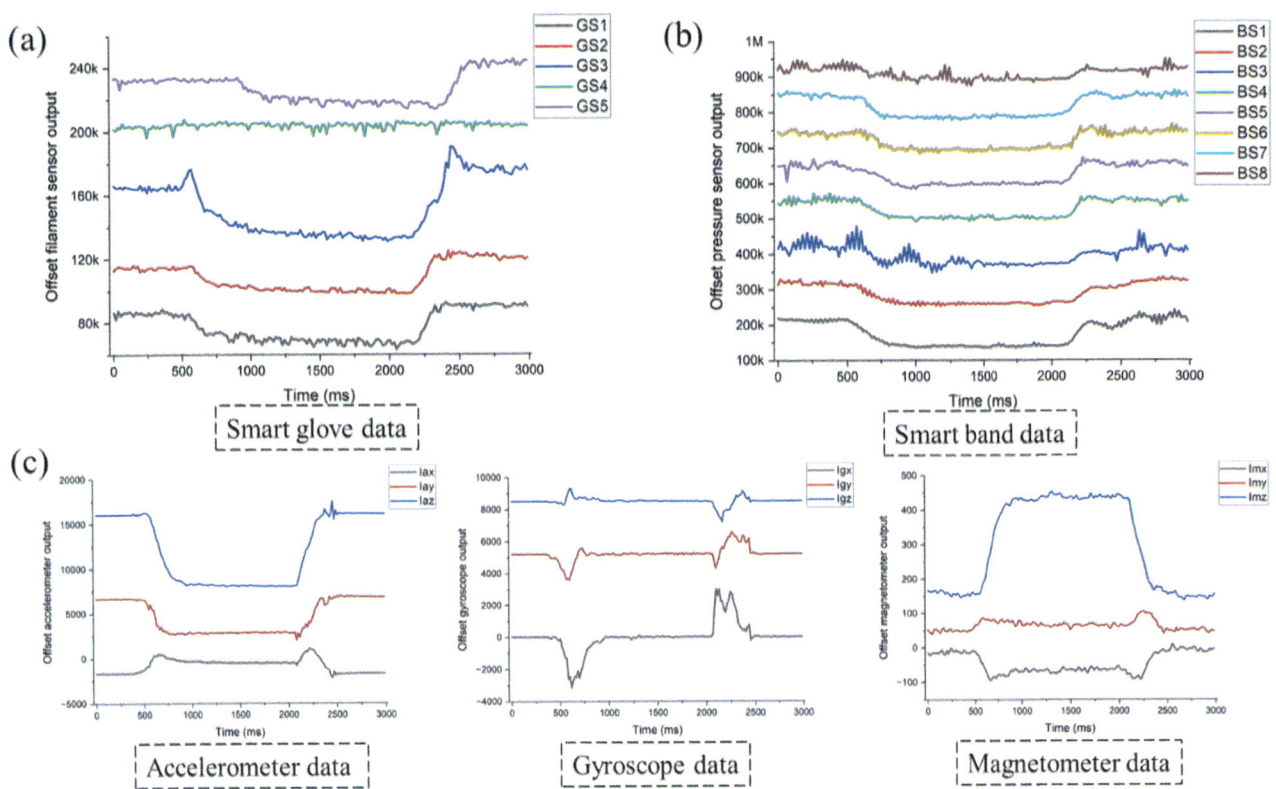

Fig. 4 Graphs showing the raw data obtained from the sensor fusion module for **a** smart glove, **b** smart band, and **c** IMU

sensor response can be observed for GS1, GS2, and GS3, and a moderate change is seen in GS5 which is the sensor on the thumb while no change in the sensor response is seen in GS4 which is the index finger. Likewise, Fig. 4b represents the data of eight sensors in the smart band for gesture G1. Although a distinct relation between the individual sensor and the finger position cannot be determined by this data, it is evident that all eight sensors are responsive to the performed gesture. Similarly, the data from the three sensor components of the IMU is shown in Fig. 4c. It can be observed that both the accelerometer and the magnetometer exhibit significant and stable sensor response for the performed gesture G1. Yet again, despite the visible sensor response, a direct interpretation that correlates the individual finger movement and the data of these two sensors is not possible. Furthermore, it can be observed that the data of the gyroscope does not provide a stable signal but rather impulses at the start and stop of the gesture. This implies that a gyroscope might not be a suitable sensor solution for the recognition of static gestures.

B. *Evaluation of the Performance of the Smart Glove*

Further analysis of the acquired data was conducted by generating a data matrix for all the sensors in the band corresponding to the performed gestures. The entry in the matrix is a relative change in the measured sensor output computed by performing a difference between the rest position data (at 250 ms) and gesture position data (at 1500 ms). A bar graph is generated from the resultant matrix as shown in Fig. 5a which represents the sensitivity of filament sensors in the glove for the performed gestures. It can be observed that a significant change among the five sensors and across the different performed gestures is evident. It appears that the gesture G5 has a negligible difference among the five sensors, however, in this particular gesture, all fingers are open which means none of the five sensors are activated and hence the result. This implies that the smart glove solution with filament sensors has the potential to capture all the finger movements corresponding to the performed ten gestures. In addition, a radar chart of the data matrix is presented in Fig. 5b which further provides a closer insight into the significance of each sensor attached to a finger corresponding to the movement of the finger. It is also clear

that except for G5 all the sensors exhibit different levels of sensitivity for the performed gestures.

C. *Evaluation of the Performance of the Smart Band*

A similar analysis is performed for smart band sensors by populating a similar data matrix and generating a bar graph as shown in Fig. 6a. Unlike the smart glove, only five out of the ten gestures were significantly detected by the smart band. The smart band sensor data show moderate variation in the sensor response for gestures G5 to G9. These are the gestures where the hand is either fully open or one of the fingers and thumb is closed. More investigation must be conducted to substantially validate the lack of sensitivity of the smart band for this particular gesture combination. The radar chart of the data is shown in Fig. 6b which presents a similar pattern with moderate deviations for all the sensors across gestures. It is evident that the classification of the gestures using a smart band technology is challenging because the technology relies on the subtle movement of muscles that are interlinked with all the fingers. However, in combination with a machine learning algorithm, the band could provide noticeable accuracy in classifying these gestures.

D. *Evaluation of the Performance of IMU*

Finally, a similar data matrix was populated for the 3 sensing elements of the IMU and the corresponding bar graph and radar plots are presented in Fig. 7. It can be observed that both the accelerometer and magnetometer present clear and

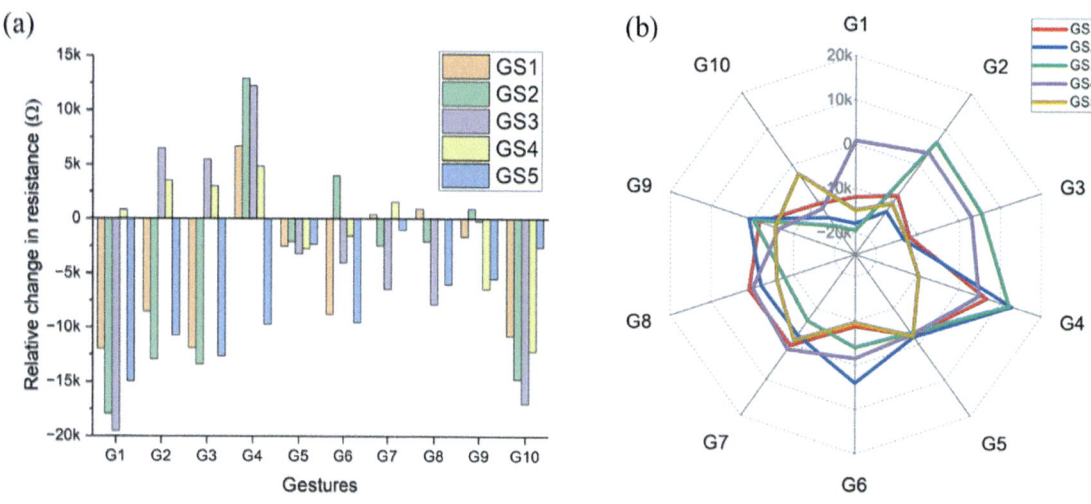

Fig. 5 **a** Performance of the sensors in the glove and **b** radar chart presenting the significance of the sensor in each finger against the performed gestures

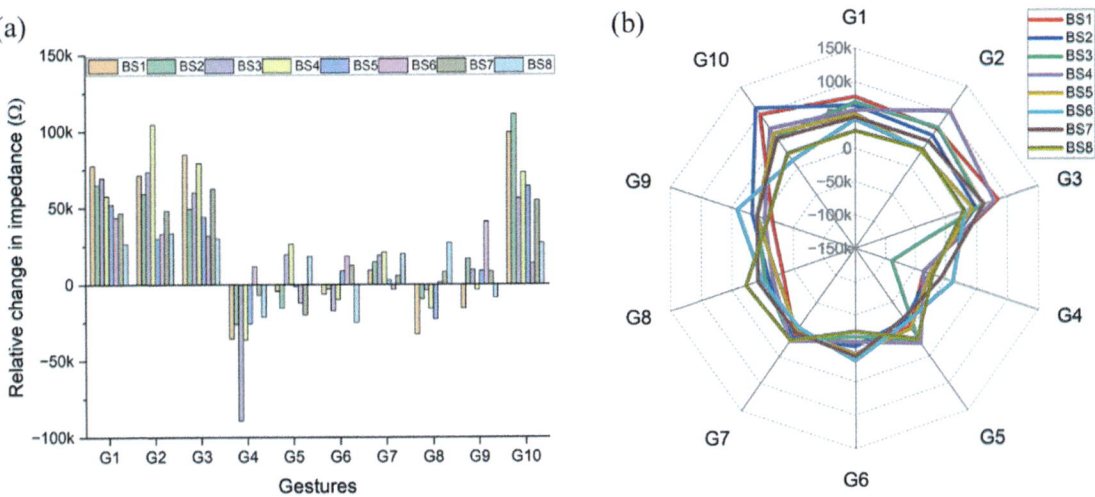

Fig. 6 **a** Performance of the sensors in the smart band and **b** radar chart presenting the significance of each sensor against the performed gestures

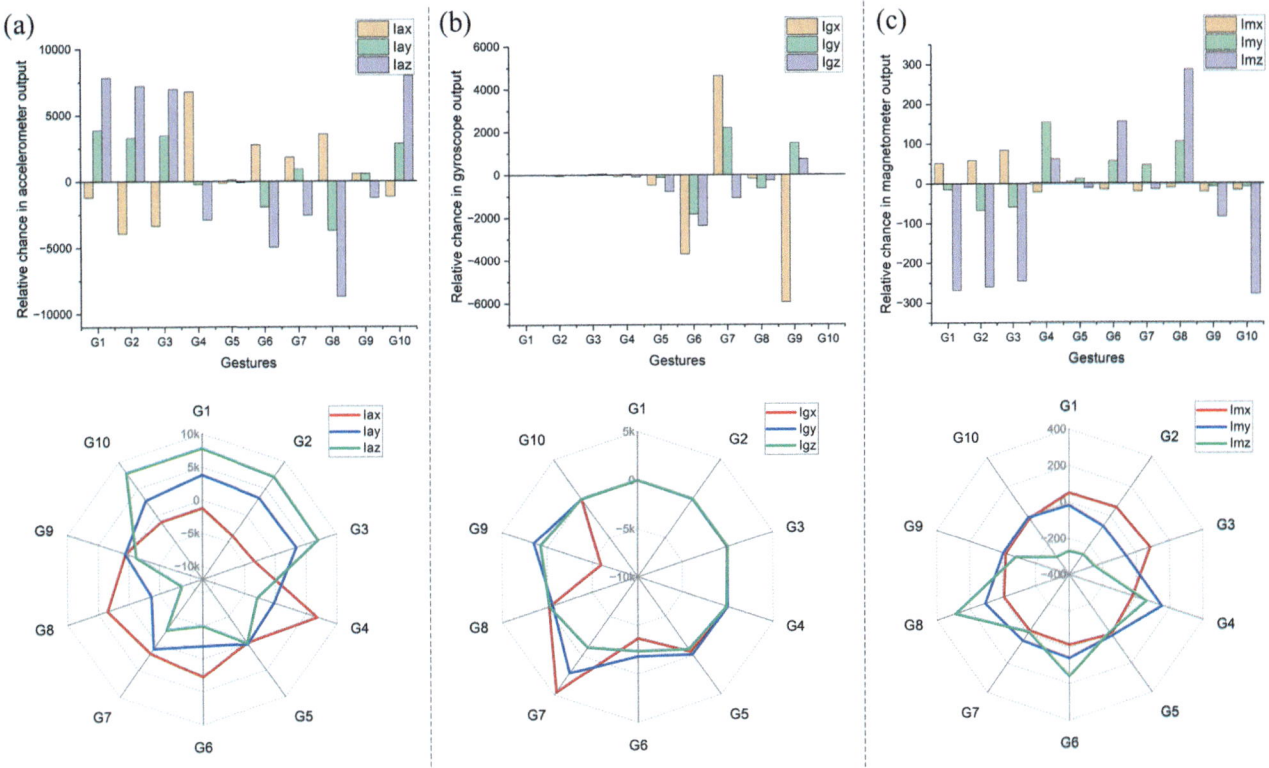

Fig. 7 Performance of the individual sensing component and the corresponding radar chart presenting the significance against the performed gesture for **a** accelerometer, **b** gyroscope, and **c** magnetometer

distinguishable sensor responses for the performed gestures. Interestingly, similar to the smart glove, these two sensors show a negligible change for gesture G5 where all fingers are open. Although it appears from Fig. 7b that the gyroscope provides valuable information for some of the gestures, it is arbitrary, unstable, and not repeatable across different trials. Despite the poor performance, the gyroscope might be highly valuable for studies dealing with dynamic gestures. As concluded for the smart band, a direct interpretation and correlation of the IMU data with the performed gesture are not possible and the need for a machine learning algorithm is evident.

4 Conclusion

In this paper, a multisensor measurement framework has been demonstrated for the recognition of ten ASL numerical sign language gestures. Three different sensor technologies including a smart glove with nanocomposite filament strain sensors, a smart band with nanocomposite pressure sensors, and an IMU sensor are developed into a wearable solution that can be attached to the forearm of the user. The system was worn by a healthy male volunteer and the recorded data by the LabVIEW interface has been analyzed to evaluate the performance of the multisensor framework for gesture recognition. The results have provided valuable insights into each sensor technology. The smart glove with five filament strain sensors on each finger is more than sufficient to recognize the chosen gestures for a single test subject without the need for any sophisticated machine learning algorithms. The smart band has exhibited significant sensitivity for 50% of the chosen gestures and a need for a machine learning algorithm is evident. A similar conclusion has been driven to the IMU sensors except for the gyroscope, which is not suitable for static gesture recognition procedures. Moreover, in future investigations with more gestures and more test subjects the need for machine learning algorithm is evident for accurate recognition of gesture with the multisensor measurement framework.

Acknowledgements Funded by the Deutsche Forschungsgemeinschaft (DFG, German Research Foundation)—Project-ID 416228727—SFB 1410.

References

1. Martinez-Hernandez, U., Metcalfe, B., Assaf, T., Jabban, L., Male, J., & Zhang, D. (2021). Wearable Assistive Robotics: A Perspective on Current Challenges and Future Trends. *Sensors, 21*(20), 6751.
2. Rajendran, D., Ben Atitallah, B., Ramalingame, R., Quijano Jose, R. B., & Kanoun, O. (2021). Ultra thin nanocomposite in-sole pressure sensor matrix for gait analysis. *In Advanced Sensors for Biomedical Applications* (pp. 33–45). Springer, Cham.
3. Teixeira, E., Fonseca, H., Diniz-Sousa, F., Veras, L., Boppre, G., Oliveira, J., & Marques-Aleixo, I. (2021). Wearable devices for physical activity and healthcare monitoring in elderly people: A critical review. *Geriatrics, 6*(2), 38.
4. Gaurav, S. (2021). Teleoperation, human intent prediction and imitation learning methods for collaborative robots (Doctoral dissertation, University of Illinois at Chicago).
5. Guo, L., Lu, Z., & Yao, L. (2021). Human-machine interaction sensing technology based on hand gesture recognition: A review. *IEEE Transactions on Human-Machine Systems.*
6. Zhang, Q., Wang, D., Zhao, R., & Yu, Y. M. (2019). Enabling end-to-end sign language recognition with wearables. *In Proceedings of the 24th International Conference on Intelligent User Interfaces, Marina del Ray, CA, USA* (pp. 17–20).
7. Fleming, A., Stafford, N., Huang, S., Hu, X., Ferris, D. P., & Huang, H. H. (2021). Myoelectric control of robotic lower limb prostheses: A review of electromyography interfaces, control paradigms, challenges and future directions. *Journal of neural engineering, 18*(4), Article 041004.
8. Ben Atitallah, B., Barioul, R., Ghribi, A., Bouchaala, D., Derbel, N., & Kanoun, O. (2020, July). Electrodes placement investigation for hand gesture recognition based on impedance measurement. In *2020 17th International Multi-Conference on Systems, Signals & Devices (SSD)* (pp. 1173–1177). IEEE.
9. Ben Atitallah, B., Rajendran, D., Hu, Z., Ramalingame, R., Quijano Jose, R. B., Veiga Torres, R. D., Kanoun, O. (2021). Piezo-resistive pressure and strain sensors for biomedical and tele-manipulation applications. Advanced Sensors for Biomedical Applications, 47–65.
10. Nowak, M., Eiband, T., & Castellini, C. (2017) Multi-modal myocontrol: Testing combined force-and electromyography. In *2017 International Conference on Rehabilitation Robotics (ICORR)* (pp. 1364–1368). IEEE
11. Jiang, S., Gao, Q., Liu, H., & Shull, P. B. (2020). A novel, co-located EMG-FMG-sensing wearable armband for hand gesture recognition. *Sensors and Actuators A: Physical, 301*, Article 111738.
12. Jiang, S., Lv, B., Guo, W., Zhang, C., Wang, H., Sheng, X., & Shull, P. B. (2017). Feasibility of wrist-worn, real-time hand, and surface gesture recognition via sEMG and IMU sensing. *IEEE Transactions on Industrial Informatics, 14*(8), 3376–3385.
13. Wajahat, M., Lee, S., Kim, J. H., Chang, W. S., Pyo, J., Cho, S. H., & Seol, S. K. (2018). Flexible strain sensors fabricated by meniscus-guided printing of carbon nanotube–polymer composites. *ACS Applied Materials & Interfaces, 10*(23), 19999–20005.
14. Ko, J., Jee, S., Lee, J. H., & Kim, S. H. (2018). High durability conductive textile using MWCNT for motion sensing. *Sensors and Actuators A: Physical, 274*, 50–56.
15. Bressem, J. (2013). 70. A linguistic perspective on the notation of form features in gestures. In Volume 1 (pp. 1079–1098). De Gruyter Mouton.

Understanding the Capabilities of FMG and EMG Sensors in Recognizing Basic Gesture Components

Giuseppe Sanseverino, Dominik Krumm,
Rajarajan Ramalingame, Chintan Malani, Rim Barioul,
Olfa Kanoun, and Stephan Odenwald

Abstract

Gestures are one of the most intuitive ways that humans use to interact with others or convey information. The idea that hand gestures could facilitate human–machine interaction has recently gained increasing interest among researchers. Various technologies have been investigated, providing both visual and sensor-based gesture recognition. While camera-based solutions suffer from the constraint of specific and expensive laboratories, wearable sensor-based solutions allow lower costs and higher flexibility, enabling gesture recognition even in public spaces. Although several solutions are available in the literature, most of them focus on specific sensor principles and specific gestures. The aim of this work is to recognize basic gesture components, defined as primary elements that compose more complex gestures, using both force myography (FMG) and electromyography (EMG), and to highlight their strengths and weaknesses. This will provide the foundation for the recognition of more complex human upper limb movements. To this end, a laboratory study was conducted with ten participants. FMG signals were collected by means of a wearable sensor network consisting of an instrumented smart band with eight pressure sensors and a wireless datalogger. EMG data were acquired using three commercial sensors. The recorded data were analyzed using k-nearest neighbor classifier and extreme learning machine algorithms. The results showed that the data recorded using FMG had higher accuracy in recognizing the ten different static hand gestures studied compared to the EMG data.

1st International Conference Hybrid Societies, 2023.

G. Sanseverino (✉) · D. Krumm · C. Malani · S. Odenwald
Department of Sports Equipment and Technology, Chemnitz
University of Technology, Chemnitz, Germany
e-mail: giuseppe.sanseverino@mb.tu-chemnitz.de

D. Krumm
e-mail: dominik.krumm@mb.tu-chemnitz.de

C. Malani
e-mail: chintan.malani@mb.tu-chemnitz.de

S. Odenwald
e-mail: stephan.odenwald@mb.tu-chemnitz.de

R. Ramalingame · R. Barioul · O. Kanoun
Professorship of Measurement and Sensor Technology, Chemnitz
University of Technology, Chemnitz, Germany
e-mail: rajarajan.ramalingame@etit.tu-chemnitz.de

R. Barioul
e-mail: rim.barioul@etit.tu-chemnitz.de

O. Kanoun
e-mail: olfa.kanoun@etit.tu-chemnitz.de

Keywords

Human–Machine interaction · Gesture recognition · Body-Attached sensor networks · FMG · EMG

1 Introduction

A. *Body-Attached Sensor Networks for Gesture Recognition*

Gestures are one of the most common and effective ways humans communicate with each other. In their review, Mitra and Acharya [1] describe gestures as expressive and meaningful body motions that aim to convey information or interact with the environment. Among all the different kinds of gestures, hand gestures are of great interest in the field of human-machine interaction (HMI). For example, Guo et al. [2] state in their paper that gestures are simpler and more intuitive than traditional HMI methods thanks to their high degree of differentiation, flexibility, and efficiency in conveying information. They also provide a comprehensive review of the technologies used for gesture recognition, identifying both visual and sensor-based solutions. Vision-based solutions include standard RGB cameras, RGB depth sensors such as Kinect (Microsoft Corporation, Washington, USA), and more complex marker-based motion capture systems such as Vicon (Vicon Motion Systems Ltd, Oxford, UK). The aforementioned technologies have demonstrated their performance in several studies. For example, Ren et al. [3] achieved a mean accuracy of over 90 % in recognizing ten different static hand gestures with an RGB depth camera. However, vision-based techniques can involve very complex and expensive setups. In addition, they are often tied to specific laboratories, which limits testing conditions. In contrast, sensor-based technologies also provide good results without the need for dedicated laboratories. This makes them cheaper and more flexible, and also opens up the possibility of performing gesture recognition in public spaces.

Different sensor principles can be used for gesture recognition, of which electromyography (EMG) and force myography (FMG) are among the most common, as they can be easily integrated into wearable devices. In addition, arrays of FMG or EMG sensors can be interconnected to form Body-Attached Sensor Networks (BASNs). For example, Qi et al. [4] used a sleeve equipped with eighteen EMG electrodes to recognize nine different static hand gestures and achieved 95 % recognition accuracy. On the other hand, Ramalingame et al. [5] achieved 93 % accuracy in recognizing ten different static hand gestures using FMG. Specifically, a BASN was developed consisting of a smart band equipped with eight pressure sensors worn on the forearm and a wireless datalogger.

Given the good results that can be obtained with both sensor principles, the aim of this paper was to investigate the differences in recognition accuracy when using EMG and FMG and to highlight strengths and weakness of both methods. An experimental test was conducted with ten participants using two different BASNs to recognize ten static gestures.

B. *Basic Gesture Components*

Most of the available works on static hand gesture recognition are focused on postures related to sign language alphabets, e.g., numbers or letters in American Sign Language. However, in this study, the focus is on Basic Gesture Components (BGCs), which represent the fundamental element for complex gestures. Ten BGCs were included in this study. Table 1 provides the descriptions and pictograms [6] of the selected BGCs.

2 Methods

A. *Sensor Network*

The BASN used for FMG consisted of a custom-made smart band and a miniature wireless datalogger [7] for signal acquisition. For the synthesis of the pressure sensors to be integrated

Table 1 Descriptions and pictograms of basic gesture components (bgcs) considered in the study

BGC	Description	Pictogram	BGC	Description	Pictogram
1	Finger 2 stretched		6	Finger 1 to 3 stretched	
2	Fist		7	Finger 1 and 5 stretched	
3	Flat hand		8	Finger 1 to 5 crooked	
4	Finger 2 to 5 flapped down; Finger 1 stretched		9	Finger 1 to 4 stretched	
5	Finger 1 to 5 connected		10	Finger 2 and 5 stretched	

into the smart band, multi-walled carbon nanotubes (CNTs, purchased from Nanocyl, Belgium) were pre-dispersed in the organic solvent tetrahydrofuran (THF) using ultrasonication and magnetic stirring. The prepared CNT dispersion was then combined with polydimethylsiloxane (PDMS purchased from Dow Crowning, Germany) and the dispersion procedure was continued using again ultrasonication and magnetic stirring. The crosslinking agent was added to the prepared nanocomposite at a ratio of 10:1 to complete the synthesis process. To fabricate the pressure sensor, the nanocomposite material was poured into a mold and cured in an oven at 120 °C for 2 hours. Eight such cured sensors with a diameter of 2 cm were then combined with inter-digital electrode structures and integrated into a stretchable medical band to realize the smart band shown in Fig. 1. These pressure sensors operate on the principle of resistance change [8] or capacitance change [9] under applied pressure. As the pressure applied to the surface of the sensor increases, more conductive paths are promoted by the MWCNTs in the PDMS polymer, resulting in a decrease in the measured resistance of the sensor.

Three Trigno Avanti Sensors (Delsys Inc., Natick, MA, USA) were used to acquire EMG data in this study. Each of these commercial sensor units consisted of an EMG sensor and a 6-axis IMU. The three sensors communicated wirelessly with a Trigno base station (Delsys Inc., Natick, MA, USA), which was connected to a PC for data acquisition.

B. *Test Protocol*

Ten healthy volunteers, five females and five males, aged 22 to 52 years, participated in this study and provided

Fig. 1 Body-Attached Sensor Network for FMG. **a** Smart band instrumented with eight pressure sensors; **b** miniature wireless datalogger for signal acquisition

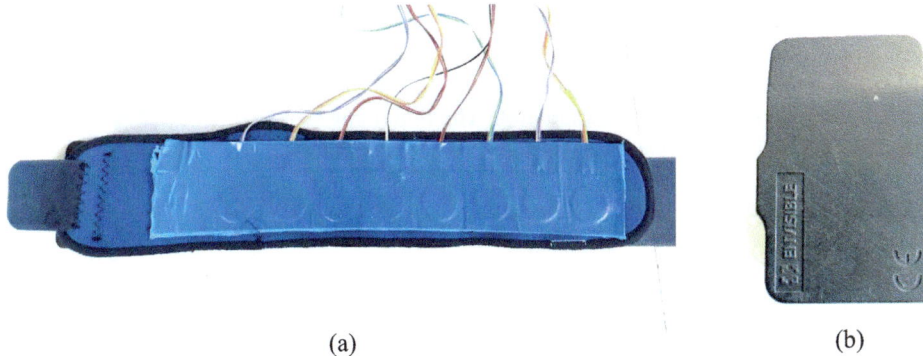

(a) (b)

written informed consent. The study was carried out in agreement with the Declaration of Helsinki and approved by the institute's ethics committee (reference #V-331-15-GJ-Sensor-13052019).

In preparation for the test, participants were asked to wear the smart band on the upper part of the right forearm while the wireless datalogger was attached to the upper arm. The three EMG sensors were each attached superficially with double-sided adhesive tape to the muscle belly of the right biceps brachii, flexor carpi radialis and extensor digitorum, as these muscles are responsible for a large portion of the movements of the forearm and the hand. Figure 2 shows the sensor set-up used.

Participants were asked to perform all BGCs five times in randomized order according to the following protocol: (i) start from the rest position (standing still with arms relaxed along the body) and perform the gesture; (ii) hold the gesture until the test operator gives the stop signal; (iii) return to the rest position. For the smart band, an acquisition window of 3 s per gesture, starting from the moment the person leaves the rest position, was considered. Each EMG acquisition lasted 5 s and started one second before the start of the acquisition with the smart band. The signals acquired included the movements performed by the participant to leave the rest position and perform the gesture, as well as the holding of the gesture.

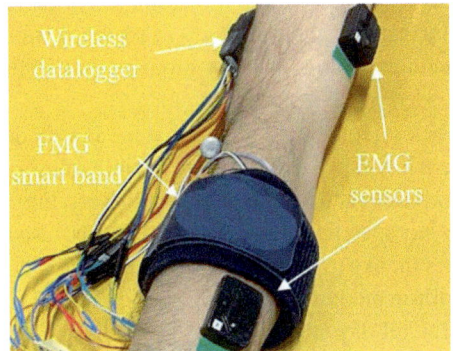

Fig. 2 Sensor set-up used in the test

C. *Data Analysis*

Raw sensors data, stored as csv files, were imported into MATLAB (The MathWorks, Natick, MA, USA). Using the information contained in the filenames, the data were programmatically annotated with respect to the person, gesture, and repetition parameters. The 500 annotated EMG signals were trimmed to a uniform size of exactly 3 s (6000 x 3 data points per measurement). Signals were filtered using a 5th order Butterworth low-pass filter with a cutoff frequency of 10 Hz and a delay of zero. Thirteen features were extracted for each EMG sensor. The EMG features were mean absolute value, variance, root mean square (RMS), waveform length, mean frequency, median frequency, Shannon entropy, integrated EMG, kurtosis, skewness, root sum of square level, maximum absolute value, and the simple square integral. Thus, a total of 39 features were characterizing one BGC.

To recognize the ten different BGCs based on the EMG data, a k-nearest neighbor classifier was used as a classification model. The model used 8 nearest neighbors, a city-block distance with a squared inverse distance weighing function, and standardized predictors, i.e., the model centered and scaled each column of the predictor data with the column mean and standard deviation. To train the model and test its validity, the measurements were divided into a training set and a test set. Repetitions 1 to 4 (n = 400) were used for training and repetition 5 (n = 100) for testing the classification model. The percentage success rate of the trained classification model in correctly predicting the test data was calculated as 1 minus the ratio of the number of correct predictions to the number of all predictions times 100. Based on the true and predicted BGCs, a confusion matrix chart was created to evaluate how the currently selected classifier performed in each class.

The FMG signals were analyzed using an extreme learning machine (ELM) algorithm as previously done by Ramalingame et al. [5]. Five features were extracted for each sensor signal in the time domain, namely: minimum, maximum, mean, median, and RMS. The 40 features were used as input to the feature selection algorithm proposed by Barioul et al. [10], named Random Walk Binary Gray Wolf Optimizer. The features thus selected were used to train and test an ELM with 3000 hidden nodes and a sigmoid activation function. The ELM was trained with 5-fold cross validation, using 80 % of the collected data for training and validation. The remaining 20 % of the collected data were used to test the trained model.

3 Results

Figure 3 shows the confusion matrix created by the k-nearest neighbor classifier for the EMG signals. The success rate for the prediction based on the test data was 53%.

The Random Walk Binary Gray Wolf Optimizer delivered the best features to be considered for each FMG sensor. These are reported in Table 2. The ELM trained with these features produced the confusion matrix shown in Fig. 4 and achieved an overall testing accuracy of 86%.

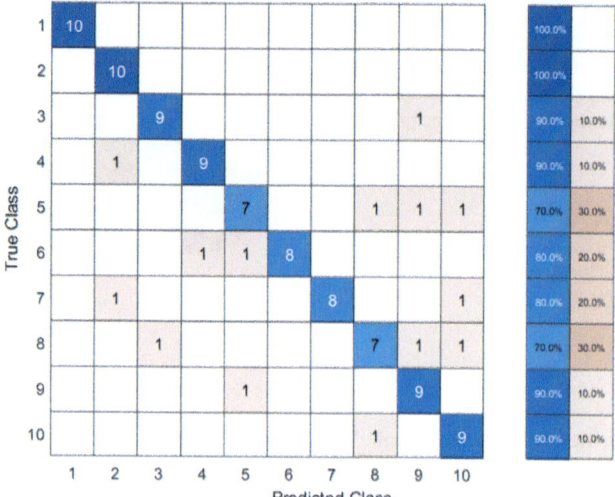

Fig. 4 Confusion matrix generated by the ELM algorithm for the FMG signals

4 Discussion and Conclusion

Data analysis showed that higher accuracy can be achieved by FMG as compared to EMG and therefore it is more suitable for the recognition of the BGCs used in this study. In this regard, it should be noted that the poorer results associated with EMG could be biased by the positioning of the sensors. The BASN developed for FMG covers the entire circumference of the forearm and ensures that the volume changes of all muscles responsible for hand movements are captured by one of the eight pressure sensors that compose the smart band. In contrast, the EMG sensors used could only detect muscle activation of two of the muscles in the forearm. This limitation should be addressed in future studies by using additional EMG sensors in the measurement set-up. This could lead to higher accuracy in the recognition of static hand gestures in accordance with the results of Qi et al. [4]. Another limitation of the test was the small number of participants. A larger training data set with more participants may improve the quality of data analysis for both FMG and EMG. Even though FMG achieved good accuracy, it was still lower than accuracy achieved by Ramalingame et al. [5] in previous similar tests, so further refinement of the test protocol and FMG BASN may be needed in the future.

Nevertheless, it is worth noting that the BASN developed for FMG is a less expensive and more flexible solution than the EMG and is easier and more convenient to wear. Furthermore, it requires no specific preparation and can be used outside laboratories thanks to its connection to a wireless datalogger that has built-in memory for storing data and sufficient battery capacity for all-day acquisitions. Vice versa, even though EMG sensors can be connected to a portable device (e.g., smartphone) via Bluetooth, they are much more expensive,

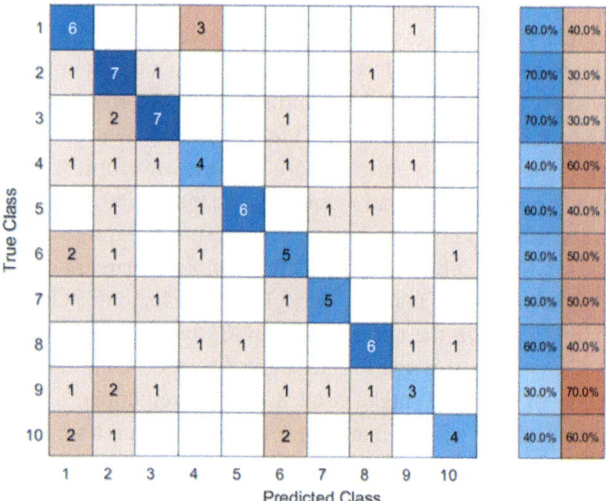

Fig. 3 Confusion matrix generated by the k-nearest neighbor classifier for the EMG signals

Table 2 Selected features to be considered for the training of extreme learning machine for each FMG sensor obtained with the Random Walk Binary Gray Wolf Optimizer

Sensor number	Selected features
1	Median values
2	All
3	Median values
4	None
5	Maximum and median values, RMS
6	Minimum and mean values
7	None
8	All

less comfortable to wear, and require prior preparation, such as the need to shave the skin before use.

Acknowledgements Funded by the Deutsche Forschungsgemeinschaft (DFG, German Research Foundation)—Project-ID 416228727—SFB 1410.

References

1. Mitra, S. & Acharya, T. (2007) Gesture recognition: a survey. *IEEE Trans. Syst., Man, Cybern. C, 37*(3), pp. 311–324, https://doi.org/10.1109/TSMCC.2007.893280.
2. Guo, L., Lu, Z., & Yao, L. (2021). Human-Machine interaction sensing technology based on hand gesture recognition: a review. *IEEE Trans. Human-Mach. Syst., 51*(4), 300–309. https://doi.org/10.1109/THMS.2021.3086003
3. Ren, Z., Meng, J., & Yuan, J. (2011) Depth camera based hand gesture recognition and its applications in Human-Computer-Interaction. In *2011 8th International Conference on Information, Communications & Signal Processing*, pp. 1–5, https://doi.org/10.1109/ICICS.2011.6173545
4. Qi, J., Jiang, G., Li, G., Sun, Y., & Tao, B. (2020). Surface EMG hand gesture recognition system based on PCA and GRNN. *Neural Computing and Applications, 32*(10), 6343–6351. https://doi.org/10.1007/s00521-019-04142-8
5. Ramalingame, R., et al. (2021). Wearable smart band for american sign language recognition with polymer carbon nanocomposite-based pressure sensors. *IEEE Sens. Lett., 5*(6), 1–4. https://doi.org/10.1109/LSENS.2021.3081689
6. Bressem, J. (2013) A linguistic perspective on the notation of form features in gestures. In Handbücher zur Sprach- und Kommunikationswissenschaft / Handbooks of Linguistics and Communication Science (HSK) 38/1, Müller, C., Cienki, A., Fricke, E., Ladewig, S., McNeill, D., & Tessendorf, S., Eds.: DE GRUYTER, pp. 1079–1098.
7. Hill, M., Hoena, B., Kilian, W., & Odenwald, S. (2016). Wearable, modular and intelligent sensor laboratory. *Procedia Engineering, 147*(3), 671–676. https://doi.org/10.1016/j.proeng.2016.06.270.RR1
8. Ramalingame, R., et al. (2019). Flexible piezoresistive sensor matrix based on a carbon nanotube PDMS composite for dynamic pressure distribution measurement. *J. Sens. Sens. Syst., 8*(1), 1–7. https://doi.org/10.5194/jsss-8-1-2019
9. Ramalingame, R., Lakshmanan, A., Müller, F., Thomas, U., & Kanoun, O. (2019). Highly sensitive capacitive pressure sensors for robotic applications based on carbon nanotubes and PDMS polymer nanocomposite. *J. Sens. Sens. Syst., 8*(1), 87–94. https://doi.org/10.5194/jsss-8-87-2019
10. Barioul, R., Raju, R. M., Varghese, S., Kanoun, O. (2022) Random walk binary grey wolf optimization for feature selection in sEMG based hand gesture recognition. In *2022 IEEE 9th International Conference on Computational Intelligence and Virtual Environments for Measurement Systems and Applications (CIVEMSA)*, pp. 1–6, https://doi.org/10.1109/CIVEMSA53371.2022.9853687.

Feature Selection for Hand Gesture Recognition Using Six FSR Sensors Bracelet

Sajeda Al-Hammouri, Rim Barioul, Khaldon Lweesy, Mohammed Ibbin, and Olfa Kanoun

Abstract

Gesture recognition plays a critical role in applications ranging from gaming and remote manipulation to assistive technologies for individuals with disabilities. Force myography (FMG) has recently emerged as a promising approach for gesture recognition due to its strong classification capabilities. In this study, eleven American Sign Language (ASL) gestures were classified using FMG signals. To improve classification accuracy, three advanced swarm-based feature selection algorithms: the Binary Gray Wolf Optimizer (BGWO), Binary Grasshopper Optimization Algorithm (BGOA), and Binary Hybrid Gray Wolf Particle Swarm Optimizer (BGWOPSO), were evaluated as wrapper methods for an Extreme Learning Machine (ELM) classifier. Ten volunteers contributed data using a bracelet equipped with six force-sensitive resistor (FSR) sensors. Results show that the BGWOPSO-based feature selection approach significantly outperformed the other methods, increasing ELM classification accuracy from 55.82% to 91.36%.

S. Al-Hammouri (✉) · K. Lweesy · M. Ibbin
Biomedical Engineering Department, College of Engineering, Jordan
University of Science and Technology, Irbid, Jordan
e-mail: sfalhammouri18@eng.just.edu.jo

M. Ibbin
e-mail: mohib@just.edu.jo

R. Barioul · O. Kanoun
Professorship of Measurement and Sensor Technology, Technische
Universität Chemnitz, 09126 Chemnitz, Germany
e-mail: rim.barioul@ieee.org

O. Kanoun
e-mail: olfa.kanoun@etit.tu-chemnitz.de

K. Lweesy
Electrical, Computer, and Biomedical Engineering Department,
College of Engineering, Abu Dhabi University, P.O. Box 59911 Abu
Dhabi, United Arab Emirates
e-mail: klweesy@just.edu.jo

Keywords

Extreme learning machine · Force myography · FSRs ·
Hybrid gray wolf particle swarm

1 Introduction

Gesture recognition is increasingly gaining attention as a method for human–computer interaction (HCI) in which the machine can understand the human body language where the ability to recognize hand movements is crucial for a variety of applications, including gaming [1], producing prosthetic limbs for amputees, controlling devices [2], and sign language recognition [3]. Different methods, which may be divided into vision-based and sensor-based methods, can be used to identify hand gesture. Despite the fact that the vision-based methods have been widely utilized for gesture recognition, they have certain drawbacks, including occlusion [4], sensitivity to the background, light, and reflected objects. On the other hand, the sensor-based methods including the surface electromyography (sEMG), which measures the electrical bio signals produced during muscle contractions, has been widely investigated in the area of gesture recognition. However, this method still has some drawbacks that limit its usage as it is affected by sweat and humidity [5]. Also, due to the noisy nature of the EMG signal, it needs a lot of pre-processing and feature extraction processes [6].

Alternatively, another low-cost, non-invasive method that records the volumetric changes resulting from muscle contractions is called force myography (FMG) has been widely used recently as it has many advantages over EMG signals in terms of its need for less preparation of the skin, and it is less affected by skin impedance and sweat [7, 8]. In addition, the FMG is distinguished by its stability and durability against outside electrical noise. Different sensors can be used to collect the FMG with a majority to the resistive polymer-thick-film (RPTF) sensors that include two different types of sensors: Force-sensitive resistors (FSR) sensors and Flexiforce® sensors [9] due to their low cost, small size, flexibility, widespread availability, and simplicity of integration into mobile devices [10].

This work aims to improve the classification accuracy of eleven gestures by enhancing the performance of ELM by selecting the best features and optimizing its input weights since the randomness in initializing the input weights is the cause of uncertainty problems. This can be done with the help of swarm intelligence algorithms which were created using natural patterns found in the interactions of live things, such bird flocks, wolfs, and grasshoppers. In this study, the BGWO, BGOA, and BGWOPSO are investigated as optimizers and feature selection methods duo their strong global search ability and robust performance, thereby the aim is not to invent a new algorithm, but to find out an algorithm, which significantly improves the performance of the ELM in terms of convergence behavior and accuracy. Therefore, the focus is on increasing the accuracy while maintaining a minimal number of sensors to ensure user comfort.

2 Related Work

In order to distinguish only three grasping tasks, reaching, grasping, and moving an object, a band with six FSR sensors was utilized to gather FMG signals from eight volunteers in [11]. They achieved classification accuracy rates of 97.1% and 96.7%, respectively, using SVM and linear discriminant analysis (LDA). A support vector machine (SVM) classifier was used to detect only four particular hand gestures (fist, palm, down, and up) in [12], where five FSR sensors were utilized to capture raw FMG signals from forty volunteers. They achieved 60% accuracy. While [13], whose raw FMG data was gathered from ten healthy people utilizing eight FSR sensors band, was only able to distinguish six hand movements (grasping, rotating, shaking, pushing, pulling, and opening) with almost 93% testing accuracy. In [14] they classified six grasping tasks obtained from eleven healthy people with an accuracy of 87.7% using a leap motion and a band with 10 FSR sensors. This demonstrates that even with standard artificial intelligence algorithms, a high degree of accuracy may be achieved for a relatively small number of movements. In [15], the data were gathered from 10 healthy participants using a band with eight FSR sensors, the numbers from (0–9) in American Sign Language (ASL) were classified using extreme learning machine (ELM) with 90% classification accuracy. A wristband with four FSR sensors was utilized in [16] to collect the FMG signal and classify nine gestures from the ASL, obtaining 89.65% testing accuracy but they only tested their work on one volunteer.

Different features were extracted from the FMG signal such as mean absolute value (MAV), root mean square (RMS), linear fit, parabolic fitting (PF), autoregressive model (AR) for classifying grasping and no grasping, RMS, slope, mean-mode difference, and slope sign count [17].

One of the most challenging and important steps after feature extraction is feature selection (FS) which aims to reduce data dimension by removing unnecessary, irrelevant, and redundant information to improve machine learning performance [18]. The effectiveness of these algorithms comes from their ability to find the best solution in a reasonable time [19]. For example, the GWO algorithm is better than the evolutionary algorithm and particle swarm optimization (PSO) because it has fewer parameters and is simpler to implement. However, like other stochastic algorithms, it had some drawbacks such as it is easily trapped in the local optimal value and provides a poor convergence behavior at exploitation [20]. From here other new population-based algorithms that are inspired by the behavior of grasshopper swarms in nature were introduced in [21] in which it is characterized by special adaptive mechanism that is smoothly balances exploration and exploitation which makes this algorithm able to outperform other techniques such as PSO, GA, GWO and differential evolution (DE). But, this algorithm also suffers from slow convergence speed and falls into local optimum while solving specific problems. Then different studies in the literature have shown that using a hybrid algorithm by combining two or more algorithms can result in much better performance than using each algorithm alone. For example, in a binary version of the hybrid algorithm built of PSO and GWO called BGWOPSO was developed to determine the best feature subset. It was tested on 18 standard UCI benchmark data sets and the results demonstrate that it performs better than GWO and PSO in terms of accuracy, selection time, and the number of selected features.

As it can be seen, there are several algorithms in the literature but here in this work, BGWO, BGOA, and BGWOPSO were selected since they are considered from the best available algorithms, as they proved their effectiveness compared to other evolutionary algorithms.

3 Materials and Method

In this section, the hardware design for the used device for the data collection will be discussed followed by a detailed description of pre-processing and feature extraction process.

3.1 Hardware Design and Data Collection

The design of our bracelet is composed of 6 FSRs (Interlink FSR 402) embedded in the inner side of a velcro band with

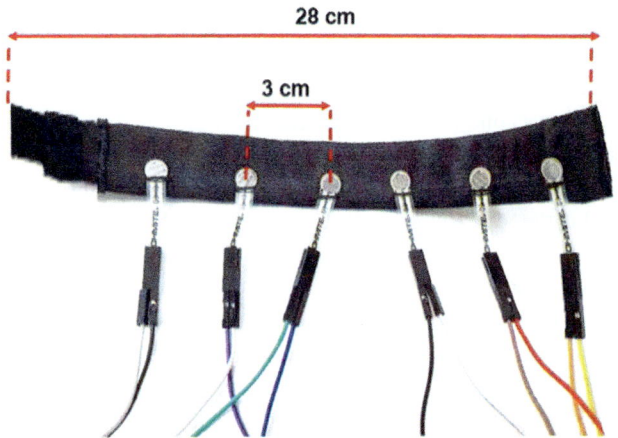

Fig. 1 Six-FSR sensor band used for FMG data acquisition. The sensors are evenly distributed along the 28 cm strap, with each sensor spaced 3 cm apart

a length of 28 cm wrapped around the volunteer's forearm. The sensors were evenly distributed with a distance of 3 cm as shown in Fig. 1. All of these sensors were connected via voltage divider circuits to two teensy boards to reduce the acquisition time. Each board has two analogs to digital converters (ADC) and each one of them is used to convert the value of two sensors. All six sensors were connected to a double-sided printed circuit with dimensions of 5.2 cm × 7.2 cm.

In this study, the raw FMG signals were collected from ten healthy participants (5 males, 5 females, height: 165.5 ± 7.4 cm, weight: 65.3 ± 12.3 kg, wrist circumference: 16.4 ± 2.3 cm) aged between 25 and 35 years old, all of whom were told about the experimental methods before participating. The data recorded by the control board was transferred to a desktop computer using a USB connection, and the collected data was saved in an excel file using Tera-term software for evaluation. The subjects were asked to perform 11 gestures from ASL (numbers from 0 to 10) following a video that shows how to make the gestures in a detailed and clear way. All the performed gestures are shown in Fig. 2. Also, the rest position was recorded. One hundred fifty samples were recorded for each gesture, where each gesture was per- formed ten times. After each trial, there was a brief relaxing time of about one minute and 5 minutes between different gestures for subjects to avoid muscle fatigue.

3.2 Pre-processing and Feature Extraction

Many studies have shown that during the rest state, there is some pressure that can be noticed due to the armband fastening. Thus, to remove the effect of this pressure and to remove any interference from muscles in the rest state, the raw signals were pre-processed by subtracting the minimum signal values recorded at rest from the raw signals. Another method is proposed here in our study called the Z method to get rid of the remaining pressure. In this method, the mean of the rest signal was subtracted from the original data and then divided by the standard deviation of the rest signal, as shown in Eq. 1.

$$Z = \frac{X - \mu_{rest}}{\sigma_{rest}} \tag{1}$$

where X represents the raw signal value, μ_{rest} and σ_{rest} are the mean and the standard deviation for the rest signal, respectively. In this work, only six-time domain features were extracted: MAV, RMS, AR, minimal, and maximal values and mean. Normalizing the extracted features is one of the most important steps in machine learning, especially since most of the features have different range values. Therefore, making sure that no feature dominates the others is necessary, and this can be achieved using the Min-Max normalization method, which is one of the most used methods to normalize data to the range [0 1] because it ensures that all features have the same scale, unlike the z-score normalization. Equation 2 shows the Min-Max normalization formula where x represents the data that needs to be normalized, min is the minimal sample value, and max is the maximal sample value.

$$Y = \frac{x - min}{max - min} \tag{2}$$

4 Extreme Learning Machine for Classification

ELM is a single-layer feedforward neural network (SLFN) introduced by Huang in 2004 [22]. As its name implies, it is composed of only one hidden layer on its architecture that connects the input layer to the output layer. So it is composed

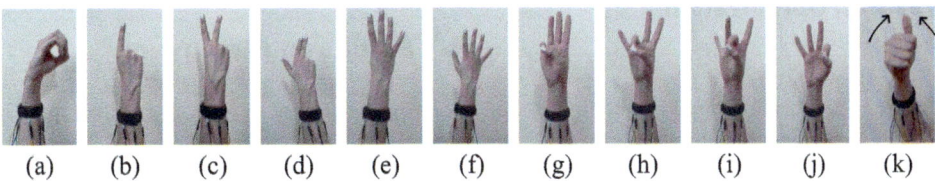

(a) (b) (c) (d) (e) (f) (g) (h) (i) (j) (k)

Fig. 2 Performed ASL number signs: (a) Zero, (b) One, (c) Two, (d) Three, (e) Four, (f) Five, (g) Six, (h) Seven, (i) Eight, (j) Nine, (k) Ten (dynamic gesture)

of three main layers. ELM is considered a fast training algo- rithm because the weights (W) and the bias that connect the input layer to the hidden neurons are initialized randomly only for the first time and remain constant without the need to be tuned during the training phase [23]. The output weights that connect the hidden neurons to the output layer are calcu- lated analytically by a one-step calculation through Moore– Penrose (MP) generalized inverse [24]. ELM exceeds some available machine learning algorithms such as artificial neural networks which usually use backpropagation algorithms to tune the weight parameters. Despite the popularity of these algorithms, they had some drawbacks in slowing down the learning speed, being trapped in local minima, and are consid- ered a time-consuming process [25]. In addition, many studies showed that ELM was faster than SVM with good predictive performance [26, 27].

Despite all of its attractive properties, the random initial- ization of the input weight caused poor generalization in reaching the optimal results easily [28].Thereby, swarm intel- ligence can also be used to optimize the weights of ELM such as in [29–31] the GWO, PSO, and GOA were used to optimize input weights and biases of ELM. The following subsections are showing some theoretical explanations about the used optimization algorithms in this paper such as BGOA, BGWO, and BGWOPSO.

4.1 Binary Grasshopper Optimization Algorithm (BGOA)

It is an optimization algorithm inspired by grasshopper's movement patterns and their search for food. It can be repre- sented as a set of solutions that are randomly produced to build an artificial swarm, according to a mathematical model proposed in [21] to simulate grasshopper behavior. After that, all the candidate grasshoppers are evaluated using a fitness function, where the best agent found is considered as the target grasshopper. Then this leader grasshopper starts to attract the other grasshoppers in the swarm to move toward him. The original version of GOA is used to solve continuous prob- lems but for feature selection, the continuous version must be converted to binary and the simplest way to do that is the sigmoid function, which was used in this study.

4.2 Binary Gray Wolf Optimization Algorithm (BGWO)

GWO is one of the recent swarm intelligence techniques that was proposed in 2014 [32]. This algorithm mimics the social hierarchy of intelligence of gray wolves in hunting and their social interaction in nature. The Social Hierarchy of Gray wolves' pack is divided into four groups: alpha, beta, delta,

and omega. Alpha is the most powerful and dominant wolf and is responsible for making decisions, determining the place and time for rest, and leading the entire pack in hunting, migration, etc. The second level of the hierarchy of gray wolves is beta, which helps alpha make decisions or any other pack activities. They are considered the best candidates to take the pack's lead if the alpha wolf is ill or dead. The third level is delta, which must obey alpha and beta, but they dominate omega wolves. In the mathematical model for the GWO. It is always assumed that alpha (α) is the fittest or the best solution, beta (β) and delta (δ) are the second and the third-best solutions obtained so far. While the rest of the candidate solutions are assumed to be omega (ω). Since the location of the prey is unknown in the search space, in order to simulate the hunting process of the gray wolves mathematically, alpha, beta and delta are considered to have the best knowledge about the location of the prey. Thus, they oblige and guide the omega wolves to update their positions according to the position of the best search agents $X_{Alpha}, X_{Beta}, X_{Delta}$. For solving binary problems a binary version of GWO was proposed in [33] using the sigmoid function.

4.3 Binary Gray Wolf with Particle Swarm Optimizer (BGWOPSO)

In this algorithm, each particle in PSO is represented by a position and velocity vector, in which each particle has its own intelligence and searches a search space centered on the best solution it has discovered so far. Particles are also aware of the best location that all particles (as a swarm) have dicovered till now. The velocity and positions are updated according to the following equations in order to combine GWO and PSO

$$v_i^{k+1} = W * \left(v_i^k + c_1 r_1 (x_1 - x_i^k) + c_2 r_2 (x_2 - x_i^k) + c_3 r_3 (x_3 - x_i^k) \right)$$
(3)

where W is the positive inertia weight, x_1, x_2, and x_3 are the best three solutions, r is a random number between [0, 1], and c represents the acceleration coefficients. This technique, which combines the advantages of both GWO and PSO, has been shown to be effective in controlling the trade-off between exploitation and exploration during optimization iterations which made it superior to other available algorithms

5 Results and Discussion

In this section, the proposed BGWOPSO-ELM is verified on the collected FMG data where the classification accuracy was used as performance criteria and its performance was compared to other well-known used algorithms in the litera- ture such as BGOA and BGWO. The data were divided into

Table 1 Experimental parameter configuration for the algorithms employed in this study

Parameters	Value
Number of search agents	10
Number of iteration	100
Number of runs	50
Alpha in fitness function	0.99
Beta in fitness function	0.01
a in GOA	[0 2.079]
l in GOA	1.5
f in GOA	0.5
c1, c2, c3 in PSO	0.5
ELM hidden nodes	400
Number of initial features	36

Table 2 Classification performance and selected feature counts of BGWOPSO-ELM compared with baseline algorithms

Name	Train accuracy	Test accuracy	Average number of features
ELM only	59.3 ± 0.013	55.82 ± 0.002	All features (36)
BGWOPSO	97.89 ± 0.001	91.36 ± 0.009	32
BGOA	82.84 ± 0.007	78.48 ± 0.018	30.2
BGWO	84.3 ± 0.011	76.45 ± 0.012	32.6

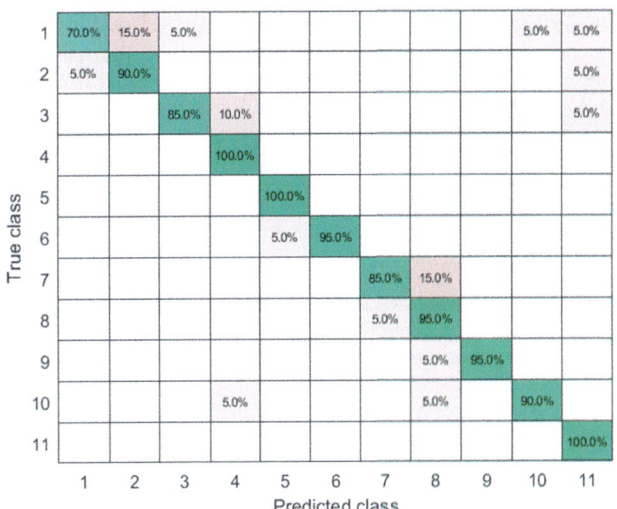

Fig. 3 Confusion matrix for the classification of 11 hand gestures using the BGWOPSO model. Diagonal cells represent correct classifications, while off-diagonal cells indicate misclassifications

two groups in which 80% of the data is used for training and 20% for testing. All algorithms used in this study are implemented using MATLAB software and executed on an Intel Core i7 machine. The maximum number of iterations in every run is set to 100. For a proper comparison, the same parameters were used to compare all algorithms as shown in Table 1. The results for classifying numbers are shown in Table 2, it can be noticed that using ELM only without any optimization algorithms in 6 sensors data gives poor results which are not sufficient to be implemented in any hardware system, this reduction in accuracy can be due to the randomness of ELM weight initialization and the effect of reducing the number of sensors to only 6 while having a large number of gestures. To reach the main objective of this work which aims to detect and recognize different gestures from the ASL with the highest classification accuracy using only six FSR sensors, three algorithms were used to optimize the ELM classifier. The comparison between different algorithms in optimizing the ELM is shown in Table 2 where it can be noticed that BGWOPSO managed to achieve the highest results with 97.89% for training accuracy and 91.36% for testing accuracy and this confirms its ability to select the best features, in addition, to optimize the randomness in initializing the input weights

for the ELM by removing the unwanted weights causing poor prediction accuracy or slowing the response of ELM because, in this algorithm, the PSO aids the GWO optimizer by managing its exploration and exploitation using inertia constant to enhance the convergence speed and attain the optimal global optima.

The confusion matrix for classifying numbers from 0 to the dynamic gesture 10 is shown in Fig. 3 where it can be noticed that classifying numbers (5,7,8) represented by (6,8,9) on the confusion matrix can reach 95% as classification accuracy with 100% specificity for three gestures (3, 4, and the dynamic gesture 10). In addition, it was able to recognize numbers (1, 9) with good classification accuracy reaching 90%. While discriminating numbers (2 and 6) achieved 85%.

6 Conclusion and Future Work

In this study, we investigated three algorithms as a feature selection wrap-per based on ELM and tested on FMG signal collected from ten volunteers performing the ASL numbers using a band with only 6 FSR sensors. The Z method was used to get rid of the excess pressure which remains during the rest posture due to attaching the band around the wrist. Six-time domain features were extracted followed by feature selection and optimization process for the ELM using three different optimization algorithms such as BGWO, BGOA, and BGWOPSO. The results showed that the BGWOPSO achieved a classification accuracy of 91.36% outperforming BGOA and BGWO. In the future, a wearable ASL translator with an embedded BGWOPSO wrapper based on ELM is planned.

Acknowledgements This research is funded by the Deutsche Forschungsgemeinschaft (DFG, German Research Foundation)—Project-ID 416228727—SFB 1410 and the German Academic Exchange Agency (DAAD) within the project PROFILE (grant number 57612192).

References

1. Gupta, M., Rohini, R., Reddy, P., Prakash, P. B., & Kumar, K. (2016). Gesture controlled Metal detection Land Rover. *International Journal of Engineering Trends and Technology, 21*(05), 229–231.
2. Xie, R., Sun, X., Xia, X., & Cao, J. (2015). Similarity matching-based extensible hand gesture recognition. *IEEE sensors journal, 15*(6), 3475–3483.
3. Vaitkevičius, A., Taroza, M., Blažauskas, T., Damaševičius, R., Maskeliunas, R., Woźniak, M. (2019). Recognition of american sign language gestures in a virtual reality using leap motion. *Applied Sciences 9*(3) 445.
4. Maqueda, A. I., del Blanco, C. R., Jaureguizar, F., García, N. (2016) Temporal pyramid matching of local binary subpatterns for hand-gesture recognition. *IEEE Signal Processing Letters 23*(8) 1037–1041.
5. Ha, N., Withanachchi, G. P., & Yihun, Y. (2019). Performance of forearm fmg for estimating hand gestures and prosthetic hand control. *Journal of Bionic Engineering, 16*(1), 88–98.
6. Fujiwara, E., Suzuki, C. K. (2018) Optical fiber force myography sensor for identification of hand postures. *Journal of Sensors 2018*.
7. Connan, M., Ruiz Ramírez, E., Vodermayer, B., Castellini, C. (2016) Assessment of a wearable force-and electromyography device and comparison of the related signals for myocontrol. *Frontiers in Neurorobotics 10* (2016) 17.
8. Nissler, C., Mouriki, N., & Castellini, C. (2016). Optical myography: Detecting finger movements by looking at the forearm. *Frontiers in neurorobotics, 10*, 3.
9. Xiao, Z. G., & Menon, C. (2019). A review of force myography research and development. *Sensors, 19*(20), 4557.
10. Delva, M. L., Lajoie, K., Khoshnam, M., & Menon, C. (2020). Wrist-worn wearables based on force myography: On the significance of user anthropometry. *BioMedical Engineering OnLine, 19*(1), 1–18.
11. Sadarangani, G. P., Menon, C. (2017) A preliminary investigation on the utility of temporal features of force myography in the two-class problem of grasp vs. no-grasp in the presence of upper-extremity movements. *Biomedical Engineering Online 16* (1) 1–19.
12. Chen, Y., Liang, X., Assaad, M., Heidari, H. (2019) Wearable resistive-based gesture-sensing interface bracelet. *In: 2019 UK/China Emerging Technologies (UCET), IEEE, 2019*, pp. 1–4.
13. Anvaripour, M., Saif, M. (2018) Hand gesture recognition using force myography of the forearm activities and optimized features. *In: 2018 IEEE International Conference on Industrial Technology (ICIT), IEEE*, pp. 187–192.
14. Jiang, X., Tory, L., Khoshnam, M., Chu, K., Menon, C. (2018) Exploration of gait parameters affecting the accuracy of force myography-based gait phase detection. *In: 2018 7th IEEE International Conference on Biomedical Robotics and Biomechatronics (Biorob), IEEE*, pp. 1205–1210.
15. Ben Atitallah, B., Abbasi, M. B., Barioul, R., Bouchaala, D., Derbel, N., Kanoun, O. (2020) Simultaneous pressure sensors monitoring system for hand gestures recognition, *In: 2020 IEEE Sensors, IEEE*, pp. 1–4.
16. Barioul, R., Ghribi, S. F., Derbel, H. B. J., Kanoun, O. (2020) Four sensors bracelet for american sign language recognition based on wrist force myography. *In: 2020 IEEE International Conference on Computational Intelligence and Virtual Environments for Measurement Systems and Applications (CIVEMSA), IEEE*, pp. 1–5.
17. Islam, M. R. U., Waris, A., Kamavuako, E. N., & Bai, S. (2020). A comparative study of motion detection with fmg and semg methods for assistive applications. *Journal of Rehabilitation and Assistive Technologies Engineering, 7*, 2055668320938588.
18. Beni, G., Wang, J. (1993) Swarm intelligence in cellular robotic systems. *In: Robots and Biological Systems: Towards a New Bionics?, Springer*, pp. 703–712.
19. Talbi, E. -G. (2009) Metaheuristics: from design to implementation, *74, John Wiley & Sons*.
20. Long, W. (2016) Grey wolf optimizer based on nonlinear adjustment control parameter. *In: 2016 4th International Conference on Sensors, Mechatronics and Automation (ICSMA 2016), Atlantis Press*, pp. 643–648.
21. Saremi, S., Mirjalili, S., & Lewis, A. (2017). Grasshopper optimisation algorithm: Theory and application. *Advances in Engineering Software, 105*, 30–47.
22. Huang, G.-B., Zhu, Q.-Y., Siew, C.-K. (2004) Extreme learning machine: a new learning scheme of feedforward neural networks. *In: 2004 IEEE international joint conference on neural networks (IEEE Cat. No. 04CH37541), 2, IEEE*, pp. 985–990.
23. Huang, G.-B., & Chen, L. (2007). Convex incremental extreme learning machine. *Neurocomputing, 70*(16–18), 3056–3062.
24. Zhang, X., Yang, Z., Cao, F., Cao, J., Wang, M., & Cai, N. (2020). Conditioning optimization of extreme learning machine by multi-task beetle antennae swarm algorithm. *Memetic Computing, 12*, 151–164.
25. Cao, W., Gao, J., Ming, Z., Cai, S. (2017) Some tricks in parameter selection for extreme learning machine, in: IOP conference series: materials science and engineering, *261, IOP Publishing*, p. 012002.
26. Zhang, R., Huang, G.-B., Sundararajan, N., & Saratchandran, P. (2007). Multicategory classification using an extreme learning machine for microarray gene expression cancer diagnosis. *IEEE/ACM transactions on Computational Biology and Bioinformatics, 4*(3), 485–495.
27. Zhang, L., Zhang, D., Tian, F. (2016). Svm and elm: Who wins? object recognition with deep convolutional features from imagenet. *In: Proceedings of ELM-2015 Volume 1, Springer*, pp. 249–263.
28. Alencar, A. S., Neto, A. R. R., & Gomes, J. P. P. (2016). A new pruning method for extreme learning machines via genetic algorithms. *Applied Soft Computing, 44*, 101–107.
29. Shariati, M., Mafipour, M. S., Ghahremani, B., Azarhomayun, F., Ahmadi, M., Trung, N. T., Shariati, A. (2020). A novel hybrid extreme learning machine–grey wolf optimizer (elm-gwo) model to predict compressive strength of concrete with partial replacements for cement. *Engineering with Computers 1–23*.
30. Han, F., Yao, H.-F., & Ling, Q.-H. (2013). An improved evolutionary extreme learning machine based on particle swarm optimization. *Neurocomputing, 116*, 87–93.
31. Yu, C., Koopialipoor, M., Murlidhar, B. R., Mohammed, A. S., Armaghani, D. J., Mohamad, E. T., Wang, Z. (2021) Optimal elm–harris hawks optimization and elm–grasshopper optimiza-

tion models to forecast peak particle velocity resulting from mine blasting. *Natural Resources Research* 1–16.

32. Mirjalili, S., Mirjalili, S. M., & Lewis, A. (2014). Grey wolf optimizer. *Advances in Engineering Software, 69*, 46–61.

33. Emary, E., Zawbaa, H. M., Grosan, C., Hassenian, A. E. (2015) Feature subset selection approach by gray-wolf optimization. *In: Afro-European Conference for Industrial Advancement*, Springer, pp. 1–13.

Design and Preliminary Testing of a Shoulder Exoskeleton Based on a Soft Bellow Actuator

Stanislao Grazioso, Teodorico Caporaso,
Benedetta M. V. Ostuni, and Antonio Lanzotti

Abstract

This paper describes the design and preliminary testing of a soft exoskeleton for the shoulder which supports the adbuction/adduction movements. The soft exoskeleton is based on a single soft bellow actuator, a pneumatically actuated system which is composed by multiple consecutive chambers that expands when compressed air is inflated till providing a desired motion. In this work we present the design, analysis, prototyping and testing of the soft bellow actuator as well as its preliminary integration into a first version of bellow-based soft exoskeleton for the shoulder.

Keywords

Soft exoskeletons · Soft actuators · Soft robotics

1 Introduction

Robotic exoskeletons are wearable mechanical structures that enhance the power of a human [1]. To date, most of current design of robotic exoskeleton transmit forces from the robot to the human using rigid link and structures [1]. Despite being very reliable and allowing the transmission of high forces, these concepts suffer from low wearability. Indeed, they are usually bulky structures that can be seen as armor for the humans. Over the last years, robotic exoskeleton that transmit forces through cable–driven system or pneumatically actuated chambers have emerged as a good alternative which enhance the wearability, reduce the overall weight and improve the comfort for the humans. These systems are referred to as soft exoskeletons [2].

Multiple concepts have been proposed for assistance of shoulder movements, mostly based on pneumatic actuation. In [3], the authors have developed a soft wearable robot for the shoulder supporting abduction and horizontal flex- ion/extension on two separate pneumatic actuators. The evolution of this system is represented by the work in [4], where the authors have developed an inflatable bifurcated actuator which better distributes the forces under the upper limb. A different design approach for assistance of the shoulder indeed uses an external exomuscle [5]. This solution, although being effective in rehabilitation settings, might be less effective in everyday life. Among the available pneumatic actuators for the development of soft exoskeletons, bellows technology has recently emerged as a promising solution. A good example of a soft robotic exoskeleton based on bellow technology is represented by [6], where the authors have developed a concept for assistance of elbow flexo-extension movements.

In this paper we report the design and preliminary testing of a bellow-based soft wearable robot for assistance of adbuction/adduction movements of the shoulder, particularly suited for overhead tasks [7]. Section 2 illustrates the concept design of the exoskeleton. In Sect. 3, we report the design and prototyping, analysis and testing of the actuator. The preliminary realization and testing of the exoskeleton based on the soft bellow actuator is indeed reported in Sect. 4. Finally, conclusions are given in Sect. 5.

2 Concept

The concept of bellow-based soft exoskeleton is illustrated in Fig. 1. It is based on a suit where two soft bellow actuators (one for each arm) are placed between the upper arm and the chest. The system is intended to provide assistance to the movements of shoulder abduction/adduction. To select the level of assistance, a regulation system at the wrist level is foreseen. The soft bellow actuators are placed on the suit through flexible plates and adjustment straps. They have a suitable

S. Grazioso (✉) · T. Caporaso · B. M. V. Ostuni · A. Lanzotti
Department of Industrial Engineering, University of Naples Federico II, Naples, Italy
e-mail: stanislao.grazioso@unina.it

© The Author(s) 2026
B. Meyer et al. (eds.), *Hybrid Societies*, Advances in Science, Technology & Innovation,
https://doi.org/10.1007/978-3-032-03488-5_11

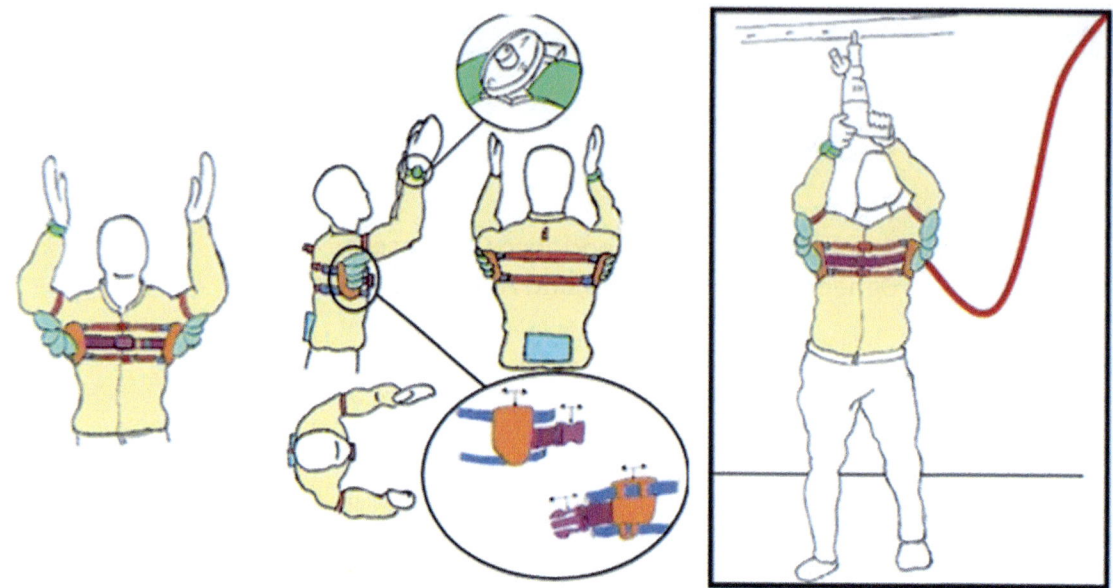

Fig. 1 Concept of a bellow-based exoskeleton for assistance of overhead tasks

Fig. 2 CAD concept of the exoskeleton based on the soft bellow actuator

connector for compressed air, to allow the connection to the pneumatic tubing coming from a pressure-regulated circuit. The movements of the flexible plates on the nylon strips can potentially allow the bellows to assist, beside the abduction/adduction movements, also the flexo/extension movements of the shoulder.

The CAD model of the bellow-based soft exoskeleton is reported in Fig. 2. The model includes the suit, the bellows, the flexible plates between the bellows and the suit, the nylon straps and clips. The suit of the model is realized as a three-dimensional surface starting from a CAD model of a human being, as scanned by our 3D body scanner [8].

3 Realization of the Soft Bellow Actuator

3.1 Prototyping

The soft bellow actuator is realized with connected and sealed multiple chambers made of TPU double coated Nylon material. The fabrication steps are shown in Fig. 3. First, rectangles of TPU double coated Nylon material are cutted (1); after, we cut two rectangles, one for connection of the inlet valve, one for connection of multiple chambers (2); the third step is to attach the inlet valve on the rectangle sheet (3); after, two fabric rectangles are heat sealed together along all sides (4); now, there is the step of connecting together the various air chambers with double-sided tape (5); finally, the last step is joining the top end of the inner tubes together (6).

3.2 Analysis

The analysis of the soft bellow actuator has been conducted within the Ansys environment. We model the bellow as a series of flat rectangular surfaces, on which holes are made to allow the inner chambers to connect with each other.

An explicit dynamic analysis was carried out, given the high deformations that the material is subject to. The bellow was model considering TPU double coated Nylon material, with material properties taken from [9]. The mesh consists of shell elements. The internal constraints were given so

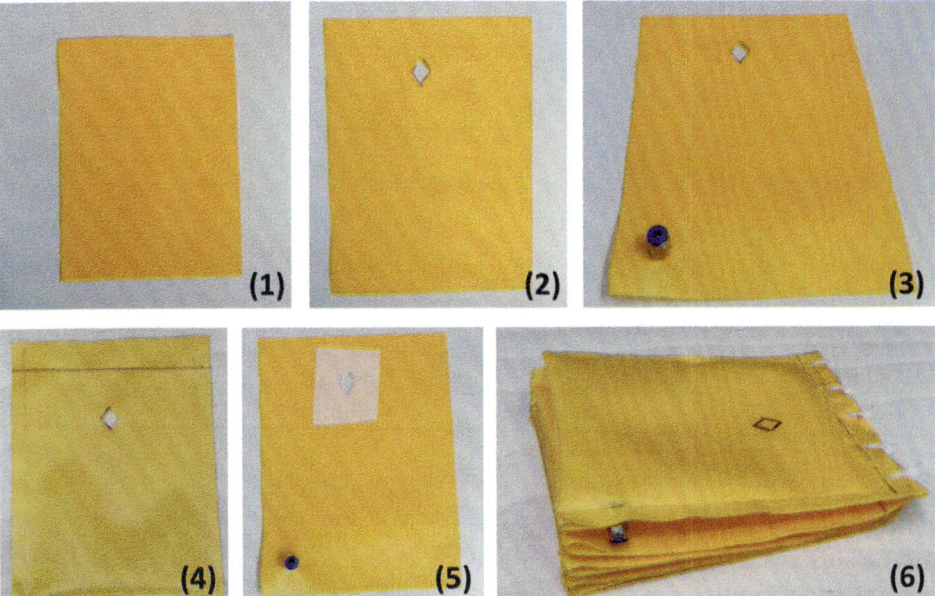

Fig. 3 Fabrication steps of the soft bellow actuator. (1) cutting of rectangles of material; (2) drill inner holes to join the various air chambers; (3) attach the air inlet to one of the holes; (4) heat seal two fabric rectangles along all sides; (5) connect the various air chambers together with double-sided tape by matching the previously created holes; (6) join the top end of the inner tubes together using sewing or ties

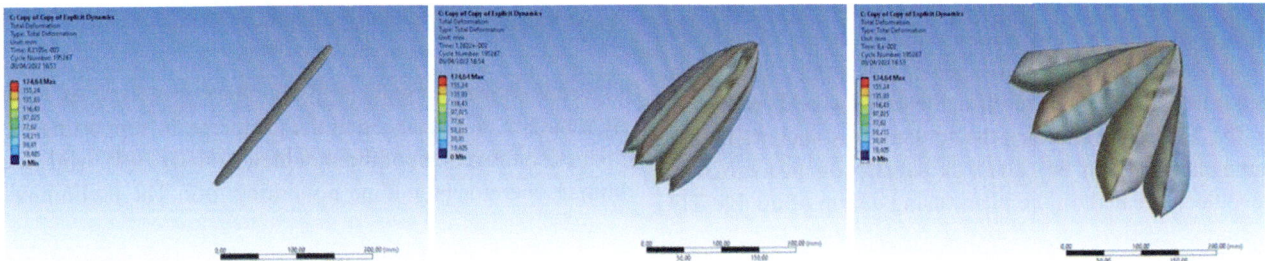

Fig. 4 Finite element analysis of the soft bellow actuator

that the surfaces were coupled two by two to form the air chambers. The load was assigned as a pressure on the inner surfaces of the inner tubes. A hinge on the short side was used as a constraint. The load and constraint conditions are such that the air chambers open like a bellows. The screenshots of the analysis are reported in Fig. 4. It is important to consider that the results of the analysis are purely qualitative. Indeed, the aim was only to understand the dynamics of opening of the bellow.

3.3 Testing

The tests on the soft bellow actuator were performed on a test stand which simulates the abduction movement of the shoulder. It consists of an L-shaped wooden structure where a wooden beam (with length equal to that of the human arm) is hinged. The beam has a hook where it is possible to place weights that generate a torque equal to the one generated by the arm. In addition, in correspondence of the hinge is placed a goniometer that allows a quick measurement of the abduction angle reached by the actuator. The tests on the soft bellow actuator (with six chambers) are shown in Fig. 5, where compressed air is inflated within the bellow till reaching a maximum angle set equal to 90° (input pressure equal to 19 psi). The experiments have been performed at ASTRO Lab, the laboratory of Joint Lab IDEAS of University of Naples Federico II devoted to advancements in soft robotics.

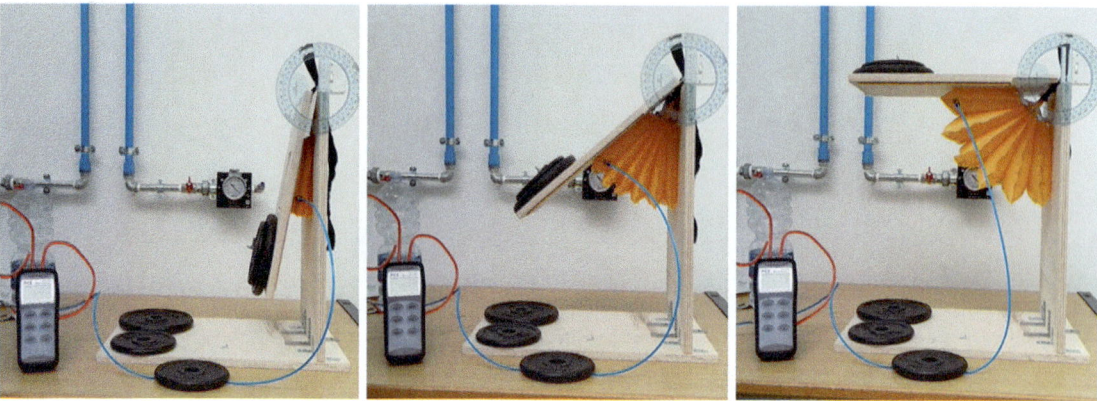

Fig. 5 Tests of the pneumatically actuated soft bellow actuator

4 Preliminary Realization of the Soft Exoskeleton

4.1 Assembly and Integration

Apart from the soft bellow actuators, the full exoskeleton is composed by other components, as illustrated in Fig. 6. In particular, we have realized two TPU plates for the connection of the bellows with the human body. To fix the actuator to the arm and the torso we have used nylon straps with clips to allow an easy and quick adjustment and wearing of the system. Regarding the fixing on the arm, the straps were directly sewn on the TPU plate and then the actuator was anchored on the same plate (see Fig. 6e). To fix the actuator on the torso, loops have been created on the plate (using a strip of double TPU

coated nylon) inside which the straps can pass. Thanks to these "rails," the actuator is free to move on the torso of the operator and allow her/him a greater range of movement (abduction/adduction and, eventually, flexo/extension). For the prototype, we did not realize the suit, and therefore we have used a system of straps to support the actuator during the use.

4.2 Preliminary Wearability and Usage Tests

Figure 7 shows the realized preliminary prototype as worn by a human. We have tested the prototype in supporting the abduction movement of the shoulder, from 0 to 90°, while the human operator is holding a working tool. The preliminary

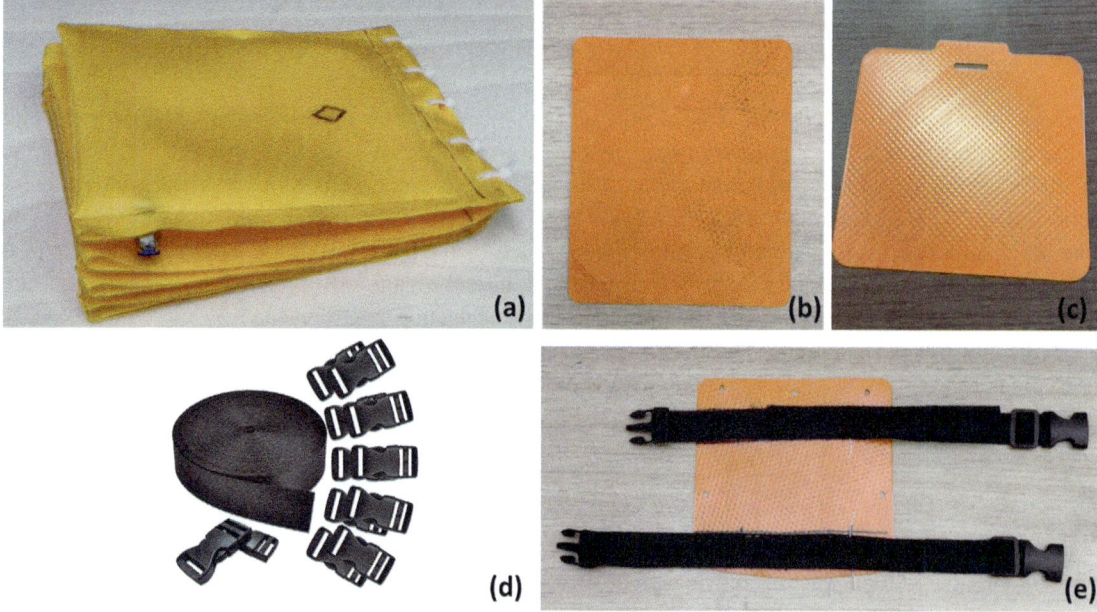

Fig. 6 Components of the prototype of exoskeleton. **a** = soft bellow actuator; **b** = TPU plate for the arm; **c** = TPU plate for the chest; **d** = Nylon straps and clips; **e** = TPU plate with nylon straps

Fig. 7 Exoskeleton worn by a human on a shoulder abduction movement while carrying a manual tool

tests are very promising in terms of comfort from the human side, but in progress is a study to quantify the wearability and usability of the device.

5 Conclusion

A novel concept of a soft exoskeleton based on bellow technology is presented in this paper to support the shoulder adbuction/adduction movements. The concept might be used for the development of assistive devices that can support people in performing overhead tasks, namely tasks with arms above head level. In this work, we have presented the design, modeling, analysis and prototyping of a single soft bellow actuator and a first integration into a soft exoskeleton for the shoulder. The soft bellow actuator is able to move in the range 0 to 90 degrees, while not changing its shape if subject to the external load of the arm and of the tool. This is quite promising toward the further exploitation of this technology for the realization of simple but yet effective assistive devices that can be used in industrial, rehabilitation and also daily living tasks.

References

1. De Looze, M. P., Bosch, T., Krause, F., Stadler, K. S., & O'sullivan, L. W. (2016). Exoskele- tons for industrial application and their potential effects on physical work load. *Ergonomics, 59*(5), 671–681.

2. Panariello, D., Grazioso, S., Caporaso, T., Gironimo, G.D., Lanzotti, A. (2021). A detailed analysis of the most promising concepts of soft wearable robots for upper–limb. *In: International Conference on Design, Simulation, Manufacturing.* pp. 71–81.

3. O'Neill, C. T., Phipps, N.S., Cappello, L., Paganoni, S., Walsh, C. J. (2017). A soft wearable robot for the shoulder: Design, characterization, and preliminary testing. *In: 2017 International Conference on Rehabilitation Robotics.* pp. 1672–1678

4. O'Neill, C., Proietti, T., Nuckols, K., Clarke, M. E., Hohimer, C. J., Cloutier, A., Lin, D. J., & Walsh, C. J. (2020). Inflatable soft wearable robot for reducing therapist fatigue during upper extremity rehabilitation in severe stroke. *IEEE Robotics and Automation Letters, 5*(3), 3899–3906.

5. Simpson, C., Huerta, B., Sketch, S., Lansberg, M., Hawkes, E., & Okamura, A. (2020). Upper extremity exomuscle for shoulder abduction support. *IEEE Transactions on Medical Robotics and Bionics, 2*(3), 474–484.

6. Thalman, C. M., Lam, Q.P., Nguyen, P. H., Sridar, S., Polygerinos, P. (2018) A novel soft elbow exosuit to supplement bicep lifting capacity. *In: 2018 IEEE/RSJ International Conference on Intelligent Robots and Systems (IROS).* pp. 6965–6971. IEEE.

7. Panariello, D., Grazioso, S., Caporaso, T., Palomba, A., Di Gironimo, G., Lanzotti, A. (2022). Biomechanical analysis of the upper body during overhead industrial tasks using electromyography and motion capture integrated with digital human models. *International Journal on Interactive Design and Manufacturing* pp. 1–20

8. Grazioso, S., Selvaggio, M., & Di Gironimo, G. (2018). Design and development of a novel body scanning system for healthcare applications. *International Journal on Interac- tive Design and Manufacturing (IJIDeM), 12*(2), 611–620.

9. Nguyen, P. H., Mohd, I. I., Sparks, C., Arellano, F. L., Zhang, W., Polygerinos, P. (2019). Fabric soft poly-limbs for physical assistance of daily living tasks. *In: 2019 International Conference on Robotics and Automation (ICRA).* pp. 8429–8435. IEEE

Observational Learning in Humans and Machines

Claudia Voelcker-Rehage, Fred H. Hamker, Javier Baladron, Julian Rudisch, Torsten Fietzek, and Julien Vitay

Abstract

Observational learning is referred to as a change in performance following the observation of others. With respect to motor learning, an observed action is known to facilitate motor learning mediated by brain processes that are involved during both, the observation and the execution of a certain task. Observational learning in humans has inspired robotic researchers as it may alleviate the necessity to explicitly program robots or require robots to extensively search for a suitable solution. Further, observational learning may become a central aspect of future hybrid societies where robots closely interact with humans. Here, we summarize the current state of the art in observational motor learning in humans and robots with a focus on upper extremity tasks. Further, we briefly outline a roadmap for better understanding observational motor learning by means of building brain-inspired neuronal models. Bringing together these lines of research not only advances human movement science but, in the long run, may contribute to new programming approaches in robots that facilitate human–robot interaction.

Keywords

Motor learning · Mirror neurons · Robots · Machine learning · Computational neuroscience

Contributed equally: Claudia Voelcker-Rehage and Fred H. Hamker.

C. Voelcker-Rehage · J. Rudisch (✉)
University of Münster, Münster, Germany
e-mail: julian.rudisch@uni-muenster.de

C. Voelcker-Rehage
e-mail: claudia.voelcker-rehage@uni-muenster.de

F. H. Hamker · J. Baladron · T. Fietzek · J. Vitay
Chemnitz University of Technology, Chemnitz, Germany
e-mail: fred.hamker@informatik.tu-chemnitz.de

J. Baladron
e-mail: javier.baladron@usach.cl

T. Fietzek
e-mail: torsten.fietzek@informatik.tu-chemnitz.de

J. Vitay
e-mail: julien.vitay@eodyn.com

J. Baladron
Departamento de Ingeniería Informática, Universidad de Santiago de Chile, Santiago, Chile

1 Introduction

From an evolutionary point of view, flexibility and adaptivity are essential for the survival and success of organisms. Learning involves processes that develop with experience, which change an individual's behaviors and internal states. Brain evolution made different forms of learning possible that rely on exploratory behavior. However, this is aided by methods, such as instructions, (error) feedback, appropriate task constraints, and the observation of others. Learning from others is of interest to researchers across different disciplines such as social and developmental psychology, sport and exercise science, biology and robotics, mainly because it improves goal-directed behavior.

In observational learning, typically a co-learner serves as a learning model for [1, 2]. Here, we focus on learning movements or motor skills through observational learning in humans and robots, but less on social skills, with the aim that a further understanding of how we learn from others not only advances brain science and human movement science, but also the field of robotics. If robots and machines were also able to learn from others, it would significantly alter how humans interact with such embodied technologies in future hybrid societies.

2 Observational Learning in Humans

Motor learning is vital for human existence as it enables us to produce novel skills that may be required for survival (e.g., learning how to throw a javelin to kill an animal) or for social life (e.g., learning to play a musical instrument). The processes leading to motor learning are complex and may occur through various pathways, involving different structures and subsystems of the sensorimotor control system and leading to relatively persistent changes in skill level [3]. The primary method through which we acquire novel skills is through physical practice. In addition, motor learning can also occur through mental practice as well as through observation of a model (observational learning). Observational learning is frequently used to supplement physical practice, and it has been repeatedly shown to facilitate motor learning.

Observational (motor) learning can be defined as any enduring change in motor performance following the observation of an action [4]. This definition is extremely broad and it integrates an abundance of different types of actions, ranging from simple and short-duration (e.g., reach and grasp actions) to complex and long-duration (e.g., a set of dance moves). Consequently, mechanisms that are involved when learning actions from the observation of others may differ depending on the type of action to be learned as well as the skill level of the learner.

The observation of a skill is thought to trigger the action representations that are activated during the physical execution of the action [5]. It has also been shown that the observation of the action during practice can generate similar neural activations and, under certain conditions, lead to similar adaptations in the learner as physical practice. Experiments investigating the effects of schedules of practice [6, 7], sensory information [8] and relative feedback frequency [9, 10] have shown that variables affecting learning through physical practice tend to affect observational learning in a similar way. Therefore, observational learning has often been attributed to the mirror neuron system [4, 5, 11, 12]. Studies on the human mirror neuron system locate activity related to action observation primarily in the ventral premotor and inferior parietal cortex [4, 5]. However, how someone learns from the observation of another action and what brain structures and neural circuits of the extended mirror neuron system are involved depends on the type of task, the intentions, motivation, as well as the skill level of the learner [4]. When observing an action, the learner can either extract and learn a novel or more effective strategy or coordination pattern to solve a task, or the learner can extract and learn subtle adaptations to coordination patterns required to meet specific coordination constraints [3, 13]. A toddler, for example, may observe their parents pulling down the door handle to open a door, thus learning that this is a necessary part of the task to open a door. Likewise, it may be difficult to find more effective (or effi-

cient) solutions for certain tasks in daily life as well as sports without the observation of someone else displaying an effective model, e.g., to hold a pen in a precision grip instead of a power grip, to jump over a pole using a Fosbury flop instead of the straddle technique or to follow a certain route that one has not considered before, to successfully climb a wall.

Moreover, observational learning may even be beneficial compared to other learning approaches. For example, watching a task solution or a skill acquisition process might allow the observer to engage in cognitive activities akin to the learner, including performance evaluation, error detection, and consideration of potential corrective responses [14, 15]. Therefore, the observer might pick up action strategies or coordination patterns that could be used to "solve" the task (differently) [16, 17]. In this vein, it has been argued that action observation may give the learner unique opportunities either to extract important information concerning appropriate coordination patterns and subtle requirements of the task or to evaluate the effectiveness of strategies that would be difficult, if not impossible, if he or she were to prepare and execute an impending movement concomitantly [18]. From that perspective, observational practice offers the learner a chance to conduct processing that could not occur simultaneously with physical practice [13].

As mentioned above, the type of task and skill level influence observational learning. Studies have revealed that observation of an unfamiliar motor task (e.g., a dance sequence) elicits higher activation in the dorsolateral prefrontal cortex (dlPFC) as compared to familiar tasks [19]. There are, however, also findings from movement imagery indicating that one must have a well-established motor representation of a skill for this to be translated into a motor/internal pattern of brain activity. If not, brain activation will be a visual/external pattern [20]. When an action is observed which is already mastered by the learner, he or she can adapt and internally simulate the sensorimotor consequences of the task, without actually executing it [21]. Thus, existing skills can be refined, or subtle adaptations to changing constraints can be learned.

To sum up, when internally simulating an action (e.g., during observation), both the motor commands necessary for coordination as well as the expected sensory consequences (forward model) are simulated without physically performing the task. Learning thus occurs through shaping or strengthening the sensorimotor networks, similar to physical practice [19]. This type of learning may be predominant, for example, when observing a sequence of dance steps or observing a climber finding their way to reach the top of a boulder, if the observer is already skilled for the purpose of mentally rehearsing and improving these movements [19].

Three different strategies of observational learning with respect to the observer's internal model have been discussed [22]. Vicarious reward learning describes the ability to predict and observe the outcomes of others. This information can be used by the observer to infer behavioral changes. Emulation allows the observer to infer the goal or the intention of the observed person. Emulation requires an already functional world model of the observer. When this world model is fed with the inferred goal, the problem can be mentalized and a behavioral response can be computed to achieve the desired outcome. Imitation is rather applied in the case of a poor world model, when even the inferred goal of the observed person does not allow the observer to fully mentalize the problem such that no outcome can be reliably predicted. Under this condition, the imitation strategy is applied where the observer imitates an obvious or frequent behavior. In its simplest form imitation learning does not even require access to the consequences, and value is given to those actions being observed.

A core domain of experimental studies in observational learning is upper limb actions. Two important fields of application are rehabilitation of brain injuries [23] or professional training for surgery [24]. In addition, many investigations of observational learning in robots focus on upper limb reach and grasp actions (see Section 1.3). The following section, therefore, provides a short overview of observational learning studies in healthy adults with the goal to serve as a model for observational learning in robots, potentially controlled by brain-inspired neural architectures. Observational learning studies in healthy individuals typically aim to investigate principles of motor learning and neural correlates thereof. Studies focusing on the upper limb typically employ one of two different types of actions: sequencing tasks and visuomotor adaptation tasks [4]. In motor sequencing tasks, the learner reproduces a specific sequence of movements (e.g., reaching to press different buttons). Changes in performance with observational or physical practice, i.e., an increase in reaction time or a decrease in the error rate, are attributed to motor learning. It has been shown that observing another actor and physically practicing a sequencing (serial reaction time) task leads to similar performance improvements in reaction time [25]. In motor adaptation tasks, one aims to investigate how observational learning affects the adaptation of the sensorimotor system to changing constraints. This is typically done through visual or force-field perturbations in which participants have to push a manipulandum to an end position while the manipulandum or the observed trajectories are deflected. Following some practice, the sensorimotor system adapts to the deflection and counteracts it. Adaptations of the sensorimotor system to novel force fields are linked to an updating of the internal model [26], i.e., a forward model of the inverse dynamics of the system relating muscle activations to changes in the end effector. Visual observation of a dynamic force field is, however, virtually impossible, and, as such, one cannot update an internal model from the observation of another actor

who already has adapted to a novel force field. Learning from observation is rather only possible when the learner observes the process of the actor adapting to the novel force field [27]. In addition to evidence that the cerebellum may implement an internal model [28], measures of functional connectivity suggest that the cerebellum might also mediate the learning process of sensorimotor adaptation from observing an actor adapting to a novel force-field through practice [29]. After an update of the internal model, typically so-called aftereffects are found in the absence of force-field perturbations, leading to a deflection of the movement trajectories in the opposite direction [26]. When visuomotor adaptation is mediated by observational learning, however, aftereffects are absent [30], indicating that learning does not occur through an updating of such an internal model, but rather through the acquisition of explicit knowledge about the changed constraints.

In sum, observational learning is important for skill acquisition, complementing physical practice in situations where an understanding of novel tasks is required by means of updating and strengthening the internal representations of an action. The means by which an action or skill is learned through observation, as well as the involved neural pathways, depend on the task to be learned, the skill level, motivation, and intentions of the learner.

3 Observational and Imitation Learning in Robots

Despite the pioneering review on imitation learning in humanoid robots in 1999 [31], true observational or imitation learning in robots is still unsolved. Thus, present approaches may more conservatively be called "learning from demonstrations" (LfD). Robot learning from demonstrations alleviates the necessity to explicitly program robots by decomposing the desired task into functional blocks and allows them to autonomously generalize the observed behavior to novel situations [32]. Most works follow a behavioral cloning approach, where the desired motor commands are directly provided to the robot by an expert, and a function approximator (e.g., a neural network) learns to reproduce and generalize the sensorimotor pattern. This, however, raises the correspondence problem [33], as the human expert and the robot share neither the same perceptual space (especially when the robot faces the human) nor the action space (muscles vs. actuators). Most strategies circumvent the correspondence problem by either (1) learning an explicit mapping between the human and robot joints and relying on human motion data to generate the demonstrations [34], or (2) having the human demonstrator activate the robot joints through direct manipulation or teleoperation with haptic devices [35]. Learning solely from the observation of human demonstrations raises important challenges, as only the visual consequences of the demonstration are available to the robot and, moreover, they can be extremely different when the same action (e.g., grasping an object) is performed by the expert or by the robot (change of reference frame). A lot of work in developmental robotics has tackled this problem by taking inspiration from the primate mirror neuron system [11] to map observations into the robot's actuator space, but these approaches are still limited in terms of applicability [36, 37].

Another problem with the present robotic approaches to imitation or observational learning is that the robot's policy cannot become better than the demonstrations. An important area of research in robotics nowadays is the use of reinforcement learning methods to learn efficient policies through trial and error by maximizing the sum of rewards received in the long term. After the initial breakthroughs in video games [38] or board games such as Go [39], deep reinforcement learning (DRL) became central to autonomous robotics as it allows robots to learn visuomotor policies in an end-to-end manner, i.e., without manually engineering visual features or creating motor primitives. DRL relies on the exploration of the state-action space by the agent, which can take a long time, as the agent has to discover the consequences of its actions in the long term without additional task knowledge. One method to alleviate this issue is to learn from demonstrations, replacing the initial exploration with expert trajectories, thus avoiding exploring useless regions of the state-action space. This approach, now labeled as offline RL [40], still suffers from the same correspondence problem, as the expert demonstrations must be in the same state-action space as the agent. However, the learned policy might ultimately become better than the demonstrations if the agent is allowed to further explore.

Imitating the state-action pairs provided by the demonstrations is not the only option to integrate observational learning into reinforcement learning. In many domains, the hardest part when designing an artificial system is not how the system should perform a task, but formalizing what the task actually is. In reinforcement learning, the task is defined by the reward function, which tells the system at each time step how good or bad the action was, although in most practical cases reinforcement is sparse. The difficulty in defining the reward function also raises safety issues [41]: the reward function that will be maximized by the system might not be what the programmer intended. For example, when training a robot arm to stack Lego bricks, the programmers designed the reward function as being high when the bottom face of a brick is higher than the ground [42]. The DRL agent solved this problem by simply flipping all bricks upside-down, effectively accumulating rewards but missing the point of the task.

Inverse reinforcement learning (IRL) [43] is a method that allows one to derive a reward function from observations. Instead of imitating the state-action pairs of a demonstration, an IRL system tries to infer which reward is actually optimized by the demonstrator—assuming that the demonstrator is an optimal actor. Each state-action pair, even those not shown by the demonstrator, can then be mapped to a reward function through a function approximator [44]. This estimated reward function allows learning a task using the goal actually intended by the demonstrator, not the possibly perverse incentive implied by an engineered reward function. Inverse RL fills the void between imitation learning/learning from demonstrations and observational learning; importantly, IRL does not suffer as much from the correspondence problem (the actions of the demonstrator can be different from the agent's as long as the intent is the same). The importance of IRL for observational learning is further reinforced by the observation of similar mechanisms in the human brain [45].

An important challenge in observational learning is the quality of the demonstration data [46]. On the one hand, the demonstration quality is limited due to possible non-optimal task performance of the demonstrator. Sasaki and Yamashina [47] tackled this problem with a behavioral cloning approach. On the other hand, the perception data is prone to sensory noise and occlusion. To solve this issue, researchers have used a Bayesian IRL method and applied it to an online sorting task [48]. Since in such real-world tasks low-level learning of robot motion is already complex due to many degrees of freedom, task primitives have been introduced [49].

Like IRL, which learns the decision policy and therefore a more abstract goal, Chung et al. [50] applied goal-based imitation learning by implementing a Bayesian learning model. Since the model learns by inferring the hidden goal of the demonstrator instead of simply copying the observed trajectory, the correspondence problem is also less critical. Alternatively, Park et al. [51] used a recurrent neural network and estimated a parametric bias in the input of the network to distinguish between the primary goal in imitation and the means to reach the goal. Agents are trained using backpropagation through time, to reach two goals in different ways. For imitation, the desired motor behavior is given to the model, which in turn is used to estimate a parametric bias, and the model generates a movement.

Carfi et al. [52] depict the challenges of hand-object interaction in humans and robots, giving first general definitions for object interaction tasks and highlighting the importance of perception of learning by demonstration in robot manipulation tasks. Finally, Vollmer et al. [53] emphasize the effect of robot learner feedback on the behavior of the demonstrating human, which should be considered in the design of robot observational learning when robots shall learn from humans.

In summary, observational learning in robots can be implemented at the sensorimotor level (learning from demonstrations, either purely supervised or optimized by reinforcement learning), but it suffers from the mapping problem. It can also be implemented at the cognitive/motivational level by learning to infer the underlying reward function (inverse RL) or goal (goal-directed imitation learning). In both cases, the quality of demonstrations is crucial to the success of the learning process.

4 Understanding Observational Learning by Models of Brain Function

In order to better understand observational learning, one has to not only explore its function but also identify the relevant brain structures involved and implement computational models. Those models can be tested on similar tasks to those applied to human subjects. Model simulation data can then be compared to experimental data.

Reinforcement learning relies on pathways formed by the multiple loops of the cortex with the basal ganglia. Phasic changes in the level of the neurotransmitter dopamine have been shown to encode a reward prediction error during instrumental conditioning. These circuits could therefore also be involved in observational learning. Dopamine has a strong effect on the basal ganglia, modulating both plasticity and neural activation. The neural activity in the striatum, an input nucleus of the basal ganglia, has been related to observational prediction errors [54–56].

Initial theories regarding the organization of the multiple cortico-basal ganglia loops followed the concept of parallel and segregated closed loops, each selecting actions independently [57, 58]. Modern approaches, however, favor a hierarchical organization, in which loops do not necessarily compete with each other in order to control behavior [59–64]. Hierarchical control has been already actively discussed in the context of prefrontal cortex (PFC) organization [65], but only recently in basal ganglia research. Different possible interaction mechanisms have already been identified, such as overlapping cortico-striatal projections [66], striato-nigro-striatal spirals [67] or the cortico-thalamo-striatal pathways [68]. Furthermore, the advantages of hierarchical motor control have also been discussed in the context of synthetic systems [69].

The three types of observational learning (vicarious reinforcement-learning, action imitation, and emulation) have been mainly studied from the perspective of the segregated circuit hypothesis. Following this framework, Charpentier et al. [70] proposed the existence of an arbitration mechanism to control the competition between an action imitation and an action emulation learning system.

To our knowledge, little has been said about observational learning starting from a modern hierarchical perspective. In the following paragraphs, we discuss how the different strategies of observation learning could be understood within a framework of a hierarchically organized cortex-basal ganglia network outlined in [71] and [72]. We mainly propose that the different types of observational learning correspond to changes at different levels of the hierarchy.

In the model proposed by Baladron and Hamker [71], goal-directed behavior results from a cascade of decisions made by the multiple cortex-basal ganglia loops. A ventral loop, including the nucleus accumbens and other limbic structures, selects a goal depending on current needs and environmental information. This goal contains information that allows obtaining a specific type of reward (food or drink), e.g., a particular location that has to be checked. Then, loops through the dorsomedial striatum select an abstract plan involving states that need to be reached in order to obtain the desired goal. Finally, loops through the dorsolateral striatum select appropriate actions or movements to reach the selected states. The model further includes cortico-thalamic pathways that bypass the slow processing through the loops implementing habitual actions.

Emulation requires understanding and copying the goal or components of the abstract plan of the observed agent. We believe, therefore, that emulation depends on the initial ventral loops that represent values. Observation can bias the activation of these loops and later influence the selection of an output goal sent to the following dorsomedial loops. Information regarding the outcome of actions already gathered in the following loops through previous direct experience with each action can then be exploited. The observed action does not need to be copied, as a change in the ventral selection would force the following loops in the hierarchy to create their own plan to reach the goal. Thus, emulation works by setting goals within an already well-trained model.

Imitation learning may operate between the dorsomedial and dorsolateral loops that map environmental states to the actions required to achieve them. Although initially these state-action associations are created by self-executing actions, they could also be influenced by observation. If, after observation the connection between the observed action and the observed outcome state is increased then when the corresponding state is set as a goal for the dorsolateral loop, a bias toward the recently observed action will be produced. Such a change does not require copying the observed agent's goal but only learning about their outcomes and updating an outcome-action map. Thus, imitation learning is more explorative and works also with an uninformed model, but due to a less predictable outcome, the imitated action might not directly lead to the desired outcome.

A third mechanism by which observation could influence the action selection in a hierarchical model is through the update of cortico-thalamic shortcuts. In the original model [71], these pathways directly link preceding stimuli to actions and are trained and monitored by the basal ganglia. Training however could also be made through observation. In the original model learning in these connections is very slow, so an additional mechanism would be required to increase learning speed during observation.

One of the main features of the model from [71] is that each loop computes its own prediction error signal used for learning. Supported by recent data (see [73] for a review), we have suggested that the dopamine signals deliver specific information to each loop. We propose, therefore, that the different mechanisms described above could be controlled through different changes in the dopamine levels reaching each loop. If the goal of the observed subject is recognized by cortical areas, then a phasic dopamine increase in the ventral loop could enforce learning at initial levels of the hierarchy. If the outcome of the action executed by the observed subject is novel or unexpected to the observer, a phasic increase in dopamine could be triggered in the dorsal domains, enhancing the learning of a new map between states and actions. Interestingly, both changes could happen at the same time, enabling both goal and outcome learning.

Some studies have already searched for the neural signatures of observational prediction errors. By means of fMRI studies, Burke et al. [54] related changes in activity in the ventral prefrontal cortex and ventral striatum to the computation of observational outcome prediction error. They did not find a neural signature of observational learning in the dorsal striatum. Cooper et al. [55], however, in a different experiment, found similar reward prediction error-related activation in the dorsal striatum after experiential and observational instrumental conditioning. Both experiments had therefore conflicting results, with one associating observational learning with ventral areas and the other with dorsal areas. This deviation could be due to the differential activation of the mechanisms described above. Further, Kobza and Bellebaum [56] observed activations of the dorsal striatum in both observational and active learners. Active learners though had a stronger activation when an action outcome was unpredicted, but not when a stimulus led to a different reward than its current predicted value.

The basal ganglia could also be involved in the process of goal or outcome recognition. Emulation requires comparing the information stored in the action-outcome mapping learned in the multiple basal ganglia loops with what is being observed, something possible if observed and own actions outcomes are represented in the same neural space (as in the well-studied mirror system [74]). A simple well-placed dopamine phasic increase could then activate learning of the observed behavior.

Another type of task used to study observational learning is motor adaptation, as outlined in the section on human observational learning. Here, participants are initially trained to perform an arm movement to reach a goal, and then the environment is perturbed, usually by introducing a visual rotation or a force field. Participants tend to slowly adapt their movements to cope with the new environmental conditions. Then, if the perturbation is suddenly removed, an after-effect is observed: participants tend to over-compensate and slowly return their movement to their initial condition. In an alternative approach, participants do not directly face the perturbed environment but instead observe an external agent solving the task. Participants who only observe actions successfully learn the new appropriate movement however they do not present an after-effect [30, 75], suggesting that a different mechanism is used for observational than for direct learning. We hypothesize that motor adaptation through self-experience involves an implicit process mainly at the level of the cerebellum [76, 77], while observational learning occurs through the basal ganglia.

5 Conclusion

We provided a brief overview of observational motor learning in humans and machines. In addition to the correspondence problem, solutions also face an intention recognition problem. These challenges have to be met by any successful approach to observational learning. In our opinion, interdisciplinary research that integrates human movement science, neuroscience, neurocomputational models, and robotics is necessary to further advance the field. In particular, brain-inspired neural models, if they address the challenges outlined above, may help in further understanding this magnificent ability to learn.

Acknowledgements This research has been supported by the German Research Foundation (DFG, 416228727)—SFB 1410 Hybrid Societies.

References

1. Granados, C., & Wulf, G. (2007). Enhancing motor learning through dyad practice: Contributions of observation and dialogue. *Research quarterly for exercise and sport, 78*(3), 197–203.
2. McNevin, N. H., Wulf, G., & Carlson, C. (2000). Effects of attentional focus, self-control, and dyad training on motor learning: Implications for physical rehabilitation. *Physical Therapy, 80*(4), 373–385.
3. Schmidt, R. A., Lee, T. D., Winstein, C. J., Wulf, G., and Zeaznik, H. N. (2019). *Motor control and learning* (6th ed.). Human Kinetics.
4. Ramsey, R., Kaplan, D. M., & Cross, E. S. (2021). Watch and learn: The cognitive neuroscience of learning from others' actions. *Trends in Neurosciences, 44*(6), 478–491.
5. Rizzolatti, G., & Craighero, L. (2004). The mirror-neuron system. *Annual Review of Neuroscience, 27*(1), 169–192. PMID: 15217330.
6. Blandin, Y., Proteau, L., & Alain, C. (1994). On the cognitive processes underlying contextual interference and observational learning. *Journal of motor behavior, 26*(1), 18–26.

7. Wright, D. L., Li, Y., & Coady, W. (1997). Cognitive processes related to contextual interference and observational learning: A replication of blandin, proteau, and alain (1994). *Research Quarterly for Exercise and Sport, 68*(1), 106–109.

8. Shea, C. H., Wulf, G., Park, J.-H., & Gaunt, B. (2001). Effects of an auditory model on the learning of relative and absolute timing. *Journal of motor behavior, 33*(2), 127–138.

9. Badets, A., & Blandin, Y. (2004). The role of knowledge of results frequency in learning through observation. *Journal of motor behavior, 36*(1), 62–70.

10. Badets, A., & Blandin, Y. (2005). Observational learning: Effects of bandwidth knowledge of results. *Journal of motor behavior, 37*(3), 211–216.

11. di Pellegrino, G., Fadiga, L., Fogassi, L., Gallese, V., & Rizzolatti, G. (1992). Understanding motor events: A neurophysiological study. *Experi- mental Brain Research, 91*(1), 176–180.

12. Fabbri-Destro, M., & Rizzolatti, G. (2008). Mirror neurons and mirror systems in monkeys and humans. *Physiology, 23*(3), 171–179.

13. Wulf, G., Shea, C., & Lewthwaite, R. (2010). Motor skill learning and performance: A review of influential factors. *Medical education, 44*(1), 75–84.

14. Adams, J. (1986). Use of the models knowledge of results to increase the observers performance. *Journal of Human Movement Studies, 12*(2), 89–98.

15. Black, C. B., & Wright, D. L. (2000). Can observational practice facilitate error recognition and movement production? *Research quarterly for exercise and sport, 71*(4), 331–339.

16. Hodges, N. J., & Franks, I. M. (2002). Modelling coaching practice: The role of instruction and demonstration. *Journal of sports sciences, 20*(10), 793–811.

17. Hodges, N. J. & Franks, I. M. (2004). Instructions, demonstrations and the learning process: Creating and constraining movement options. In *Skill Acquisition in Sport*, pages 169–198. Routledge.

18. Shea, C. H., Wulf, G., & Whltacre, C. (1999). Enhancing training efficiency and effectiveness through the use of dyad training. *Journal of motor behavior, 31*(2), 119–125.

19. Cross, E. S., Kraemer, D. J., Hamilton, A. F. d. C., Kelley, W. M., & Grafton, S. T. (2008). Sensitivity of the Action Observation Network to Physical and Observational Learning. *Cerebral Cortex*, 19(2):315–326.

20. Olsson, C.-J., Jonsson, B., Larsson, A., & Nyberg, L. (2008). Motor representations and practice affect brain systems underlying imagery: An fmri study of internal imagery in novices and active high jumpers. *The open neuroimaging journal, 2*, 5.

21. Jeannerod, M. (2001). Neural simulation of action: A unifying mechanism for motor cognition. *NeuroImage, 14*(1), S103–S109.

22. Charpentier, C. J., & O'Doherty, J. P. (2018). The application of computational models to social neuroscience: Promises and pitfalls. *Social Neuroscience, 13*(6), 637–647.

23. Borges, L. R., Fernandes, A. B., Dos Passos, J. O., Rego, I. A. O., & Campos, T. F. (2022). Action observation for upper limb rehabilitation after stroke. *Cochrane Database of Systematic Reviews*, (8).

24. Harris, D., Vine, S., Wilson, M., McGrath, J. S., LeBel, M., & Buckingham, G. (2018). Action observation for sensorimotor learning in surgery. *Journal of British Surgery, 105*(13), 1713–1720.

25. Heyes, C., & Foster, C. (2002). Motor learning by observation: Evidence from a serial reaction time task. *The Quarterly Journal of Experimental Psychology Section A, 55*(2), 593–607.

26. Shadmehr, R., & Mussa-Ivaldi, F. A. (1994). Adaptive representation of dynamics during learning of a motor task. *Journal of neuroscience, 14*(5), 3208–3224.

27. Mattar, A. A., & Gribble, P. L. (2005). Motor learning by observing. *Neuron, 46*(1), 153–160.

28. Wolpert, D. M., Miall, R. C., & Kawato, M. (1998). Internal models in the cerebellum. *Trends in cognitive sciences, 2*(9), 338–347.

29. McGregor, H. R., & Gribble, P. L. (2015). Changes in visual and sensory-motor resting-state functional connectivity support motor learning by observing. *Journal of neurophysiology, 114*(1), 677–688.

30. Ong, N. T. and Hodges, N. J. (2010). Absence of after-effects for observers after watching a visuomotor adaptation. *Experimental Brain Research*, pages 325–334.

31. Schaal, S. (1999). Is imitation learning the route to humanoid robots? *Trends in Cognitive Sciences, 3*(6), 233–242.

32. Ravichandar, H., Polydoros, A. S., Chernova, S., & Billard, A. (2020). Recent Advances in Robot Learning from Demonstration. *Annual Review of Control, Robotics, and Autonomous Systems, 3*(1), 297–330.

33. Nehaniv, C. L. (2007). Nine billion correspondence problems. In C. L. Nehaniv & K. Dautenhahn (Eds.), *Imitation and Social Learning in Robots, Humans and Animals: Behavioural, Social and Communicative Dimensions* (pp. 35–46). Cambridge University Press.

34. Kim, S., Kim, C., You, B., & Oh, S. (2009). Stable whole-body motion generation for humanoid robots to imitate human motions. In *2009 IEEE/RSJ International Conference on Intelligent Robots and Systems*, pages 2518–2524.

35. Babič, J., Hale, J. G., & Oztop, E. (2011). Human sensorimotor learning for humanoid robot skill synthesis. *Adaptive Behavior, 19*(4), 250–263.

36. Dawood, F., & Loo, C. K. (2016). View-Invariant visuomotor processing in computational mirror neuron system for humanoid. *PLoS ONE, 11*(3), Article e0152003.

37. Wermter, S., Palm, G., & Elshaw, M., editors (2005). Biomimetic neural learning for intelligent robots—intelligent systems, cognitive robotics, and neuroscience, volume 3575 of *Lecture Notes in Computer Science*. Springer.

38. Mnih, V., Kavukcuoglu, K., Silver, D., Rusu, A. A., Veness, J., Bellemare, M. G., Graves, A., Riedmiller, M., Fidjeland, A. K., Ostrovski, G., Petersen, S., Beattie, C., Sadik, A., Antonoglou, I., King, H., Kumaran, D., Wierstra, D., Legg, S., & Hassabis, D. (2015). Human-level control through deep reinforcement learning. *Nature*, 518(7540):529–533.

39. Silver, D., Huang, A., Maddison, C. J., Guez, A., Sifre, L., van den Driessche, G., Schrittwieser, J., Antonoglou, I., Panneershelvam, V., Lanc- tot, M., Dieleman, S., Grewe, D., Nham, J., Kalchbrenner, N., Sutskever, I., Lillicrap, T., Leach, M., Kavukcuoglu, K., Graepel, T., & Hassabis, D. (2016). Mastering the game of Go with deep neural networks and tree search. *Nature*, 529(7587):484–489.

40. Levine, S., Kumar, A., Tucker, G., & Fu, J. (2020). Offline reinforcement learning: tutorial, review, and perspectives on open problems. arXiv:2005.01643.

41. Garc´ıa, J., & Ferna´ndez, F. (2015). A comprehensive survey on safe reinforcement learning. *The Journal of Machine Learning Research, 16*(1):1437–1480.

42. Popov, I., Heess, N., Lillicrap, T., Hafner, R., Barth-Maron, G., Vecerik, M., Lampe, T., Tassa, Y., Erez, T., & Riedmiller, M. (2017). Data-efficient Deep Reinforcement Learning for Dexterous Manipulation. arXiv:1704.03073.

43. Ng, A. Y., & Russell, S. J. (2000). Algorithms for Inverse Reinforcement Learning. In *Proceedings of the Seventeenth International Conference on Machine Learning*, ICML '00, pages 663–670, San Francisco, CA, USA. Morgan Kaufmann Publishers Inc.

44. Arora, S., & Doshi, P. (2020). A Survey of Inverse Reinforcement Learning: Challenges, Methods and Progress. (arXiv:1806.06877).

45. Collette, S., Pauli, W. M., Bossaerts, P., & O'Doherty, J. (2017). Neural computations underlying inverse reinforcement learning in the human brain. *eLife*, 6:e29718.

46. Argall, B. D., Chernova, S., Veloso, M., & Browning, B. (2009). A survey of robot learning from demonstration. *Robotics and Autonomous Systems, 57*(5), 469–483.

47. Sasaki, F., & Yamashina, R. (2021). Behavioral cloning from noisy demonstrations. In *International Conference on Learning Representations*.

48. Suresh, P. S. & Doshi, P. (2021). Marginal MAP Estimation for Inverse RL under Occlusion with Observer Noise.

49. Bentivegna, D. C., & Atkeson, C. G. (2003). A Framework for Learn- ing from Observation Using Primitives. In G. Goos, J. Hartmanis, J. van Leeuwen, G. A. Kaminka, P. U. Lima, & R. Rojas (Eds.), *RoboCup 2002: Robot Soccer World Cup VI* (Vol. 2752, pp. 263–270). Springer.

50. Chung, M.J.-Y., Friesen, A. L., Fox, D., Meltzoff, A. N., & Rao, R. P. N. (2015). A Bayesian developmental approach to robotic goal-based imitation learning. *PLoS ONE, 10*(11), Article e0141965.

51. Park, J.-C., Kim, D.-S., & Nagai, Y. (2018). Learning for goal directed actions using RNNPB: developmental change of "what to imitate." *IEEE Transactions on Cognitive and Developmental Systems, 10*(3), 545–556.

52. Carf'ı, A., Patten, T., Kuang, Y., Hammoud, A., Alameh, M., Maiettini, E., Weinberg, A. I., Faria, D., Mastrogiovanni, F., Aleny'a, G., Natale, L., Perdereau, V., Vincze, M., & Billard, A. (2021). Hand-Object Interaction: From Human Demonstrations to Robot Manipulation. *Frontiers in Robotics and AI*, 8:714023.

53. Vollmer, A. -L., Mu¨hlig, M., Steil, J. J., Pitsch, K., Fritsch, J., Rohlfing, K. J., & Wrede, B. (2014). Robots Show Us How to Teach Them: Feedback from Robots Shapes Tutoring Behavior during Action Learning. *PLoS ONE*, 9(3):e91349.

54. Burke, C. J., Tobler, P. N., Baddeley, M., & Schultz, W. (2010). Neural mechanisms of observational learning. *Proceedings of the National Academy of Sciences of the United States of America*, pages 14431–14436.

55. Cooper, J., Dunne, S., Furey, T., & O'Doherty, J. (2012). Human dorsal striatum encodes prediction errors during observational learning of instrumental actions. *Journal of Cognitive Neuroscience*, pages 106–118.

56. Kobza, S., & Bellebaum, C. (2015). Processing of action- but not stimulus-related prediction errors differs between active and observational feedback learning. *Neuropsychologia*, pages 75–87.

57. Redgrave, P., Rodriguez, M., Smith, Y., Rodriguez-Oroz, M., Lehericy, S., Bergman, H., Agid, Y., DeLong, M., & Obeso, J. (2010). Goal-directed and habitual control in the basal ganglia: implications for parkinson's disease. *Nature Review Neuroscience*, pages 760–772.

58. Yin, H., & Knowlton, B. (2006). The role of the basal ganglia in habit formation. *Nature Review Neuroscience, 6*, 464–476.

59. Baldassarre, G., Caligiore, D., & Mannella, F. (2013). Computational and robotic models of the hierarchical organization of behavior. chapter The Hierarchical Organization of Cortical and Basal-Ganglia Systems: A Computationally-Informed Review and Integrated Hypothesis. Springer, Berlin.

60. Dezfouli, A., & Balleine, B. (2012). Habits, action sequences and reinforcement learning. *European Journal of Neuroscience*, pages 1036–1051.

61. Dezfouli, A., & Balleine, B. W. (2013). Actions, action sequences and habits: Evidence that goal-directed and habitual action control are hierarchically organized. *Plos Computational Biology*, pages 1–14.

62. Rusu, S., & Pennartz, C. (2020). Learning, memory and consolidation mechanisms for behavioral control in hierarchically organized cortico-basal ganglia systems. *Hippocampus*, pages 73–98.

63. Yin, H. (2014). *Neurobiology of alcohol dependence, chapter Cortico-basal ganglia networks and the neural substrates of actions*. Academic Press.

64. Yin, H. (2016). *The Basal Ganglia novel perspectives on motor and cognitive functions, chapter The basal ganglia and hierarchical control in voluntary behavior*. Springer.

65. Badre, D., & Nee, D. (2018). Frontal cortex and the hierarchical control of behavior. *Trends in Cognitive Science*, pages 170–188.

66. Averbeck, B., Lehman, J., Jacobson, M., & Haber, S. (2014). Estimates of projection overlap and zones of convergence within frontal-striatal circuits. *Journal of Neuroscience*, pages 9497–9505.

67. Haber, S., Fudge, J., & McFarland, N. (2000). Striatonigrostriatal pathways in primates form an ascending spiral from the shell to the dorsolateral striatum. *Journal of Neuroscience*, pages 2369–2382.

68. Haber, S. & Calzavara, R. (2009). The cortico-basal ganglia integrative network: The role of the thalamus. *Brain Research Bulletin*, pages 69–74.

69. Merel, J., Botvinick, M., & Wayne, G. (2019). Hierarchical motor control in mammals and machines. *Nature Communications*.

70. Charpentier, C. J., Iigaya, K., & O'Doherty, J. P. (2020). A neurocomputational account of arbitration between choice imitation and goal emulation during human observational learning. *Neuron, 106*(4), 687-699.e7.

71. Baladron, J., & Hamker, F. H. (2020). Habit learning in hierarchical cortex—basal ganglia loops. *European Journal of Neuroscience, 52*(12), 4613–4638.

72. Scholl, C., Baladron, J., Vitay, J., & Hamker, F. H. (2022). Enhanced habit formation in Tourette patients explained by shortcut modulation in a hierarchical cortico-basal ganglia model. *Brain Structure and Function, 227*, 1031–1050.

73. Collins, A., & Saunders, B. (2020). Heterogeneity in striatal dopamine circuits: Form and function in dynamic reward seeking. *Journal of Neuroscience Research*, pages 1046–1069.

74. Dinstein, I., Thomas, C., Behrmann, M., & Heeger, D. J. (2008). A mirror up to nature. *Current Biology*, pages R13–R18.

75. Lim, S. B., Larssen, B. C., & Hodges, N. J. (2014). Manipulating visual–motor experience to probe for observation-induced after-effects in adaptation learning. *Experimental Brain Research*, pages 789–802.

76. Donchin, O., Rabe, K., Diedrichsen, J., Lally, N., Schoch, B., Gizewski, E. R., & Timmann, D. (2012). Cerebellar regions involved in adaptation to force field and visuomotor perturbation. *Journal of Neurophysiology, 107*, 134–147.

77. Rabe, K., Livne, O., Gizewski, E. R., Aurich, V., A. Beck, D. T., and Donchin, O. (2009). Adaptation to visuomotor rotation and force field perturbation is correlated to different brain areas in patients with cerebellar degeneration. *J Neurophysiol*, 101:1961–1971.

Using Eye Tracking to Aid the Design of Human–Machine Interfaces (HMIs) in Industrial Applications

Alexandra Kuschnereit, Alexandra Bendixen⬡, Dominic Mandl, and Wolfgang Einhäuser⬡

Abstract

In the vision of a "hybrid society," the interaction between humans and embodied digital technologies is envisaged to be as smooth and seamless as human–human interaction. In the foreseeable future, however, humans will undoubtedly continue to interact with many machines through designated human–machine interfaces (HMIs). Designing such HMIs for production settings presents a particular challenge, as users may vary in their experience and expertise. Moreover, challenging issues of a specific HMI are often difficult to verbalize and therefore hard to obtain through user report and expert interviews alone. To make interactions through HMIs smooth, it is therefore crucial to evaluate HMIs with state-of-the-art technologies that do not require explicit report. Here, we demonstrate the use of eye tracking for HMI design in an industrial production setting, where smooth human–machine interaction is particularly critical to ensure safe and efficient operation. Using a real-life example, we illustrate how eye tracking allows dissociating users' difficulties to find a particular interaction item ("search") from their challenges in

realizing that it is indeed the item to be operated ("verification"). We argue that this distinction is crucial for (re-) designing HMIs to optimize usability and that the usefulness of eye tracking extends beyond the specific context to human–machine interaction in general.

Keywords

Gaze · Eye tracking · Search · Human–machine interface (HMI) · Production · Usability · User experience (UX)

1 Introduction

Increasing digitization and the ubiquity of easy-to-operate interfaces have driven an increasing demand for intuitive human-machine interfaces (HMIs) in industrial contexts and production settings. Additional demands arise from increasing personalization (i.e., adaption to the customer) of machinery, where one-fits-all training programs become unfeasible—yet, machines need to be operated efficiently and safely without the need for regular intervention by the manufacturer. Consequently, questions of usability and user experience (UX) have become of topical interest in HMI design for production settings. The concept of usability typically follows the definition of the norm ISO-9421:11 ([1], as cited in [2]) to encompass effectiveness, efficiency and user satisfaction. Most often, usability and UX are assessed by standardized questionnaires, with the "System Usability Scale" (SUS [3]) being one of the most abundantly used in the industrial context. However, when such questionnaires are applied only after the interaction with the device, there is the risk of cognitive biases, such as applying most weight to recent events (recency bias). If they are, however, applied at multiple time points during the interaction, the questionnaires themselves may interfere with the UX and potentially also hamper effectiveness and efficiency of device use. Hence, unobtrusive, continuous measures of human-machine

1st International Conference Hybrid Societies, 2023.

A. Kuschnereit · D. Mandl
HA DE SD, HAHN Automation GmbH, Rheinböllen, Germany
e-mail: a.kuschnereit@hahnautomation.com

D. Mandl
e-mail: d.mandl@hahnautomation.com

A. Bendixen
Cognitive Systems Lab, Chemnitz University of Technology, Chemnitz, Germany
e-mail: alexandra.bendixen@physik.tu-chemnitz.de

W. Einhäuser (✉)
Physics of Cognition Group, Chemnitz University of Technology, Chemnitz, Germany
e-mail: wolfgang.einhaeuser-treyer@physik.tu-chemnitz.de

B. Meyer et al. (eds.), *Hybrid Societies*, Advances in Science, Technology & Innovation, https://doi.org/10.1007/978-3-032-03488-5_13

interaction are highly desirable. Such measures would also be beneficial when testing a select group of users, especially when the sample needs to be limited to employees for confidentiality, safety or practicality considerations. This may bring additional biases, such as a reluctance to report lacks in understanding or a generally positive bias toward the company's HMI design. While some of these biases may be mitigated by other methods [4]—which include explicit feedback given through thinking aloud [5], experience sampling [6] or interviews of experienced users as well as usability inspection [7]—non-report-based measurements may present a particularly efficient means to obtain bias-free assessments of usability and UX issues from a diverse user sample. Here, we will illustrate such a non-report approach using eye-tracking data as our key measure.

Human-machine interaction is typically characterized by sequential sampling of (visual) information from a display and acting accordingly. In the case of a touchpad display HMI, as we consider herein, this includes the direction of visual attention to areas on the display that bear task-relevant information and to buttons that need to be operated. Hence, the interaction can be viewed as a series of visual search processes. Visual search for sufficiently complex items usually invokes the sequential allocation of selective visual attention to areas of the display [8]. As the direction of gaze typically follows the attentional focus [9], eye tracking is ideally suited to characterize the interaction with an HMI. In basic research, visual search is often subdivided into three stages [10]: initiation, scanning, and verification. In the context of HMIs, this translates to the initial planning of the interaction, the actual search for the interaction item, and realizing that the found item is indeed the one that needs to be operated. Eye tracking allows separating the latter two stages from each other: actual search for the task-relevant item (corresponding to scanning) and realizing that the item presents the correct option (corresponding to verification). In this paper, we will use example cases from an actual HMI used in production settings, to illustrate why this distinction is crucial when optimizing HMI design. This will also make the case why the use of eye tracking can provide a true benefit over considering behavioral data alone.

The use of eye tracking and attention models for interface design so far has had a strong focus on web applications [11–13], in particular with respect to menu layout and placement (e.g., [14, 15]). These cases benefit from the comparably stable spatial configuration, with the user maintaining a largely fixed position relative to the screen, and from typically constant ambient light levels. In applied fields, eye tracking has widely been used in traffic scenarios, typically with a focus on the driver and their state (for reviews: [16, 17]), as well as for operating other vehicles such as aircrafts [18], ships [19], or agricultural machines [20]. Although the HMI design or even its optimization are not typically at the focus of these studies, usually the interaction with the HMI constitutes an essential part of the research protocol. Most recently, there has also been a focus on the gaze behavior of pedestrians, typically asking the question how they should interact with future autonomous cars as part of the research motivation ([21, 22] for review). This is often carried out with the vision that the HMI to be designed is the interface between car and pedestrian to replace driver-pedestrian interaction.

In production contexts, however, eye-tracking studies are scarce. One challenge might be the varying light levels incident on the eye when turning from a machine to the HMI and back, which pose a two-fold challenge to videooculographic devices: first, the sensors or camera systems themselves may be sensitive to incident light, second, light-induced changes in pupil size may dramatically affect the eye tracker's precision and accuracy [23]. Moreover, compared to the aforementioned settings of machine operation, in production contexts users are often highly mobile and switch between operating the machine directly, acquiring or delivering material or information to other users or production units, and using the HMI. Hence, eye tracking in production contexts often has restricted itself to assess specific aspects or phases of operation—for example, quantifying cognitive demands during a particular production step by measuring fixation durations [24]. Alternatively, studies have reduced the interaction to the HMI—without the actual production machine attached (e.g., [25]). Here, we use mobile eye tracking while users operate an actual physical machine, solving 20 pre-defined subtasks. Solving each task involves receiving instructions from outside the HMI, operating the HMI and—for some subtasks—interacting directly with the machine. As a consequence, users are comparably mobile, and different phases and modes of interaction with the machine can be identified. For the present paper, we used gaze data to automatically identify those periods in which the user interacted with the HMI (as compared to—for example—reading the task instructions or interacting directly with the machine). Of these interaction periods, we chose two prototypical subtasks to illustrate how mobile eye-tracking fosters the identification of specific challenges in HMI operation. This demonstrates how mobile eye tracking can be applied in production contexts to improve usability and UX.

2 Methods

A. *Setup*

The setup consisted of a sorting machine that is typically used for training purposes, the corresponding HMI, which controlled the machine and could be operated through a 9' touch screen (about 19.8 x 11.7 cm), and a stand on which the tasks were presented (Fig. 1, left). To maintain

Fig. 1 Setup and task. Left: Setup (outside view), the panel with the task, the HMI and the sorting machine are visible from left to right; middle: user turning the task instruction to the next task; right: user solving the task by the operating the HMI

constant ambient light levels throughout, all internal and external windows to the room were blinded. The experimenter was located behind the user and monitored the function of the equipment. Gaze direction was recorded throughout the experiment at 100 Hz using a mobile eye-tracking device (Tobii Pro Glasses 2; Tobii SE, Danderyd, Sweden). The eye tracker was calibrated using the manufacturer's procedure, and calibration was validated by asking the user to fixate the edges of the HMI display before and after the main experiment. The position of the HMI display in the user's field of view was determined from a head-centered video recorded at 25 Hz with the scene camera integrated in the eye-tracking glasses.

B. *Procedure*

Users were asked to complete a total of 20 tasks using the HMI. Each task was presented on a laminated ISO A4 paper to the left of the HMI. On a go signal by the experimenter, the user turned the page of the instruction stack (Fig. 1, middle), read the instruction of the subsequent task, and then proceeded to solve the task using the HMI (Fig. 1, right). If the task required a response by the user (e.g., "What is the name of the item currently produced?"), this response was given verbally. Upon completion of one task, the experimenter instructed the user to continue with the next task. Errors were corrected by the experimenter only if they could influence the behavior of the HMI in a later task. If the users indicated that they were unable to complete a task, they were helped by the experimenter. Only tasks completed correctly without help are considered as solved correctly. The users had no time limit for completion and could re-read the instruction of the current task when needed.

The tasks included simple, single-step operations like setting the language of the display, tasks that required information to be gathered from the current display, tasks that required a specific action, tasks that required a sequence of actions to get the information needed, as well as tasks that required the direct interaction with the machine itself based on displayed information. The tasks were arranged in a fixed sequence. The starting configuration of a task depended on the

previous actions, but it was ensured (if needed through experimenter intervention) that these were identical in all users (except the display of date and time, which was current). After task completion users filled in two questionnaires to assess their affinity for technology and to have them rate the HMI. As these data were primarily collected for comparison to future versions of the HMI, they are not considered further in the present manuscript.

All procedures were approved by the local IRB (Ethikkommission TU Chemnitz; case no.: 101518697).

C. *Participants*

Thirty-one volunteers completed the experiment. Four additional volunteers were excluded prior to data collection, as they failed to meet one of the inclusion criteria (normal vision or corrected-to-normal vision with a correction between −3dpt and +5dpt, which was achieved by adding lenses to the eye-tracking device, as well as normal color vision) or if the eye-tracking device could not achieve a stable calibration. One of the 31 participants had to be excluded after data collection but prior to any data analysis, as it became evident only after completion that one of the inclusion criteria had not been met. Accordingly, 30 users were included in the analysis. All users were male employees or trainees of the company (HAHN Automation, Rheinböllen, Germany) and spanned the age groups between 16 and 55 years about uniformly (16–25 years: 6, 26–35 years: 10, 36–45 years: 10, 46–55 years: 4). Prior to data collection, participants were assigned to one of three groups based on their expertise:

- "Novices": These users (N = 9) had no prior experience with the specific HMI, with the HMI layout nor with the sorting machine to be programmed. However, they did have engineering knowledge and training, albeit not with the type of machine used. Members of this group can be considered representative of a typical customer eventually working with the device.
- "Intermediate": These users (N = 10) have no experience with the specific HMI or machine, but have experience with similar machines, such that they can be expected to

Fig. 2 Example frame with automatic screen detection. **A** The video frame with the detection result overlaid (cyan: detected screen, magenta: encompassing rectangle); the brightness of the frame has been adjusted for this figure to brighten the area outside the screen; for realistic brightness settings see right panel of Fig. 1; **B** thresholded frame; **C** morphological closing on panel B; **D** convex hull of C (note the slight difference to panel C in the bottom left of the detected area)

have a higher familiarity with the basic terminology than the users in the novices group.

- "Experts": These users (N = 11) are familiar with the type of machine and the specific HMI as they have prior experience using it, but were not involved in designing or programming the HMI.

D. *Analysis*

All data analyses were performed in MATLAB (Mathworks, Natick, MA, USA) version 9.6.0 (R2019a) or higher, including the Image Processing Toolbox (version 10.4).

(1) *Behavioral Data*

For each subtask, the "interaction time" was defined as the time from the moment the HMI display came into the scene camera's field of view to the end of the interaction (i.e., until providing the response, until conducting the required action or until abandoning the subtask). The time between turning the page and looking to the display (i.e., the first reading of the instruction) was considered preparation time. The "total interaction time" for the experiment was defined as the time from the start of the interaction time of the first subtask to the end of the last subtask.

(2) *Detection of the HMI Display*

To reference the eye-tracking data, which were recorded relative to the head-centered video, to world coordinates, we automatically determined if the HMI screen ("display") was present in the video and if so its position in the video frame. To this end, we make use of the fact that the screen typically was the brightest object in the scene, if it was visible (Fig. 2A). Hence, we first thresholded each frame to retain the 10% brightest pixels (Fig. 2B). On the resulting binary image we determined the morphological closing using a disc of 10 pixels radius as structural element. In the resulting image (Fig. 2C), we determined the convex hull of the largest connected area (Fig. 2D). The smallest quadrilateral enclosing this hull was considered the screen (Fig. 2A, cyan). As the

screen was rarely rotated more than a few degrees relative to the scene camera, for simplicity, we used the smallest encompassing rectangle (Fig. 2A, magenta) of the hull as the screen for all analyses reported here. Since this procedure was applied to all frames, including those in which the screen was not visible, in a second step, we used the following heuristics to determine whether the detected object was indeed the screen: the area of the encompassing rectangle (Fig. 2A, magenta) did not differ more than 20% from the area of the quadrilateral (Fig. 2A, cyan), the aspect ratio of the rectangle was between 0.53 and 0.63, and the mean pixel value of the area exceeded the surround by at least 120%.

The interaction onset of a subtask was determined as the first time the screen came into view and remained in view for at least 25 subsequent frames, after the previous subtask had been completed and the instruction for the current subtask had been viewed. The interaction onsets were verified manually; rare cases of incorrect detections (e.g., because the instruction sheet was mistaken for the screen by the detection algorithm) were corrected.

(3) *Eye-Tracking Data*

For eye-tracking analysis we first determined the area of the display that was relevant for the current subtask, i.e., the button to be pressed or the area where the task-relevant information could be retrieved. We then measured for each frame of the scene-camera video, whether at least one eye-tracking sample recorded during this frame fell within this "relevant display area". In this case, we considered the relevant display area as fixated during this frame.[1] Using these data, we split the interaction time for each subtask into two periods: "scanning time" and "verification time". We defined the "scanning time" as the time from the start of the interaction to the first frame in which the relevant display area was fixated. We defined the "verification time" as the time from the end of scanning time to the end of the interaction time. Together, scanning time and verification time sum to the interaction

---

[1] If no eye-data sample is available or the HMI screen is not detected for a particular frame, the display is considered not fixated for this frame. This may lead to a slight underestimation of actual fixations on the display.

time. The "scanning percentage" was defined as scanning time divided by interaction time (x 100%).

To illustrate the spatial distribution of gaze direction relative to the HMI display, we present gaze maps in addition. These were computed by splitting the rectangle encompassing the detected HMI screen (Fig. 2, magenta) into 58 lines and 100 columns, and assigning each valid eye-tracking sample to the closest of these bins.[2] The resulting 2D histogram was then smoothed by a Gaussian kernel (SD: 5 pixel) to visualize the fixation density relative to the HMI display. When comparing gaze maps to the quantitative analysis above, it is important to note that smoothing the maps for visualization can make it appear as if the relevant display area is fixated, even if the gaze sample is actually on a neighboring location.

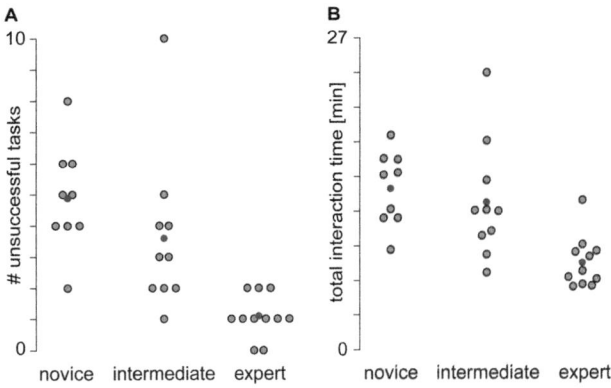

Fig. 3 Overall performance. **A** Number of errors; that is, the number of tasks that were not solved or not solved correctly. **B** Total interaction time. Each gray point denotes an individual, red point denotes group mean

(4) *Statistical Analysis*

In all analyses, we compared the interaction times between three groups using a 1-factor Analysis of variance (ANOVA) with group as the only factor. We expected that the number of correctly solved tasks increases and the total interaction time decreases with expertise level. If we found a main effect of the factor group, we analyzed by follow-up independent-sample t-tests between which pairs of groups there were differences. For the analysis of gaze data, we mostly focused on extreme individual cases (see below) and did therefore not assess group differences.

3 Results

A. *Overall Performance*

First, we determined the overall number of errors, that is, the number of subtasks (out of 20) not solved correctly (Fig. 3A). We found that novices made 4.9 (SD: 1.7) errors on average, users with intermediate expertise 3.6 (SD: 2.5) errors, and experts 1.1 (SD: 0.70) errors. There was a main effect of group on the number of errors ($F_{(2,27)} = 11.9, p < .001$) with significant pairwise difference between novices and experts ($t(18) = 6.80, p < .001$) as well as between experts and intermediate-experience users ($t(19) = 3.15, p = .005$), but not between novices and intermediate-experience users ($t(17) = 1.28, p = .217$). Similarly, the average total interaction time (Fig. 3B) was largest in novices (mean: 837s; SD: 192s), smallest in experts (451s; 134s) and in-between for the intermediate expertise group (766s; 304s). Again, there was a main effect of group ($F_{(2,27)} = 9.01, p = .001$), with significant pair-

wise differences between any pair of groups (all $t > 7.48$, all $p < .001$). This shows that the overall performance is consistent with the group assignment. It should also be noted that in the intermediate group there is substantial variability both in terms of errors and interaction time, presumably reflecting the higher heterogeneity in this group compared to the other groups.

B. *Example Case 1: Difficulty Finding Required Information*

In this example subtask, users were asked to read the cycle time from the HMI. This information is visible in the lower middle of the display (red box in Fig. 4, hereafter "relevant area") without need for pressing any button. All users solved this task eventually and we found no group differences for the interaction time in this task ($F_{(2,27)} = 0.64, p = .534$). However, there was a striking variability in interaction time within the groups, in particular for the intermediate and expert group (Fig. 5).

Analyzing the gaze data, we found that out of the 30 users, 23 (6/9 novices, 9/10 intermediate, 8/11 experts) looked at the area where the information was displayed. Despite the variability in overall interaction time, most of these users responded quickly once they had looked at the relevant display area. The average time to respond after first looking at the relevant area (i.e., the verification time) for these 23 users was 7.3s. Although there was considerable variability (standard deviation (SD) of verification times: 16.9s), which was mainly caused by extreme cases (range: 0.72s–73.9s), more than half of the users had verification times below 2s (median: 1.9s). This implies that once users had found the information, they could use it quickly. This is also evident in the scanning percentage: the relative time spent searching the information makes up nearly half the interaction time (mean: 43.2%, SD: 24.5%), albeit with a huge inter-individual variability (range: 2.1%—97.6%).

---

[2] While the aspect ratio of the screen is close to 16:9 (1.78), the fact that it appears slightly rotated in the video makes 100:58 (1.72) a more typical ratio observed for the rectangle. The error of using the rectangle rather than the actual screen outline is negligible given the kernel size used to smooth the gaze map.

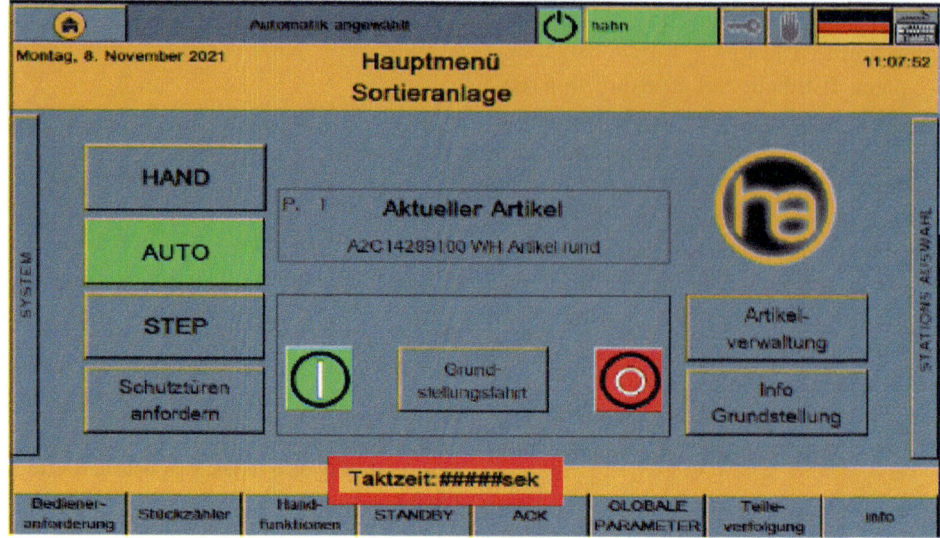

Fig. 4 Example 1—Display. The cycle time is given in the yellow strip close to the menu items (indicated by the red box, which is not present in the actual display)

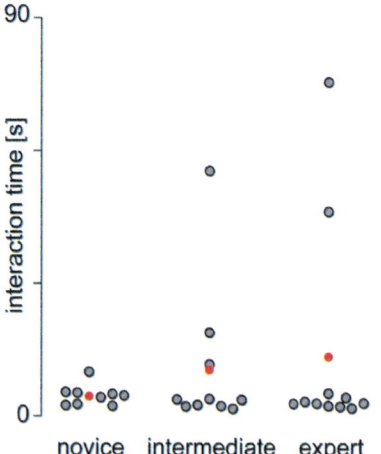

Fig. 5 Example 1—Interaction time. Notation as in Fig. 3

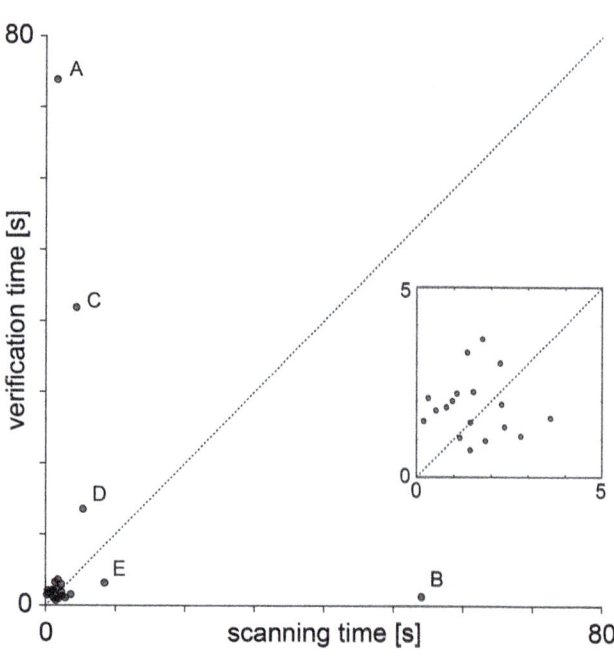

Fig. 6 Example 1—Scanning vs. verification. Users who took longer than 10s to complete the task are labeled A–E in descending order of their interaction time. Inset shows zoomed-in plot from 0 to 5s in each dimension

We pay particular attention to the five users who needed more than 10s to complete the task, as these "slow"[3] users are most relevant to identify potential HMI design issues for the specific item needed for this subtask. When directly comparing scanning time and verification time (Fig. 6), it is evident that for two of the slow users (marked "B" and "E" in Fig. 6) the bulk of interaction time went into scanning. This is, they had to look for the information for a substantial time (scanning time: 54.1s and 8.5s), but then responded quickly (verification time: 1.3s and 3.2s). Conversely, two other slow users ("A" and "C") looked at the relevant display area early (scanning time: 1.6 and 4.4s) and then took exces-

sively long to respond (verification time: 73.9s and 41.8s). The fifth slow user ("D") fell in-between with a scanning percentage of 28.4% (scanning time: 5.4s, verification time: 13.5s), though still closer to "A" and "C" than to "B" and "E."

To illustrate this point further, we consider the gaze maps of the slow users (Fig. 7). We observe that the user with the longest interaction time (A) spent substantial time searching the menu buttons at the bottom, presumably searching for an alternative screen to switch to rather than retrieving the

---

[3] Note that we use the term "slow user" for brevity of presentation for the users who took longer than 10 s in this particular subtask; this is not meant to imply any statement about individual user characteristics.

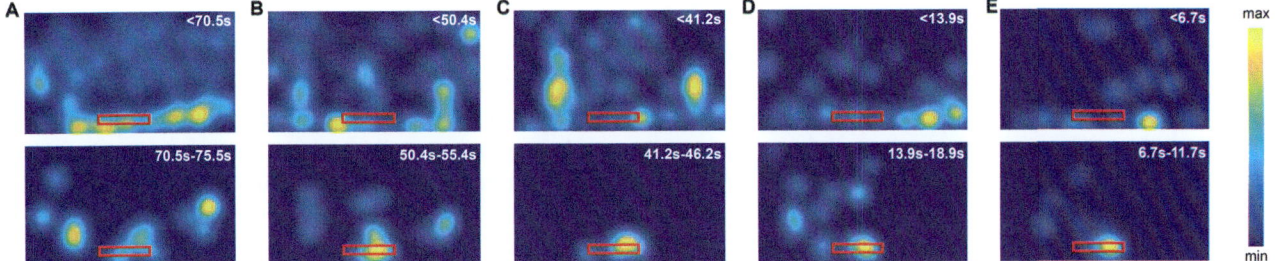

Fig. 7 Example 1—Gaze maps. Gaze maps of the five users who took longer than 10s to complete the task sorted by interaction time; top: data up to five seconds before the response, bottom: last five seconds before the response. Red box indicates approximate location of cycle time information. Panels "A"– "E" match the user identifiers of Fig. 6 Maps are scaled individually according to colorbar on the right

information from the current screen (despite looking at the information early). The other user with excessive verification time (C) in his initial glance at the relevant area looked to its right, presumably not capturing the word ("Taktzeit"—cycle time) on the left but only the time in seconds as such, and then continued search the larger menu items. Hence, even for the two users with the longest verification times, it is likely that there was a substantial search going on after they had looked at the relevant information early on without perceiving it.

Taken together, these data suggest that most of the users who took longest for task completion were not primarily hampered by misunderstanding the task or identifying the information once they had found the relevant display area. Rather, they needed to search for the relevant area, presumably because it was neither conspicuously placed on the display nor made stick out (salient) by visual means.

Interestingly, seven users (3 novices, 1 intermediate, 3 experts) responded correctly without looking at the relevant display area. While in some cases there might be technical issues preventing a correct detection of the screen in the very moment they looked at the relevant area, or the rele-

vant area was occasionally covered by the user's hand with them looking slightly above, for some of these cases, it is likely that they retrieved the information from memory. This shows another advantage of using eye tracking for assessing real-world HMIs: unlike in constrained experimental settings, in reality there are often multiple ways by which a user can obtain a particular piece of information. Behavioral data alone would not be able to distinguish those ways from those intended by the experimenter, and could thus lead to erroneous interpretations of HMI usability.

C. *Example Case 2: Failure to Recognize the Correct Button Due to Unclear Labeling*

The first example presented a case where challenges in HMI usage arose mostly from users' difficulties to find the relevant information, a challenge that could be met by making relevant information visually more conspicuous (salient) or placing it in a different location. As second example, we consider a subtask in which users had to acknowledge an (experimentally induced) equipment malfunction at the HMI.

Fig. 8 Example 2—Display. The "ACK" button is the fifth out of the 8 buttons at the bottom of the display (indicated by the red box, which is not present in the actual display). The error message is shown on the top (red box)

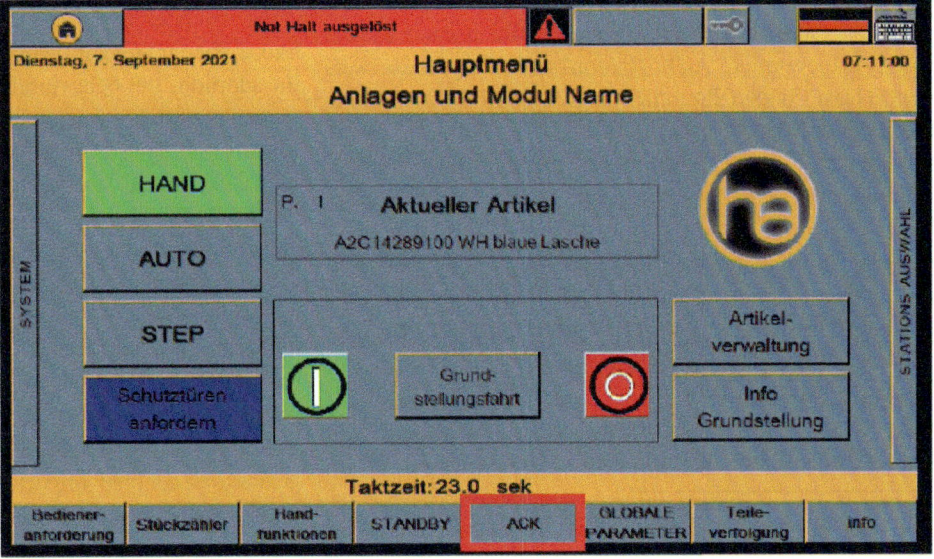

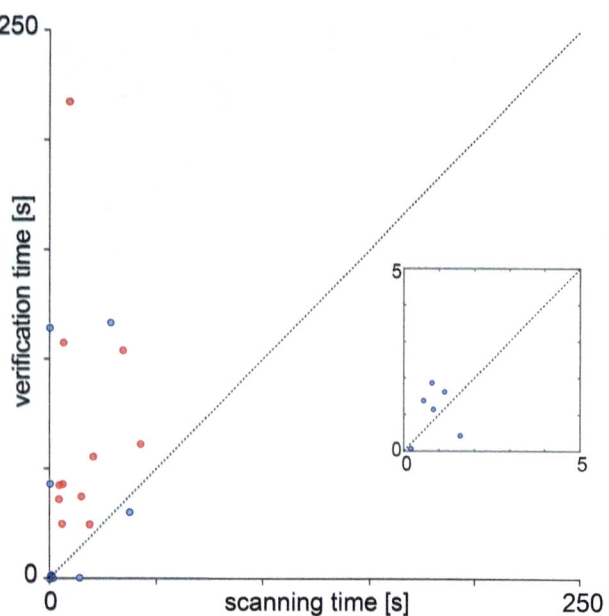

Fig. 9 Example 2—Scanning versus verification. Red: users who did not manage to complete the task; blue: users who successfully completed the task and looked at the button. Inset shows zoomed-in plot from 0 to 5s in each dimension

The corresponding button was labeled "ACK" (for "acknowledge", Fig. 8), although the language of the HMI was set to German and the instruction was given in German ("Quittiere die Störung" [=Acknowledge the malfunction]). Only two out of nine (22.2%) novices successfully solved the task and six out of ten (60.0%) in the intermediate group, while all eleven experts solved the task successfully. Moreover, the experts all solved the task quickly (range: 0.28s–14.5s, mean: 2.6s, sd: 4.1s), while there was substantial variability in the other two groups (novices: 5.5s and 67.9s; intermediate: range: 0.36s–145.4s, mean: 51.2s, sd: 64.0s).

To uncover the potential reasons for failure in task completion, we consider the eye-tracking data. All eleven users who failed to complete the task ("unsuccessful users") looked directly at the correct ("ACK") button—on average these users spent at least[4] 0.98s (SD: 1.58s; range: 0.04s to 5.52s) looking directly at it. In contrast, of the 19 users who correctly solved the task ("successful users"), there were eight (novices: 1/2; intermediate: 2/6; experts: 5/11) who did not look directly at the button.

We again separated the data into scanning time and verification time (Fig. 9). The unsuccessful users had an average scanning time of only 15.1s (SD: 12.8s, median: 9.1s, range: 4.3s–42.4s), compared to their average verification time of 68.9s (SD: 56.9s, median: 43.3s, range: 25.0s–217.7s). That is, nearly all found the correct button reasonably quickly, but then failed to realize that it is the one to be pressed. More-

over, all unsuccessful users spent more time on verification than on searching (mean scanning percentage: 20.2%, SD: 13.3%, median: 18.6%, range: 4.0%–42.5%). Inspecting the gaze maps of the unsuccessful users (Fig. 10) shows some variation: some users did engage in a search for alternative buttons, but included the "ACK" button in it (e.g., Fig. 10H, J) without ever pressing it; some apparently tried to retrieve information from the error message in the upper left of the display, spending most of their interaction time looking there (e.g., Fig. 10C, G); and some spent considerable time looking at a particular display area unrelated to the task (e.g., Fig. 10D, K). That is, only few unsuccessful users engaged in a typical search pattern, and those who did, dwelled considerable time on the "ACK" button. In sum, unsuccessful completion of this task is not a consequence of unsuccessful search.

For the successful users, the relation of scanning to verification time (Fig. 9, blue data points) is more diverse, with some spending most of their interaction time searching, but some showing extreme verification times up to nearly two minutes (upper left in Fig. 9)—well in and above the range of unsuccessful users.

Taking the data of successful and unsuccessful users together shows clearly that finding the correct information is not the major challenge in this task. Instead, users fail (or need an excessively long time) to complete the task in many cases because they do not recognize that the button they are looking at, is indeed the correct button to press.

4   Discussion

In the present study, we used an example from industrial practice to illustrate the benefits of using eye-tracking in HMI design: We first used behavioral data (interaction time and whether a response was correct) to verify that on a group-level performance across all tasks was related to expertise. This validates that the tasks chosen are representative of those encountered in everyday interactions with the HMI. In addition, the behavioral data allows us to identify individual users who were challenged by a specific subtask, irrespective of their group. We then focus eye-tracking analysis on the gaze data of those users to identify the features of the HMI that may have hampered their performance. In doing so, we make an important distinction between the (i) time needed to find the task-relevant information (scanning time) and (ii) the time to realize that this indeed is the relevant information (verification time). This distinction is crucial for redesigning the HMI appropriately, as in the former case, the location of the information needs to be made more conspicuous, in the latter, the information itself needs to be adapted. Importantly, the two factors cannot easily be distinguished based on behavioral data alone. This illus-

---

[4] This is a lower bound, as there can be periods where gaze or screen are undetected.

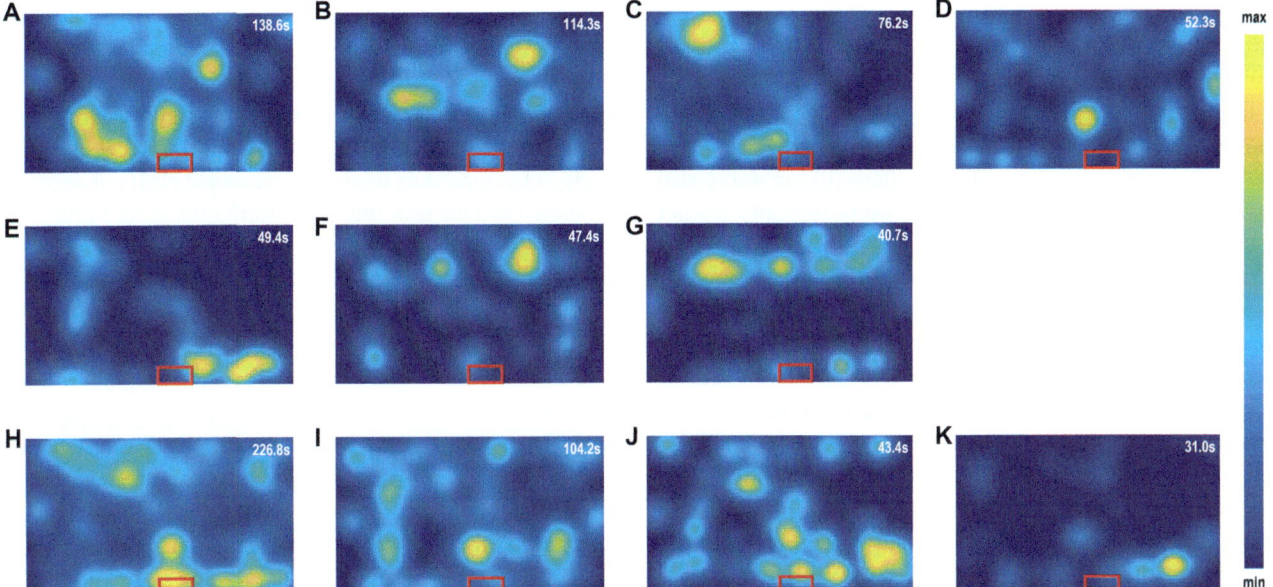

Fig. 10 Example 2—Gaze maps. Gaze maps of gaze positions relative to the screen of the eleven users who did not manage to complete the task. Maps are scaled individually according to colorbar on the right; novices in panels **A–G**, intermediates in panels **H–K**, sorted within each group in descending interaction time, which is given in the upper right. Red box indicates approximate button position

trates how eye tracking leads to additional benefits in HMI design.

The distinction between scanning and verification is particularly important, as in the HMI literature, the accessibility of information to the user is most often related primarily to its physical salience, i.e., how much the information-bearing item differs from its surrounding (see [26] for a review on the general concept of visual salience). This is most clearly stated for signals to be treated with high priority (i.e., warnings and alerts), which are supposed to be "as salient … as possible" [27], larger, higher in contrast, and different in color [28]. This is put to the extreme in the "dullscreen" principle, where items within normal operation range are presented in grayscale and color is exclusively reserved for alerts ([29], but see [30] for a recent re-assessment). The lack of salience contributes to scanning time, and our example 1 illustrates a case where the need to search extensively may arise from the relevant location lacking salience (black text on a yellow background, while isolated colorful items on neutral background are also present, Fig. 4). However, focusing HMI design on salience and choosing a conspicuous location alone falls short: salience does not affect the verification time, which—as illustrated in particular in our example case 2—can be a main contributor to delays or even failures in executing a task: If users do not realize that "ACK" is the correct button, because the information is unclear and misleading, they will be reluctant to press it—no matter how salient it is. Notably, it is not evident that querying users would reveal a distinction between scanning and verification, as statements along the line "I could not find the correct button" are well conceiv-

able, but equally ambiguous as behavioral data. More structured interviews might capture specific examples, but still will miss verification failures not considered a priori. In contrast, by tracking the user's gaze, the inability to find the information can easily be distinguished from the failure to realize its function.

Using eye tracking in HMI design complements the more common approach to assess the usability and UX by questionnaires. Questionnaires provide a standardized tool to assess the overall impression of the user and are therefore particularly useful to assess overall user satisfaction and UX, in particular in early stages of HMI (re-)design. However, the identification of specific challenges or weaknesses of an HMI poses a challenge for questionnaire-based methods: if questionnaires are applied only rarely, e.g., once after the test, they can provide only an overall impression. In turn, if they are applied frequently, they interfere with the UX and possibly the usability, i.e., with their objects of measurement. Eye tracking, in contrast, allows a continuous, unobtrusive assessment of the user's interaction with the HMI. Eye tracking is therefore particularly well-suited to identify specific issues interfering with usability and to optimize detailed aspects in later stages of HMI design.

As compared to questionnaires, eye-tracking circumvents the general difficulty for users to remember and verbalize challenges and solution strategies they experienced during particular subtasks or with particular display items. In addition, it overcomes the issue that users, often recruited from within the company producing the HMI, may be biased toward positive statements and thereby (inadvertently) downplay issues they may have encountered. In this respect, using gaze data comes with the additional advantage that it can be fully anonymized prior to analysis, such that concerns about using the data to assess individual performance, which again may bias sample selection and data, are alleviated.

Unlike purely screen-based interactions (like in web-based applications), in industrial contexts users typically do not spend all their productive time interacting with the HMI. Rather they interact with other production sites, provide or retrieve instructions or material, operate or monitor the machine directly, etc. in addition to using the HMI. When assessing HMIs in a realistic setting, it is therefore crucial to distinguish times used to operate the HMI from other phases to avoid wrong conclusions about the HMI interactions. Indeed, different groups (experts vs. novices) may differ in the time interacting with the HMI, even if the overall task duration is the same [25]. Here, the use of the mobile eye-tracking device also allows this distinction between receiving instructions, operating the HMI (the phase the present study focuses on), and operating the machine directly without the HMI.

We use eye tracking with a comparably small sample size per group. Rather than focusing on averages and group differences, we advocate putting emphasis on extreme cases across the whole sample (i.e., irrespective of expertise group): identify outliers based on behavioral data, and then use the gaze data to identify *potential* causes of challenges or advantages these users experience when solving the particular task. It could be argued that larger samples are desirable, but this is often not feasible in industrial contexts: there often is only a comparably small number of users who can potentially be included in a study, and those may have highly variable expertise, which again may be different from the target group (e.g., prospective customers). Resorting to expert interviews (querying a very small, select group in detail about the HMI), as is often done in industrial contexts, also suffers from this issue and is also prone to some of the aforementioned biases. Hence, we consider eye tracking an ideal means to arrive at meaningful hypotheses about issues and challenges imposed by an HMI with a comparably small and variable sample of users. Nonetheless, it is evident that if these data are used to redesign an HMI, it is highly advisable to directly test the new version against the old version ("A-B testing") to check whether the issues are solved and no unforeseen new issues arise. For this phase of design validation, eye tracking again can be a useful tool to uncover effects not evident in behavior alone and to achieve meaningful insight without large effort or excessive samples sizes (cf. [25]).

When interpreting eye-tracking data, it is important to keep in mind that looking at (or near) an item cannot always be equated with actually perceiving it. This phenomenon of "looking without perceiving" ("inattentional blindness" [31]), and similar failures of visual awareness are well-known and widely researched in applied fields, such as driving [32, 33],

aviation [34] or medical diagnostics [35]. In our present study we also see the reverse effect: some participants who do not need to look at some piece of information to retrieve it, presumably because they have memorized it earlier and now simply recollect it. First, this shows that when using purely behavioral data to assess real-world HMIs, it needs to be ensured that the task-relevant information is not retrievable elsewhere. Second, if users retrieve information from memory because it is not sufficiently conspicuous, this also causes the risk that changes in the information go unnoticed as the item retrieved from memory dominates.

In the present setting, we use eye tracking for offline analysis to improve HMI design. Especially in conjunction with virtual or augmented-reality settings, it might be of interest to have a continuous closed-loop monitoring of gaze direction to provide automated assistance whenever the user engages in an unusually long gaze interaction with the display. Similarly, HMIs themselves might partially be controlled by gaze to free the user's hands or to avoid display contact, for example in medical settings, hazardous environments or cleanrooms. The development of affordable, yet sufficiently accurate eyetracking devices, as pushed forward in the context of assistive and rehabilitation technology (e.g., [36, 37]), renders a wide-spread use of such technology conceivable in the near future.

Most contemporary commercial eye-tracking devices use the position of the pupil as at least one part of the information to determine gaze direction. Such systems have the additional benefit that pupil size is measured "for free". Although varying light levels are an issue, pupil size can in principle be used to assess a variety of cognitive functions ([38] for a review), including cognitive load and effort ([39] for a review), which in turn may allow inferences about the HMI design [40]. Moreover, pupillometry can be combined with eye-tracking data to improve closed-loop eye-tracking-based interfaces [41]. Besides eye tracking and pupillometry, electroencephalography [42] and electrocardiography (ECG) or heart rate measurements [43], have frequently been suggested to complement questionnaire data for a continuous assessment of the user's state. Eye tracking goes beyond assessing the user's state, since it provides spatially and temporally specific information about the user's interaction with the HMI. Moreover, eye tracking does not require averaging across multiple events or prolonged interaction epochs, but allows analyzing brief single episodes of interaction. In addition, eye tracking can readily be combined with these techniques and thereby be complemented by other physiological measures. Finally, eye tracking has become easily accessible and applicable, with affordable mobile solutions being available and wide-range applications in production context at least conceivable. While we deliberately chose a production setting as a real-world testbed for our demonstration, many of these considerations will generalize to other domains in which

HMIs are used in dynamic, real-world environments—that is, far beyond the design of web interfaces, where eye tracking is already widely used [11–15]. The particular distinction between scanning and verification is likely of relevance in all situations in which a multitude of actions can be selected from and the available options are neither self-evident nor overtrained; that is, in all cases where users with limited experience operate complex HMIs. We therefore recommend a three-step process for gaze-based evaluation of HMIs: (i) Validate that tasks are sufficiently complex and representative of everyday interactions with the HMI, for example, by verifying a group-level dependence of behavioral performance on expertise, (ii) identify those combinations of items and individuals that show extreme performance values and thus suggest particular challenges for HMI use, (iii) for these combinations, identify whether the challenge primarily arises from scanning or from verification costs.

Although our example cases clearly demonstrate how eye tracking can be a useful tool for assessing HMIs in production and manufacturing, several caveats apply. Quickly changing light levels can interfere with the measurement precision and accuracy, by affecting the cameras directly or by affecting pupil size [23]. Especially if high precision is needed to distinguish nearby interaction targets, this needs to be incorporated to avoid misinterpretation of the data. On a more general level, the apparent ease of use of eye-tracking devices provokes incorrect application and overconfidence in the data (see [44] for a recent survey), for example, by measuring data outside the calibrated region, ignoring physical shifts of the device over time causing potentially substantial errors (cf. [45]) or interpreting interpolated missing data as prolonged fixations. The latter becomes particularly relevant, when excessive fixation durations (or dwell times) are used as diagnostic information for users experiencing difficulties, as has recently been proposed for a production setting, thermal spraying [24]. Moreover, when tracking in real-world settings, it becomes a challenge to retrieve the gaze target in the real world from the head-centered coordinate system in which eye data are measured, without the need for substantial manual annotation. In the present study, this was the rationale to restrict analysis to the interaction phase (where the screen is readily detectable from the scene camera), and to defer the analysis of preparation phases and the interaction with the machine as such.

A comparably novel concept in production, which may alleviate several of the aforementioned challenges for real-world eye tracking, is the creation of digital twins in virtual reality (VR). This approach allows testing machines or certain aspects of them without having to build or reconfigure the actual device [46]. This aligns well with the use of gaze tracking, as several commercial virtual reality systems now provide eye-tracking functionality. In sufficiently detailed VR settings and with an appropriate control strategy [47], gaze behavior in VR closely mimics that of the real world [48]. When using head-mounted displays, this comes with the additional advantage that the position of all elements in the virtual setting are precisely known, such that gaze can readily be referenced to the actual objects of interest. Moreover, light levels can be controlled such that the aforementioned confounding factors are limited. While it remains to be shown (similar to [48]) that gaze data obtained in VR have high validity for real-world HMI operation, it is likely that combining the digital twin concept in VR with gaze tracking will further accelerate rapid prototyping and re-designs of HMIs in production in the near future.

Acknowledgements The work was funded in part by the Deutsche Forschungsgemeinschaft (DFG, German Research Foundation) project ID 416228727—SFB 1410, projects A04 and C01.

References

1. ISO/TC 159/SC 4 Ergonomics of Human-System Interaction (Technical Committee), "Ergonomic requirements for office work with visual display terminals (VDTs) – Part 11: Guidance on usability," International Organization for Standardization, (1998). https://www.iso.org/standard/63500.html
2. Sauro, J., & Lewis, J. R. (2016) Quantifying the user experience: Practical statistics for user research. *Morgan Kaufmann.*
3. Brooke, J. (1996) SUS—A quick and dirty usability scale. In Usability Evaluation in Industry (1st ed.), CRC Press. https://doi.org/10.1201/9781498710411.
4. Nielsen, J. (1993). *Usability Engineering.* Academic Press.
5. van Someren, M., Barnard, Y. F., & Sandberg, J. (1994). *The think aloud method: A practical approach to modelling cognitive processes.* Academic Press.
6. Larson, R., & Csikszentmihalyi, M. (2014) The experience sampling method. *In Flow and the Foundations of Positive Psychology.* Springer, Dordrecht. https://doi.org/10.1007/978-94-017-9088-8_2.
7. Nielsen, J. (1994) Usability inspection methods. *In Conference Companion on Human Factors In Computing Systems,* pp. 413–414.
8. Treisman, A. M., & Gelade, G. (1980). A feature-integration theory of attention. *Cognitive Psychology, 12*(1), 97–136. https://doi.org/10.1016/0010-0285(80)90005-5
9. Deubel, H., & Schneider, W. X. (1996). Saccade target selection and object recognition: Evidence for a common attentional mechanism. *Vision Research, 36*(12), 1827–1837. https://doi.org/10.1016/0042-6989(95)00294-4
10. Malcolm, G. L., & Henderson, J. M. (2009) The effects of target template specificity on visual search in real-world scenes: Evidence from eye movements. *J. Vis., 9*(11), article 8. https://doi.org/10.1167/9.11.8.
11. Ehmke, C., & Wilson, S. (2007) Identifying web usability problems from eyetracking data. *British HCI conference 2007, University of Lancaster,* UK. https://openaccess.city.ac.uk/id/eprint/3917/1/Ehmke-final.pdf.
12. Kaspar, K., Ollermann, F., & Hamborg, K. C. (2011) Time-dependent changes in viewing behavior on similarly structured web pages. *J. Eye Mov. Res., 4*(2), article 4. https://doi.org/10.16910/jemr.4.2.4.

13. Thoma, V., & Dodd, J. (2019) Web usability and eyetracking. *In Eye Movement Research. Studies in Neuroscience, Psychology and Behavioral Economics*, pp. 883–927, Springer, Cham. https://doi.org/10.1007/978-3-030-20085-5_21.

14. McCarthy, J. D., Sasse, M. A., Riegelsberger, J. (2004) Could i have the menu please? an eye tracking study of design conventions. In O'Neill, E., Palanque, P., & Johnson, P. (eds.) *People and Computers XVII — Designing for Society*, pp. 401–414, Springer, London. https://doi.org/10.1007/978-1-4471-3754-2_25.

15. Leuthold, S., Schmutz, P., Bargas-Avila, J. A., Tuch, A. N., & Opwis, K. (2011). Vertical versus dynamic menus on the world wide web: Eye tracking study measuring the influence of menu design and task complexity on user performance and subjective preference. *Computers in Human Behavior, 27*(1), 459–472. https://doi.org/10.1016/j.chb.2010.09.009

16. Grüner, M., Ansorge, U. (2017) Mobile eye tracking during real-world night driving: A selective review of findings and recommendations for future research. *J. Eye Mov. Res., 10*(2), article 1. https://doi.org/10.16910/jemr.10.2.1.

17. Kapitaniak, B., Walczak, M., Kosobudzki, M., Jóźwiak, Z., Bortkiewicz, A. (2015) Application of eye-tracking in drivers testing: A review of research. *Int. J. Occup. Med. Environ. Health, 28*(6), pp. 941–954. https://doi.org/10.13075/ijomeh.1896.00317.

18. Ziv, G. (2016). Gaze behavior and visual attention: A review of eye tracking studies in aviation. *The International Journal of Aviation Psychology, 26*(3–4), 75–104. https://doi.org/10.1080/10508414.2017.1313096

19. Hareide, O. S., & Ostnes, R. (2017). Maritime usability study by analysing eye tracking data. *Journal of Navigation, 70*(5), 927–943. https://doi.org/10.1017/S0373463317000182

20. Prati, E., Grandi, F., & Peruzzini, M. (2021) Usability testing on tractor's HMI: a study protocol. *In International Conference on Human-Computer Interaction*, pp. 294–311, Springer, Cham. https://doi.org/10.1007/978-3-030-78092-0_19.

21. Lehet, D., Novotný, J. (2022) Assessing the feasibility of using eye-tracking technology for assessment of external HMI. *In 2022 Smart City Symposium Prague (SCSP)*, pp. 1–6, IEEE. https://doi.org/10.1109/SCSP54748.2022.9792547.

22. Lévêque, L., Ranchet, M., Deniel, J., Bornard, J. C., & Bellet, T. (2020). Where do pedestrians look when crossing? A state of the art of the eye-tracking studies. *IEEE Access, 8*, 164833–164843. https://doi.org/10.1109/ACCESS.2020.3021208

23. Drewes, J., Masson, G. S., & Montagnini, A. (2012) Shifts in reported gaze position due to changes in pupil size: Ground truth and compensation. *In ACM Symposium on Eye Tracking Research and Applications*, pp. 209–212. https://doi.org/10.1145/2168556.2168596.

24. Bocklisch, F., Paczkowski, G., Zimmermann, S., & Lampke, T. (2022). Integrating human cognition in cyber-physical systems: A multidimensional fuzzy pattern model with application to thermal spraying. *J. Manufact. Sys., 63*, 162–176. https://doi.org/10.1016/j.jmsy.2022.03.005

25. Walper, D., Kassau, J., Methfessel, P., Pronold, T., & Einhäuser, W. (2022) Optimizing user interfaces in food production: gaze tracking is more sensitive for AB-testing than behavioral data alone. *In ACM Symposium on Eye Tracking Research and Applications*, pp. 1–4. https://doi.org/10.1145/3379156.3391351.

26. Itti, L., & Koch, C. (2001). Computational modelling of visual attention. *Nature Reviews Neuroscience, 2*(3), 194–203.

27. Wogalter, M. S., Conzola, V. C., & Smith-Jackson, T. L. (2002). Research-Based Guidelines for warning Design and Evaluation. *Applied Ergonomics, 33*, 219–230. https://doi.org/10.1016/S0003-6870(02)00009-1

28. Laughery, K. R. (2006). Safety communications: Warnings. *Applied Ergonomics, 37*(4), 467–478. https://doi.org/10.1016/j.ifacol.2018.08.474

29. Lew, R., Ulrich, T. A., Boring, R. L. (2017) Nuclear reactor crew evaluation of a computerized operator support system HMI for chemical and volume control system. *In International Conference on Augmented Cognition*, pp. 501–513, Springer, Cham, https://doi.org/10.1007/978-3-319-58625-0_36.

30. Boring, R. L. (2021) When dullscreen is too dull. *In International Conference on Applied Human Factors and Ergonomics*, pp. 493–501, Springer, Cham. https://doi.org/10.1007/978-3-030-80624-8_62.

31. Mack, A. (2003). Inattentional blindness: Looking without seeing. *Current Directions in Psychological Science, 12*(5), 180–184. https://doi.org/10.1111/1467-8721.01256

32. Kennedy, K. D., & Bliss, J. P. (2013) Inattentional blindness in a simulated driving task. *In Proceedings of the human factors and ergonomics society annual meeting, 57*(1), pp. 1899–1903, SAGE Publications

33. Murphy, G., & Greene, C. M. (2016). Perceptual load induces inattentional blindness in drivers. *Appl Cog Psychol., 30*(3), 479–483. https://doi.org/10.1002/acp.3216

34. White, A., O'Hare, D. (2022) In plane sight: Inattentional blindness affects visual detection of external targets in simulated flight. *Appl. Ergon., 98*, article 103578. https://doi.org/10.1016/j.apergo.2021.103578.

35. Drew, T., Võ, M. L. H., & Wolfe, J. M. (2013). The invisible gorilla strikes again: Sustained inattentional blindness in expert observers. *Psychological Science, 24*(9), 1848–1853. https://doi.org/10.1177/0956797613479386

36. Subramanian, M., Park, S., Orlov, P., Shafti, A., & Faisal, A.A. (2021) Gaze-contingent decoding of human navigation intention on an autonomous wheelchair platform. *In 10th International IEEE/EMBS Conference on Neural Engineering (NER)*, pp. 335–338, IEEE. https://doi.org/10.1109/NER49283.2021.9441218.

37. Sunny, M. S. H., Zarif, M. I. I., Rulik, I., Sanjuan, J., Rahman, M. H., Ahamed, S. I., & Brahmi, B (2021) Eye-gaze control of a wheelchair mounted 6DOF assistive robot for activities of daily living. *J. NeuroEng. Rehab., 18*(1), pp. 1–12. https://doi.org/10.1186/s12984-021-00969-2.

38. Einhäuser, W. (2017) The pupil as marker of cognitive processes. In Zhao Q. (ed.), Computational and Cognitive Neuroscience of Vision, pp. 141–169, Springer, Singapore. https://doi.org/10.1007/978-981-10-0213-7_7

39. van der Wel, P., & van Steenbergen, H. (2018). Pupil dilation as an index of effort in cognitive control tasks: A review. *Psychonomic Bulletin & Review, 25*(6), 2005–2015. https://doi.org/10.3758/s13423-018-1432-y

40. Köles, M. (2017). A review of pupillometry for human-computer interaction studies. *Period. Polytechnica Elec. Engin. Comp. Sci., 61*(4), 320–326. https://doi.org/10.3311/PPee.10736

41. Ehlers, J., Strauch, C., Georgi, J., & Huckauf, A. (2016). Pupil size changes as an active information channel for biofeedback applications. *Appl. Psychophysiol. Biofeedb., 41*(3), 331–339. https://doi.org/10.1007/s10484-016-9335-z

42. J. Kohlmorgen, G. Dornhege, M. Braun, B. Blankertz, K. R. Müller, G. Curio, … and W. Kincses, "Improving human performance in a real operating environment through real-time mental workload detection," in Toward Brain-Computer Interfacing, pp. 409–422, 2007.

43. H. Yang, Y. Wang, and R. Jia, "Dashboard Layout Effects on Drivers' Searching Performance and Heart Rate: Experimental Investigation and Prediction," Front. Pub. Health, vol. 10, article 813859, 2022. https://doi.org/10.3389/fpubh.2022.813859

44. K. Holmqvist, S. L. Örbom, I. T. Hooge, D. C. Niehorster. R. G. Alexander, R. Andersson, and R. S. Hessels, "Eye tracking: empirical foundations for a minimal reporting guideline," Behav. Res. Meth., pp. 1–53, 2022. https://doi.org/10.3758/s13428-021-01762-8

45. K. Kopiske, D. Koska, T. Baumann, C. Maiwald, and W. Einhäuser, "Icy road ahead - rapid adjustments of gaze-gait interactions during perturbed naturalistic walking," J. Vis., vol. 21, no. 8, article 11, 2021. https://doi.org/10.1167/jov.21.8.11

46. Kritzinger, W., Karner, M., Traar, G., Henjes, J., & Sihn, W. (2018). Digital Twin in manufacturing: A categorical literature review and classification. *IFAC-PapersOnLine, 51*(11), 1016–1022.

47. Feder, S., Bendixen, A., Einhäuser, W. (2022) A hybrid control strategy for capturing cognitive processes in virtual reality (VR) in a natural and efficient way. *In 2022 IEEE International Conference on Computational Intelligence and Virtual Environments for Measurement Systems and Applications (CIVEMSA)*, pp. 1–6, IEEE. https://doi.org/10.1109/CIVEMSA53371.2022.9853646.

48. Drewes, J., Feder, S., & Einhäuser, W. (2021) Gaze during locomotion in virtual reality and the real world. *Front. Neurosci., 15*, article 656913. https://doi.org/10.3389/fnins.2021.656913.

Predicting Object Weights from Giver's Kinematics in Handover Actions

Lena Kopnarski, Laura Lippert, Claudia Voelcker-Rehage, Daniel Potts, and Julian Rudisch

Abstract

Handover actions describe the action when an object is handed over from one actor (human/robot) to another. A requirement for a smooth handover action is precise coordination between the two actors in space and time. Part of a handover action are reach and grasp movements. In order to be able to perform adequate reach and grasp movements, precise models regarding the object properties are necessary, only then anticipatory grip force scaling can take place. It is possible that receivers in handover actions observe the giver during object manipulation in order to estimate the object weight more accurately. Knowledge about the change in kinematics due to object weight in handover actions can be used to improve human–robot interactions by providing robots with better weight estimation through prediction based on human kinematics. The aim of this study was to investigate whether predictions about the object weight can be achieved from the kinematics of the giver in a handover action. Furthermore, the aim was to analyze which joint angles are particularly suitable for classifying the object weight (i.e., are most influenced by the object weight).

Keywords

Classification · Handover · Kinematics · Object weight · Pattern recognition

Contributed equally: Lena Kopnarski and Laura Lippert.

L. Kopnarski · C. Voelcker-Rehage · J. Rudisch
Institute of Sport and Exercise Sciences, University of Münster, Münster, Germany
e-mail: lena.kopnarski@uni-muenster.de

L. Lippert (✉) · D. Potts
Faculty of Mathematics, Chemnitz University of Technology, Chemnitz, Germany
e-mail: laura.lippert@math.tu-chemnitz.de

1 Introduction

Joint handover actions are performed almost every day. Whether it is handing over the stapler to a colleague at work, the glass of wine to the partner at home or to exchange cash in the supermarket. In order to carry out a smooth handover action, both actors need to coordinate precisely in space and time [1]. Handover actions are a joint action that can be divided into individual sub-actions. Thus, a handover action includes reaching and grasping as well as carrying/transporting an object comparable to a replacement task.

For reach, grasp and manipulation tasks, precise scaling of grip and load forces is necessary. Motor control research has shown that grip and load forces are already planned in anticipation [2]. A suitable grip force must be large enough to overcome the load force and prevent the object from slipping out of the fingers, but must also avoid being too large, so that the object is not crushed or the person is not getting fatigue. The necessary grip and load forces depend on both the intended action and the object properties. Accordingly, the successful scaling of the grip and load forces depend a lot on the accuracy of the estimates of the object properties. A handover action is a time-critical action for both actors at the moment of object transfer. While both actors are in physical contact with the object, the giver reduces and the receiver increases grip forces rapidly [3]. Therefore, precise anticipatory grip force scaling is of particular importance. Estimation of object properties (such as weight) is usually based on previous experience and knowledge [4]. Furthermore, it is also possible that information about the weight of an object is additionally obtained from the observed kinematics of another person lifting or moving the object [5]. This means that heavy objects can influence the kinematics of movement differently than light objects (e.g., the joint angle configuration). When an object is handed over from one person to another, the receiver can observe this movement and obtain information to create an accurate forward model on the receiver's side.

© The Author(s) 2026
B. Meyer et al. (eds.), *Hybrid Societies*, Advances in Science, Technology & Innovation,
https://doi.org/10.1007/978-3-032-03488-5_14

An accurate estimation of the object weight also plays a major role for the grip force scaling of robots [6]. If robots are confronted with the task of grasping and transporting different, unknown objects, it is necessary that they can estimate the object weight in order to produce a suitable grip force. So far, predicting the object weight before the robot has physical contact with the object is a major challenge that has already been addressed by different approaches such as image recognition [7] or thermography [8]. In the context of hybrid societies, the human–robot handover scenario plays a central role. As a the receiver, predicting the weight of the object through the kinematics of the giver could allow the robot to anticipate appropriate grip forces, even if it is an unknown object. This would give a big advantage over approaches like image recognition, where object classes have to be learned first and therefore can never cover the variety of everyday objects.

In this study, we, therefore, want to investigate how changes in the weight of an object can be identified from the motion kinematics of a giver during a han-dover task. Furthermore, we investigate which joint angle contributes most to the classification of object weights in order to find out which kinematic characteristic of the giver should be considered most in a handover movement in order to provide the most reliable prediction. Hence, we recorded kinematics in handover actions in which the weight of the object to be handed over was varied. The aim was to classify the kinematics (time-profiles of joint angles) of the giver using a support vector machine (SVM) and thus to predict the object weight.

2 Methods

2.1 Participants

Forty healthy subjects (31 female) aged 22.0 4.3 years participated in the experiment, thus data were collected from a total of 20 dyads in the handover experiment. All subjects had normal or corrected to normal vision, no psychiatric or neurological disorders, and no orthopedic upper limb impairments. According to the Edinburgh Handedness Inventory [9], 39 subjects were right-handed and one subject was ambidextrous. This study was approved by the Ethics Committee of the Chemnitz University of Technology, Faculty of Behavioral and Social Sciences, on July 12, 2019—number V-343-17-CVR-SFB A01-24062019.

2.2 Materials

A passive-marker based optical motion capture system (Vicon Motion Systems Ltd, Oxford, UK) with 10 cameras (5 Vantage, 5 Vero) was used to record subject motions

at a sampling frequency of 100 Hz. Sixteen spherical reflective markers with a diameter of 6.4 mm were used for the upper body (head, trunk, shoulders, right arm). We used a marker set based on the Plug-in Gait Model [10]. The following joint angles were extracted for the right arm: shoulder (flexion/extension, abduction/adduction, internal/external rotation), elbow (flexion/extension), and wrist (flexion/extension, internal/external rotation, ulnar/radial deviation).

Two different self-constructed, 3D printed test objects were used. Test objects included transducers for the measurement of grip forces (not used for this study) and 5 infrared LEDs which enabled tracking in the Vicon system. Two different test objects, differing in size, were used. Both objects had an identical base body (8cm × 8cm × 8cm), which contained both the LEDs and the possibility of attaching weights inside the object. The grasping surfaces, which differ in size and distance from each other (5cm × 5cm × 5cm; 8cm × 8cm × 8cm) between the two objects, were located above the base body and were arranged one above the other (see Fig. 1). The lower (blue) grasping surfaces were used by the giver, the upper (yellow) by the receiver. Three different object weights were used so that the handover object weighed 400 g in the light condition, 700 g in the medium condition and 1000 g in the heavy condition (weight conditions were the same for both object sizes).

2.3 Procedure

Participation in the study consisted of two test sessions with about 7 to 14 days in between. The first session consisted of different motor and sensory tests of the right hand, a questionnaire and a replacement task (no joint action, not considered

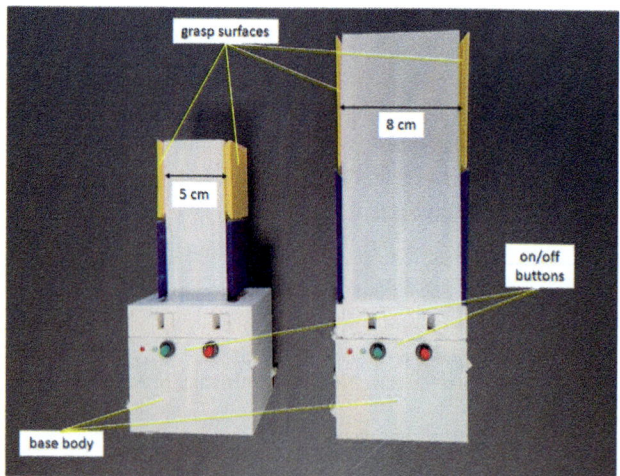

Fig. 1 Small (left) and big (right) object. Weights can be embedded in the base body

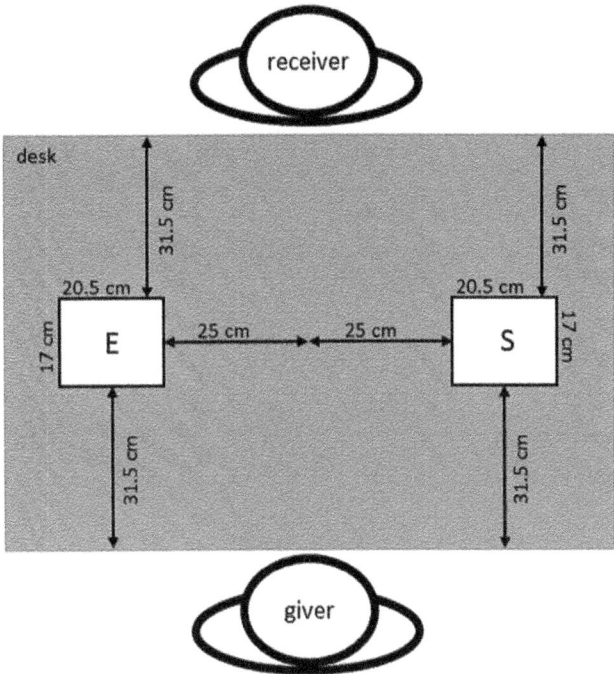

Fig. 2 Experimental set up

further here). In the second session, two subjects sat opposite each other at a table. At the start of a trial, the object was placed on a foam pad (17 cm × 20.5 cm) fixed centrally to the table on the right-hand side of the giver (see Fig. 2). The subjects were instructed to perform a handover action as natural as possible. After an acoustic signal, the giver grasped the object at the lower grasping surfaces (blue) and handed it over to the receiver, who grasped it at the upper grasping surfaces (yellow). The receiver then placed the object on a foam pad on the other side of the table (see Fig. 2), which ended the trial. Subjects performed handover actions with all six object configurations, with object size blocked and object weight presented in a pseudo-randomized order. Each subject was assigned either the role of the giver or the receiver, with the assignment being swapped halfway through the trials. Across all trials, each condition was performed 10 times (2 roles × 2 object sizes × 3 object weights), resulting in four blocks and 120 trials. In total, our data set has $M = 2400$ trials (40 givers, with 60 trials each).

2.4 Analyses

2.4.1 Data Preprocessing

The used data for our approach are times series, denoted by $\alpha_i(t)$ for the joint angle $i = 1, \ldots 7$ belonging to the three wrist, three shoulder and one elbow angles. In order to compare the time series, they were cut to their individual start and end time stamp. The starting point is the time after which the giver has grasped the object. This is defined by the moment

when the velocity of the object exceeds a threshold of 0.02 m/s for the first time. The end point is reached when givers and receivers wrist have minimal distance. Consequently, due to differences in the movement, the time series have different lengths for different trials. This can be fixed by a time normalization, where we use for every trial instead of the given time stamps $t_1, \ldots, t_e$ the normalized time stamps

$$0, \frac{t_2 - t_1}{t_e - t_1}, \frac{t_3 - t_1}{t_e - t_1}, \ldots, \frac{t_{e-1} - t_1}{t_e - t_1}, 1.$$

Hence, we modeled every angle by a function depending on the time,

$$\alpha_i(t), \quad t \in [0, 1], \quad i = 1, \ldots 7, \tag{1}$$

which we have given at discrete time points and has length 1. Missing data was interpolated linearly and some trials were discarded due to too many missing values, which reduced the total amount of 2400 trials. After filtering due to data recording errors we received $M = 2256$ trials. The time series $\alpha_i(t)$ are plotted in Fig. 3 for some random trials from one person.

2.4.2 Prediction Procedure

The aim was to extract the important information from the time series $\alpha_i(t)$ which classify the objects weight and to find out which joint angles are important for the classification. Each time series $\alpha_i(t)$ can be viewed as a smooth function defined on $[0, 1]$, which allows for a decomposition in basis functions. A useful basis in this case is the half-period-cosine basis, which allows us for every angle i to approximate the time series well by

$$\alpha_i(t) \approx \sum_{k=0}^{n-1} a_k^{(i)} \cos(\pi k t) =: \tilde{\alpha}_i(t), \quad i = 1, \ldots 7. \tag{2}$$

The half-period cosine basis is a good choice for the approximation of non-periodic functions, since in this case the decay rate is $\mathcal{O}(n^{3/2})$ and the coefficients $a_k^{(i)}$ can be calculated easily and fast from the time series at discrete points by using the discrete cosine transform (DCT), see [12, Chap. 6] [11].

Hence, we described the time series $\alpha_i(t)$ by $n = 8$ coefficients $a_k^{(i)}$ with $k = 0, \ldots, 7$, which is a very compressed expression in contrast to the full time series. Furthermore, $n = 8$ is a reasonable choice between over- and underfitting. For smaller n the error between $\alpha_i(t)$ and $\tilde{\alpha}_i(t)$ is too big, whereas for bigger n the approximation fits to much the noise in the measured data. The approximated time series $\tilde{\alpha}_i(t)$ are plotted as dashed lines in Fig. 3. Another advantage of this procedure is that the sum of cosine functions smooths out the measurement inaccuracies, which led to noisy data in the

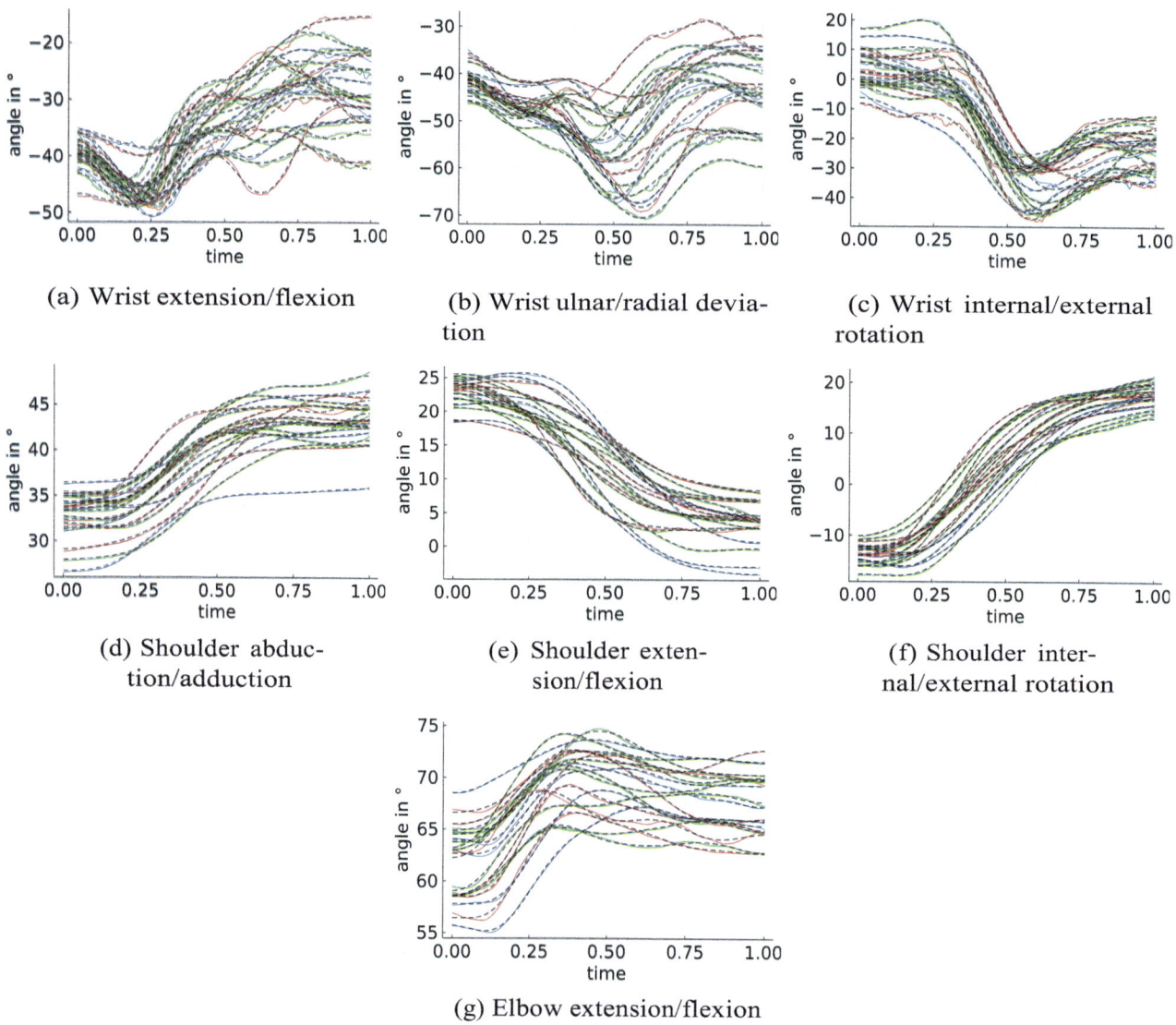

(a) Wrist extension/flexion

(b) Wrist ulnar/radial deviation

(c) Wrist internal/external
rotation

(d) Shoulder abduc-
tion/adduction

(e) Shoulder exten-
sion/flexion

(f) Shoulder inter-
nal/external rotation

(g) Elbow extension/flexion

Fig. 3 A subset of the time series $\alpha_i(t)$ (solid) of one person for the 7 joint angles, together with the approximated time series $\tilde{\alpha}_i(t)$ (black, dashed). The colors belong to the light (blue), medium (red), and heavy (green) object

original time series. Each trial has $n = 8$ Cosine-coefficients for each of the considered 7 joint angles. This can be seen as data compression in comparison the raw time series, since there are only 56 degrees of freedom for each trial.

We have the label vector $\mathbf{y} \in \{1, 2, 3\}^M$, which assigns 1, 2 and 3 to the trials with light, medium and heavy object, respectively. When we use all 7 joint angles for the prediction, we receive a matrix $\mathbf{X} \in \mathbb{R}^{56 \times M}$ containing the coefficients $a_k^{(i)}$ for every trial. For the classification we use Julia's SVM, which is contained in LIBSVM in the Machine Learning package. Different strategies for a train/test split and cross-validation (CV) are possible in order to give a measure how good our classification is. For cross-validation, we first split all M trials randomly 80/20 in train- and test data (CV all trials). And in a second variant we choose randomly the trials from

8 persons as test set and the trials from all other persons as training set, which is also an approximately 80/20-splitting (CV person-wise). In all cases, we did the CV 10 times, resulting in 50 classification tasks in total, from which we average the classification rate. In general, we standardize the values of $\mathbf{X}$ belonging to the test trials by a Z-transform, which transforms the mean and the variance of every column of $\mathbf{X}$ to zero and one, respectively. This is necessary, since the coefficients are scaled differently. For the prediction of the trials in the test set, we have to transform the values in $\mathbf{X}$ belonging to the test trials by the same transformation like the training set.

We want to study the influence of the 7 different joint angles for the classification. Therefore we do the previously described classification using only the joint angles in the subset $\mathbf{u} \subset \{1, \ldots, 7\}$. This means, we only use the coefficients $a_k^{(i)}$ with $i \in \mathbf{u}$ for the classification. We denote the classification rate by $cv(\mathbf{u})$. The *Shapley values* are a common tool for describing feature contributions. They were introduced in [12] for game theory and more recently also used for approximation theory, [13, 14].

Our variables are the 7 different joint angles. Given any subset $\mathbf{u} \subset \{1, \ldots, 7\}$ of the joint angles, the value that subset creates on its own is its explanatory power. We use here the classification rate as a way to measure explanatory power. Shapley showed that there is a unique valuation ϕ, that satisfies some reasonable axioms. Using our classification rate $cv(\mathbf{u})$, these values are defined by

$$\phi_i = \frac{1}{7} \sum_{\mathbf{u} \subseteq \{1,\ldots,7\} \setminus \{i\}} \binom{6}{|\mathbf{u}|}^{-1} (cv(\mathbf{u} \cup \{i\}) - cv(\mathbf{u})), \quad (3)$$

where $cv(\emptyset) := 0$. The values ϕi give some notion for importance of the joint angles for classification task.

3 Results

Our aim was to predict the object weight classification from the cosine coefficients $a_k^{(i)}$ of the time series belonging to the time series of the joint angles.

Considering all joint angles, we achieved a classification rate of 0.683 (CV all trials) and 0.567 (CV person-wise). We plotted in Fig. 4 some coefficients $a_k^{(i)}$ for $k = 0, 1$. One can see that there are some person specific behaviors, which means that some coefficients $a_k^{(i)}$ slightly differ for the persons, independent of the objects weight.

The mean classification rate which we reached for different subsets u and the different CV strategies are summarized in Table 1. There we present the subsets u, for which the highest classification rates are possible. The best classification rate 0.68 is reached involving all joint angles. Additionally, using for instance only one of the joint angles results in a low classification rate: Predicting the weight only from wrist ulnar/radial deviation gives prediction rate 0.354, since 3 different classes have to be predicted, this is no meaningful prediction.

One further question was the influence of the different joint angles to the classification rate. We calculated the Shapley values (3) for the two different CV strategies and plot the results in Fig. 5. The wrist ulnar/radial deviation is least necessary for the classification. Whereas, the shoulder internal/external rotation and elbow extension/flexion are import angles for the classification.

4 Discussion

In this study, we investigated how the weight of an object in a handover task can be predicted by the joint angles of the giver's active arm. For this purpose, a discrete cosine transform and a support vector machine (SVM) were used to classify the different weights.

The results of the cross-validation (CV) show that a prediction of the object weight is possible, whereby a higher classification rate is achieved when the data of a subject is included in both the training and the test data set (CV all trials). We attribute this to the individuality of the movement. Looking at Fig. 4, we notice that the data are clustered by person. This individual influence factor on movement has also already been shown in several studies [15–18]. The influence factor of individuality possibly also affects the classification performance. In other words, if an SVM is trained with the kinematic data of one person, a more reliable prediction of the kinematics of the same person is obtained than the prediction with kinematic data of another person.

Furthermore, by determining the Shapley value, it could be shown that especially the shoulder rotation and elbow movement is influenced by the object weight and therefore makes an important contribution to the prediction of the object weight. Wrist ulnar/radial deviation provides the least amount of explanation in object weight prediction.

It has already been shown that people have the ability to estimate the weight of an object by observation while another person is grasping, transporting or manipulating the object [19, 20]. Efforts to find out on which kinematic characteristics these judgements depended yielded the result that mainly the duration of the lifting movement is used to make such judgements about the object weight [5]. The results of our study extend these findings and allow the use of kinematic data for weight prediction by using the joint angles alone without explicitly determining the lifting duration.

A limitation of our experimental setup is that we recorded the giver kinematics in a very controlled environment, as the start position of the object was always the same and the grasping position was only varied by the two different object sizes. This results in a relatively low variance of the giver movements. This contrasts with real handover actions in everyday life, where the object can always be at a different starting position and orientation, resulting in a greater kinematic variance of the giver.

Therefore, we suggest that in further experiments the starting position could be varied. Moreover, a variation of the weight classes would be interesting. In this study we investigated the three weight classes 400 g, 700 g, 1000 g, which could be varied both in distance and in rage, possibly resulting

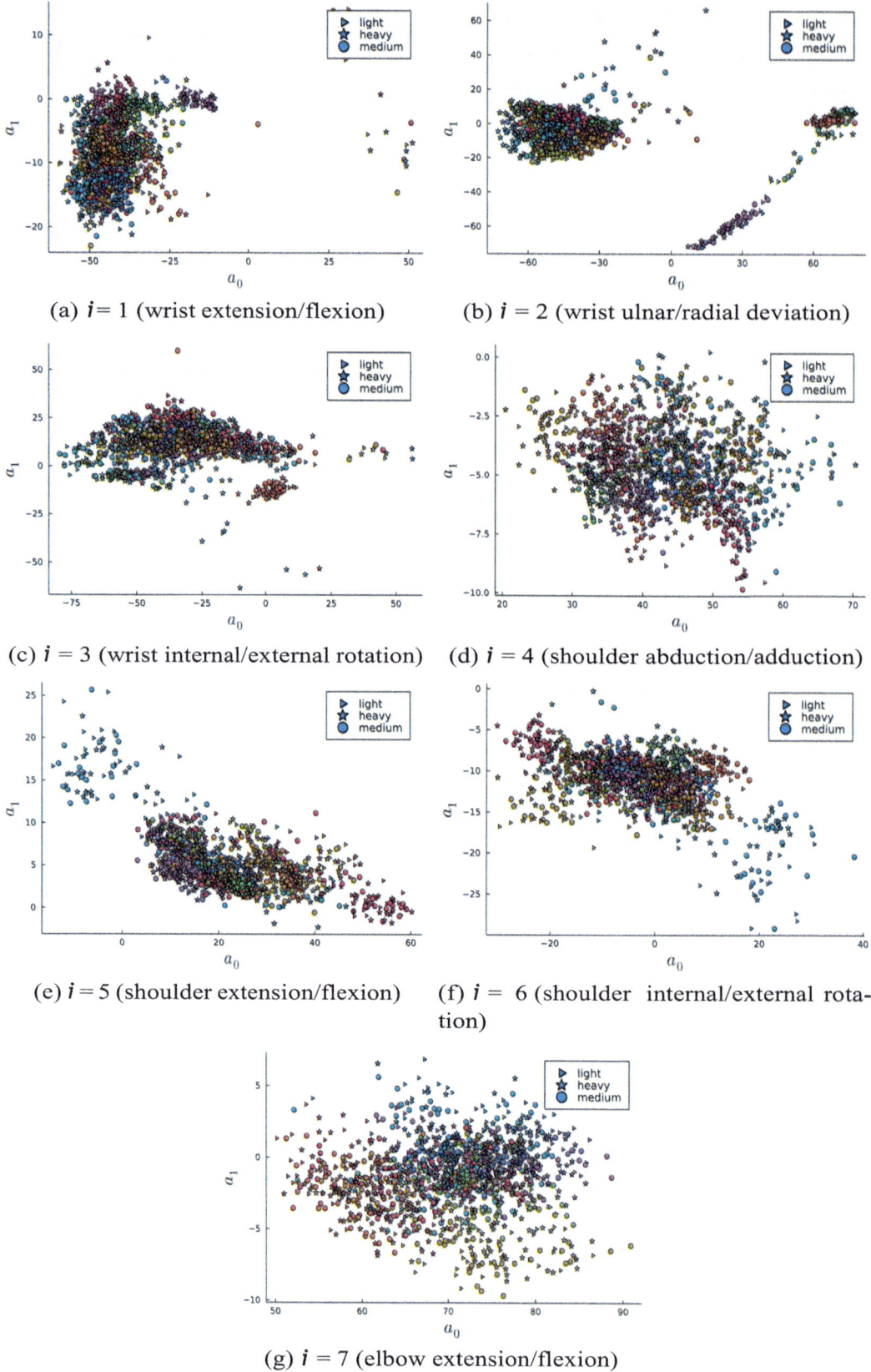

(a) $i = 1$ (wrist extension/flexion)

(b) $i = 2$ (wrist ulnar/radial deviation)

(c) $i = 3$ (wrist internal/external rotation)

(d) $i = 4$ (shoulder abduction/adduction)

(e) $i = 5$ (shoulder extension/flexion)

(f) $i = 6$ (shoulder internal/external rotation)

(g) $i = 7$ (elbow extension/flexion)

Fig. 4 Coefficients $a_k^{(i)}$ for $k = 0, 1$ for the different angles and all trials. The different colors belong to trials from different persons. The shape belongs to the object's weight

Table 1 Classification rates of predicting the objects weight from subset u of joint angles using different cross-validation strategies

$u\backslash$ CV	All trials	Person-wise
{1, 2, 3, 4, 5, 6, 7}	0.683	0.567
{1, 3, 4, 6, 7}	0.680	0.578
{1, 3, 4, 5, 6, 7}	0.679	0.579
{3, 4, 5, 7}	0.678	0.578
{2, 4, 5, 7}	0.676	0.573
{2, 3, 4, 5, 6, 7}	0.676	0.603
{3, 4, 6, 7}	0.675	0.613
{1, 4, 5, 6}	0.675	0.575
{1, 2, 4, 5, 6, 7}	0.674	0.587
{4, 6, 7}	0.664	0.619
{4, 5, 6, 7}	0.670	0.614
{5, 6, 7}	0.659	0.606
$\vdots$	$\vdots$	$\vdots$
{3}	0.482	0.421
{4}	0.447	0.437
{2}	0.391	0.354

Mapping of the angles
1 → Wrist extension/flexion
2 → Wrist ulnar/radial deviation
3 → Wrist internal/external rotation
4 → Shoulder abduction/adduction
5 → Shoulder extension/flexion
6 → Shoulder internal/external rotation
7 → Elbow extension/flexion

Fig. 5 Shapley values with the classification rate as explanatory power and two different cross-validation strategies

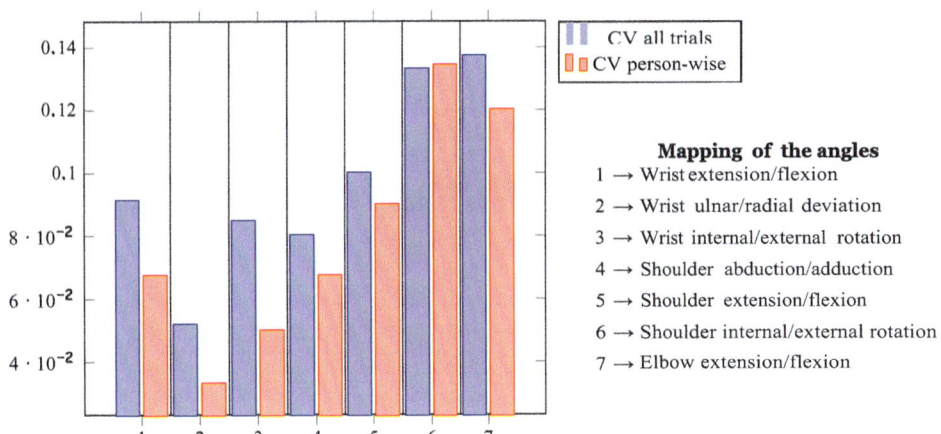

Mapping of the angles
1 → Wrist extension/flexion
2 → Wrist ulnar/radial deviation
3 → Wrist internal/external rotation
4 → Shoulder abduction/adduction
5 → Shoulder extension/flexion
6 → Shoulder internal/external rotation
7 → Elbow extension/flexion

in different classification accuracies. By changing the weight classes, it is conceivable that in the future not only classifications but also weight estimations will be possible through the analysis of kinematic data.

The SVM is widely used for classification tasks. Therefore, also here good results were expected. Furthermore, one has to choose specific parameters from the time series for the classification, since the raw time series are noisy. And, furthermore, the joint angles at specific times are not robust against changes in the absolute position: The same movement from a slightly different starting position can lead to completely different joint angles.

In joint handover actions, observation and prediction (about the intentions of the other but also about the object properties) play an important role. This is why it is necessary that robots in hybrid societies are also able to achieve this, so that human actors can interact intuitively and smoothly. The approach described in this study can contribute to improving human–robot interactions in hybrid societies by having the robot predict object properties through observation.

Acknowledgements This research has been supported by the German Research Foundation (DFG, 416228727)—SFB 1410 Hybrid Societies.

References

1. Sebanz, N., Bekkering, H., & Knoblich, G. (2006). Joint action: bodies and minds moving together. *Trends in Cognitive Sciences*, 10(2):70–76.

2. Hermsdörfer, J., Li, Y., Randerath, J., Goldenberg, G., & Eidenmüller, S. (2011). Anticipatory scaling of grip forces when lifting objects of everyday life. *Experimental Brain Research*, 212(1):19–31

3. Mason, A. H., & Mackenzie, C. L. (2005). Grip forces when passing an object to a partner. *Experimental Brain Research*, 163(2):173–187.

4. van Polanen, V., & Davare, M. (2015) Sensorimotor Memory Biases Weight Perception During Object Lifting. *Frontiers in Human Neuroscience*, 9.

5. Hamilton, A., Joyce, D. W., Flanagan, R., Frith, C. D., & Wolpert, D. M. (2007). Kinematic cues in perceptual weight judgement and their origins in box lifting. *Psychological Research*, 71(1):13–21.

6. Copot, D., Ionescu, C., Nascu, I., & De Keyser, R. (2016). Online weight estimation in a robotic gripper arm. In *2016 IEEE International Conference on Automation, Quality and Testing, Robotics (AQTR)*, pages 1–6.

7. Standley, T., Chen, D., Sener, O., & Savarese, S. (2017). image2mass: Estimating the mass of an object from its image. page 10.

8. Aujeszky, T., Korres, G., Eid, M., & Khorrami. F. (2019). Estimating Weight of Unknown Objects Using Active Thermography. *Robotics*, 8(4):92. Number: 4 Publisher: Multidisciplinary Digital Publishing Institute.

9. Oldfield, R. C. (1971). The assessment and analysis of handedness: The Edinburgh inventory. *Neuropsychologia*, 9(1):97–113.

10. Vicon, M. S. (2020). Vicon nexus plug-in gait reference guide.

11. Plonka, G., Potts, D., Steidl, G., & Tasche, M. (2018) *Numerical Fourier Analysis*. Applied and Numerical Harmonic Analysis. Birkhäuser.

12. Shapley, L. (1952). A value for n-person games. *Technical report*. DTIC Document.

13. Owen, A. B. (2014). Sobol' indices and shapley value. *SIAM/ASA Journal on Uncertainty Quantification, 2*(1), 245–251.

14. Sundararajan, M., & Najmi, V. (2020) The many shapley values for model explanation. In H. D. III and A. Singh, editors. in *Proceedings of the 37th International Conference on Machine Learning*, volume 119 of *Proceedings of Machine Learning Research*, pages 9269–9278. PMLR, 13–18.

15. Bednarik, R., Kinnunen, T., Mihaila, A., Fränti, P. (2005). Eye-Movements as a biometric. In *Proceedings of the 14th Scandinavian conference on Image Analysis*, SCIA'05, pages 780–789, Berlin, Heidelberg. Springer-Verlag.

16. Bekemeier, H., Maycock, J., & Ritter, H. (2019). What Does a Hand-Over Tell?—Individuality of Short Motion Sequences. *Biomimetics, 4*(3).

17. Cunado, D., Nixon, M. S., & Carter, J. N. (1997). Using gait as a biometric, via phase-weighted magnitude spectra. In Bigu¨n, J., Chollet, G., & Borgefors, G., editors, *Audio- and Video-based Biometric Person Authentication, Lecture Notes in Computer Science*, pages 93–102, Berlin, Heidelberg,. Springer.

18. Girges, C., Spencer, J., O'Brien J. (2015). Categorizing identity from facial motion. *Quarterly Journal of Experimental Psychology*, 68(9):1832–1843

19. Rizzolatti, G., Fadiga, L., Fogassi, L., & Gallese, V. (1999). Resonance behaviors and mirror neurons. *Archives Italiennes De Biologie*, 137(2–3):85–100.

20. Sciutti, A., Patane, L., Nori, F., & Sandini, G. (2014). Understanding object weight from human and humanoid lifting actions. *IEEE Transactions on Autonomous Mental Development, 6*(2), 80–92.

A Parametric Model to Assess Minnesota Dexterity Test with IMU

Marvin Rehm, Giuseppe Sanseverino,
Teodorico Caporaso, Antonio Lanzotti,
Stephan Odenwald, and Alois Pichler

Abstract

Evaluations of injuries and the assessment of residual dexterity is carried out by clinicians by means of standardized tests. Several dexterity tests have been developed and all of them require the test administrator to sit in front of the subject and visually assess the motions performed. Automatic tools can reduce the time needed for the evaluation and can provide additional useful information. For many of the available dexterity tests, automated evaluation methods have been developed. However, for the Minnesota dexterity test, despite being widely used, just one automated solution based on multiple depth sensor cameras was developed. This study aims to provide a wearable and flexible method to automatically evaluate the outcomes of Minnesota dexterity tests. The proposed methodology consists of a parametric model that is capable of processing acceleration data collected with an IMU attached to the centre of mass of the subject's dominant hand. The developed model was successfully validated against a subject study with ten participants.

Keywords

Manual dexterity · Biomechanics · Wearable devices · Automatic assessment · Parametric model

1 Introduction

1.1 Dexterity Tests and Their Assessment

Dexterity tests are used by clinicians to measure the accuracy of hand and finger movements under controlled conditions. They can help to identify and evaluate certain forms of brain damage or musculoskeletal disabilities. Several standardized tests are available in literature, each of them has a slightly different focus. The Nine-Hole Peg Test [4], for example, is mainly used to measure finger dexterity, the Box and Block Test [7] does not require high precision and measures unilateral gross manual dexterity, while the Minnesota Manual Dexterity Test (MMDT) [6] evaluates subject's ability to grab and place objects quickly and is mostly used for the evaluation of injuries. Above-mentioned dexterity tests are in general visually assessed by clinicians, what requires a certain amount of time that can be reduced by using automatic tools. With regard to the MMDT, the time taken by the subject to completely fill in the board is the only parameter used for the evaluation. In literature, some studies were found that focus on the automatic evaluation of dexterity tests. Temporiti et al. [9] proposed a performance assessment of Nine-Hole Peg Test using optical cameras, Onã et al. [8] used RGB-Depth camera solution to automatically assess the Box and Block test, while Zhang et al. [10] used wearable sensors for the automatic assessment of the Box and Block test. Only one work was found that focused on the automatic evaluation of MMDT in the literature. Caporaso et al. [2] proposed a tool

M. Rehm · A. Pichler
Faculty of Mathematics, Chemnitz University of Technology, Chemnitz, Germany
e-mail: rehm.marvin@outlook.de

A. Pichler
e-mail: alois.pichler@math.tu-chemnitz.de

G. Sanseverino (✉) · S. Odenwald
Department of Sports Equipment and Technology, Chemnitz University of Technology, Chemnitz, Germany
e-mail: giuseppe.sanseverino@phil.tu-chemnitz.de

S. Odenwald
e-mail: stephan.odenwald@mb.tu-chemnitz.de

T. Caporaso · A. Lanzotti
Fraunhofer JL IDEAS, Department of Industrial Engineering, University of Naples Federico II, Naples, Italy
e-mail: teodorico.caporaso@unina.it

A. Lanzotti
e-mail: antonio.lanzotti@unina.it

© The Author(s) 2026
B. Meyer et al. (eds.), *Hybrid Societies*, Advances in Science, Technology & Innovation,
https://doi.org/10.1007/978-3-032-03488-5_15

for the automatic assessment of this test based on multiple RGB-Depth cameras. Even if good results were achieved by the authors, the proposed test set-up is still very complex. This paper aims to introduce an IMU-based tool for the automatic assessment of MMDT that is easy to use and guarantees higher flexibility with regard to vision-based solutions.

1.2 Parametric Models

When executing the MMDT, the centre of mass (CoM) of the hand per- forms a characteristic trajectory. To automatically assess the test, the simplest approach would be to double integrate the acceleration data obtained with the IMU. Unfortunately, the noise associated with the collected accelerations cause this approach to fail. Machine learning approaches are also conceivable. Ahmed et al. [1], for example, attempt to use such approaches to classify pedestrian movements on the base of acceleration data. Attempts have also been made to recognize human activity by evaluating IMU data using deep learning, cf. [5].

The evaluation of MMDT with wearable devices was not investigated before. In this case, machine learning approaches would not guarantee a good accuracy due to the limited amount of training data. Therefore, the idea is to approximate the available data with a parametric model, information thus obtained can be used to identify the movements.

A parametric model is well suited for this case because it is not affected by the considerable recording errors present in the available data. In fact, there is no need to build the model on the data, but it is sufficient to use the characteristic trajectory associated with the MMDT, while measured data are only used to estimate the parameters (period length, e.g.).

2 Methods

2.1 Test Protocol

The Minnesota Manual Dexterity Test consists of a rectangular board with 60 holes equally distributed on 15 columns and 60 objects with cylindrical shape that fit into the board holes. The participant is required to sit in front of the board with his dominant hand placed on a predefined spot on the desk. When the test starts, the subject leaves the rest position and starts filling the board. Figure 1a illustrates the test set-up. Accelerations are measured at the approximate centre of mass of the subject's hand by means of a 6 DoF IMU (Dialogg, Envisible GmbH, Chemnitz, Germany). The accelerometer sampling rate was set to 100 Hz and the axes are oriented as shown in Fig. 1b. An RGB camera ($1920 \times 1080p$, 30fps) was used to record reference videos of each test session.

Tests were conducted at the ERGOS lab of the Fraunhofer Joint Lab IDEAS in Naples. Data from 10 voluntary healthy (25 ± 3 years) subjects were collected. An additional test was performed where one subject was deliberately asked to commit a common mistake defined by Caporaso et al. [2] as reverse placing. Later this allows to test the effectiveness of the developed methodology in identifying this kind of error. All participants provided writ- ten informed consent. The study was in accordance with the Declaration of Helsinki.

2.2 Model Development

The steps followed to automatic assess the MMDT are reported in Fig. 2. In particular the assessment goes through three phases: (i) definition of the region of interest that goes from the start to the end of subject's hand movement; (ii) identification of the motion profile of the CoM of the hand; (iii) analysis of obtained data to identify possible mistakes committed by the subject.

The phase (i) enables the calculation of the actual test duration for each subject. In fact, once the signal is trimmed, the actual duration is equal to the number of frames that compose the time-series divided by the acquisition rate of the sensor.

The test is executed in a coordinate system that is fixed to the desk and defined as in Fig. 1b. However, during the test execution, the subject's hand changes its orientation and this leads to a deviation of the accelerometer axes to the reference. This issue is overcome by operating a conversion of these coordinate systems. In fact, rotation matrices, called *Givens-rotations* can serve this purpose. These are defined as

Fig. 1 Scheme of Minnesota manual dexterity test set-up (**a**) and accelerometer axes disposition (**b**)

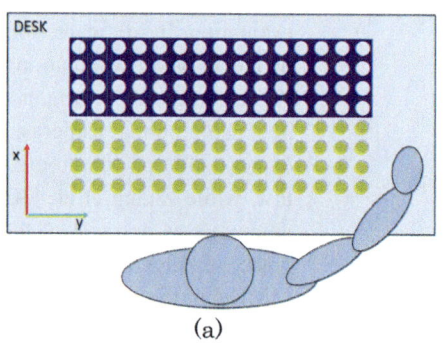

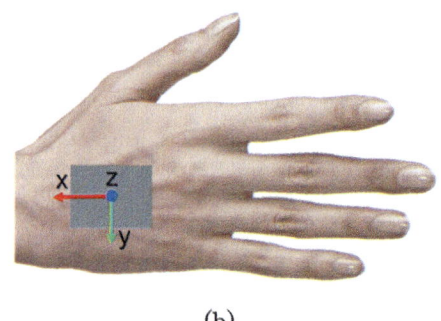

(a)   (b)

Fig. 2 Flowchart showing the steps followed to get to the automatic assessment of Minnesota dexterity test

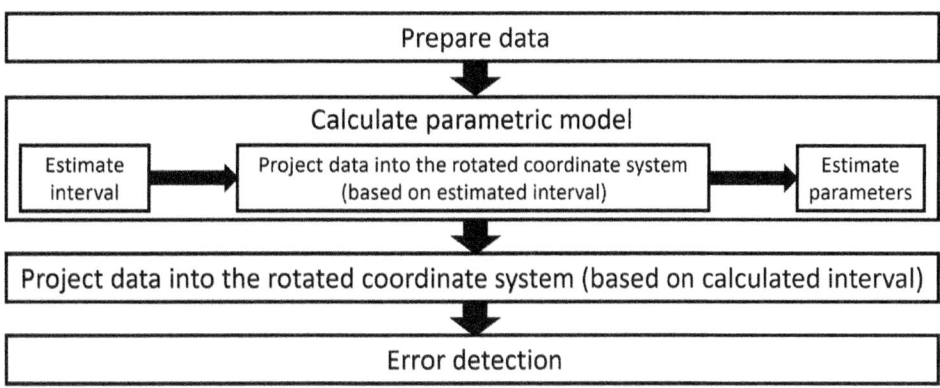

$$\begin{pmatrix} \cos(\omega) & \sin(\omega) \\ -\sin(\omega) & \cos(\omega) \end{pmatrix} \begin{pmatrix} x_{\text{sensor}} \\ y_{\text{sensor}} \end{pmatrix} = \begin{pmatrix} x_{\text{reference}} \\ y_{\text{reference}} \end{pmatrix} \quad (1)$$

where ω is the rotation angle. Since the movement in the $y$reference direction is minimal, the idea is to minimize it over ω. This consideration leads to the optimization problem

$$\min_{\omega} \sum_{i \in I} \left| -\sin(\omega) x_{\text{sensor}}^i + \cos(\omega) y_{\text{sensor}}^i \right| \quad (2)$$

where I is the time interval of interest. The rotation angel is calculated separately for each of the 15 columns. For this purpose, the assumption is used that the rotation angel is constant within a column. This assumption corresponds to observations made from subjects.

Equation (2) results in the optimal rotation angle ω, which can be used to calculate the rotated acceleration data in the direction $x$reference.

The next step is to define a model for the accelerations. Direct modelling is not possible, because the theoretical pattern for the acceleration is unknown. However, the movement has a clear characteristic in the $x$reference direction, which is easy to model. The following parametric model for the trajectory then is considered:

$$g(t|A, \alpha) := -A(1 - \alpha t) \sin(8\pi t), \quad (3)$$

where

$$f(t|T, \varphi, A, \alpha) := g\left(\frac{t - \varphi}{T} - \left\lfloor \frac{t - \varphi}{T} \right\rfloor \middle| A, \alpha \right). \quad (4)$$

The function g, globalized by the function f, models the trajectory traced by the centre of mass of the subject's hand to complete one column of the Minnesota Test board.

The parameters of the model are T, φ, A and α. T is the so-called period length, which corresponds to the time the subject needs for one column. The intercept φ is used for a right or left shift of the model, to match the peaks of the measured data and the model. The parameter A scales the amplitude.

The amplitudes of the four peaks within a column decrease, modelled by the factor $(1 - \alpha t)$. That is, the parameter α controls this effect.

Because of the relationship between trajectory and acceleration, it is now possible to obtain a parametric model for the acceleration, by differentiating the model obtained for the trajectory two times with respect to the time t; that is

$$g''(t|A, \alpha) = A(16\pi\alpha \cos(8\pi t) - 8\pi(8\pi\alpha t - 8\pi) \sin(8\pi t)), \quad (5)$$

and

$$f''(t|T, \varphi, A, \alpha) = \frac{1}{T^2} g''\left(\frac{t - \varphi}{T} - \left\lfloor \frac{t - \varphi}{T} \right\rfloor \middle| A, \alpha \right). \quad (6)$$

The next step consists in estimating the parameters (T, φ, A, α) by minimizing the error function. This leads to the minimization problem

$$\min_{T, \varphi, A, \alpha} \sum_{i \in I} \left(f''(t_i|T, \varphi, A, \alpha) - a_i \right)^2 \quad (7)$$

where a_i is the acceleration measured in correspondence with the centre of mass of the subject's hand, in the reference coordinate system. The *L-BFGS-B* method, a quasi-Newton method that supports box constraints [11], is employed to solve problem (7).

The algorithm then calculates the parameters (T, φ, A, α) for the 15 columns one after the other. Starting from the beginning of the time series, the interval for the first column is estimated generously, then it is possible to estimate the rotation angel for this interval. This allows to calculate the acceleration in the $x$reference direction. Afterwards, the parameters are calculated using the above-mentioned optimization method. In particular, the so-called period length T is thus obtained. The starting point of the second column can then be calculated, allowing to estimate an interval for the second column. This procedure is then repeated for the first seven columns.

The parameters for the last seven columns are calculated in the same way, but starting from the end of the time series. For the eighth column, there is then an interval for the optimization, from the end of the seventh column and the start of the ninth one. This procedure is used to keep the effect of error propagation as low as possible. A parametric model for the entire movement is thus obtained.

Since the start and end points of the columns are known, the rotation angles can be recalculated to raise the accuracy. Subsequently, it is possible to calculate the acceleration data in $x$reference direction, using the rotation angles. Before these are evaluated, they are smoothed using a moving average filter, with a window of 55 frames, which has proven itself in tests.

In order to evaluate the movement, the upper and lower peaks are searched in the data and assigned to the respective column, using the parameters obtained by the model.

Caporaso et al. [2] have shown that most common mistakes committed by the subject lead to specific patterns in the trajectory data. This makes it possible to assess the correctness of test execution and the possible presence of errors. The same approach could be used here, but the noise associated with acceleration data makes it more complex.

An error function is introduced for a general assessment of the approximation. Since the identification of the columns are particularly important for the assessment of the dexterity test, the period lengths are used as an indicator for the quality of the approximation through the parametric model. The best possible result would be that the period lengths ($T_i, i \in \{1, \ldots, 15\}$) would sum up to the total length of the time series. This consideration results in

$$ e := \left| n - \sum_{i=1}^{15} T_i \right|, \tag{8} $$

where n is the length of the time series in seconds.

3 Results

Figure 3 displays modelled and measured accelerations on top of each other, showing a good agreement of the two.

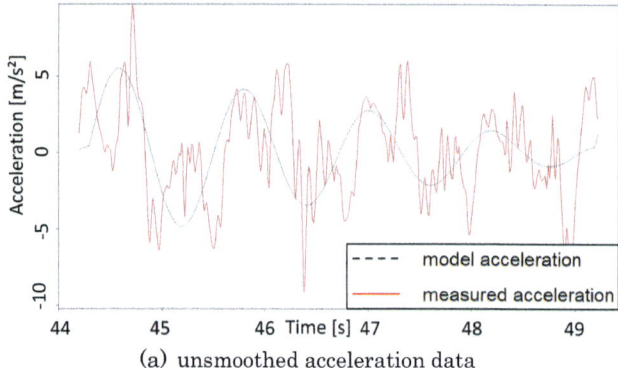

(a) unsmoothed acceleration data

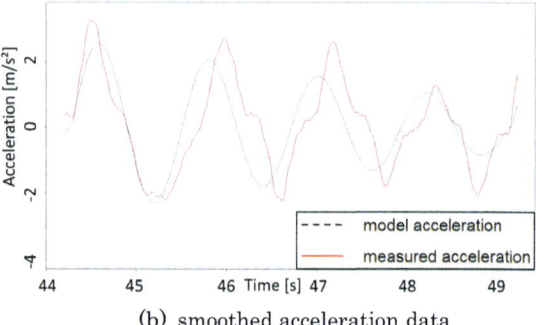

(b) smoothed acceleration data

Fig. 3 Plot of accelerations in *xreference* direction with focus on column 9 of the test board for subject 1 (red) and corresponding parametric model (black)

The approximation errors calculated for the ten subjects can be found in Table 1.

4 Discussion and Conclusions

In conclusion, it can be stated that the developed method is capable of assessing the Minnesota Dexterity test using acceleration data collected with an IMU. Results show a good estimate for the test duration, that represents the main parameter for its evaluation. In fact, the absolute error (equal to 0.88 ± 0.78 s) is comparable to the intra-individual variability for MMDT found by Desrosiers et al. [3]. In this pilot study, the obtained graphs, showing an estimate of the trajectory traced by the subject's hand, are assessed visually.

Table 1 Estimated test duration and its approximation error with regards to the actual test duration

Subject	1	2	3	4	5	6	7	8	9	10
Actual duration [s]	79.20	93.15	89.01	89.31	71.57	75.17	100.83	86.28	85.11	90.52
Calculated duration [s]	79.42	91.01	89.12	87.41	72.56	73.74	101.14	85.96	85.05	91.89
Absolute error [s]	0.22	2.14	0.11	1.90	0.99	1.43	0.31	0.32	0.06	1.37
Relative error[a] [%]	0.28	2.29	0.13	2.13	1.39	1.89	0.30	0.37	0.07	1.52

[a]The actual duration corresponds to 100%.

Nevertheless, once the peaks in the graphs are identified, a methodology like the one introduced by Caporaso et al. [2] can be applied to recognize possible mistakes committed by the subjects. However, this would require more complex algorithms to account for approximation introduced by the model. In comparison with vision-based solutions for the evaluation of dexterity tests, the proposed methodology guarantees much lower costs and higher flexibility thanks to the use of wearable devices.

Acknowledgements The present work is funded by the Deutsche Forschungsgemeinschaft (DFG, German Research Foundation)Project-ID 416228727—SFB 1410. The subject test is part of the project SAFE-WORKERS (Support in Ability and Function Evaluation for Workers) funded by Fraunhofer Joint Lab IDEAS and INAIL—Direzione regionale della Campania.

References

1. Ahmed, M. U., Brickman, S., Dengg, A., Fasth, N., Mihajlovic, M., & Norman, J. (2019). A machine learning approach to classify pedestrians' event based on IMU and GPS. *International Journal of Artificial Intelligence*, 17(2):154–167.
2. Caporaso, T., Sanseverino, G., Krumm, D., Grazioso, S., D'Angelo, R., Di Gironimo, G., Odenwald, S., & Lanzotti, A. (2022). Automatic outcomes in minnesota dexterity test using a system of multiple depth cameras, In: Gerbino, S., Lanzotti, A., Martorelli, M., Mirálbes Buil, R., Rizzi, C., Roucoules, L. (eds) Advances on Mechanics, Design Engineering and Manufacturing IV. JCM 2022. Lecture Notes in Mechanical Engineering. Springer, Cham.
3. Desrosiers, J., Rochette, A., Hebert, R., & Bravo, G. (1997). The minnesota manual dexterity test: Reliability, validity and reference values studies with healthy elderly people. *Canadian Journal of Occupational Therapy*, 64(5), 270–276.
4. Feys, P., Lamers, I., Francis, G., Benedict, R., Glenn, P., La Rocca, N., Hudson, L. D., Rudick, R., & Consortium, M. S. O. A. (2017). The nine-hole peg test as a manual dexterity performance measure for multiple sclerosis. *Multiple Sclerosis Journal*, 23(5), 711–720.
5. Hou, C. (2020). A study on imu-based human activity recognition using deep learning and traditional machine learning. In *2020 5th International Conference on Computer and Communication Systems (ICCCS)* (pp. 225–234).
6. LafayetteInstruments (1998). *The complete Minnesota dexterity test. Examiner's manual*. Lafayette Inc.
7. Mathiowetz, V., Volland, G., Kashman, N., & Weber, K. (1985). Adult norms for the box and block test of manual dexterity. *The American Journal of Occupational Therapy*, 39(6), 386–391.
8. Oña, E., Sánchez-Herrera, P., Cuesta-Gómez, A., Martinez, S., Jardón, A., & Balaguer, C. (2018). Automatic outcome in manual dexterity assessment using colour segmentation and nearest neighbour classifier. *Sensors*, 18(9), 2876.
9. Temporiti, F., Mandaresu, S., Calcagno, A., Coelli, S., Bianchi, A. M., Gatti, R., & Galli, M. (2022). Kinematic evaluation and reliability assessment of the nine hole peg test for manual dexterity. *Journal of Hand Therapy*.
10. Zhang, Y., Chen, Y., Yu, H., Lv, Z., Shang, P., Ouyang, Y., Yang, X., & Lu, W. (2019). Wearable sensors based automatic box and block test system. In *2019 IEEE SmartWorld, Ubiquitous Intelligence & Computing, Advanced & Trusted Computing, Scalable Computing & Communications, Cloud & Big Data Computing, Internet of People and Smart City Innovation (SmartWorld/SCALCOM/UIC/ATC/CBDCom/IOP/SCI)* (pp. 952–959).
11. Zhu, C., Byrd, R. H., Lu, P., & Nocedal, J. (1997). Algorithm 778: L-BFGS-B: Fortran subroutines for large-scale bound-constrained optimization. *ACM Transactions on Mathematical Software*, 23(4), 550–560.

A Novel Framework for the Generation of Synthetic Datasets with Applications to Hand Detection and Segmentation

Tom Uhlmann, Amin Dadgar, Felix Weigand, and Guido Brunnett

Abstract

We introduce a configurable framework to generate synthetic datasets featuring human characters for the training of neural networks. We demonstrate how our framework can generate a vast amount of annotated images showing a human and its hands in a multitude of poses and with varying backgrounds. Furthermore, we conduct a specific formation of synthetic hand images to train convolutional neural networks for detecting and/or segmenting hands in real scenarios in a novel way. The generated dataset aims to enhance the previously successful method of exploiting the invariancy concept with slightly more expensive rendering techniques. That is, we keep the number of subjects, poses, scenes, and other costly factors at the minimum level while increasing the diversity of the data. Our dataset features ten different human characters in various poses captured from a multitude of camera angles. We make our framework open source and publish the dataset consisting of 90,000 images.

Keywords

Machine learning · Synthetic dataset · Hand detection · Hand segmentation · Configurable graphical framework · Open source · Neural network

1 Introduction

With the improvements in hardware capabilities that enabled the design of deep networks and the booming of data generated, neural networks regained weight in solving problems after several decades of decline, this time with some crucial algorithmic improvements [1, 2]. The research and industrial communities are attracted heavily to the benefits this technology offers on a wide range of tasks, such as detection, segmentation, tracking, and estimation, required by many applications such as surveillance, human–computer interactions, and autonomous systems (e.g., driving).

A general trend in machine learning is the preference for increasing the depth of networks. However, as the number of layers increases, overfitting impedes the path to achieving the desired performance. This is especially the case if the amount of data does not increase accordingly. Thus, the ability of the model to generalize to unseen data is impaired. One immediate solution to mitigate this issue is to employ larger datasets with more diversity that could effectively narrate the missing information. However, the larger the dataset is, the more laborious, time-consuming, and error-prone the labeling process will be. These costs are especially high if the number of parameters is high (e.g., datasets of the human body or hands which contain high degrees of freedom). To tackle those impediments, a meaningful marriage between the realms of computer graphics and machine learning could be an effective strategy.

Methods of generative computer graphics offer the possibility to generate an ample amount of annotated training data in an automated or semi-automated fashion with a controlled parameter setting and reliable annotations. That is also beneficial regarding data protection issues that are a concern with material captured in real-life scenarios. However, to assist the networks in generalizing better, employing highly diverse image sets with complex rendering techniques is crucial to bridge the information gap between real and synthetic data. Nevertheless, these techniques can in turn be costly to employ. Considerations such as a large number of poses, subjects, scenes, and various lighting conditions in a random fashion (randomization techniques) are a few costly aspects contributing to the complexity.

We present a configurable framework based on freely available software to create synthetic images for training

T. Uhlmann (✉) · A. Dadgar · F. Weigand · G. Brunnett
Chemnitz University of Technology, Chemnitz, Germany
e-mail: uhto@hrz.tu-chemnitz.de

© The Author(s) 2026
B. Meyer et al. (eds.), *Hybrid Societies*, Advances in Science, Technology & Innovation,
https://doi.org/10.1007/978-3-032-03488-5_16

convolutional neural networks with a focus on scenes containing human models. We will show how to configure and use this novel framework to produce a large dataset with relatively low cost and manual effort, yet high diversity for human hand segmentation and detection.

The rest of the paper is organized as follows. In the next section, we will review the literature focusing on synthetic dataset generation of human models and their hands. Then we will present our novel framework. This is followed by a more detailed view of hand recognition and the presentation of our generated dataset. Then we will show some results and discuss limitations and future work in the last section.

2 Previous Work

The quantity and quality of a dataset are of paramount importance regarding the choice of the training set for machine learning. If these two factors are considered appropriately regarding the data before the learning phase, they can prevent the crucial obstacles of over- and underfitting. Processes on the models' structure such as dropout and regularization will assist in addressing overfitting, however, they are performed within the learning phase [3]. Overfitting describes a state of the neural network where it is specifically tailored to the input data and unable to generalize to new inputs. On the other side, underfitting means a state where the network is not suited for the task, resulting in many mistakes. A suitable dataset should contain enough samples for training and testing, but also it ought to possess enough diversity to cover most situations and prevent overfitting. Synthetic data provides the benefit of controlling all parameters present in the dataset, which assists us in avoiding certain features not being over or underrepresented in a dataset. Additionally, we can create an infinite amount of samples that can be annotated automatically. This property makes these methods very attractive for dataset generation.

Several techniques to generate synthetic datasets have been published in the scientific literature [4], however, they fall into two main categories. One class of approaches aims to train a neural network that will generate new images of the same class. In that direction, Kar et al. [5] developed a framework called Meta-Sim and trained a neural network that modifies the attributes of a scene graph and a 3D renderer to produce the final labeled image. Karras et al. [6] illustrated an approach to create an architecture for generative adversarial networks (GANs) which automatically separates high-level features. Then, the variables are stochastically altered, and new synthetic images are generated. They used their framework to create a highly diverse dataset of human faces. A similar approach was proposed by Zhang et al. [7] that used a GAN to create massive datasets from a few annotated samples. They directed their work on segmenting objects in images.

However, besides occasional unstable (never converging) behaviors of GAN approaches, they produce samples with limited diversity, as the generated data exhibits the same statistics as the training set. Variance, that is of particular significance when employing synthetic images, is therefore not sufficiently pronounced.

The other mainstream approach is to explore a concept known as domain randomization [8]. There, one considers randomization of the parameter set and a 3D renderer, where the statistics of variables in a dataset are estimated and replicated. In other words, the underlying idea is to randomly consider a wide range of scenes, poses, lighting conditions, and camera viewpoints, expecting to reach an acceptable level of richness in the synthetic data, by which the model would generalize to many real-world scenes. In that context, Prakash et al. [9] created a synthetic dataset for 2D bounding box detection of cars. However, this approach is also costly [10] albeit the easiness of the tasks of simple geometrical object detection [11, 12]. For substantial variations of rendering parameters such as texture, scene, background objects, and models, complex realistic rendering techniques must be considered. Additionally, considerable challenges are yet to be addressed especially for more complicated tasks such as hand detection or segmentation.

Modules that extract the distinguishing features which represent hands for detecting, tracking, and localizing them in the scene and segmenting hands from the cluttered and natural background are central in many intelligent systems. However, technical challenges like occlusions, cluttered backgrounds, lighting conditions, and technological factors such as the number and type of cameras, required frame rates, and accuracy all play a significant role in the overall performance of those systems. Various approaches have been proposed by researchers addressing some or all of the above issues. Image processing and computer vision approaches such as Fourier based techniques [13], motion-based methods [14], active contour model-based techniques [15], temporal-based approaches [16], optimization methods [17, 18], depth-based technologies [19], and deep convolutional neural network frameworks [1, 2] are some successful examples.

Recently, an approach proposed by [20] aimed to detect hands in real scenes using a neural network that is trained on synthetic images. However, unlike the two main categories of data generation, they sought to exploit a different well-known notion called the invariancy concept. They considered a specific method of synthetic data generation to train a neural network for hand detection. That is, instead of employing costly domain randomization methods, they generated a training set with two simple uniform background colors, a simple empty scene configuration, texturing, lighting condition, and shading. Additionally, their dataset comprises only one (right) hand model, and no other body parts existed in the scene. However, to compensate for the tremendous gaps that exist between their dataset and a conventional real set, and to achieve the desired diversity in the hand's shape, they rotated the hand model about two axes which causes the most amount of alteration in the shape. Those axes were in-plane-axes (e.g., x and y), and they let the out- plane-axis rotation and translations get inferred by the powerful invariancy concept intertwined in neural network technologies, since the hand's outer appearance does not change.

Inspired by this work, we aim to tie the exploitation of domain randomization and invariancy concepts in a hybrid way to enhance them one step forward. That is, we slightly increase the complexity of backgrounds, scenes, and lighting conditions and also employ the whole body. However, this slight addition in complexities does not fulfill the real-synthetic data gap. Therefore, we couple these complexities and exploit them further using the invariancy concept similar, but not identical, to Dadgar's [20] work. More specifically, instead of rotating the hand or the subject about in-plane-axes, we rotate the camera around the subject. Therefore, we can maximize the diversity of resulting 2D backgrounds and poses' shapes while keeping the costly factors such as the number of designed scenes, subjects, poses, and lighting conditions low.

3   Synthetic Dataset Generation Framework

Our framework (see Fig. 1) employs a free, easy to use, and high-quality rendering software, named Blender [21], as its central unit and couples it with Python API for automatic configuration and scene generation. It takes two configuration of parameters stored as a plain text file as inputs and a set of human models created with MakeHuman [22]. A Python script takes those inputs and generates n scenes, with $n \in \mathbb{N}$

being the number of images contained in the final dataset. By rendering the resulting scenes, we get one image each. Additionally, we output another image containing the annotations in form of color masks. The details of our process are as follows.

To create a dataset, we define a set of parameters that comprises environment, camera, background, model, and pose. This set of parameters determines how each sample is generated. The parameters of the environment are described as all settings that remain constant for an entire dataset. The most influential aspects of the environment are the employed 3D scene, image resolution, render engine, and what annotations are created. We use an arcball camera that always points to a specific point of our model (e.g., toward the chest) so that all essential features are visible. Hence, the parameters of the camera are the radial distance as well as the polar and azimuth angle. By modifying these parameters, we can view our object from various perspectives. For the background, we employ high-dynamic range (HDR) images and image-based lighting. This makes it easy to cover various lighting situations without a complex and error-prone light setup and results in properly illuminated pictures. For the background, we vary the used HDR image only. The illumination of the character changes accordingly based on the background image, as shown in Fig. 4, where the same scene is viewed from different camera angles. If we did not utilize image-based lighting, we would have to create a light setup for each background scene, which is time-consuming and complicated since it requires significant manual effort. To cover a wide range of human models, we use the free software MakeHuman [22], which can create rigged human characters with a wide range of properties such as sex, age, clothing, and body proportions. The models are created beforehand, and we tend them to cover a wide range of attributes without containing a specific bias. That is, we consider female and male characters equally, with various

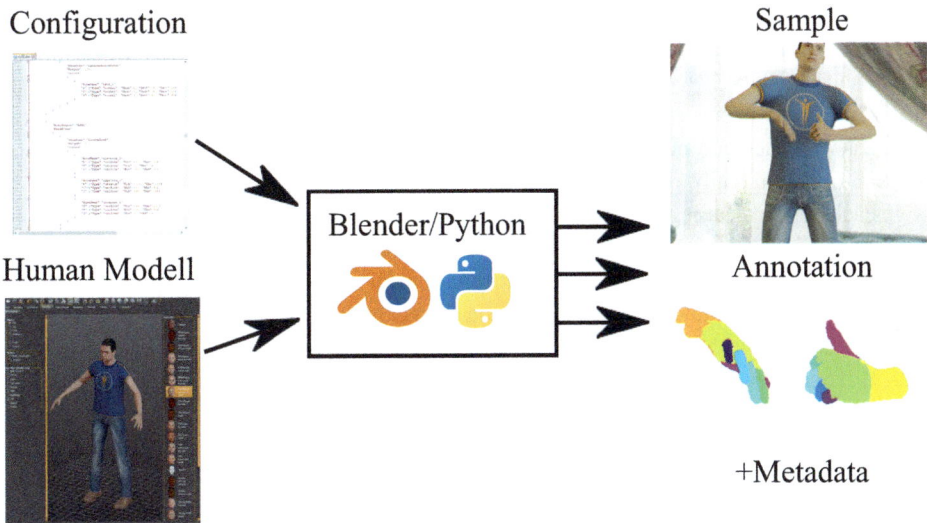

Fig. 1 Our framework uses a set of human models and a configuration file as inputs and creates a large set of annotated samples for those. We use Blender's powerful Python API and rendering capabilities to generate the samples from the inputs automatically

Configuration

Sample

Human Modell

Blender/Python

Annotation

+Metadata

Fig. 2 This figure shows a
defined base pose for the hand
and three automatically generated
variations of this base pose

Base Pose   Variation 1   Variation 2   Variation 3

Fig. 3 Two sample images from
our dataset (top) and their
corresponding annotation image
(bottom). The annotation shows
the different segments of the
hands in unique colors

skin colors and clothing styles when producing our dataset. However, since our framework is configurable, one can reuse it with alternative properties of interest, such as skin color, gender, age, clothing style, or add new models. Then, one can create a different image set if the generated images do not exhibit the desired traits targeting a specific application.

The most cumbersome part of the parameter definition is setting up the poses. It is infeasible to consider all poses a human body can undergo, since the number of joint angle combinations is numerous. To manage this complexity, we use a two-stage approach. First, we separate the model into several regions. These are the chest, legs, arms, head, and hands, and we assume that these individual regions can be posed independently. This assumption is feasible, because the skeletal sub-chain of each region can be considered independent. Next, we define a set of standard poses for each region which can be varied automatically within defined joint angle ranges and anatomically reasonable limits. In addition, we assign each pose a meaningful name, in order to make editing by the user easier. Figure 2 shows the base pose "pointing" for a hand and some automatically generated variations of this base pose. To create a new pose, we select a base pose in the first stage according to a decision tree inspired by Prakash et al. [9]. Then, we vary each base pose according to a normal distribution. With this two-stage approach, we can control the complexity of the human skeleton, while creating believable poses with high variation. The range of variation for each pose has to be defined carefully to avoid unnatural poses and

self-intersections of the model. Currently, this is up to the user.

To obtain the annotations, we separate the mesh into different parts by grouping together the respective vertices of the mesh. That highly depends on the scenario the dataset is intended to be created for. A reasonable separation could be the body regions introduced earlier. For our use case, we separated the hands into various regions. Then, we assign each defined vertex group a unique color and generate, for each sample, an additional image containing the color-coded parts, which show the pixels that define their masks within the image in unique colors. Reliable and accurate annotations can be per- formed in an automated fashion by using this approach. Figure 3 shows two samples from our hand recognition dataset and their respective annotations. It is evident, that manual annotations would be highly time-consuming and would never reach the accuracy we achieve with our approach.

4 Hand Recognition Dataset

For our hand recognition dataset, we extended the idea of Dadgar et al. [20] by including the whole human body and both hands. Though the previous dataset successfully trained the YOLO network [1] on the task of tracking. A direct extension of that approach to the more difficult task of hand segmentation seems unsatisfactory. More precisely, we tailored the dataset and its annotation for the Mask-RCNN

Fig. 4 This figure shows a 180-degree view of a single model and pose in four-degree steps. Note that the lighting condition changes for each view, but the subject is still lit correctly

network training [2] to segment the hand. Though in many scenes, the hands were segmented, however, there was a considerable number of false positives, too.

Addressing that failure was the main inspiration for us to slightly increase the complexity of the dataset's backgrounds, scenes, lighting conditions, and models for later testing the dataset in segmentation and/or even hand pose estimation networks. To that, we use our novel configurable graphical frame-work to generate synthetic data for human characters. With our configurable framework, we constructed a dataset that features ten different human characters in various poses from a multitude of camera angles. We sample around the 'up'-axis only in 2-degree steps, hence we have 90 perspectives for each pose and model, covering a 180-degree radius. The number of backgrounds, scenes, and lighting conditions is equal to the number of subjects and in total, the dataset consists of 90,000 images. To create the actual images, Blender provides two physically-based render engines. A ray tracer called Cycles and a rasterizer called Eevee, which both use physically-based materials. We chose the Eevee rasterizer as it provides all features we need for our setup, requires about 10 times less time to generate an image, and offers a comparable quality. The rendering engine could be changed easily, since our framework is compatible with both. It took about 30 h to render the whole dataset with a consumer grade computer.

Though following a similar strategy as the previous work, there is a considerable conceptual difference in the generation of this dataset, too: Previously, there existed no body parts except for a hand. So one could easily rotate the hand around the x–y axes. Here, however, we have the whole body. Therefore, rotation of the hand is limited to the anatomical constraints that

the body compels to the hand movements. Thus we rotate the camera around the subject's circular vicinity and let the dataset be as diverse as possible in the terms of the outer shape of both hands. Additionally, we rotate the camera merely around the "up" axis of the body.

The benefits of such a dataset are: (1) We cover the whole subjects' body and two hands, which let the network learn the differences between the hands and other similarly-colored parts. (2) By rotating the camera, the whole background, scene, and lighting changes. Therefore, we gain further diversity on other factors, too.

Figure 3 shows two examples of our highly diverse dataset and their respective annotations. A 180-degree view's example of a single pose and model can be seen in Fig. 4. For the sake of clarity, we only included every second sample thus each sample illustrates a four-degree step. In these images, the changes in lighting condition, due to the image-based lighting, is clear while the subject is appropriately visible with enough contrast. Additionally, the backgrounds and the scenes of each image are distinct from each other. These points demonstrate the second benefit of our approach.

5 Results

We conduct a simple experiment to evaluate the suitability of the samples for recognition using an existing neural network. To that end, we created 1000 samples with a resolution of 800×600 pixels and tested them with the widely used pose detection software OpenPose [23]. We manually checked the detection performance by grouping several types of misdetections, and then we counted their occurrences. As shown

Table 1 Misdetections using OpenPose [23] on a sample dataset of 1000 images created with our framework

Error type	Count	Misdetection rate (%)
Model not detected	0	0
Background detected as body part	3	0.31
Body parts falsely detected	21	2.14
Body parts not detected	34	3.47
Both hands not detected	10	1.02
One hand not detected	81	8.27
Hand not fully detected	74	7.55

in Table 1, we mainly encountered errors when the images contained extreme body poses, such as both arms raised high above the head or the samples with low contrast situations. The overall misdetection rate is immensely low if we only view the body part scenario, and it is low if we consider the misdetection of the hands. The misdetection of hand features is mainly caused by their low resolution in the created images. Our experiment illustrates that the generated images can be detected with a high degree of confidence by an existing network and, therefore, could be employed as a training set for neural networks.

Using our framework, we created a dataset of 90,000 images with ten different human characters, five female and five male, and a resolution of 800×600 pixels which is particularly suited for hand recognition using a convolutional neural network. The samples show the hands in higher resolution compared to our sample dataset while also showing the rest of the body.

We publish our framework open source for everyone to use on GitHub (https://github.com/CrossForge/Synthetic HumanDatasetGenerator), and we also make our hand recognition dataset available for research purposes on Zenodo (https://doi.org/https://doi.org/10.5281/zenodo.7415618).

6   Discussion and Future Work

We developed a generative and configurable graphical framework to create synthetic datasets of human characters suitable for training neural networks based on open-source software. Our framework has the major advantage of cross-platform compatibility since it uses Blender as the central unit, which is available on all typical operating systems. It is configurable, extensible, and simple to use, which makes it easy to adapt to many use cases with human posable characters.

When considering synthetic data to train neural networks, one requires a large amount of data with considerable variance to let the network generalize well to the task and prevent overfitting. In that context, we illustrated the advantages of our framework in generating training data for hand recog-

nition. Our framework enabled us to create an arbitrary amount of high-quality samples with well-defined properties with a significant amount of control. We illustrated its capabilities by creating a novel annotated dataset for hand recognition extending previous work. Although the generated samples are not as natural as real photographs, the annotations are much more accurate and reliable, with suitable masks and meta-data. Additionally, they do not raise any concerns regarding data privacy. Finally, we evaluated the suitability of the dataset in an experiment for employing it in a neural network. In future works, we intend to add an automatic method to detect and exclude poses that contain self-occlusions, and we want to validate the benefits of our dataset by using it to train and test a neural network for hand recognition purposes.

Acknowledgements This research was funded by the Deutsche Forschungsgemeinschaft (DFG, German Research Foundation)—Project-ID 416228727—CRC 1410.

References

1. Redmon, J., Divvala, S., Girshick, R., & Farhadi, A. (2016). You only look once: Unified, real-time object detection. In *IEEE Conference on Computer Vision and Pattern Recognition (CVPR)*

2. He, K., Gkioxari, G., Dollar, P., & Girshick, R. (2020). Mask R-CNN. *IEEE Transactions on Pattern Analysis and Machine Intelligence, 42*(2), 386–397.

3. Jabbar, H., & Khan, R. Z. (2015). Methods to avoid over-fitting and under-fitting in supervised machine learning (comparative study). *Computer Science, Communication and Instrumentation Devices, 70.*

4. Dandekar, A., Zen. R. A., & Bressan, S. (2018). A comparative study of synthetic dataset generation techniques. In *International Conference on Database and Expert Systems Applications* (pp. 387–395). Springer.

5. Kar, A., Prakash, A., Liu, M.-Y., Cameracci, E., Yuan, J., Rusiniak, M., Acuna, D., Tor-ralba, A., & Fidler, S. (2019). Meta-sim: Learning to generate synthetic datasets. In *Proceedings of the IEEE/CVF International Conference on Computer Vision* (pp. 4551–4560).

6. Karras, T., Laine, S., & Aila, T. (2019). A style-based generator architecture for generative adversarial networks, in *Proceedings of the IEEE/CVF Conference on Computer Vision and Pattern Recognition* (pp. 4401–4410).

7. Zhang, Y., Ling, H., Gao, J., Yin, K., Lafleche, J.-F., Barriuso, A., Torralba, A., & Fidler, S. (2021). Datasetgan: Efficient labeled data factory with minimal human effort. In *Proceedings of the IEEE/CVF Conference on Computer Vision and Pattern Recognition* (pp. 10145–10155).

8. Tobin, J., Fong, R., Ray, A., Schneider, J., Zaremba, W., & Abbeel, P. (2017). Domain randomization for transferring deep neural networks from simulation to the real world. In *IEEE-International Conference on Intelligent Robots and Systems (IROS)* (pp. 23–30).

9. Prakash, A., Boochoon, S., Brophy, M., Acuna, D., Cameracci, E., State, G., Shapira, O., & Birchfield, S. (2019). Structured domain randomization: Bridging the reality gap by context-aware synthetic

data. In *2019 International Conference on Robotics and Automation (ICRA)* (pp. 7249–7255). IEEE.

10. Shrivastava, A., Pfister, T., Tuzel, O., Susskind, J., Wang, W., & Webb, R. (2017). Learning from simulated and unsupervised images through adversarial training. In *Proceedings of the IEEE Conference on Computer Vision and Pattern Recognition* (pp. 2107–2116).

11. Jaderberg, M., Simonyan, K., Vedaldi, A., & Zisserman, A. (2013). Synthetic data and artificial neural networks for natural scene text recognition. In *IEEE International Conference on Computer Vision (ICCV)* (pp. 569–576).

12. Tremblay, J., Prakash, A., Acuna, D., Brophy, M., Jampani, V., Anil, C., To, T., Cameracci, E., Boochoon, S., & Birchfield, SD. (2018). Training deep networks with synthetic data Bridging the reality gap by domain randomization. In *Proceedings of the IEEE Conference on Computer Vision and Pattern Recognition Workshops* (Vol. 2018, pp. 1082–1090).

13. El-ghazal, A., & Belkasim, S. (2007). A new shape signature for Fourier descriptors. *Computer, 1*, 161–164.

14. Luo, Q., Kong, X., Zeng, G., & Fan, J. (2010). Human action detection via boosted local motion histograms. *Machine Vision and Applications, 21*(3), 377–389.

15. Kronfeld, T., Brunner, D., & Brunnett, G. (2010). Snake-based segmentation of teeth from virtual dental casts. *Computer-Aided Design, 7*, 1–10.

16. Just, A., & Marcel, S. (2009). A comparative study of two state-of-the-art sequence processing techniques for hand gesture recognition. *Computer Vision and Image Understanding, 113*(4), 532–543.

17. Spruyt, V., Ledda, A., & Geerts, S. (2010). Real-time multi-colour space hand segmentation. In *Proceedings—International Conference on Image Processing, ICIP* (pp. 3117–3120).

18. Spruyt, V., Ledda, A., & Philips, W. (2014). Robust arm and hand tracking by unsupervised context learning. *Sensors (Switzerland), 14*(7), 12023–12058.

19. Sharp, T., Keskin, C., Robertson, D., Taylor, J., Shotton, J., Kim, D., Rhemann, C., Leichter, I., Vinnikov, A., Wei, Y., Freedman, D., Kohli, P., Krupka, E., Fitzgibbon, A., & Izadi, S. (2015). Accurate, robust, and flexible real-time hand tracking. In *ACM Conference on Human Factors in Computing Systems (CHI)*, pp. 3633–3642.

20. Dadgar, A., & Brunnett, G. (2020). SaneNet: Training a fully convolutional neural network using synthetic data for hand detection. In *SAMI 2020—IEEE 18th World Symposium on Applied Machine Intelligence and Informatics, Proceedings* (pp. 251–256).

21. B. O. Community. (2018). *Blender—A 3D modelling and rendering package.* Blender Foundation, Stichting Blender Foundation, Amsterdam.

22. M. O. Community. (2021). *Makehuman—Open source tool for making 3d characters.* http://www.makehumancommunity.org

23. Cao, Z., Simon, T., Wei, S.-E., & Sheikh, Y. (2017) Realtime multi-person 2d pose estimation using part affinity fields. In *Proceedings of the IEEE Conference on Computer Vision and Pattern Recognition* (pp. 7291–7299).

A Computational Study on Supervised and Unsupervised Gait Data Segmentation

Christoph Helmberg(iD), Tobias Hofmann(iD), Dominik Krumm(iD), and Stephan Odenwald(iD)

Abstract

This paper addresses the identification of patterns and change points in high-dimensional time series data recorded by body-attached sensor networks. We discuss different approaches to state detection in a supervised and in an unsupervised learning scenario. Within this context, we provide empirical evidence that our methods are capable of identifying relevant motion patterns in complex time series. Among other conceivable applications, we are particularly motivated by the idea that recognized motion change points serve as valuable information for embodied digital technologies interacting in public spaces.

Keywords

Time series data · Segmentation · Sparsest cut · Nearest centroid

MSC 2020

90C35 · 05C90 · 62H30

C. Helmberg · T. Hofmann (✉)
Department of Mathematics, Chemnitz University of Technology, Chemnitz, Germany
e-mail: tobias.hofmann@math.tu-chemnitz.de

C. Helmberg
e-mail: helmberg@math.tu-chemnitz.de

D. Krumm · S. Odenwald
Department of Sports Equipment and Technology, Chemnitz University of Technology, Chemnitz, Germany
e-mail: dominik.krumm@mb.tu-chemnitz.de

S. Odenwald
e-mail: stephan.odenwald@mb.tu-chemnitz.de

1 Introduction

The foot is the last link in the kinematic chain of human locomotion [19]. The internal forces generated by the muscles during the execution of a movement are transmitted to the ground via the foot. By recording these plantar loads it is possible to draw certain conclusions about the person's current activity. For example, if one wants to have continuous and always up-to-date information about the current state of an agent in public space, recording the pressure under the foot provides a sound data basis for this. This data can be used, for example, to try to predict and recognize human interactions in public space. The term interaction is described by [15] as the process in which people react to each other and coordinate their behavior (movements) with regard to a common interest.

The identification of specific events during human gait through the use of wearables, i.e., body-attached sensors, has recently been used for diagnosis and therapy, especially in the health sector. For example, an ambulatory monitoring system using a smartwatch was able to continuously track fluctuations in Parkinson's disease [17]. It has also been shown that plantar pressure data can be used to detect and predict freezing of gait, a walking disorder in advanced stages of Parkinson's disease [21]. In another study, wearable inertial sensors were shown to be very sensitive in detecting gait disorders in patient with multiple sclerosis, even in the early stages when global scales do not provide clinical information [21]. In summary, these tools allow very accurate analyses and help to diagnose diseases such as Parkinson's disease, cerebral palsy, multiple sclerosis and stroke [1].

Notwithstanding the promising results of these studies, the gait data to be analyzed are very heterogeneous, high-dimensional and time-dependent [18], making their analysis non-trivial. For this reason, statistical or machine learning methods are usually employed. For linear or less complex problems, statistical techniques or so-called data reduction techniques are usually used. Statistical techniques can be applied to gait models, for example, to study the effects of different independent variables on dependent variables [18]. Multivariate statistical techniques such as principal component analysis (PCA) and linear decrement analysis (LDA) can be used to represent gait data and identify linear relationships [14]. The goal of machine learning techniques such as supervised (classification-based) and unsupervised (cluster-based) learning is to develop algorithms that either learn from experience in the form of labeled data or automatically discover useful patterns from given data points [18].

Our paper aims to discuss different approaches for state detection in a supervised and in an unsupervised learning scenario. We demonstrate empirically that our methods are able to identify relevant motion patterns in complex time series from plantar pressure data. Our approach for the supervised setting provides well-interpretable results and is computationally efficient, but it is tied to scenarios in which relevant motion patterns to be classified are already known. The unsupervised approach represents an application-independent methodological contribution to recognizing similar states in high-dimensional time series; it is related to optimization techniques for change point and outlier detection. A recent overview on offline change point detection can be found in [22], while [6] summarizes approaches to online outlier detection. While, we mainly consider the offline setting, the approach presented here is intellectually closer to those described in [6].

2 Data Acquisition Methodology

One male person (age 33 years, height 1.87 m, body mass 85 kg) voluntarily participated in the pilot study and gave written informed consent. The study was conducted in accordance to the Declaration of Helsinki. Three body-attached sensor networks (BASNs) were used to collect gait data during the experiment. The BASNs each consisted of a small, lightweight and wearable sensor node[1] [8]. Two BASNs also had a pressure insole (Smart footwear sensors/HD 002, IEE, Echternach, LUX) connected to them. The sensor node has a processor that enables data acquisition and preprocessing, a wireless data logger, and an integrated 3-axis IMU (accelerometer and gyroscope) sensor. The pressure insoles were operated through the sensor node using a plug-in connector. Measurements with the BASN can be performed without external devices (e.g., PC, smartphone, etc.), and the collected data can be stored on the internal memory and later downloaded to any PC using the appropriate software (Envisible sensors). Before the experiment, pressure measurement insoles of size XL were inserted between the midsole and

insole of the participant's shoes (size $44\frac{1}{2}$ EUR). The sensor nodes were attached to the lateral side of each shoe with adhesive tape. After ensuring that the BASNs were functioning correctly, the participant moved to the starting position for the experiment and started the recording with a sampling rate of 100 Hz. According to the study protocol, the test person first walked through a straight corridor of about 35 m in length. He then entered the stairwell through a door and descended 7 floors with two flights of 8 stairs each. He then walked back into the corresponding corridor and climbed 7 floors. There he walked straight ahead again through the original corridor. Thereafter, he went down two floors. From there, he walked straight back through the corresponding corridor and descended 2 more floors. He turned around and climbed 4 floors again, arriving at the original floor at his starting position. Finally, he walked about 10 m straight ahead through the original corridor, turned around and walked 10 m back to the starting, respectively, ending point.

The BASN data are read out to an *envisible* file with tab-delimited values. This file was used without further preprocessing for the unsupervised experiments. For the supervised setting, the data are segmented into individual strides. This segmentation is based on the summed pressure of the left foot calculated from the eight channels of the respective pressure measurement insoles. We determine *left foot contact* events after periods in which no pressure is recorded. Small pressure values that are below the precision range of the sensors are initially set to zero. Our idea for identifying foot events is similar to the algorithm available on OSF [12], but ours is less nuanced because we intend to evaluate our data analysis approaches with as little preprocessing as possible.

3 Supervised Learning by Nearest Centroids

We consider motion data given by a sequence $(y_i)_{i \in [T]}$ where $[T]$ is a short-hand for $\{1, \ldots, T\} \subset \mathbb{N}$ and $y_i \in \mathbb{R}^n$ for $i \in [T]$. For the test sequence on which our empirical investigations are based, we have $T = 32,452$ data points y_i of dimension $n = 34$. Each data point represents a 10 ms time frame, amounting to a total sequence duration of nearly 5.5 min. The information stored in the time series vectors y_i originates from 34 channels of three body-attached pressure, accelerometer, and gyroscope sensors captured as described in the previous section. Furthermore, we consider a finite set of labels L as well as a map $\ell : \{y_1, \ldots, y_T\} \rightarrow L$ which assigns each time series vector one of the given labels. This map represents the *true* state of motion a person is in. For the test series at hand, the set L contains five states: self-paced walking, reduced-paced walking, turning, climbing stairs, and descending stairs. The respective labels were assigned by hand when the test walk was recorded. A schematic drawing of the time series

---

[1] The sensor node is currently a research product. However, it will be commercially avail- able in the next future (https://www.envisible.de/).

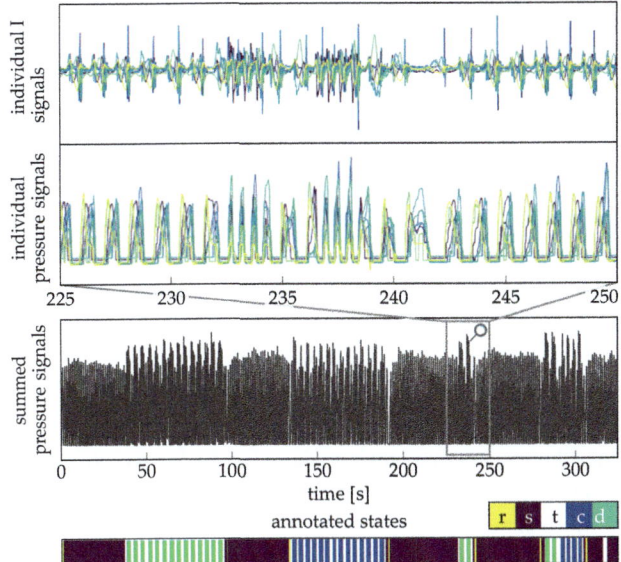

Fig. 1 Human movement time series. All data are displayed in normalized form. The bottom line diagram provides the annotated states *reduced-paced walking* (r), *self-paced walking* (s), *turning* (t), *climbing stairs* (c), and *descending stairs* (d)

and the given annotation is displayed in Fig. 1. For better visibility, only a selection of channels, stemming from the left foot's sensor, is shown there. At the very bottom is a diagram that illustrates the annotated states of motion over time. Corresponding summed foot pressure signals are shown in the chart above the bottom line. At the top and underneath, for a smaller time series interval, there are the profiles from six acceleration and eight individual foot pressure signals, respectively.

In this section, the time series vectors y_i are preprocessed such that for each coordinate their average is zero and their standard deviation is one. Another fairly natural preprocessing operation when dealing with human movement data is to divide a walk into individual strides. When doing this as described at the end of the previous section, we obtain a monotonously increasing sequence of time points $(t_s)_{s \in [S]}$, where $S \in \mathbb{N}$ is the number of detected strides and $t_s \in [T]$ are points in time when a stride is finished.

When observing new gait information from moving subjects, a reasonable data analysis question is to ask, based on conducted observations, in what state of motion a subject is in. So suppose that we are given labels for some time series vectors y_i and we are interested in predicting the label of strides for which the state is unknown. There are many machine learning methods for such supervised data classification tasks. Possible approaches include the nearest neighbor method [3], support vector machines [20], or neural networks [4], in many variants, and there are more sophisticated standardization schemes than the one we used [10]. Yet, we shall outline a simple, interpretable and effective procedure to classify the strides of the human movement data at hand,

which essentially is based on the nearest centroids method [13].

Our approach starts by gaining information about each stride from the individual time series vectors a stride comprises. Setting $t_0 = 0$ and using the absolute value component wise as in $|y_i| = \left(|(y_i)_1|, \ldots, |(y_i)_n| \right)^\top$, we determine the vector of strides attributes

$$a'_s = \sum_{\substack{i \\ i \in \{t_{s-1}+1, \ldots, t_s\}}} |y_i| \quad (\in \mathbb{R}^n)$$

for each stride $s \in [S]$. Taking the absolute value in this sum for example has the effect that we do not distinguish between turning left and turning right. This is in line with the annotation given with the test data, where no distinction is made between left and right turns. The strides lengths in seconds provide further useful information. Those are given by the numbers $t_s - t_{s-1}$, which we append to a'_s to obtain $a_s \in \mathbb{R}^{n+1}$ for $s \in [S]$. The idea we describe here could be extended, at higher computational costs, for example by determining Fourier coefficients for the curves the vectors $y t_{s-1} + 1, \ldots, y t_s$ describe for a stride $s \in [S]$. Such ideas are described by [11] to predict object proper- ties from movement kinematics. Not only the attributes, but also the labels we associate with a stride need an aggregation procedure, since the labels given to the individual time series vectors y_i might not perfectly fit to the strides segmentation given by $(ts)_{s \in [S]}$. We assign a stride s the label $\ell(s)$ which is most frequent among $\ell(y t_{s-1} + 1), \ldots, \ell(y t_s)$. Learning on training data $S' \subset [S]$ is then done by computing the centroids

$$y_c := \frac{1}{|\{s \in S' : \ell(s) = c\}|} \sum_{\substack{s \in S' \\ \ell(s)=c}} a_s$$

for each class $c \in L$. The prediction for some new data point $y \in [S] \backslash S'$ is then performed by determining

$$\arg\min_{c \in L} \| y_c - y \|,$$

where we chose $\|.\|$ as the Euclidean norm.

To test this procedure empirically, we run 1000 trials, where we split the given strides uniformly at random into training and test data. From each class in L, 2/3 of the strides are taken for training and 1/3 of them are taken for evaluating predictions. The results are shown in Fig. 2. We almost always recognize when the test person walks or descends stairs and detect slowed movement with high probability. The detection of turning and climbing movements is not that reliable, with a rate of 62% and 77%, respectively. Those two states are sometimes confused. However, it is to be expected that such an effect occurs, since it is sometimes somewhat difficult to

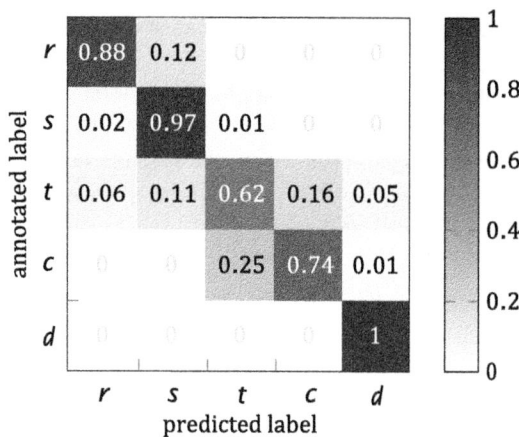

Fig. 2 Confusion matrix showing the ratio between predicted and annotated states, based on our nearest centroid classifier

distinguish between these movements even when annotating the data. All in all, for little computational effort, we obtain a well-interpretable method that allows quite reliable predictions. A weakness of our approach is that it is tightly bound to a pre-segmentation into individual strides. If we wish to handle more complex motion patterns, that involve multiple steps or strides or no clearly identifiable steps at all, we need broader methodological concepts. We shall focus on such in the following section.

4 Unsupervised Segmentation by Sparsest Cuts

The task discussed here is to identify and cluster consecutive segments of the high-dimensional time series $(y_i)_{i \in [T]}$ that correspond to similar movements in order to support their later manual classification as walking, climbing stairs, etc. For actual computations the y_i are first transformed so that for each coordinate the average is zero and estimating the covariance matrix over all coordinates results in the identity; no further preprocessing is applied.

For reasons to be explained later, the unsupervised approach for identifying data segments that correspond to similar movements in (y_i) will work on time window versions $\overline{y}_i = (y_i^\top, y_{i+1}^\top, \ldots, y_{i+m-1}^\top)^\top \in \mathbb{R}^{nm}, i \in [T - m + 1]$, where the lengths m will be selected algorithmically starting with $m = 1$ and for each choice of m the approach first forms a graph representation $G^{(m)} = (V^{(m)}, E^{(m)})$ of the time series $(\overline{y}_i)$. The nodes $v \in V^{(m)}$ of the graph may be thought of as arising from a discretization of the state space that the time series traverses. Concretely, each node $v \in V^{(m)}$ is a point $v \in \mathbb{R}^{nm}$. Each point $\overline{y}_i$ of the time series is mapped to the node $v \in V^{(m)}$ that is closest to it via $v^{(m)} : [T - m + 1] \to V^{(m)}, i \mapsto |\text{argmin}_{v \in V^{(m)}} \|v - \overline{y}_i\|,$ where $\|\cdot\|$ denotes the Euclidean norm. For the untransformed

original data with $m = 1$ this results in the Mahalanobis-norm. Ties in this assignment are assumed to appear with probability zero and need to be resolved consistently otherwise. The edge set is defined as

$$E^{(m)} = \left\{ \left\{ v^{(m)}(i), v^{(m)}(i + 1) \right\} : \right.$$
$$\left. v^{(m)}(i) \neq v^{(m)}(i + 1), i \in [T - m] \right\}.$$

Further, we introduce vertex weights $w_v^{(m)} = |\{i \in [T - m + 1] : v = v^{(m)}(i)\}|$ for $v \in V^{(m)}$ and edge weights $w_e^{(m)} = |\{i \in [T - m] : e = \{v(i), v(i + 1)\}\}|$ for $e \in E^{(m)}$, i.e., the vertex weights count the number of times the respective node is assigned and edge weights count the number of transitions from one node to the next within the time series. Geometrically the nodes are the centers of Voronoi cells and the weights count how often the time series *crosses the boundary* between two consecutive cells forming an edge. If edge weights are considered as edge multiplicities, the time series' walk through the graph forms an Eulerian trail by construction (with node weights counting loops). In particular, periodically repeated movements correspond to running through the same sequence of nodes repeatedly up to small deviations due to small changes in movement that cause slightly different node assignments for similar movements. Therefore repeated sequences should form strongly connected clusters within the graph that should be identifiable by weighted sparsest cuts (see, e.g., [9] for a discussion of several sparsest cut variants). A weighted sparsest cut of a graph $G = (V, E)$ with node weights $(w_v)_{v \in V}$ and edge weights $(w_e)_{e \in E}$ is any bipartition (V_1, V_2) of the node set $V = V_1 \dot\cup V_2$ minimizing

$$\frac{w_E(V_1, V_2)}{\min\{w_{V_1}, w_{V_2}\}},$$

where $w_E(V_1, V_2) = \sum_{e \in \{\{u,v\} \in E : u \in V_1, v \in V_2\}} w_e$ is the weight of the cut and $w_{V_1} = \sum_{v \in V_1} w_v$ (likewise wV_2) is the weight of the nodes in V_1 (in V_2). The induced subgraphs $G[V_1]$ and $G[V_2]$ will be called the shores of the cut.

Observation 1 *Consider a connected graph with positive node and edge weights. If in a weighted sparsest cut one shore consists of several connected components, then each component itself induces a weighted sparsest cut with the same weight ratio. In particular, there is a weighted sparsest cut that separates the nodes into two connected components.*

Proof Pick a sparsest cut and let it partition the node set V into V_1 and V_2 with (weighted) cut value $w_E(V_1, V_2)$. Suppose V_1 consists of two nonempty node sets V_1', V_1''

with $V_1 = V_1' \cup {}^\cdot V_1''$ and $E(V_1', V_1'') = \emptyset$. Put $a = wE$ $(V_1', V_2) > 0$ (by connectedness), $b = wV' > 0$, $c = wE(V_1'', V2) > 0$, $d = wV'' > 0$ with no edge linking the two. Suppose, for contradiction $\frac{a}{b} \neq \frac{c}{d}$, w.l.o.g., $\frac{a}{b} < \frac{c}{d}$, and let $g = w_{V_2}$. If $b > d + g$ then $b + d > g$ and $\frac{a}{\min\{b,d+g\}} = \frac{a}{d+g} < \frac{a+c}{g} = \frac{a+c}{\min\{b+d,g\}}$ contradicts the choice of the sparsest cut. On the other hand for $b \leq d + g$ the relation

$$\frac{a}{b} < \frac{c}{d} \Leftrightarrow \frac{a}{b} < \frac{a+c}{b+d} \qquad (1)$$

gives $\frac{a}{\min\{b,d+g\}} = \frac{a}{b} < \frac{a+c}{b+d} \leq \frac{a+c}{\min\{b+d,g\}}$ which again contradicts the sparsest cut property. Therefore $\frac{a}{b} = \frac{c}{d} = \frac{a+c}{b+d}$. Furthermore $b \leq d + g$, because for $b + d > g$ the relative cut value would then exceed the first if $b + d < g$ or the second if $d < b + g$. Therefore $(V_1', V_1'', V2)$ and $(V_1'', V_1', V2)$ are both weighted sparsest cuts. ∎

The use of sparsest cuts for successively splitting sequences of almost identi- cal movements is motivated by the following exactness result in the case of tracking sequences of unperturbed, well discernable movements/states that may be, and are mostly, carried out in periodic fashion.

Theorem 1 *Suppose a time series records a sequence* $(v_i)_{i \in [T]} \in V^{[T]}$ *within some discrete set V where this sequence itself is generated by a sequence of states* $(s_i)_{i \in [\ell]} \in S^{[\ell]}$ *for some finite set S and $\ell \in \mathbb{N}$, so that each appearance of a state $s \in S$ appends to (v_i) exactly the same sequence* $(v_1^s, \ldots, v_{n_s}^s) \in V^{n_s}$ *of length ns. Assume further each state appears at least as often consecutively to itself in (s_i) as it is adjacent to other states, and that there is a $1 \leq \underline{\kappa} \leq \min\{n_s : s \in S\}$ so that each subsequence of (v_i) of length $\underline{\kappa}$ belongs to exactly one $s \in S$ or to a unique transition from s to s' for some $s, s' \in S$. Given* $\underline{\kappa} \leq \kappa \leq \min\{n_s : s \in S\}$ *let the graph $G\kappa = (V\kappa, E\kappa)$ have nodes $V_\kappa = \{(v_i, \ldots, v_{i+\kappa-1}) : i \in [T - \kappa + 1]\}$ and edges* $\{\{(v_i, \ldots, v_{i+\kappa-1}), (v_{i+1}, \ldots, v_{i+\kappa})\} : (v_i, \ldots, v_{i+\kappa-1}) \neq (v_{i+1}, \ldots, v_{i+\kappa}), i \in [T - \kappa]\}$ *with node and edge weights counting their appearance in (v_i). For $s \in S$ let $V_K(s) = \{v \in V_K \text{ vis a subsequence of } (v^s, \ldots v^s, \ldots v^s, \ldots v^s)\}$.*

Given $\overline{V} \subset V\kappa$ call a state $s \in S$ uncut in the induced subgraph $G_K[\overline{V}]$ if $V_K(s) \cap \overline{V} \in \{\emptyset, V_\kappa(s)\}$ and call a cut $(V', \overline{V}\backslash V')$ in $G_K[\overline{V}]$ in $G_K[\overline{V}]$ separating states if every connected component in $G_K[V']$ and and $G_K[\overline{V}\backslash V']$ has all states uncut but contains at least one $V_\kappa(s) = $ for some $s \in S$ In a connected induced subgraph $G_K[\overline{V}]$ that has all states uncut and contains at least two states $(|\{s \in S : \emptyset \neq V_k(s) \subset \overline{V}\}| \geq 2)$ a sparsest cut cuts into an uncut state only if the state's weight accounts for at least half of the weight $w_{\overline{V}}$, otherwise it separates states.

Proof It will be convenient to collect for s, $s' \in S$ with $s = s'$ transition nodes $V_K(s, s') = \{v \in V_K : v \text{ is a subsequence of } (v_1^s, \ldots, v_{n_s}^s, v_1^{s'}, \ldots, v_{n_s}^{s'})\}$.

By the assumptions on (vi), $\underline{\kappa}$ and κ, the sets $V_K(s)$, $V_K(s, s')$ are disjoint for s, $s' \in S$ with $s = s'$ and cover all nodes of V_κ. W.1.og. $V_\kappa(s) = \emptyset$ for $s \in S$. For each state $s \in S$ let $h_s = |\{i \in [\ell - 1] : ((s_i = s) \vee (s_{i+1} = s)) \wedge (s_i = S_{i+1})\}|$ denote the number of times s is adjacent to some other state in (si) and let $h_s^c = |\{i \in [\ell - 1] : s_i = s = s_{i+1}\}|$ denote the number of times s follows s consecutively. By assumption $h_s^c \geq h_s \geq 1, w_{K(s)} \geq h^c \kappa$ and each consecutive execution appends at least κ nodes of $V_\kappa(s)$ (not necessarily all of them distinct), so $w_{V_K(s)} \geq h_s^c \kappa$.

Let $G_K[\overline{V}]$ be connected with all states uncut and let it contain at least two states. The sparsest cut value in $G_K[\overline{V}]$ may be bounded from above by $\frac{1}{k}$ as follows. Among all states $s \in S$ with $V_K(s) \subset V$ let $\underline{s}$ be one for which taking $\underline{V} = \{v \in \bar{V} : v \in V_K(s) \vee v \in V_K(s, s') \text{ for some } s' \in S \text{ with } V_K(s') \cap \bar{V} = \emptyset\}$ minimizes the ratio $w_{E_\kappa}(\underline{V}, \overline{V}\backslash\underline{V}) / \min\{w_{\underline{V}}, w_{\overline{V}\backslash\underline{V}}\}$. Let $h_{\underline{s}}'$ be the number of times $\underline{s}$ is adjacent to other states within V^-. Due to consecutive repetitions, appears at least $h_{\underline{s}}^c \geq h_{s_-} \geq h_{\underline{s}}'$ times with at least k $V_\kappa(s)$ - nodes κ $V\kappa(s)$-nodes within the time series, so $w_{\underline{V}} \geq \kappa h_{\underline{s}}^c$. Likewise, each state s in $\overline{V}$ contributes at least κ nodes within $\overline{V}$ for each time it is adjacent to $\underline{s}$, thus $w_{\overline{V}_\lambda \backslash V} \geq \kappa h_s$. Therefore the value of this cut ratio is bounded from above by $h_{\underline{s}} / \min\{kh_{\overline{s}}^c, kh_{\underline{s}}'\} \leq \frac{1}{k}$ which gives rise to the desired bound.

Suppose now $G_K[\overline{V}]$ has a sparsest cut $(V', \overline{V}\backslash V')$ that does not separate states. W.l.o.g., each shore $G_K[V']$ and $G_K[\overline{V}\backslash V']$ is connected by Observation 1.

Then there is one shore $\hat{V} \in \{V', \overline{V}\backslash V'\}$ with $\hat{V} \subset_{s,s' \in S, s=s'} V_K(s, s')\emptyset = (V_K(s) \cap \hat{V}) = V_K(s)$ for some $s \in S$. In words, the sparsest cut must separate a set of transition nodes or cut an edge running between nodes that uniquely belong to the same state.

Consider first a cut that separates a set of transition nodes. For any pair of distinct states $s, s' \in S$ with $V_k(s, s') = \emptyset$ these transition nodes form a connected component in the graph induced by all transition nodes, because by assumption on κ and the time series there is a node in $V\kappa(s)$ directly preceding and one node in $V_\kappa(s')$ directly following the nodes of $V_\kappa(s, s')$. Any subsequence of nodes in $V_\kappa(s, s')$ consists of at most κ 1 nodes. Thus, each transition contributes two cut edges and a weight of at most κ 1, yielding a cut ratio of value at least $2/(\kappa$ 1). By Obs. 1, the upper bound above and (1) an isolated node set of this type cannot be contained in a sparsest cut, i. e., any such node set also includes an adjacent node that uniquely belongs to a state. Therefore either a full state is included or a state is cut.

Assume now that $(V', \overline{V} \backslash V')$ cuts a state $s \in S$ with $V_K(s) \subset \overline{V}$ whose weight accounts for at most half of the weight of $\overline{V}$. W.l.o.g. let $w_{V'} \le w_{\overline{V} \backslash V'}$. Put $\omega_s = w_{V_K(s)}$ for the weight of s, $\omega_s = w_V \cap V_K(s)$ for the contribution to the lighter shore $G_K[V']$, and let $h'_s = w_{E_K}(V \cap' V_K(s), V \langle V_K(s))$ denote the weighted number of adjacent nodes in $G_K[V']$. By assumption ω_s is at most half the weight of $\overline{V}$, so h_s / ω_s is an upper bound on the sparsest cut within $\overline{V}$. Let $a' = w_{E_K}(V', \overline{V} \backslash V')$ denote the value of the corresponding weighted cut in $\overline{V}$ and $b' = wv'$ the node weight of the smaller shore in $\overline{V}$ so that a'/b' is the sparsest cut value in $\overline{V}$. Let $V'' = V' \backslash V_K(s)$, $a'' = w_{E_K}(V'', \overline{V} V'')$ $b'' = w''_v$ 'so that a''/b'' is the corresponding cut ratio where instead of cutting s only the h'_s edges into s are cut (the nodes of s are shifted so that s is part of the larger shore; $V'' = \emptyset$ because otherwise $(V_K(s), \overline{V} \cdot \backslash \cdot V_K(s))$ renders a sparser cut). Note that $a' \ge a'' - h'_s + 2h^c_s$, because in each consecutive appearance of s at least two edges must be cut. Furthermore $b' = b'' + \omega'_s$. Then

$$\frac{a'' - h'_s + 2h^c_s}{b'' + \omega'_s} \frac{a'}{-} h_{\underline{s}}. \le b' \le \omega_s.$$

Due to $h_s \le h_s \le h^c_s$ (the latter by assumption), $\omega'_s < \omega s$ and (1) we have

$$\frac{a'' + h_s}{b'' + \omega_s} < \frac{a'' + 2h^c_s - h'_s}{b'' + \omega'_s} \le \frac{h_s}{\omega_s} \Rightarrow$$
$$\frac{a''}{b''} \Rightarrow \frac{h_s}{\omega_s} < \frac{2h^c_s - h'_s}{\omega'_n} \Rightarrow \frac{a''}{b''} < \frac{a'}{b'}.$$

This is in contradiction to $\frac{a'}{b'}$ giving the weighted sparsest cut value. ∎

This theorem justifies to partition the node set V_k recursively by sparsest cuts in order to separate different states.

It is worth noting, that a state that corresponds to a cycle without self-intersections in G_k will be cut into two connected components. If there are h direct successions of the state within the time series these two parts will appear interleaved for h times when cut for the first time. This should make detection of states within the clusters easier.

The structural result motivates the application of sparsest cut to graphs whose nodes correspond to a discretization of rolling time windows of the time series and whose edges reflect the transitions between these. Even for rather small κ forming G_k exactly as in the theorem might lead to huge vertex sets V_k and even small data perturbations would cause many nodes to be visited only once. A more practicable approach seems to be to choose a manageable size k of the graph in advance and to compute for each desired time window length m a new discretization $V^{(m)}$ with $k = |V^{(m)}|$ nodes. Nested sparsest cut is then applied to the graph $G^{(m)}$ ($\kappa = 1$) whose nodes are built for time window length m.

The choice of the rolling time window length m should be guided by the conditions of the theorem on the choice of κ. On the one hand m should be large enough to render the node-state correspondence unique (each time window has to contain some characteristic part), on the other it will be difficult to separate states whose duration is significantly below m. In this experimental study we follow a rather involved approach of solving the problem repeatedly for successively larger m. For the current academic purposes this allows to study the development of the segmentation as m grows. Starting with $m = 1$ we compute a discretization $V^{(m)}$ with k nodes via k-means, compute approximately nested sparsest cuts on the corresponding graph $G^{(m)}$ and then choose the next larger m by evaluating the statistics of nodes of the current $V^{(m)}$ that reappear regularly in the time series. In the nested approach the next connected component to partition is selected as the one which induces the longest uninterrupted time window in the time series.

Finding a sparsest cut in a graph is an NP-complete problem, hence we resorted to heuristics. A possible starting point could be a semidefinite re- laxation of sparsest cut [2]. In our experiments it was easier to work with a semidefinite node-weighted bisection relaxation [16], where the smaller side has to have at least one fifth of the total weight. The relaxation was solved by the spectral bundle method [7]. For each intermediate semidefinite solution and for several random Goemans–Williamson-roundings [5] the associated sorting of the nodes was used as seed for improvement heuristics searching for sparsest cuts.

The following experiments are intended to illustrate the relevance of the theorem even for practical data and to provide some intuition on the development and dependence of the partitioning on k and m. In particular, the code was run for $k \in \{200, 300, 500\}$. For $k = 200$ it successively generated the time window sizes $m = 1, 17, 61, 120$, for $k = 300$ it produced $m = 1, 61, 117$ and for $k = 500$ $m = 1, 62, 119$. Figure 3 displays the development of the nested bisection until 10 partitions were reached for $m = 1, m \approx 60, m \approx 120$. In each plot the horizontal axis corresponds to the time line of the time series, the vertical steps illustrate the effect of the successive development of the partitions on the classification of the time series elements. In this, a separate new color is assigned to each new connected component of the partition (the larger shore keeps the previous color) and all time steps assigned to these nodes get the nodes' color.

The plots suggest that the dependence on the rolling time window size m is significantly larger than on the number of nodes k. For small m the visual feedback seems correct, but states do not seem to be separated well. Quite differently,

Fig. 3 Development of nested approximate sparsest cut for k-means nodes where $k \in \{200, 300, 500\}$, the rolling time window size m is generated automatically

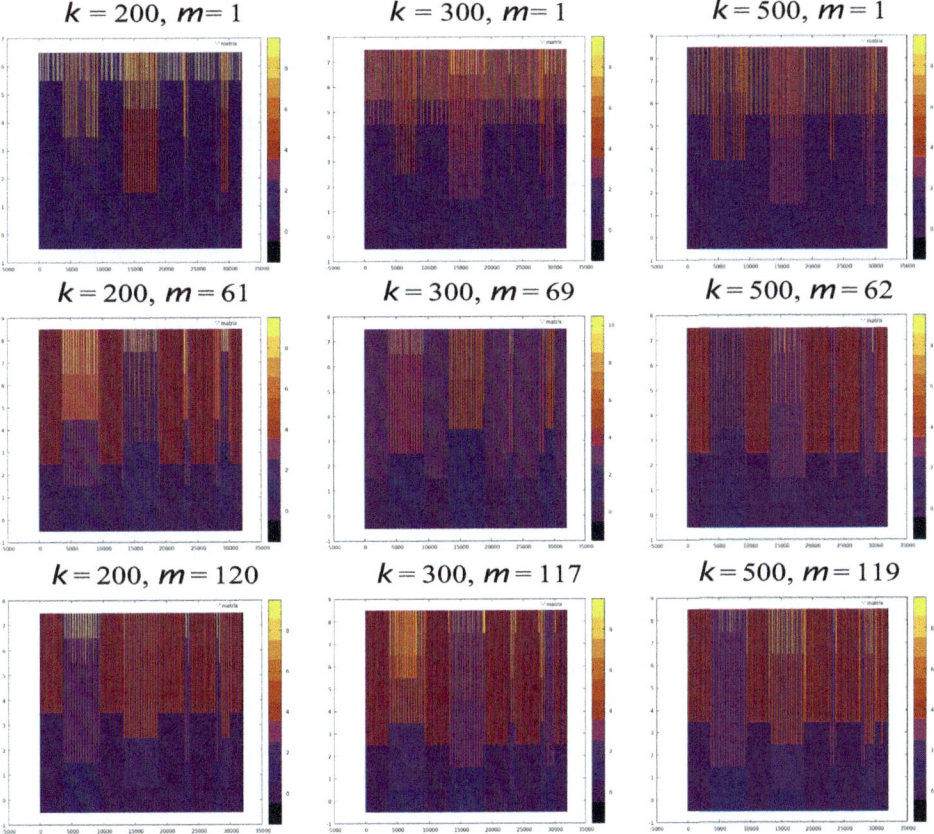

for large m the sparsest cut first separates the main states and subdivides them only later. This corresponds well to the motivating theorem, which requires the time window size to be large enough so that nodes uniquely correspond to states or transitions between two states.

Considering the specific data background for the case $m \approx 120$ we see that for $k = 200$ the code first splits off going down the stairs and then going up, for $k = 300$ it first splits off walking against using the stairs and then going down the stairs, for $k = 500$ it again first splits off walking against using the stairs but then going up the stairs. Later subdivisions seem to correspond to splitting the states into recursive substates and finally splitting the states of walking or stair climbing themselves, which results in periodic patterns. So while the major decisions are consistent, the order is somewhat arbitrary. This is to be expected in view of the difference in k with randomized k-means as discretization approach and the randomized approximate computation of sparsest cuts in the resulting graph.

In summary, for rather large m the approach seems promising for identifying the major states correctly quite independent of the choice of k. For a detailed decomposition, however, there might still be a long way to go.

5 Concluding Remarks

Theorem 1 as well as the empirical insights from Fig. 3 show that the unsupervised approach presented in Sect. 4 is able to detect states in high dimensional time series. In practice, this allows to reveal relevant recurring movement patterns one can expect to identify. For example, when analyzing gait data for health applications or encounters in public spaces, this can help tailor data analysis systems to the specifics of individual persons or machines. In this sense, the supervised and unsupervised techniques we present do not exclude each other. The unsupervised approach makes relevant motion pat- terns available for statistical analysis. The supervised approach, while being interpretable and well-suited for use in real-time scenarios, relies on certain knowledge about the scenario at hand. In Fig. 2, for example, the classes to be distinguished were already known, and we began our analysis by just assuming that the stream of data can be presegmented into individual strides. The information actually needed here is what our unsupervised methods are designed to provide.

Acknowledgements Funded by the Deutsche Forschungsgemeinschaft (DFG, German Research Foundation)—Project-ID 416228727—SFB 1410.

Conflicts of Interest The authors declare no competing interests.

References

1. Aqueveque, P., Germany, E., Osorio, R., & Pastene, F. (2020). Gait segmentation method using a plantar pressure measurement system with custom-made capacitive sensors. *Sensors, 20*, 656.
2. Arora, S., Rao, S., & Vazirani, U. V. (2009). Expander flows, geometric embeddings and graph partitioning. *Journal of the ACM, 56*(2, Article 5).
3. Cover, T., & Hart, P. (1967). Nearest neighbor pattern classification. *IEEE Transactions on Information Theory, 13*(1), 21–27.
4. Feldman, J., & Rojas, R. (2013). *Neural networks: A systematic introduction.* Springer.
5. Goemans, M. X., & Williamson, D. P. (1995). Improved approximation algorithms for maximum cut and satisfiability problems using semidefinite programming. *Journal of the ACM, 42*, 1115–1145.
6. Gupta, M., Gao, J., Aggarwal, C. C., & Han, J. (2014). Outlier detection for temporal data: A survey. *IEEE Transactions on Knowledge and Data Engineering, 25*(1), 1–20.
7. Helmberg, C., & Rendl, F. (2000). A spectral bundle method for semidefinite programming. *SIAM Journal on Optimization, 10*(3), 673–696.
8. Hill, M., Hoena, B., Kilian, W., & Odenwald, S. (2016). Wearable, modular and intelligent sensor laboratory. *Procedia Engineering, 147*, 671–676.
9. Hochbaum, D. S. (2013). A polynomial time algorithm for Rayleigh ratio on discrete variables: Replacing spectral techniques for expander ratio, normalized cut and Cheeger constant. *Operations Research, 61*(1), 184–198.
10. Kessy, A., Lewin, A., & Strimmer, K. (2018). Optimal whitening and decorrelation. *The American Statistician, 72*(4), 309–314.
11. Kopnarski, L., Lippert, L., Rudisch, J., & Voelcker-Rehage, C. (2023). Predicting object properties from movement kinematics. Brain Informatics, 10 (26).
12. Krumm, D., Sanseverino, G., & Odenwald, S. (2022, April 29). Plantar pressure and inertial data of a 7.5 km hike. https://doi.org/10.17605/OSF.IO/63HJS
13. Manning, C. D., Raghavan, P., & Schutze, H. (2008). *Introduction to information retrieval.* Cambridge University Press.
14. Phinyomark, A., Osis, S., Hettinga, B. A., & Ferber, R. (2015). Kinematic gait patterns in healthy runners. *Journal of Biomechanics, 48*, 3897–3904.
15. Poiesi, F., & Cavallaro, A. (2015). Predicting and recognizing human interactions in public spaces. *Journal of Real-Time Image Processing, 10*, 785–803.
16. Poljak, S., & Rendl, F. (1995). Nonpolyhedral relaxations of graph-bisection problems. *SIAM Journal on Optimization, 5*(3), 467–487.
17. Powers, R., Etezadi-Amoli, M., Arnold, E. M., Kianian, S., Mance, I., Gibiansky, M., Trietsch, D., Alvarado, A. S., Kretlow, J. D., Herrington, T. M., Brillman, S., Huang, N., Lin, P. T., Pham, H. A., & Ullal, A. V. (2021). Smartwatch inertial sensors continuously monitor real-world motor fluctuations in Parkinson's disease. *Science Translational Medicine, 13*.
18. Prakash, C., Kumar, R., & Mittal, N. (2018). Recent developments in human gait research. *Artificial Intelligence Review, 49*, 1–40.
19. Rosenbaum, D., & Becker, H.-P. (1997). Plantar pressure distribution measurements. Technical background and clinical applications. *Foot and Ankle Surgery, 3*, 1–14.
20. Scholkopf, B., & Smola, A. (2018). *Learning with kernels: Support vector machines, regularization, optimization, and beyond.* MIT Press.
21. Shalin, G., Pardoel, S., Lemaire, E. D., Nantel, J., & Kofman, J. (2021). Prediction and detection of freezing of gait in parkinson's disease from plantar pressure data using long short-term memory neural-networks. *Journal of NeuroEngineering and Rehabilitation, 18*, 1–15.
22. Truong, C., Oudre, L., & Vayatis, N. (2020). Selective review of offline change point detection methods. *Signal Processing, 167*, Article 107299.

Radar-Based Gait Analysis

Wolfgang Kilian, Aline Püeschel, Stefanie Doetz,
and Stephan Odenwald

Abstract

Patients with neurological diseases often suffer from motor disorders leading to changes in gait pattern and gait disorders. In order to assess the progression of the disease and the effectiveness of therapies, gait analyses are conducted. However, these are often associated with considerable time efforts for both patients and medical staff and require a high amount of space. In this paper, a system based on radar technology is evaluated for gait analysis as an alternative measurement system to the previously used pressure distribution measuring devices and optical motion capture systems. The gait pattern of a test persons were extracted from data and five gait parameters were derived. During the evaluation, the sensor position and height has been varied, and different methods of analysis have been considered. The validity of the system was investigated by quantitatively comparing the measured biomechanical parameters with those recorded by a pressure distribution measurement system. The results show that it is possible to measure medically relevant gait parameters using radar technology.

W. Kilian · A. Püeschel (✉) · S. Odenwald
Department of Sport Equipment and Technology, Technische Universität Chemnitz, Chemnitz, Germany
e-mail: alpu@hrz.tu-chemnitz.de

W. Kilian
e-mail: wkilian@hrz.tu-chemnitz.de

S. Odenwald
e-mail: odenwald@hrz.tu-chemnitz.de

S. Doetz
Technische Universität Chemnitz, Chemnitz, Germany
e-mail: stdo@hrz.tu-chemnitz.de

Keywords

Human gait · Radar · Gait analysis · Human movement assessment

1 Introduction

According to an estimation from 2020, approximately 2.8 million people worldwide live with Multiple Sclerosis (MS) [29]. This ranks MS among the most common neurological diseases in adults between the ages of 20 and 40 [27, 32]. To assess the effectiveness of a therapy as well as to evaluate the progress, various diagnostic procedures are used. An essential part is the examination of various gait parameters and gait analysis. Although gait patterns are individual, gait parameters, such as stride length, cadence, track width and speed can be used to quantify gait characteristics [12]. Changes in gait pattern of MS patients have already been detected in the early stages using gait parameters [1, 16, 23, 26]. Characteristic changes for instance can include a shortened stride length, lower cadence, greater track width and lower speed [26]. To tackle these changes more or less elaborate measuring procedures are known. Current analysis devices are divided into wearable and non-wearable systems. Wearable sensors require precise placement and are uncomfortable, which can itself impact walking parameters [9]. Markerless motion capture systems on the other side using visible or infrared light, such as Microsoft Kinect, have been widely investigated [7]. However, due to poor lighting conditions or clothing, there can be impairments [31]. Furthermore, previous gait analysis systems are location-based and are not available to general practice. The available systems require professional supervision and support, as calibration is often essential or even markers have to be attached to anatomical locations. The high time expenditure of patients and experts at the examination centres, especially in the field of MS diagnostics, can negatively affect results due to fatigue or weakness and also ties up human and spatial resources, which in turn leads to a reduced capacity of the centres. To avoid these disadvantages, a radar-based system was investigated for markerless gait analysis.

2 State of the Art

2.1 Radar Technology

Radar technology includes methods for detecting and localising objects and determining their parameters using electromagnetic waves [4, 14]. The basic structure of a radar system is shown in Fig. 1. Electromagnetic waves are emitted by an aerial (primary signal). When this radiation hits an object, part of the signal is reflected and subsequently detected by the receiver as secondary signal [4]. This signal contains information about distance, angle and speed of detected objects. By measuring the Doppler effect, object parameters can be derived [15]. Based on the frequencies of the primary and secondary signals as well as the speed of light, the radial velocity of a measuring point can be determined [6]. Due to micro-movements of individual segments, frequency modulation of the reflected signal occurs, which leads to secondary bands around the Doppler frequency and is referred to as micro-Doppler effect [2]. As a result, a frequency range with significant time dependency is generated [6]. A disadvantage of the system is the inability to display a direct visual representation of the gait pattern [5]. Nevertheless, radar technology offers some advantages compared to other methods for gait analysis. It enables measurements independent of prevailing light conditions. The radar system is not location-bound and can easily be transferred between different rooms. This offers the potential for simple implementation in the clinical setting. Since radar signals penetrate clothing, the system provides more comfort and privacy for the patient. There is also no

need for markers that can lead to changes in gait patterns, may get lost and require professional placement [15, 30].

For many years, Continuous Wave Radar (CW-Radar) has been state-of-the-art for monitoring human locomotion [10] and has already demonstrated success in analysing individual movements [19]. Moving objects can be detected by changing the reflected signal [10]. CW-Radar are thus able to detect temporal gait parameters. However, it is not possible to determine the distance of objects. Frequency Modulated Continuous Wave (FMCW) radars compensate for this disadvantage by periodically changing the frequency of the main signal, thus enabling the simultaneous measurement of velocity and distance data [15].

2.2 Radar in Gait Analysis

The application of radar technology in motion analysis mainly focuses on recognition and classification of gait [15]. Early research approaches tended to rely on deriving simple gait parameters [8], detecting the presence of a person or distinguishing human and animal gait pattern [5, 17]. More recently, the main goal is to extract temporal parameters such as cadence, speed and gait phases [18]. Advanced systems allow the differentiation of walking with and without arm-swinging [30] as well as holding objects [13, 22]. Research has been conducted on fall events in elderly persons [10, 13, 21] or pathological gait patterns in clinical settings [24, 25, 33]. Zhang derived the gait parameters walking speed, cadence, cycle duration and symmetry of movement of both legs. The accuracy of the results can be compared with those of a video camera system [33]. Sun et al. were able to derive further movement parameters such as double step length as well as physical parameters such as leg length and body height by extracting body components [28]. Using a CW radar, it is possible to classify gait into normal, pathological and assisted

Fig. 1 Basic structure of a radar system

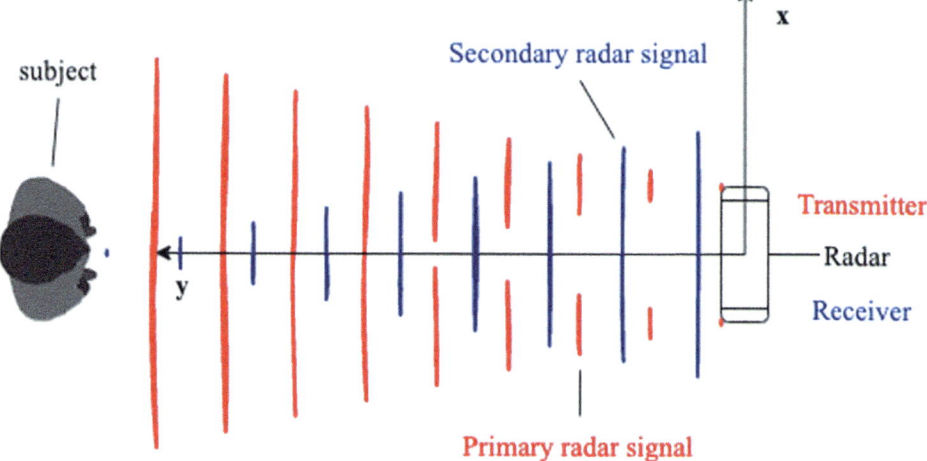

walking with an accuracy of 93.8% [25]. Individual body segments such as foot, leg and torso were identified, which is the basis for recording individual joint velocities [20]. With radar technology, it is possible to ex- amine complex properties of the human gait and to capture gait parameters. Thus, gait disorders can be recognised and gait can be distinguished from other translational gait movements such as assisted locomotion. In future, this technology offers the potential to monitor elderly people in their living environment. Moreover, it enables a prognostic diagnosis of movement restrictions of the human gait [24]. The aim of this study is to extract various spatio-temporal parameters from gait cycles using a simple and inexpensive radar sensor. In contrast to previous work, no commercial sensor system is used, but rather a platform that can be flexibly adapted to different requirements. This is particularly significant with regard to the application in the clinical environment. The evaluation of the measurement data should also be designed to be as comprehensible as possible.

3 Methods

3.1 Experimental Setup

In this paper, the FMCW radar AWR1642 BoosterPack™ with a frequency band of 77–81 GHz and a 4Rx, 2Tx antenna configuration from Texas Instruments™ was utilised. TI Gallery App was used for controlling and configuring the sensor. For the visual documentation and control of the experiments, the gait analyses were recorded with a GoPro HERO7 Black. The video data was not explicitly analysed, but merely reviewed to identify any inconsistencies in the radar data and to synchronise the radar system with Zebris FDM system. The measurements were conducted both in a living environment and in the laboratories of the Department of Sports Equipment and Technology at Chemnitz d2azsUniversity of Technology. All measurements are conducted on healthy subjects without

known neurological or orthopedic disorders. The focus of this study is the fundamental suitability of using the selected radar system for gait analysis. The initial investigation was to examine whether it is possible to assess the gait pattern of healthy people on the basis of selected parameters. Therefore, the test persons may not suffer from any diseases that influence the gait pattern. Within the framework of the work, measurements are performed on five different subjects of which three were female and two male aged 26–41 years. Different sensor positions and heights were investigated as part of the experimental testing. Furthermore, a comparative study with a pressure distribution measurement plate from Zebris Medical was conducted. The components of the setup are shown in Fig. 2.

A total of 63 measurements were conducted and evaluated with the radar system. The measurements can be subdivided into test series A–C. Three measurements were recorded and evaluated for each test situation and configuration in A and B. For test series A, these were divided into slow/fast walking + sensor in front of/behind the subject, small/large steps + sensor in front of/behind the subject. For test series B, the sensor is placed at the heights mentioned in Table 1, whereby based on the results of test series A, the sensor is always placed in front of the subject and the participant approaches it. Based on the findings of test series A and B, in C the sensor is placed at ground level and in front of the subject. A total of 3 m of the 7 m track is instrumented with pressure measurement plates. Each measuring plate has a sensor area of 1440, 560 mm. A total of 11264 sensors are distributed across the surface. The sampling frequency was 100 Hz. Five measurements are conducted on each of three different participants. Table 1 shows an overview of all test series.

The same configuration of the system is used within the individual test series and is shown in Table 2. An intermediate increase of the frame rate resulted in a computer error and was reseted.

Fig. 2 Experimental setup (**a**): 1—radar sensor, 2—camera GoPro Hero7, 3—PC; **b** height adjustable stand; **c** walking track

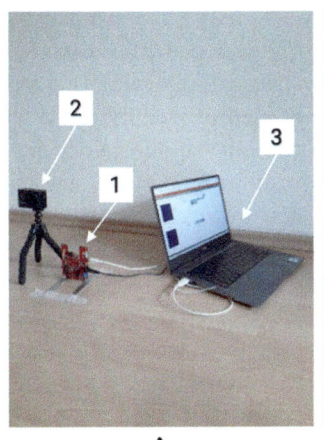

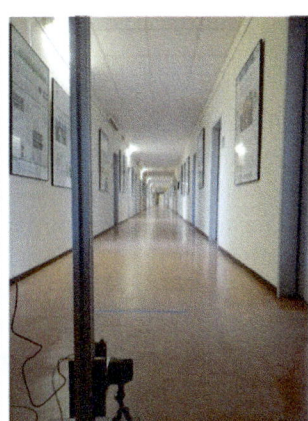

A   B   C

Table 1 Overview of all test series

Series	Numbers	Description	Distances
A	1–24	Basic suitability for gait analysis (different parameters in gait velocity, step length and position of the sensor)	Walking distance of 8.5 m, distance to radar of 1 m
B	1–24	Investigation of the influence of the radar height	Walking distance of 14 m, distance to radar of 1 m
C	1–15	Comparing measurements of the radar system and an established system for gait analysis (Zebris FDM).	Walking distance of 7 m, distance to radar of 1 m, instrumented track of 3 m

Table 2 Settings of the radar system

Description	Setting
Frame rate (fps)	10
Range resolution (m)	0.039
Maximum unambiguous range (m)	14.99
Maximum radial velocity (m/s)	2.01
Radial velocity resolution (m/s)	0.13

3.2 Data Processing and Analysis

The sensor captures two spatial coordinates and the velocity of all detected points. By calculating the distance of the detected objects, the exemplary graphic shown in Fig. 3 emerges. It includes objects without changing their distance to the sensor over time. Looking at their speed, it is obvious that these are stationary points emanating from objects such as door frames. Consequently, these objects are not relevant for gait analysis and are not considered. Furthermore, there is a point cloud (red area) in Fig. 3 that approaches the sensor over time. These measurement points reflect the subject moving towards the sensor. Using the workflow shown in Fig. 4, significant points are identified within the gait cycle. These are the basis for determining gait parameters.

The evaluation of the measurements is realised with MATLAB R2022a and is shown schematically in Fig. 4. In the first part ("blue" area), the detected measurement points are calculated and filtered. Subsequently, two different evaluation methods are used to generate distance-time profiles for the derivation of gait parameters ("red" and "green" area).

Despite previous filtering in evaluation method A1 (red area in Fig. 4), multiple objects were detected during one timestamp. Using the minimum function, the closest value to the radar sensor is selected to reduce the values per time unit to one. This method detects the front leg or foot and avoids covering relevant points with the other limb. Medians are used to smooth the graph. The determined plot with the filtering

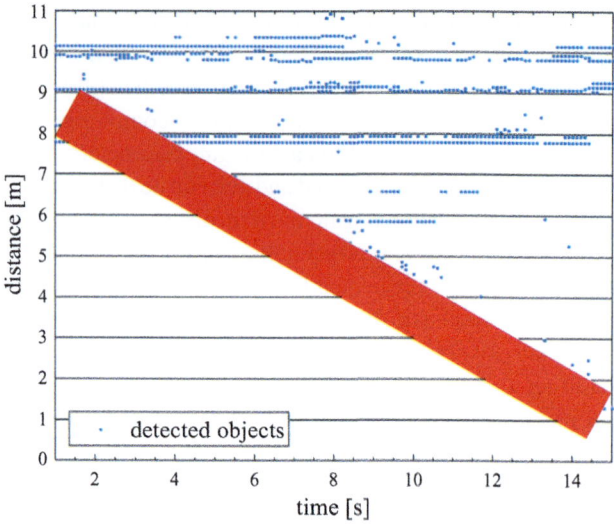

Fig. 3 Display the distance of all detected objects of a measurement over time

of A1 is shown in Fig. 5. It represents the gait cycle of the subject. To identify individual steps from the measurement data, distinctive points in the gait cycle are detected. For this purpose, the first derivation of the function is calculated as shown in Fig. 5. The extreme values of this function occur at regular intervals and are used to determine the timestamps of individual steps. Depending on the direction of the gait, maxima or minima are evaluated. The associated distances of the respective timestamps are used to identify gait parameters. Evaluation method A2 includes an additional processing of the measured values. All detected objects within a distance of 0.6 m around the graph presented in Fig. 5 are considered for further evaluation. As a result, reflections and other obstacles are eliminated. Based on the reduced number of measurement points, the object of minimal distance to the radar sensor is determined for each point in time.

4 Results

Figure 5 shows the distance-time curve derived with the data analysis presented. This has a step-like progression, which is used to identify individual steps. To determine a salient point during the gait cycle, the local minimum is identified with the help of the derivation according to time. The steps determined by this method, which are used to calculate the individual walking parameters are shown in Fig. 5 as well.

Within the research, step duration, step length, stride length, step velocity and cadence were evaluated. In preliminary tests, it was demonstrated that it is possible to record a shortened step length. During the measurement the test person was therefore asked to take small or large steps. Based on the step duration a change in the test person's walking

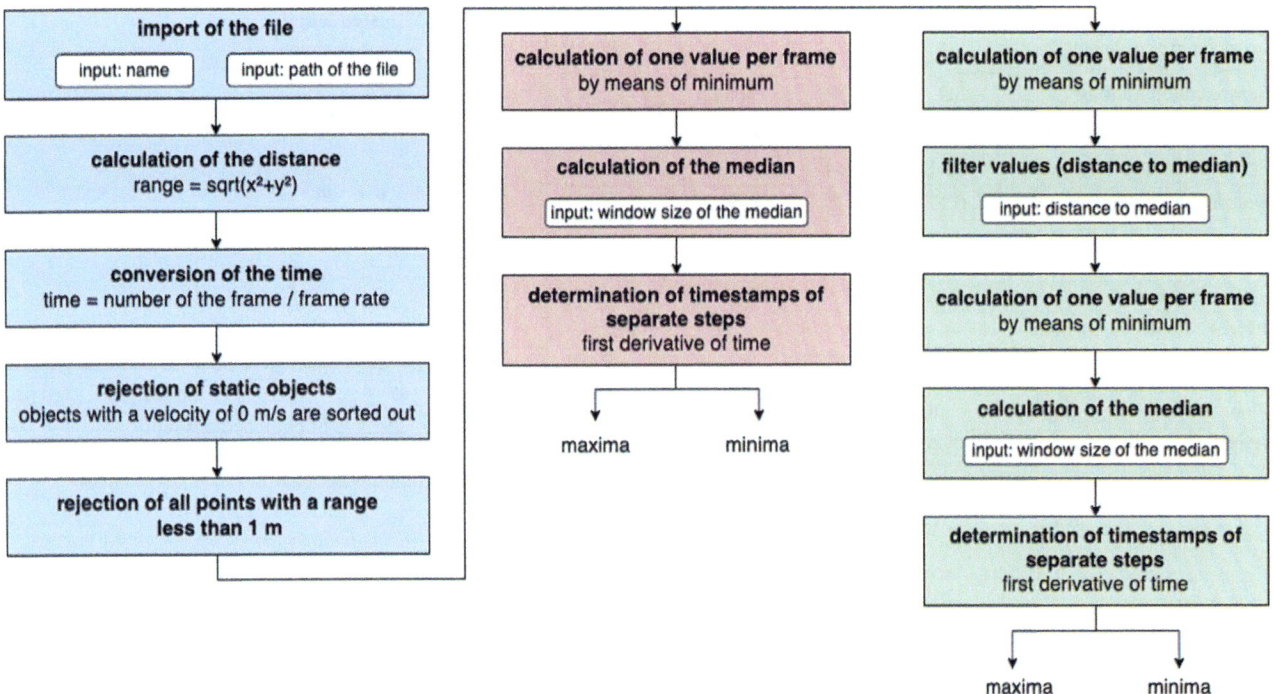

Fig. 4 MATLAB workflow

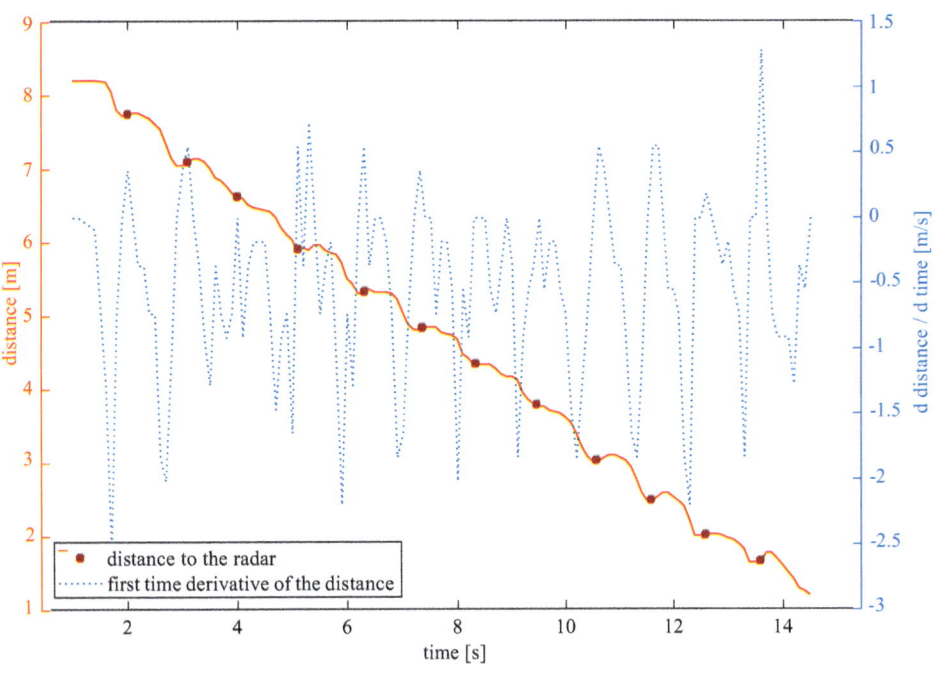

Fig. 5 Plot of evaluation method A1 and the first derivative to identify significant points in gait cycles for identification individual steps

speed could be detected with the system. The investigations indicated that the sensor should be placed in front of the test person, as the scattering based on the standard deviation, interquartile range (IQR) and span is lower compared to other positions. In addition, evaluation method 1 (red area in Fig. 4) also has lower scatter measures and is used for further procedure. In Table 3 SD, IQR and span for test series A are shown.

Based on the investigations, the sensor was placed at floor level, as this allows better step detection. The other sensor heights such as hip height, knee height and half knee height proved to be unsuitable. At the hip height of the test person, it was not possible to derive gait parameters with the developed data processing because no clear identification of the steps was possible. The data at hip level show a continuous straight line from which it is possible to narrowly derive the velocity

Table 3 Statistical evaluation of the sensor configurations in test series A concerning stride length

Sensor configurations	SD	IQR	Span
Sensor in front + method 1	**0.08**	**0.11**	**0.24**
Sensor in front + method 2	0.11	0.14	0.40
Sensor behind + method 1	0.17	0.21	0.53
Sensor behind + method 2	0.18	0.21	0.56

Table 4 Result of test series B

Height	SD	IQR	Span
Floor	**0.13**	**0.17**	**0.48**
Half-knee	0.17	0.21	0.67
Knee	0.15	0.17	0.59
Hip	n/s	n/s	n/s

Table 5 Gait parameters

Gait parameter	ICC
Step duration	0.258
Step length	**0.708**
Stride length	**0.706**
Step velocity	0.249

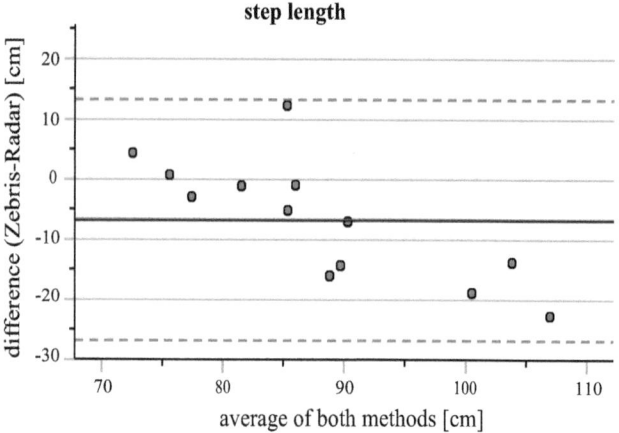

Fig. 6 Bland–Altman plot of step length

of the moving object but not single steps. The measurements at knee height and half knee height showed larger scatter measures than at floor level, which is shown in Table 4. This is supposed to be due to faulty step recognition.

Comparative measurements were carried out with pressure distribution measuring plates and the radar system. Overall 13 measurements could be evaluated. The intraclass correlation coefficient (ICC) is used to assess the correlation of the two measurement methods and determined for the individual gait parameters (see Table 5). According to Koo and Li [11], the two measurement methods show insufficient correlation regarding step duration and step velocity. Based on the ICC, the two measurement methods showed moderate correlation for the parameters step length and stride length according to Koo and Li and good correlation according to Cicchetti [3].

The Bland-Altman plot of step length (see Fig. 6) indicates that the average difference between the two measurement methods is 7.7%. Ten of the 13 measurements have a negative difference. This result implies that the radar system measures a longer step length than the Zebris Medical Force Distribution Measurement (ZebrisFDM). In this example, the distance between the upper and lower Limit of Agreement (LoA) is 40.10 cm.

5 Discussion

Independent of the evaluation method or the position of the radar sensor a basic classification of slow and fast gait pattern based on the duration of steps is possible. Considering a single gait analysis, it is obvious that the duration of the individual steps is affected by a certain variance. Therefore, it is necessary to increase the resolution of the step duration from 10 fps in order to achieve an improvement in precision. The radar sensor provides a maximum frame rate of 30 fps. However, this was not possible with the currently used technical equipment.

During the investigations, it was demonstrated that it is possible to differentiate between gait pattern with small and large steps. However, the individual gait analyses showed a partly considerable variance within a series of measurements. This may be caused by the test person himself, who varies his step length. To examine this effect, simultaneous comparative measurements were conducted with a gold standard. Furthermore, the variance of the measured values may be due to the applied data evaluation. The derived gait curve may already be imprecise since relevant points have not been considered. Another problem is the potential incorrect identification of the relevant points and thus of the steps.

The comparative investigations with both systems indicated that the radar system has a poor agreement concerning some gait parameters. One of the reasons are different approaches applied to determine the stride length. Radar systems determine step length based on the distance of individual steps from the radar sensor. Using ZebrisFDM, the step length is determined by the distance between the individual steps. In addition, the active measuring length of the systems must also be considered. Since the ZebrisFDM offers only an active measuring length of 3 m another system for comparison should be used. Likewise, the minimum distance due to the radiation angle of the radar must be ensured.

Another aspect for the lack of agreement based on the ICC between the two systems is the wearing of shoes during the measurements. Shoes were worn during examinations to obtain a realistic picture of patients' gait patterns. However, this can lead to deviations in the parameters of the pressure measurement plates. Therefore, the manufacturer recommends barefoot gait analysis. The measurements on test person 3 showed particularly pronounced differences between the two systems. Since each test subject wore shoes of his or her choice, this could be an explanation. The evaluation of the measurements is based on statistical values such as the standard deviation, IQR, ICC and Bland-Altman diagrams. In some cases, the measurements showed a pronounced variability. However, the significance of these values is limited by the small number of measurements conducted. The significance of further studies increase with the number of participants and test subjects. The dispersion of the characteristic in the population could therefore enhance agreement and validity.

6 Conclusion

The investigations confirmed that with the radar system a basic assessment of human gait pattern based on the parameter's step duration, step length, stride length, step velocity and cadence is possible within the limitations of the current system-setup. It is necessary to examine the possible extent for detecting gait changes with the present system. In this respect, it is important to check which other gait parameters can be assessed with the help of the system. Special attention on differences between the two sides of the body and the movement of the upper body is needed particularly for implementation in clinical settings. Improvements in the evaluation methodology are necessary, as the results revealed a potential for optimisation in the step recognition. Initial testing in clinical environment should follow to improve the usability of the system.

It remains to be examined whether motor deficits of neurological patients can be recorded with the help of the system. Furthermore, a test in a clinical environment is needed. Additional gait parameters need to be evaluated, especially whether it is possible to detect differences between the right and left side of the body.

Acknowledgements Funded by the Deutsche Forschungsgemeinschaft (DFG, German Research Foundation)—Project-ID 416228727—SFB 1410.

References

1. Benedetti, M. G., Piperno, R., Simoncini, L., Bonato, P., Tonini, A., & Giannini, S. (1999). Gait abnormalities in minimally impaired multiple sclerosis patients. *Multiple Sclerosis (Houndmills, Basingstoke, England), 5*(5), 363–368.
2. Chen, V. C., Li, F., Ho, S.-S., & Wechsler, H. (2006). Micro-doppler effect in radar: Phenomenon, model, and simulation study. *IEEE Transactions on Aerospace and Electronic Systems, 42*(1), 2–21.
3. Cicchetti, D. V. (1994). Guidelines, criteria, and rules of thumb for evaluating normed and standardized assessment instruments in psychology. *Psychological Assessment, 6*(4), 284–290.
4. Detlefsen, J. (1989). *Radartechnik: Grundlagen, Bauelemente, Verfahren, Anwendungen.* Springer.
5. van Dorp, P., & Groen, F. C. A. (2003). Human walking estimation with radar. *IEE Proceedings—Radar, Sonar and Navigation, 150*(5), 356.
6. Geisheimer, J. L., Marshall, W. S., & Greneker, E. (2001). A continuous-wave (CW) radar for gait analysis. In *Conference Record of Thirty-Fifth Asilomar Conference on Signals, Systems and Computers (Cat.No.01CH37256)* (Vol. 1, pp. 834–838). IEEE.
7. Gholami, F., Trojan, D. A., Kovecses, J., Haddad, W. M., & Gholami, B. (2017). A Microsoft Kinect-based point-of-care gait assessment framework for multiple sclerosis patients. *IEEE Journal of Biomedical and Health Informatics, 21*(5), 1376–1385.
8. Hornsteiner, C., & Detlefsen, J. (2008). Characterisation of human gait using a continuous-wave radar at 24 GHz. *Advances in Radio Science, 6*, 67–70.
9. Jarchi, D., Pope, J., Lee, T. K. M., Tamjidi, L., Mirzaei, A., & Sanei, S. (2018). A review on accelerometry-based gait analysis and emerging clinical applications. *IEEE Reviews in Biomedical Engineering, 11*, 177–194.
10. Jokanovic, B., & Amin, M. (2018). Fall detection using deep learning in range-Doppler radars. *IEEE Transactions on Aerospace and Electronic Systems, 54*(1), 180–189.
11. Koo, T. K., & Li, M. Y. (2016). A guideline of selecting and reporting intraclass correlation coefficients for reliability research. *Journal of Chiropractic Medicine, 15*(2), 155–163.
12. Lemhöfer, C., & Glogaza, A. (2019). Gangzyklus und Ganganalyse – Eine Einführung. *Physikalische Medizin, Rehabilitationsmedizin, Kurortmedizin, 29*(01), 11–14.
13. Li, J., Phung, S. L., Tivive, F. H. C., & Bouzerdoum, A. (2012). Automatic classification of human motions using Doppler radar. In *The 2012 International Joint Conference on Neural Networks (IJCNN)* (pp. 1–6). IEEE.
14. Ludloff, A. (2002). *Praxiswissen Radar und Radarsignalverarbeitung.* Wiesbaden; s.l.: Vieweg+Teubner Verlag.
15. Mancini, S. L., Troy, W., Hall, K. A., Wu, X., & Wang, H. (2021). Radar technology as a mechanism for clinical gait analysis: A review. *Journal of Annals of Bioengineering, 2021*(01).
16. Martin, C. L., Phillips, B. A., Kilpatrick, T. J., Butzkueven, H., Tubridy, N., McDonald, E., & Galea, M. P. (2006). Gait and balance impairment in early multiple sclerosis in the absence of clinical disability. *Multiple Sclerosis (Houndmills, Basingstoke, England), 12*(5), 620–628.
17. Otero, M. (2005). Application of a continuous wave radar for human gait recognition. In Kadar, I. (ed.) *Signal Processing, Sensor Fusion, and Target Recognition XIV* (p. 538). SPIE.
18. Prakash, C., Kumar, R., & Mittal, N. (2018). Recent developments in human gait research: Parameters, approaches, applications, machine learning techniques, datasets and challenges. *Artificial Intelligence Review, 49*(1), 1–40.

19. Quaiyum, F., Tran, N., Piou, J. E., Kilic, O., & Fathy, A. E. (2019). Noncontact human gait analysis and limb joint tracking using Doppler radar. *IEEE Journal of Electromagnetics, RF and Microwaves in Medicine and Biology, 3*(1), 61–70.

20. Ren, L., Tran, N., Foroughian, F., Naishadham, K., Piou, J. E., Kilic, O., & Fathy, A. E. (2018). Short-time state-space method for micro-Doppler identification of walking subject using UWB Impulse Doppler radar. *IEEE Transactions on Microwave Theory and Techniques, 66*(7), 3521–3534.

21. Ricci, R., & Balleri, A. (2015). Recognition of humans based on radar micro-Doppler shape spectrum features. *IET Radar, Sonar & Navigation, 9*(9), 1216–1223.

22. Ritchie, M., Ash, M., Chen, Q., & Chetty, K. (2016). Through wall radar classification of human micro-Doppler using singular value decomposition analysis. *Sensors (Basel, Switzerland), 16*(9).

23. Schmidt, R. M., Faiss, J. H., & Köhler, W. (Eds.). (2015). *Multiple Sklerose*. Elsevier Urban & Fischer.

24. Seifert, A.-K., Amin, M. G., & Zoubir, A. M. (2017). New analysis of radar micro-Doppler gait signatures for rehabilitation and assisted living. In *2017 IEEE International Conference on Acoustics, Speech and Signal Processing (ICASSP)* (pp. 4004–4008). IEEE.

25. Seifert, A.-K., Amin, M. G., & Zoubir, A. M. (2019). Toward unobtrusive in-home gait analysis based on radar micro-Doppler signatures. *IEEE Transactions on Bio-Medical Engineering, 66*(9), 2629–2640.

26. Sosnoff, J. J., Sandroff, B. M., & Motl, R. W. (2012). Quantifying gait abnormalities in persons with multiple sclerosis with minimal disability. *Gait & Posture, 36*(1), 154–156.

27. Steinlin Egli, R. (2011). *Multiple Sklerose verstehen und behandeln: [Hintergründe und Studienergebnisse, Untersuchung und Behandlung, Clinical Reasoning in Fall- beispielen]*. Springer.

28. Sun, Z., Wang, J., Sun, J., & Lei, P. (2017). Parameter estimation method of walking human based on radar micro-Doppler. *IEEE Radar Conference (RadarConf), 2017*, 0567–0570.

29. *The multiple sclerosis international federation* (3rd ed.). Atlas of MS (2020).

30. Tivive, F. H. C., Bouzerdoum, A., & Amin, M. G. (2010). A human gait classification method based on radar Doppler spectrograms. *EURASIP Journal on Advances in Signal Processing, 2010*(1).

31. Wang, D., Park, J., Kim, H.-J., Lee, K., & Cho, S. H. (2021). Noncontact extraction of biomechanical parameters in gait analysis using a multi-input and multi-output radar sensor. *IEEE Access, 9*, 138496–138508.

32. Wiendl, H., & Kieseier, B. C. (2010). *Multiple Sklerose: Klinik, Diagnostik und Therapie*. Kohlhammer Verlag.

33. Zhang, J. (2012). Basic gait analysis based on continuous wave radar. *Gait & Posture, 36*(4), 667–671.

Deep Learning for Human Activity Recognition with Plantar Pressure and Movement Data from an Instrumented Smart Sock System

Noah Zuchna, Bernd Resch, and Stephan Odenwald

Abstract

Human activity recognition (HAR) is a task which, if solved by machine intelligence, has a vast potential to improve technological advances in many different fields, medical engineering being one of the more prominent ones. Recent advances in the development of wearable sensor technologies, mobile computing, cloud computing and machine learning (ML) methods made cutting-edge technology broadly available to the scientific community. One such medical application is the therapy and monitoring optimization of subjects with Parkinson's disease (PD). Their aim is to improve the objectiveness, quality, and quantity of monitoring the course of the disease through plantar pressure and movement data collection. This may, in turn, lead medical supervisors to better informed, AI-supported decisions and new insights about PD and its therapy. However, previous research has not sufficiently investigated algorithms to assist movement data analysis to lay the fundament, on which such goals can be accomplished. Thus, this paper presents a new methodology for machine learning based time-series classification using data collected by the ENVISIBLE ParKInSock system. ParKInSock is a multi-sensor system that collects plantar pressure data with 8 highly dynamic force sensing resistor (HD-FSR) cell sensors and linear acceleration and rotational velocity data with an inertial measurement unit (IMU) on each foot, generating a total of 28 data channels. These 28 signals are minimally processed, windowed, and used as features to train two machine learning algorithms: (1) the kernel-based RandOm Convolutional KErnel Transform (ROCKET) algorithm, and (2) a long short-term memory (LSTM) type recurrent neural network (RNN). These two models are trained to perform a supervised learning classification task with classes for activities of daily living (ADL). In a first trial, using cross-validation, the mean accuracy of the ROCKET and LSTM model on the test dataset were $99.61\% \pm 0.26\%$ and $99.52\% \pm 0.22\%$, respectively. The dataset consisted of 4.4 h of five different ADL of one individual subject. In a second trial, six different ADL from three individual subjects created ~1 h of data. The average accuracy results on this dataset were $94.93\% \pm 1.49\%$ and $95.71\% \pm 1.45\%$ for the ROCKET and LTSM model, respectively.

Keywords

Deep learning · Human activity recognition · Instrumented sock · Plantar pressure data · Movement data

N. Zuchna (✉) · B. Resch
Department of Geoinformatics - Z_GIS, Paris Lodron University Salzburg, Salzburg, Austria
e-mail: noah-franz-lukas.zuchna@s2022.tu-chemnitz.de

B. Resch
e-mail: Bernd.Resch@plus.ac.at

B. Resch
Center for Geographic Analysis, Harvard University, Cambridge, MA, USA

S. Odenwald
Department of Sports Equipment and Technology, Chemnitz University of Technology, Chemnitz, Germany
e-mail: stephan.odenwald@mb.tu-chemnitz.de

1 Introduction

Human activity recognition (HAR) is a broad field of research, which is using artificial intelligence (AI), most often deep learning (DL) [1], to identify and name human movements and activities based on data generated by sensors. Using body-worn sensors is a subfield of this area, the most prevalent sensor type being the accelerometer [2]. While the fields of application are various, a dominant one is the medical sector. Here, autonomous HAR could lead to breakthrough innovations, potentially improving healthcare in clinics and at home.

© The Author(s) 2026
B. Meyer et al. (eds.), *Hybrid Societies*, Advances in Science, Technology & Innovation,
https://doi.org/10.1007/978-3-032-03488-5_19

The research of this paper addresses a specific medical application, the therapy and monitoring optimization of subjects with Parkinson's disease (PD). The goal is to improve the objectiveness, quality, and quantity of monitoring the course of the disease through plantar pressure and movement data collection and hence lead medical supervisors and AI to better informed decisions and new insights about PD and its therapy. As a next step, algorithms capable of basic HAR should be developed to assist movement data analysis and thereby lay the fundament on which such goals can be accomplished. An especially interesting field of study is the medication state of subjects with PD, with motor fluctuations being a common problem of subjects with PD. Studying these ON- and OFF-states where the medication is showing an effect or not, respectively, should help to solve some issues revolving about the different medication therapies, for example L-Dopa. Exemplary problems are the wearing-off effect over long periods of using a certain medication therapy or phenomena like short peak OFF-states during an ON-state [3]. If effects like these occur, clinicians try to adjust the medication schedule of the patient to eliminate these effects.

However, not only are these interactions with the patient and thus the chance to adapt the medication schedule based on feedback from the person usually very seldom, the feedback is also error-prone, subjective and sparse. There are existing solutions to evaluate the progression of the disease at the clinic. Being the gold standard, pressure plates can be used to confidently predict the medication state by precisely measuring gait parameters [4]. Also, other day to day movements and activities were studied during ON- and OFF-states, for example how far a person can bend over forwards. Another issue is that the condition of patients is varying every day. In summary, data from one day at the clinic can leave a lot of insights for possible improvements undiscovered. If plantar pressure and movement data from outside the clinic would also exist, the speed and quantity of the feedback could be drastically increased, opening up new opportunities for telemedicine and thereby improved life quality for PD patients and others. A first step into this direction could be, if everyday movement data could be recorded in an unobtrusive way and automatically segmented into different ADL, making it easier for clinicians to work with the data.

Thus, in this research state-of-the-art time-series classification machine learning techniques are applied to data collected by the newly developed ENVISIBLE ParKInSock, an unobtrusive, instrumented smart sock system. More specifically, the main proposition of this paper is if a LSTM model would perform well in a HAR task of classifying basic activities of daily living (ADL) given the plantar pressure and movement data of the ParKInSock. Why this main proposition was chosen shall be clarified in the related work Sect. 2. Our results demonstrate that not only the ROCKET model performs very well on the collected data, but also the LSTM model delivers highly accurate HAR with up to 100% accuracy on the test dataset.

2 Related Work

The unique selling point in this experiment is the data collected by the ENVISIBLE ParKInSock being used for HAR. While measuring plantar pressure with pressure plates is the gold standard for gait analyses in clinics, unobtrusive ways for long-period measurement, also outside of clinics, is yet a challenge to be solved. Therefore, many endeavours have been undertaken to develop insoles and socks which have plantar pressure sensing capabilities. A recent review paper by Chen et al. [5] provides an overview of the different technologies that are being developed and used for the purpose of disease monitoring and analysis, marking force sensing resistors (FSR) for plantar force sensing as a traditionally used mechanism and long short-term memory (LSTM) as a common algorithm architecture choice. Two other recent review papers [1, 2] about DL for sensor-based HAR portray clearly the feasibility of using DL for the typical pattern recognition kind task of HAR, predicting it to be the way to better model generalisation, accuracy and to be a relief upon the need to do feature engineering, being able to learn much more high-level and meaningful features automatically. A review of deep time-series clustering (DTSC) methods identifies recurrent neural networks (RNN) and LSTMs as the most commonly used techniques for time-series analysis tasks, especially in supervised learning. Paydarfar et al. [6], using piezo resistor sensor insoles with three sensors each and a RNN classifier, showed an accuracy of $87.0\% \pm 8.9\%$ on a HAR task with six activities, also suggesting further research in this direction, just like so many others reviewing and using comparable approaches [7]. The example of a recent research by Yin et al. [8] show the continuing interest in using convolutional neural networks (CNNs) and LSTMs for HAR in 2022, even during the rise of transformer models challenging state-of-the-art hybrid models as the former [9].

3 Methods

3.1 Data Collection

The two main components of the data collection are the measuring device and the mode in which the device was used to collect data from subjects. For the measuring device, each ENVISIBLE ParKInSock consists of a pressure sensing insole and a Dialogg, a data collection unit also developed by ENVISIBLE. While the insole is a non-removable part of the sock, being glued in between the outer and inner sock, the Dialogg can be easily removed for the purpose of dressing,

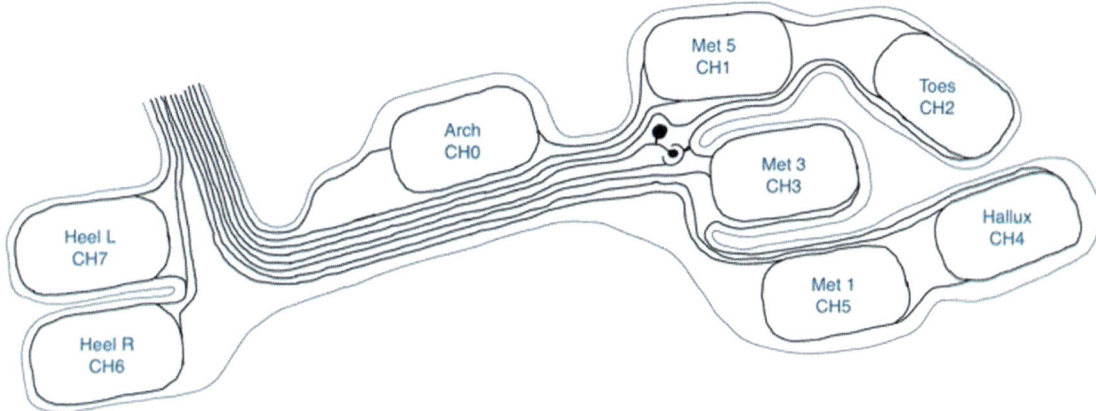

Fig. 1 Layout of the sensors composing the insole and the associated channels

and washing the sock more easily, or for using the Dialogg for other purposes. By using the 8-cell High Dynamic Force Sensing Resistor (HD-FSR) Insole by IEE Smart Sensing Solutions (cf. Fig. 1) for the left and right sock, a total of 16 pressure signal channels are connected to the two Dialogg devices which measure the output of these two insoles.

Placed at the left and right outer ankle, the two Dialoggs further collect angular velocity and linear acceleration by means of an in-device inertial measurement unit (IMU). The internal battery of the Dialoggs make measurements of about five hours on the left side and of about six hours on the right side possible. The difference is due to the fact that the left Dialogg is configured as the server, making a live connection via Wi-Fi possible and thus has higher energy consumption. Connecting to the Wi-Fi generated by the left Dialogg, one can open the Dialogg Control App and use it to configure the measurement's settings and name, start a measurement, live view the data that is being collected, label the data while it is being collected, download the data to the device, upload it to a server and more. The settings of the Dialogg give various options as to enable/disable data collection of specific sensors, measure at different frequencies, apply maxima for angular velocity and linear acceleration and several other options. All measurements were conducted with a measuring frequency of 100 Hz, the linear acceleration was set to a maximum of -4 to 4 Gn, and the angular velocity was set to a maximum of -500 to $500°/s$.

For the mode of data collection, with the objective of collecting high quality data, the measurements were conducted as follows. For the first trial with one individual subject, five ADL (biking, laying, sitting, standing, and walking) were selected which are relatively distinct from each other, but together make up the most of our daily lives. In a second trial with three subjects, these six ADL were collected: Walking downstairs and upstairs, laying, sitting, standing, and walking. Before the measurement, the task was planned, and the circumstances documented. For instance,

the route, on which the subject was to walk on, was defined and circumstances like additional weights through a backpack and different terrains (asphalt, stones, soil...) were documented. In summary, the data in the first trial was collected in various different terrains including outdoor and indoor environments, but always with the same pair of shoes or with only the sock. In the second trial, the different subjects were all wearing different shoes, but the terrain was always from the same indoor environment. The measurement was only started, when the subject was already doing the activity and was stopped again while it was still doing the activity. The study was approved by the institution's ethics committee (reference number #101,521,057) and was in accordance with the Declaration of Helsinki.

3.2 Pre-processing

For data pre-processing, analyses, and model development, the programming language Python [10] version 3.7 was used in the integrated development environment PyCharm [10], extended by the libraries Pandas [11, 12], NumPy [13] and Scikit-learn for loading and manipulating the data. For plotting the data Plotly [14] and Matplotlib [15] were used. As a backend for Keras [16] we used TensorFlow 2.0 [17]. To manage all the libraries, Anaconda [18] was used. A Python script fulfils the task of generating one big dataset out of all the individual collected measurements that were priorly sorted into folders by the activity that was recorded. At the same time, the individual files of the left and right foot are combined, and some values that are not required are dropped. Synchronising the left and right foot data is done with the help of the UNIX timestamp that is saved in the accompanying ".info" file, if this file exists. Else, the assumption is made, that both measurements started at the same time. An inner join is used to combine left with right, dropping all the values that might have been recorded while the other side was not recorded.

(1) *Reducing the Amount of Data*

As finding lower sampling frequencies that still convey all the information, necessary to make accurate classifications possible, can drastically lower the need for computational resources, and since it makes experimentation cycles a lot less resources intensive, there is an interest in finding the most efficient sampling frequency. Therefore, analysing which frequency that might be through comparing different sampling frequencies and observing if enough information is still there, was done. Resampling was done by taking only every n-th value, since that way simulates measuring at a lower frequency, while other methods, like taking the mean of n values, can only be done if the data were originally collected at a higher frequency and is only downsampled afterwards. Also, although it might be tempting, it doesn't make sense to pick different sampling frequencies for different activities, since in the final application the unlabelled input data will always come in with one specified measuring frequency. Additionally, the mechanics of the sensor cells and the resolution of the Analog–Digital-Converter used in the Dialogg create limitations for the highest possible measuring frequency and accuracy, and the precision of the measured values, respectively. This results in redundancy in the data which was reduced by rounding from eight to five decimals.

(2) *Pre-processing due to the Characteristics of the 8-Cell Insole*

The datasheet of the IEE 8-cell insole provides a typical response curve of the insole to applied pressure [19], making an approximate relation of measured Voltage values to applied pressure in millibar possible. Further, the datasheet describes the insole to have an actuation force of 250 mbar and a sensitivity range of up to seven bars. Thus, the pressure values were converted using the given curve, values below 250 mbar were set to zero and values above seven bars were set to seven.

(3) *Sliding Window*

This continuous multivariate dataset was then split into instances of 2.5 s using a sliding window approach, having been proposed to be the optimal trade-off between recognition performance and speed [20]. Moving the window over the data, a new window was only created if all data points in the window showed the same activity label, else the data points were dropped. In an example of data with a measuring frequency of 10 Hz, this gives windows of 28 × 25 values for variables and time steps, respectively. Matching the windowed dataset, a separate list of the true labels was generated, serving as the y to our X.

3.3 Development of the Classification Model

The HAR task the classification model was trained to do was to predict one of the defined labels for each given instance of a window. All classification models that are used have a neural network as the basis for their functionality. And since the target class values are integers but neural networks have a binary output by their nature, the method "to_categorical" of Keras is used to generate a binary matrix representation of the input class values.

(1) *Rocket*

The ROCKET algorithm builds upon the recent successes of CNNs for time-series classification tasks, by using random convolutional kernels to transform the raw times series data into features that are then used to train a linear classifier. The ready-to-use implementation by Sktime [21, 22] is used and therefore must only be fitted to the data. The name being derived from "RandOm Convolutional KErnel Transform," ROCKET has proven to be not only exceptional in terms of accuracy, but also to be much less time-consuming compared to its contemporary challengers [23].

(2) *LSTM*

The long short-term memory (LSTM) technique, first proposed by Hochreiter in 1997 [24], is a type of recurrent neural network (RNN) that can process sequential data and is commonly used in natural language processing and time-series forecasting tasks. Specifically, it has also been very successfully used in many attempts of doing HAR with body-worn sensors [25]. That is because the LSTM is able to extract, learn and remember features over long periods of data, in this case over many time steps. The basic architecture as proposed by Brownlee [26] is slightly adapted and is implemented with Keras for this application in the following way. The first layer is a LSTM layer with 100 units. The second layer is a dropout layer, which randomly drops out (sets to 0) a certain percentage of the input units in order to prevent overfitting. In this case, the dropout rate is 0.5, meaning half of the units will be dropped out during each training step. The third layer is a dense layer, which is a standard fully-connected layer in a neural network. This layer has 100 units and uses the rectified linear activation function. The fourth and final layer is another dense layer with units according to the number of classes, which uses the softmax activation function. This is typically used as the output layer in a classification model, as it will produce probabilities for each of the classes.

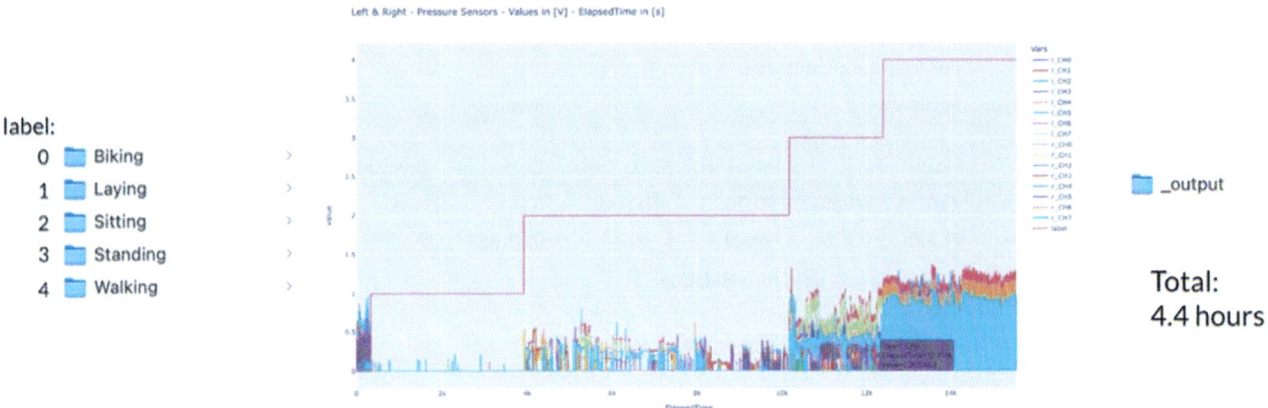

Fig. 2 Plot of the resulting output from the dataset generation step for the first dataset

4 Results and Discussion

In the dataset generation step for the first dataset from the first trial with one participating subject, the combination of eleven measurements was combined to a total of 4.4 h of recorded activities or 1,557,185 data points of 28 feature values in each data point (cf. Fig. 2).

In the second trial, by recording each of the three subjects doing each of the six ADL for approximately three minutes, it sums up to around one hour of recorded activities.

The datasets were thereafter resampled to lower frequencies and random instances from all the different activities were compared to the original frequency of 100 Hz. Going down through 50, 33, 20, 10, and 5–10 Hz was still able to capture all the activities in a way, that a human could recognise the activity at hand, at maximum needing little training to do this classification task. Thus, 10 Hz was initially picked for resampling the data for training the model. In the second trial, 33 Hz were also used to improve the model accuracy on the more challenging second dataset. 33 Hz deliver significantly more information on how the pressure curve of a footstep looks, not just giving an edgy line which roughly represents the movement but giving a rather precise idea of how the pressure distribution changes throughout every given movement. It seems that it is exactly this information, that helps the model to discern upstairs, downstairs, and normal walking from each other.

Resampling and rounding as described in Sect. 3.2(1) reduces the amount of data by ~94%, processing the raw example dataset number one from 503.8 mb to only 31.2 mb and dataset number two from 75.1 to 4.5 mb. Applying the conversion described in Sect 3.2(2) changes the plantar pressure values measured in Volt to Millibar.

After windowing the datasets, pre-processing is done, and random samples can be plotted for review. The plots gave an impression of the distinctiveness of the data in shape and statistical measures between the different activities, which gave confidence for ML methods to be successful on it.

In summary, the data collection method proposed in Sect. 3.1 yielded good results, producing high quality clean data which is adequate for model development.

In an earlier pre-test it was found that the ROCKET algorithm outperforms the MUSE algorithm [27] on a dataset with 20 min of data. Notice, that the datasets used here are different in resolution, window size, features, etc. because they were differently pre-processed than described earlier. MUSE performed higher accuracy while only using 2% of the dataset for training and the rest for validation (accuracy on the validation set—MUSE: 0.65%; ROCKET: 0.35%), but accuracy was equal when using a 66/33 train-test-split (accuracy on the validation set—MUSE: 0.87%; ROCKET: 0.87%)—the difference being that MUSE took a total of 1 h 50 min to train, while ROCKET, staying true to its name, took only 1 min 49 s to train. Therefore, ROCKET was chosen to be the benchmark performing classifier for comparison with the classification model that was the main objective to be developed in this research, the LSTM model.

For the developed LSTM model in comparison to the ROCKET algorithm, Table 1 shows the resulting values of the evaluation metrics used for six trained and validated classifiers for the described HAR task on dataset number one. The cross-validation was done using the "StratifiedKFold" function from Scikit-learn with shuffling, keeping the class proportions the same while shuffling them to get k-folds of training and validation data. Stratification and shuffling helped with getting better generalisation results. In the case of stratification this is possibly due to the obvious fact, that each trained model is getting equally sized representations of each class. Shuffling most probably brings better generalisation because the model is getting more and bigger variations between the individual training samples.

Table 1 Validation results for ROCKET and LSTM on Dataset 1

Model	0.3%[a]	6%[a]	Ten-fold cross Validation								
	of dataset used for training		Accuracy				Mean			Mean F1 score	
	Accuracy (%)		Mean	Min (%)	Max (%)	Loss	Precision	Recall	Macro	Micro
ROCKET	77.33	98.26	99.61% ± 0.26%	99.20	100.00	N/A	N/A	N/A	N/A	N/A
LSTM	64.16	95.08	99.52% ± 0.22%	99.20	99.84	0,02,852	99.52%	99.52%	99.39%	99.52%

[a] Percentage of data used for training. The rest was used for validation

Table 2 Validation and test results for ROCKET and LSTM on Dataset 2

Model	Pre-processing	Six-fold cross-validation								Train-test-split
		Accuracy			Mean			Mean F1 score		
	Sampling frequency (HZ)	Mean	Min (%)	Max (%)	Loss	Precision	Recall	Macro	Micro	Test size
ROCKET	10	95.47% ± 2.27%	91.39	98.01	N/A	N/A	N/A	N/A	N/A	N/A
ROCKET	33	95.71% ± 1.45%	93.42	97.35						
LSTM	10	93.25% ± 1.14%	92.07	95.15	0.3014	93.36%	92.95%	91.31%	93.25%	25%
LSTM	33	94.93% ± 1.49%	92.51	96.48	0.1772	95.21%	94.93%	93.18%	94.93%	25%

For the score metric results for validating and testing the models with dataset number two, confer Table 2. The pre-processing column shows to which frequency the original 100 Hz data was resampled to. Six-fold cross-validation here was also done using stratification and shuffling, again having shown significant improvements when applied. In order not to create a bias on the resulting LSTM model by doing hyperparameter tuning based on the validation results, a train-test-split is done before cross-validation and the test data is kept for the final model.

The fitting and evaluating of the ROCKET algorithm on dataset one, using ten-fold cross-validation, took ~38 min and gave results of up to 100% accuracy on the test dataset, resulting in confusion matrices that have almost only values on the diagonal and only few misclassifications that seem to be appearing without a pattern. On the more challenging dataset number two though, having lower accuracy scores, the confusion matrices are giving us more insights. Also, an expected pattern is occurring that is actually not too worrying. Since it is known from the collection of the data of climbing the stairs up and down, that in between the staircases there were even platforms where the subject always had to take a few normal walking steps to get to the next staircase, this data is expected to be unclean. Therefore, one would wish the model to classify the incorrectly as 'Upstairs' and 'Downstairs' labelled data to be classified correctly as "Walking," which is exactly what we can see in the confusion matrices in Fig. 3. On the left with a sampling frequency of 10 Hz and an accuracy of 93.8% and on the right with 33 Hz and 97.4% accuracy.

The rest of the confusion matrices shows where the model also seems to have trouble. But in these two cases all the errors made are in connection with the unclean data, probably resulting from it. In summary, the ROCKET model classifier, for both the first and second dataset, delivers great results with high accuracy.

The execution time for training and evaluating the LSTM model on the first dataset was ~1 h 1 min. On the second dataset the computation time was a bit faster, because only a six-fold cross-validation, opposed to a ten-fold one, was done, although 1000 epochs were used instead of 120. In the confusion matrices for the second dataset a similar confusion was arising, probably again from the unclean data. If the data would be clean, the chances are high, that the results would be near perfect as it is the case with the first dataset. Overall, also with the LSTM model the performance is very satisfying.

5 Conclusions

This paper proposed a methodology for analysing an HAR task on pressure and movement data generated by the ENVIS-IBLE ParKInSock measurement system. For this purpose, a complete model development and evaluation cycle was created, comparing a custom LSTM to the ROCKET algorithm. Firstly, a mode for data collection was proposed and an accompanying automated dataset generator was implemented. This makes it easy and fast to collect novel data and get a labelled ready-to-use dataset. During pre-processing, redundancy was reduced in simple but effective ways, the sensor characterisation was applied to the raw signals, and a sliding window approach was implemented. Using a sampling frequency of only 10 Hz reduced the amount of data to be processed radically, but still delivered enough information

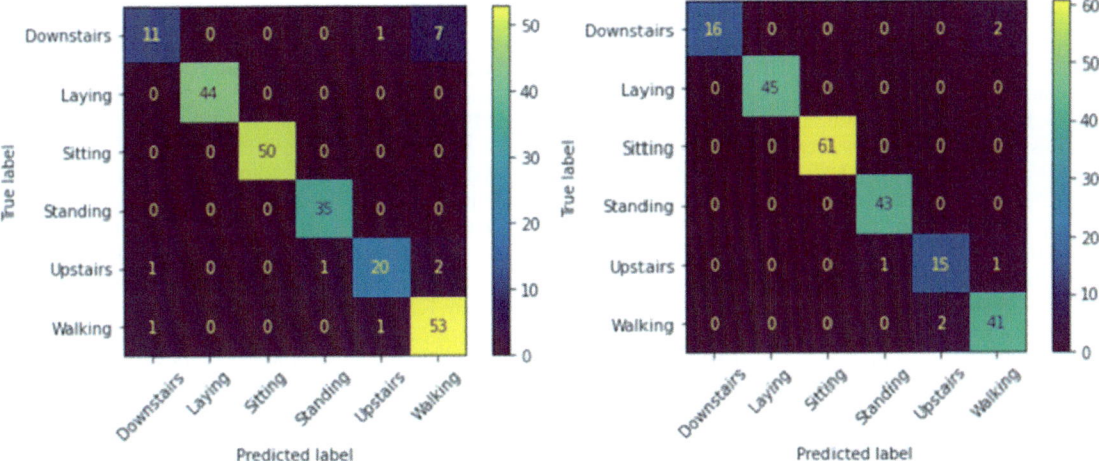

Fig. 3 Two exemplary confusion matrices from training the ROCKET model on the second dataset

for the model to deliver excellent results. When model performance is lacking due to other issues, using a higher sampling frequency can help to improve performance. Since a lack of labelled data for supervised learning is an obvious and common problem, self-limiting the available training data as shown in Table 1 gives important insights for future data collection and model development to enhance the capabilities of this and similar devices in HAR. Using the ROCKET algorithm creates a comparison to state-of-the-art multivariate time-series classifiers, which keeps outperforming the developed LSTM model. The evaluation suggests that the data has proficient quality, descriptiveness, and cleanness for the used models to deliver excellent results on. Therefore, this is a successful step towards being able to better analyse the recorded activities of subjects with PD, by segmenting the recorded ADL into the different activities.

We conclude that HAR, which is not based on IMU data but on insole plantar pressure data generated by the ENVIS-IBLE ParKInSock system, prepares the way for machine intelligence in activity and gait analysis and many other meaningful applications, not merely in the medical sector. Using the proposed methods for data collection and pre-processing should make worthwhile further studies possible, easy, and fast. Lower sampling frequencies were shown to be very effective, which is an important insight for the further development of body-worn sensor systems, because reducing the amount of data load creates a huge relief for the whole application, starting from the sensors, all the way to the final model.

Finally, the assumption that a LSTM model would perform well for HAR on plantar pressure and movement data was proven correct, delivering near to perfect results using tenfold cross-validation on the collected first dataset (99.52% ± 0.22% Mean Accuracy, 0.029 Mean Loss, 99.52% Mean Precision and 99.52% Mean Recall), although still being outperformed by the ROCKET model with even slightly

better and faster results (99.61% ± 0.26% Mean Accuracy, 99.20% Min Accuracy, 100.00% Max Accuracy). Even when using only 6% of the dataset for training, which results in 373 instances (equivalent to ~15.5 min), both models still show accurate results (LSTM: 95.08% Accuracy, ROCKET: 98.26% Accuracy). With the dataset number two it is also shown that the differences in the data due to coming from different subjects doesn't seem to create any problems. Thus, assumptions about the explanatory power of the data and the effectiveness of kernel- and LSTM-based models for HAR tasks were proven. Therefore, further research following this direction is encouraged.

In a next step, the methods that have been proven to work in this research shall be used on a bigger scale, collecting longer periods of data, from several individuals and of more ADLs. The mode of data collection will be complemented by a second unsupervised mode, where subjects wear the ParKInSock in their daily lives documenting their ADLs with the help of a diary.

References

1. Chen, K., Zhang, D., Yao, L., Guo, B., Yu, Z., & Liu, Y. (2022). Deep learning for sensor-based human activity recognition: overview, challenges, and opportunities. *ACM Computing Surveys, 54*(4), 1–40. https://doi.org/10.1145/3447744
2. Wang, J., Chen, Y., Hao, S., Peng, X., & Hu, L. (2019). Deep learning for sensor-based activity recognition: A survey. *Pattern Recognition Letters, 119*, 3–11. https://doi.org/10.1016/j.patrec.2018.02.010
3. Hardie, R. J., Lees, A. J., & Stern, G. M. (1984). ON-off fluctuations in Parkinson's disease: A clinical and neuropharmacological study. *Brain, 107*(2), 487–506. https://doi.org/10.1093/brain/107.2.487
4. Rudisch, J., et al. (2021). Agreement and consistency of five different clinical gait analysis systems in the assessment of spatiotemporal gait parameters. *Gait & Posture, 85*, 55–64. https://doi.org/10.1016/j.gaitpost.2021.01.013

5. Chen, J., et al. (2022). Plantar pressure-based insole gait monitoring techniques for diseases monitoring and analysis: A review. *Advanced Materials Technologies, 7*(1), 2100566. https://doi.org/10.1002/admt.202100566

6. Paydarfar, A. J., Prado, A., & Agrawal, S. K. (2020). Human activity recognition using recurrent neural network classifiers on raw signals from insole piezoresistors. In *2020 8th IEEE RAS/EMBS International Conference for Biomedical Robotics and Biomechatronics (BioRob)* (pp. 916–921). New York City, NY, USA: IEEE. https://doi.org/10.1109/BioRob49111.2020.9224311

7. Potluri, S., Ravuri, S., Diedrich, C., & Schega, L. (2019). Deep learning based gait abnormality detection using wearable sensor system. In *2019 41st Annual International Conference of the IEEE Engineering in Medicine and Biology Society (EMBC)* (pp. 3613–3619). Berlin, Germany: IEEE. https://doi.org/10.1109/EMBC.2019.8856454

8. Yin, X., Liu, Z., Liu, D., & Ren, X. (2022). A novel CNN-based Bi-LSTM parallel model with attention mechanism for human activity recognition with noisy data. *Science and Reports, 12*(1), 7878. https://doi.org/10.1038/s41598-022-11880-8

9. Dirgová Luptáková, I., Kubovčík, M., & Pospíchal, J. (2022). Wearable sensor-based human activity recognition with transformer model. *Sensors, 22*(5), 1911. https://doi.org/10.3390/s22051911

10. JetBrains. (2022). *PyCharm 2022.1.2 (Professional Edition)*. Retrieved Aug 02, 2022, from https://www.jetbrains.com/pycharm/

11. McKinney, W. (2010). Data structures for statistical computing in Python. In *Presented at the Python in Science Conference* (pp. 56–61). Austin, Texas. https://doi.org/10.25080/Majora-92bf1922-00a

12. Reback, J., et al. (2022). *pandas-dev/pandas: Pandas 1.4.3*. Zenodo. Retrieved June 23, 2022, from https://doi.org/10.5281/ZENODO.3509134

13. Harris, C. R., et al. (2020). Array programming with NumPy. *Nature, 585*(7825), 357–362. https://doi.org/10.1038/s41586-020-2649-2

14. Plotly Technologies Inc. (2015). *Collaborative data science*. Plotly Technologies Inc. https://plot.ly

15. Caswell, T.A., et al. (2021). *matplotlib/matplotlib: REL: v3.5.1*. Zenodo. Retrieved December 11, 2021, from https://doi.org/10.5281/ZENODO.5773480.

16. Chollet, F., et al. (2015). *Keras*. https://keras.io

17. TensorFlow Developers. (2022). *TensorFlow*. Zenodo. Retrieved May 23, 2022, from https://doi.org/10.5281/ZENODO.4724125

18. Anaconda Inc. (2016). *Anaconda navigator (Version 2.1.1)*. https://anaconda.org/

19. IEE Smart Sensing Solutions. (2022). *High dynamic force sensing resistor—Insole*. www.iee.lu

20. Wang, G., Li, Q., Wang, L., Wang, W., Wu, M., & Liu, T. (2018). Impact of sliding window length in indoor human motion modes and pose pattern recognition based on smartphone sensors. *Sensors, 18*(6), 1965. https://doi.org/10.3390/s18061965

21. Löning, M., Bagnall, A., Ganesh, S., Kazakov, V., Lines, J., & Király, F. J. (2019). *sktime: A unified interface for machine learning with time series*. https://doi.org/10.48550/ARXIV.1909.07872

22. Király, F. (2022) *alan-turing-institute/sktime: v0.13.0*. Zenodo. Retrieved July 14, 2022 form https://doi.org/10.5281/ZENODO.6832750

23. Ruiz, A. P., Flynn, M., Large, J., Middlehurst, M., & Bagnall, A. (2021). The great multivariate time series classification bake off: A review and experimental evaluation of recent algorithmic advances. *Data Mining and Knowledge Discovery, 35*(2), 401–449. https://doi.org/10.1007/s10618-020-00727-3

24. Hochreiter, S., & Schmidhuber, J. (1997). Long short-term memory. *Neural Computation, 9*(8), 1735–1780. https://doi.org/10.1162/neco.1997.9.8.1735

25. Alqahtani, A., Ali, M., Xie, X., & Jones, M. W. (2021). Deep time-series clustering: A review. *Electronics, 10*(23), 3001. https://doi.org/10.3390/electronics10233001

26. Brownlee, J. (2022). LSTMs for human activity recognition time series classification. *Machine Learning Mastery*. Retrieved August 02, 2022, from https://machinelearningmastery.com/how-to-develop-rnn-models-for-human-activity-recognition-time-series-classification/

27. Schäfer, P., & Leser, U. (2017). *Multivariate time series classification with WEASEL+MUSE*. https://doi.org/10.48550/ARXIV.1711.11343

28. van Rossum, G., & Drake, F. L. (2010). The Python language reference, Release 3.0.1 [Repr.]. In Drake, F. L. (Ed.), *Python documentation manual/Guido van Rossum*. no. Pt. 2. Hampton, NH: Python Software Foundation.

29. Pedregosa, F., et al. (2012). *Scikit-learn: Machine learning in Python*. https://doi.org/10.48550/ARXIV.1201.0490

Avoidance of Collisions Through Prospective Path Planning

Edgar Scherstjanoi and Svetlana Wähnert

Abstract

Collaborative systems in which humans and robots share a common physical environment are part of the fourth industrial revolution. Since the health integrity of humans must not be compromised, collisions are often avoided by emergency stops of the machines, which in turn have a significant impact on the throughput of automated processes. By using motion predictions, a prospective navigation of robot movements emerges. In this paper, the benefits of such a human-robot system are evaluated. Recorded motion data of a human subject are used to simulate a quasi error-free motion prediction, which is integrated as a dynamic obstacle into the path planning of virtual robots. It is shown that a simple strategy significantly reduces safety risks, with only a minor impact on the throughput of the robots.

Keywords

Collaborative systems · Mobile robots · Motion prediction · Simulation

1 Introduction

Machine learning algorithms offer a wide range of possibilities for the development of artificial intelligence. A very common example is pattern recognition in image, sound and text data, e.g. for the detection of humans, recognition of control commands or natural language processing [1]. A specific field in machine learning research concerns regression models, which can be used to generate temporal predictions based on sequential training data [2]. Regression problems can be identified in many domains, usually utilizing different types of data. Using motion capture technologies, kinematic information of human motion can also be used as training data, allowing the development of models suitable for motion prediction [3]. Despite the extensive research results, there is a lack of industrial applications, especially in regard to work activities. The reasons for this may lie in a difficult-to-estimate relationship between costs and benefits of such implementations. From an occupational science perspective, the prediction of work motion can be advantageous to create a prospective robot navigation for collaborative systems (e.g. [4]). If robots were able to make such predictions, routes could be calculated that purposefully avoid this path instead of crossing it, which might reduce the risk of collision. For a schematic representation of this idea, see Fig. 1.

Assuming that the prediction of human work activities is technically implementable, the following research questions need to be answered in order to estimate achievable benefits in relation to the costs incurred.

1. Is the risk of collision reduced when a predicted path is avoided?
2. Is a longer forecast associated with a greater reduction in collision risk associated?
3. Does the resulting detours impact the robot throughput?

This paper examines these questions, in order to motivate applied research and development of sophisticated regression models for human motion prediction. A simple environment of a collaborative human-machine workplace was simulated to determine the possible added value of using motion prediction to avoid collisions. Instead of using a specific prediction model, the human's motions were recorded in advance and transmitted to the robots during the experiment as a quasi-correct predicted trajectory. For a statistical evaluation, distances between humans and robots as well as the achieved throughput were measured.

E. Scherstjanoi (✉) · S. Wähnert
Chair of Ergonomics, Dresden University of Technology, Dresden, Germany
e-mail: edgar.scherstjanoi@tu-dresden.de

S. Wähnert
e-mail: svetlana.waehnert@tu-dresden.de

B. Meyer et al. (eds.), *Hybrid Societies*, Advances in Science, Technology & Innovation,
https://doi.org/10.1007/978-3-032-03488-5_20

Fig. 1 Schematic illustration for increasing safety in collaborative workplaces. Crossing paths might lead to collisions (left), which can be prevented by a robot navigation, which takes a human motion forecast into account (right)

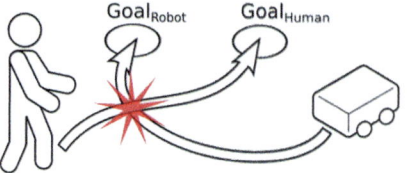
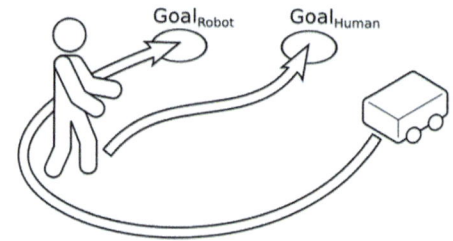

2 Methods

2.1 Simulation Environment

A two-dimensional virtual room of size 10 m × 5.5 m, with a spatial resolution of 0.1 m per gridded cell and a temporal update frequency of 60 Hz was used for this experiment. There were no static obstacles, only the boundary at the edge of the room. In this room, the motion trajectories of 4 robots and one human were simulated. A single simulation run lasted 5 min.

For an implementation of robot behavior "Robot Operating System" (ROS) [5] was utilized. This enables simulations with digital twins of autonomous driving robots, such as "Turtlebots" [6]. Common standard algorithms for localization, navigation and sensor emulation are available for vehicle operation in ROS.

The position of the human in the virtual room was measured and recorded with an inertial motion capture system ("Xsens MVN"). Actions performed included spontaneous alternations between standing and different walking speeds and directions.

While the motion of the human subject was captured beforehand, the robots acted dynamically in the virtual environment to reach an individual list of random chosen goals. A route to a single goal was determined on the one hand by the shortest path according to Dijkstra's algorithm, on the other hand by the avoidance of dynamic obstacles according to the "Dynamic Window Approach" (DWA) [7]. As dynamic obstacles the current positions of all actors were considered. Moreover, the recorded sequence of temporally following positions of the human was taken into account and used for a simulation of a trajectory prediction.

This quasi-true forecast sequence is hereinafter referred to as pseudo predicted sequence (PPS). The resulting cost map contained the information for all current inaccessible areas, including inflated areas around each obstacle. In combination with the current motion of the robot, a route to the next destination was determined. For a detailed description of robot navigation in ROS see [8].

2.2 Manipulation of PPS in Different Scenarios

The main manipulated variable was the temporal length of the PPS, i.e. the number of future positions of the human that were available to the robots. Manipulating PPS length enables the investigation according to the significance of collision reduction using motion trajectory prediction. Moreover, with different PPS lengths it is possible to examine how a longer prediction affects the behavior of a robot. Three PPS lengths were compared:

- No PPS: only the current position of the human.
- Short PPS: human position during the following two seconds.
- Long PPS: human position during the following four seconds.

Figure 2 shows exemplarily how different PPS lengths influence the dynamic navigation plans of the robots.

To ensure that results are not confounded with a randomly chosen constellation of target points, the simulation was performed in four independent scenarios (S_1–S_4), each with

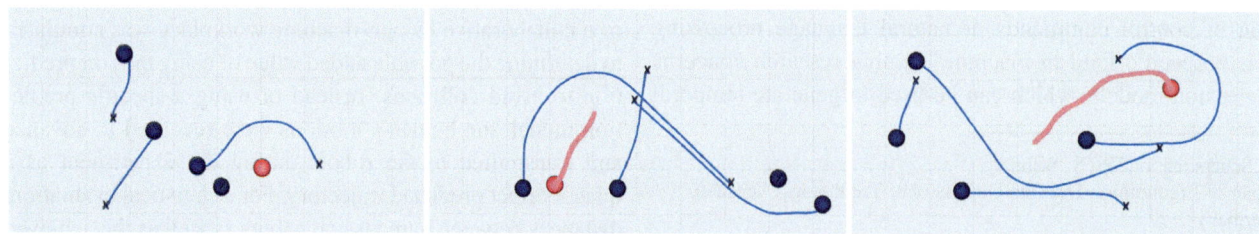

Fig. 2 ROS "RViz" top-view screenshots in which different PPS lengths affect the robot navigation. The light-red dot and trajectory represents the position and upcoming path of the human, while blue dots and trajectories show the robots positions and current navigation plans. The x's mark the current target of a robot. Left: no PPS, middle: short PPS, right: long PPS

differently chosen target points for all actors. Due to the emulated signal noise of the virtual sensors (e.g. Laser Distance Sensor), each combination of PPS length and scenario were repeated $n = 10$ times. Since the error variation between the runs can be assumed to be random, a between-subjects evaluation design was applied.

2.3 Data Collection and Analysis

The measurement of robot performance consisted of two safety variables and one efficiency variable. For the safety performance measurement it was assumed that as the distance between actors decreases, the risk of collision increases. Hence, two safety-critical zones were defined around each robot, each with a radius of 1 and 0.5 m, representing different degrees of risk of collision.

Consider $R10$ as the relative time (in %) that the human subject is within the critical zone of 1 m, respectively $R05$ within the critical zone of 0.5 m. To analyze the detours caused by the avoidance navigation planning, the number of targets reached by the robot within the 5 min run is noted as GL. For each run, the average team performance across all four robots was computed in terms of $R10$, $R05$ and GL. This resulted in 3 (PPS length) × 4 (scenario) × 10 (run) = 120 total data points per performance parameter.

3 Results

Figure 3 shows the results for $R10$, $R05$ and GL grouped by PPS length and scenario. For all three performance parameters, a two-way ANOVA with factors PPS length (No, Short and Long) and scenario S (S_1 to S_4) was conducted, where in each case only the main effect of PPS length was significant ($R10: F(2,108) = 57.31\ p< 0.001, \hat{\eta}_p^2 = 0.52; R05: F(2,108) = 69.20, p < 0.001, \hat{\eta}_p^2 = 0.56; GL: F(2,108) = 7.64, p <$

$0.001, \hat{\eta}_p^2 = 0.12$). For the levels of PPS length averaged over the scenarios, pairwise post hoc t-tests were conducted using Bonferroni adjusted alpha method. The analysis revealed that the mean values of $R10$, $R05$ and GL were significantly higher for short than for long PPS. Furthermore, the mean values of $R10$ and $R05$ were significantly higher for no PPS than for short and long PPS. For GL, no PPS had a significantly higher mean than long PPS, whereas the difference to short PPS was not significant (see Fig. 3). From these results, it can be concluded that PPS length had an effect on all three performance variables, regardless of the scenario. A short PPS length led to a reduction in the relative time the robots were within the safety-critical zones of both 0.5 m and 1 m to the human subject without significantly reducing the efficiency measure. A long PPS length led to a further reduction in these two safety-critical measures, but also to a significant reduction in efficiency.

4 Discussion and Outlook

In summary, answers to the posed research questions can be derived from the results of the experiment as follows:

1. It is possible to utilize human motion prediction as a tool to significantly reduce the collision risk in collaborative work places.
2. It is likely that a longer time forecast will also result in a more significant reduction in collision risk.
3. The resulting detours may have only a slight negative effect on the throughput of the robots. Nevertheless, it is reasonable that the influence increases as the prediction time increases.

It was shown that robot path planning, which integrates prediction of human motion, reduces collision risks. Despite the increasing risk reduction with a longer prediction sequence, this is also accompanied by a significant degradation in robot throughput. It can be concluded, that a balanced

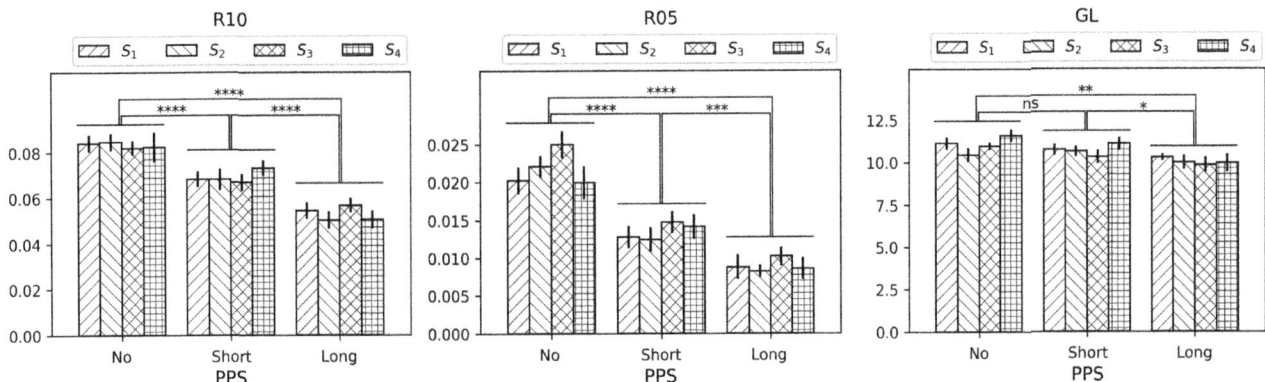

Fig. 3 Average $R10$(%), $R05$(%) and GL(#) for each scenario (S_1–S_4) and PPS length (No, Short and Long). For $R10$ and $R05$, high values indicate poor performance, while for GL, high values indicate good performance. Error bars represent the standard error. Significance levels correspond to the results of the Bonferroni adjusted post hoc t-tests (ns = not significant, * $p < 0.05$,** $p < 0.01$,*** $p < 0.001$,**** $p < 0.0001$)

parameterization and use of a prediction model leads to a better practical application.

However, the virtual experimental environment consisted of a simple, rectangular room without static obstacles and with sufficient space for alternative route planning. Typical work environments are often designed differently in this respect, e.g. with narrower corridors, doors or installed work equipment. The interpretation of the results is therefore limited. To increase the validity of the experiment, a repetition under more realistic environmental conditions would be expedient.

For navigation, a simple strategy of the robots was implemented: the avoidance of upcoming human positions (PPS). An improvement of this method, e.g. by weighting depending on a predicted time of a collision, could also contribute to the reduction of the collision risks with possibly lower load on the throughput of the robots.

The use of the PPS served as a tool to assess the potential benefits of a AI technology, which can be used in collaborative human-robot systems to improve safety. However, regression models are generally subject to error, so testing with real predictive models would be required to determine an industrially usable validity. Since many methods already exist for this purpose (e.g. [9–11]), the experiments could be repeated with such data-driven algorithms, using the methodology of the presented work as a reference.

Finally, an outlook on an experimental setup under real conditions should be mentioned. Since the simulated Turtlebots are available as real devices, it is reasonable to also investigate the behavior of a human in the presence of the mobile robots.

References

1. Sarker, I. H. (2021). Machine learning: Algorithms, real-world applications and research directions. *SN Computer Science, 2*(3), 1–21 (Springer).
2. Masini, R. P., Medeiros, M. C., & Mendes, E. F. (2021). Machine learning advances for time series forecasting. *Journal of Economic Surveys* (Wiley Online Library).
3. Mourot, L., Hoyet, L., Le Clerc, F., Schnitzler, F., & Hellier, P. (2022). A survey on deep learning for skeleton-based human animation. In *Computer Graphics Forum, 41*(1), 122–157 (Wiley Online Library).
4. Unhelkar, V. V., Lasota, P. A., Tyroller, Q., Buhai, R.-D., Marceau, L., Deml, B., & Shah, J. A. (2018). Human-aware robotic assistant for collaborative assembly: Integrating human motion prediction with planning in time. *IEEE Robotics and Automation Letters, 3*(3), 2394–2401.
5. Quigley, M., Conley, K., Gerkey, B., Faust, J., Foote, T., Leibs, J., Wheeler, R., & Ng, A. Y. (2009). ROS: An open-source robot operating system. In *ICRA workshop on open source software* (Vol. 3, No. 3.2, p. 5). Kobe.
6. Amsters, R., & Slaets, P. (2019). Turtlebot 3 as a robotics education platform. In *International Conference on Robotics in Education (RiE)* (pp. 170–181). Springer.
7. Fox, D., Burgard, W., & Thrun, S. (1997). The dynamic window approach to collision avoidance. *IEEE Robotics & Automation Magazine, 4*(1), 23–33 (IEEE).
8. Zheng, K. (2021). Ros navigation tuning guide. In *Robot Operating System (ROS)* (pp. 197–226). Springer.
9. Chiu, H.-K., Adeli, E., Wang, B., Huang, D.-A., & Niebles, J. C. (2019). Action-agnostic human pose forecasting. In *2019 IEEE Winter Conference on Applications of Computer Vision (WACV)* (pp. 1423–1432). IEEE.
10. Sofianos, T., Sampieri, A., Franco, L., & Galasso, F. (2021) Space-time-separable graph convolutional network for pose forecasting. In *Proceedings of the IEEE/CVF Inter- national Conference on Computer Vision* (pp. 11209–11218).
11. Fragkiadaki, K., Levine, S., Felsen, P., & Malik, J. (2015). Recurrent network models for human dynamics. In *Proceedings of the IEEE International Conference on Computer Vision* (pp. 4346–4354).

Sociomorphic Technologies—On the Typology of Artificial Actors

Michael R. Müller and Anne Sonnenmoser

Abstract

The central question of this paper is not whether Embodied Digital Technologies (EDTs) are to be understood as social actors by analogy with humans. The question is rather how the contingency of machine behavior is communicatively processed in design and how the mechatronics of machines is transformed into a socially accountable shape. To describe the contingency of machine behavior, we draw on Heinz von Foerster's concept of the nontrivial machine. To discuss the actor status of machines, we use the concept of "social frameworks" developed by Erving Goffman: While "natural frameworks" aim at a technical control of environmental phenomena, social frameworks operate with displays and expectations of behavior in order to provide orientation and to gain influence. Goffman's concept of social frameworks leads to the possibility of identifying different types of non-human, technically developed social actors. Each of these types is—as our comparative study shows—characterized by a different principle of organizing the relationship between man and machine.

Keywords

Artificial actors · Social displays · Sociomorphism · Anthropomorphism · Micro-sociology

M. R. Müller (✉) · A. Sonnenmoser
Institute for Media Research, Chemnitz University of Technology, Chemnitz, Germany
e-mail: michael-rudolf.mueller@phil.tu-chemnitz.de

A. Sonnenmoser
e-mail: anne.sonnenmoser@phil.tu-chmenitz.de

1 Introduction

How can we name the processes happening between humans and machines when the machines are so-called "social robots," "autonomous vehicles," or "smart assistants"? Can we call the interactions between man and machine *social interactions*? Can we speak of machines as *social actors*? [1–6]. Is it adequate to transfer such "human-related" [7] terms and categories to machines? When robots are repeatedly presented to us in humanoid form or when they are described as social, autonomous or intelligent, this suggests a corresponding transfer of concepts. At the same time, such anthropomorphisms provoke justified skepticism: not infrequently, the hybrid of wanting to imitate humans seems to be at work here, and the corresponding machines regularly fall very short of the properties promised [8, 9]. Instead of transferring concepts, the sociological motto would then be to avoid such concepts.

In the analytical examination of the development and the wealth of variants of the aforementioned types of machines, however, one quickly comes to the conclusion that in the design of such machines, role patterns and expressive figures of the social world are repeatedly taken up and behavioral sequences are produced, resulting in concrete manifestations of what Georg Simmel or Leopold von Wiese call "social forms" [10, 11]. Thus, one is faced with a paradoxical problem: an arbitrary transfer of basic theoretical concepts such as that of the social actor would level the differences between man and machine all too quickly. A rigorous avoidance of the term, on the other hand, would mask the commonalities that human-human and human-machine interaction have. Here as there, problems of pragmatic accountability of the human or machine counterpart have to be dealt with. And here as well as there, one resorts to established social forms of presentation and expression to tackle such problems.

Given this paradoxical problem, in the following, we will take an empirical path and examine the said machines from a social-centered perspective: as embodied techniques, i.e.,

B. Meyer et al. (eds.), *Hybrid Societies*, Advances in Science, Technology & Innovation,
https://doi.org/10.1007/978-3-032-03488-5_21

as preconceived phenomena or, in Alfred Schutz' words, as "first-order constructions" [12]. Hence, the question is not whether these machines are to be understood as social actors by analogy with humans [13–16]. The question is rather how the relative contingency of machine behavior is communicatively processed in design (how this contingency is thus "framed"). According to Thomas Luckmann [15], we assume that the status of the social actor is not bound to any form of humanity or human likeness—not even when human likeness is repeatedly used as a metaphor in robotics (or in theology) to assert agency. With Erving Goffman [17], we would rather like to ask whether we do not have to distinguish between *different concepts of social actors* depending on how the contingency of environmental phenomena is dealt with in social communication.

2 Sociomorphic Technology

In the empirical examination of designs, one repeatedly encounters human-like representations, but not at all exclusively. Zoomorphic and fictional designs are used as well, combined with anthropomorphic designs or simply replacing them. Analytically, this wealth of variants is by no means always easy to get to grips with. In the course of our research,[1] we therefore distinguish four basic "display-dimensions" [18].

- By the term *pictorial display* we mean the appearance of a machine (determined by its size, shape, and material nature).
- *Behavioral display* captures the style of movement of a machine and its positioning in social space.
- The term *linguistic display* refers to spoken language and the use of gestures.
- Finally, the term *digital display* captures technical image projections. Such digital displays are used, for example, for the pictographic realization of mimic expressions.

With regard to potential human-machine interactions, these displays—depending on their design and composition—fulfill different metacommunicative functions. Depending on the case, they serve

- to illustrate basic properties or skills of respective machines (we call this the *projective function*),
- to initiate and structure the exchange between man and machine (we call this the *adaptive function*),

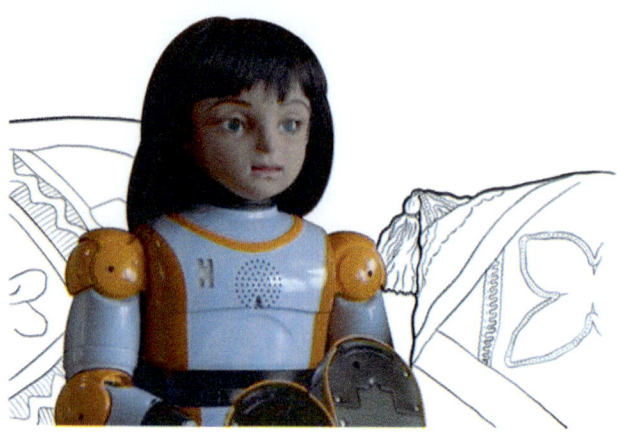

Fig. 1 So-called health robot Alice. *Source* Own archive

- to avoid or deal with breaks and crises in the course of interaction (we call this the *stabilizing function*).

In the case of the so-called health robot *Alice* (Fig. 1, cf. [19]), for example, the projective function presents itself as follows: At the level of linguistic display, the machine announces, "Hello, I am Alice, I am a health robot". With this speech act, a personal origo is designated ("I"). An explicit self-representation is initiated ("I am"). This in turn is concluded with the designation of an individual name ("Alice") and a categorical machine identity ("I am a […] robot"). At the level of the pictorial display, we find firstly a humanoid designed doll-like head segment and secondly design elements from small device design in the torso area (Fig. 1). The humanoid head segment (which is mimically movable in the original) holds out the prospect—as a design affordance—of an interpersonal exchange with this machine. The torso design, on the other hand, illustrates the technical character of the machine.

In the interaction of pictorial and linguistic display (and further display-dimensions) specific properties and functions of the machine are presented i.e., the way in which this machine can or should be encountered in a concrete situation is illustrated: It can be addressed—according to its display—like a personal being, but must nevertheless be understood as a piece of equipment. By combining design elements of different origins, properties such as personal addressability and social responsiveness are emphasized, while at the same time the idea of a comprehensive human likeness of the machine is negated.

We must dispense with further examples here. Nevertheless, analyses such as these illustrate that corresponding displays and their interplay give the respective machines a specific social shape. The machine Alice, for example, is presented as a device that has an origo and can reflect on it, that is addressable to others and can produce or influence social situations, and that can, to some extent, also recognize

---

[1] In the following, we refer to results of the research project "Social Displays. On the Accountability of Embodied Digital Technologies in Everyday Life" of the CRC 1410 "Hybrid Societies". Cf. [18].

and react to disturbances in the course of interaction. Corresponding machine designs are in this sense sociomorphic[2]: they are designs that performatively realize different social forms (such as origo-reference, presence, eye contact, shame, servitude, and other forms).

By giving machines a sociomorphic appearance, one suggests to observers a social rather than a merely technical orientation toward these machines. We will elaborate on this in the next chapter. Crucially, sociomorphic designs reflect—according to our thesis—attempts to develop concepts of machines as social actors, rather than as mere technical tools.

3 Concepts of Actors

So what are the concepts of social actors that are emerging—according to our thesis—in contemporary developments of design? In the following, we outline an initial typology (which will have to be adapted to future developments at the due time). As a point of reference (*tertium comparationis*), we base this typology on a variation of Goffman's well-known question "*What is it that's going on here?*" [17, p. 8]. In terms of human-machine interactions, this question must be: *What am I dealing with in the face of this or that machine? What do I have to expect from it?* Thus, we base our typology on the assumption that the functional and behavioral modes of non-trivial machines—at least for laymen—are structurally unclear or uncertain, i.e., in need of clarification.

This uncertainty results, among other things, from the fact—Heinz von Foerster [23] has elaborated this—that concrete behavior of non-trivial machines is not unambiguously predictable even if their basic mode of operation should be (somewhat) known. Such machines just do not work trivially, i.e., the input or cause x does not necessarily lead to the output or effect x'. There are many reasons for this characteristic of non-trivial machines: Machine learning may be one reason, but also networking with external data infrastructures or simply human remote control or ad hoc programming (cf. Fig. 2).

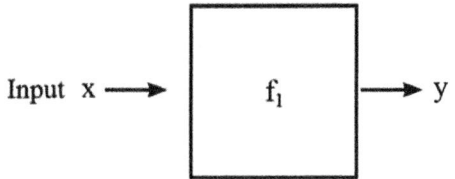

a) Trivial machine
(according to von Foerster 2019)

Input x $\longrightarrow$ f_1 $\longrightarrow$ y

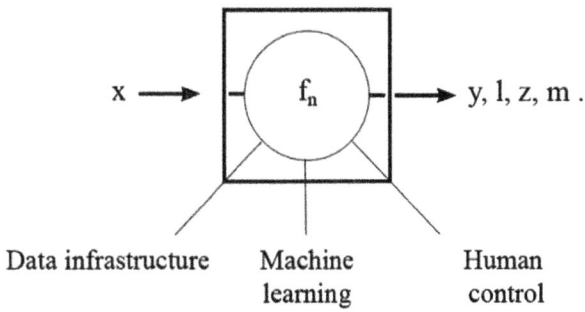

b) Non-trivial mashine

x $\longrightarrow$ f_n $\longrightarrow$ y, l, z, m ...

Data infrastructure   Machine learning   Human control

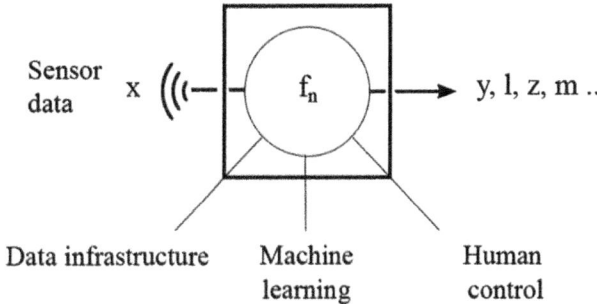

c) Non-trivial machine not at free availability

Sensor data   x $((($ f_n $\longrightarrow$ y, l, z, m ...

Data infrastructure   Machine learning   Human control

Fig. 2 Ontology of non-trivial machines. Non-triviality is not an attribute of a particular technology or particular control techniques, but refers to the relative uncertainty or ambiguity of machine behavior from an observer's perspective: although non-trivial machines are "deterministic systems, some of them are unpredictable in principle and others are unpredictable for practical reasons [...]: an output observed after a particular input will most likely not be observable with the same input at a later time" [23, p. 359] (see fig. b). According to von Foerster, the cause of such behavioral changes are changes in the "internal state" of respective machines (see above learning processes, networking, remote control): "Whereas in the trivial machine [cf. fig. a)] only one internal state is ever involved in its operation, it is precisely the change from one internal state to another that makes the non-trivial machine so intangible" [23, p. 358]. This problem of contingency, facing observers of the behavior of non-trivial machines, is exacerbated when such machines, based on their sensory systems or external data influx, not only respond (in an uncertain way) to intended inputs from their counterparts, but also process other sensory data into (uncertain) outputs (cf. fig. c): from the observer's perspective, such machines adapt to their environment even without observer-side operation, i.e., automatically.

Relevant aspect of the machine display   *Accountable characteristic*

Display of a facial scheme   Responsivity

Aesthetics of Butler   Servility

Scheme of childliness / cuteness   Juvenility

Display of autonomy   Solipsism

Fig. 3 Displays of non-trivial machines. *Source* Own illustration, own archive, cf. also: https://www.imdb.com; https://www2.deloitte.com; https://i.ytimg.com; https://www.goingelectric.de; https://www.zukunftsinstitut.de; https://2.bp.blogspot.com; https://robotsguide.com; https://image.slamtec.com; https://jobs.derstandard.at; https://www.key film.nl; https://robotsguide.com; https://image.futurezone.at; https://voi.id; https://ecomento.de; https://ecomento.de; https://www.eurotrans port.de; https://derletztefuehrerscheinneuling.com

A *first type of actor*,[3] which can be understood as an attempt to solve this problem of uncertainty, is realized by personifying presentations. Characteristic of such presentations is, among other things, the elaboration of a face or at least a facial scheme [24]. From the perspective of a human counterpart, facial schemes can be read as signs of being taken note of by a machine (in whatever form) and of being able to enter into a communicative exchange (of whatever kind) with it. Such facial schemes are affordances that enable the attribution of a certain responsivity to the respective machines. However, such presentations do not give any information about the technical mode of operation of the respective machines, nor do they make the concrete machine behavior clearly predictable. Nevertheless, such presentations allow the human counterpart to orient his own behavior toward these machines: not toward their technical function, but toward their visible responsiveness. The human counterpart can assume (for the time being) that the machine takes note of him or her (to whatever extent) and that one can communicate something to it, the machine.

Such a presentation (and interpretation) of a machine is what Goffman calls *social framing*: What will happen or what is already the case is not—as in natural framings—attributed to a "complete determinism, to complete determination," but to an "agency" [17, p. 22], which can intervene in a controlling way and which can be influenced accordingly, if necessary.

The *second type of actor* we want to discuss is diametrically opposed to the first, and is usually called an autonomous

---

[3] Our findings are based on the analysis of 282 EDT-designs. The qualitative sample was compiled and analyzed according to the principles of maximum contrast [26]. The aim of this sampling strategy was to capture as many different manifestations as possible of the phenomenon under investigation, i.e., to represent the entire spectrum of design possibilities and thus also of actor-concepts. For the data analysis, extensive, i.e. corpus-based, comparisons of pictorial displays were combined with single-case studies of the interplay of all the above-mentioned display levels. See here in more detail [18].

vehicle. If one takes a closer look at pictures like those of Fig. 3, series 4, it becomes clear what is meant by "autonomy" here: Autonomy here means (instead of the ability to use reason) the obvious absence of a human being where usually a human being is and controls—the absence of human heteronomy. Also such presentations do not give any information about the concrete operation of respective machines. And they also do not make the machine behavior unambiguously predictable. Will the machine take notice of me? Will it stop? Is it remote controlled? Does it run on digital tracks with the help of suitable sensors? Or does it move freely in space? All these questions remain unsolved. Nevertheless, presentations of autonomy have a social orientation value. For the time being, the human counterpart can assume that a machine of this kind will not take note of him or her in the way a human driver would, and that it would hardly be possible to make the machine understand anything, as it would be possible with a human driver. But it is precisely on the solipsism of such a demonstratively unsocial machine-being that one can orient oneself. Where the concept of responsivity runs the risk of suggesting possibilities of communicative exchange which are technically not given, the concept of solipsism takes the opposite path: It dissimulates such possibilities even where they may be given in rudiments.

Two further types of actors, that we will briefly discuss, are characterized by the presentation of servility and juvenility, respectively. Servility (servile subordination) is presented by means of an expressive repertoire idiomatically borrowed from the aesthetics of the butler or the waitress. Juvenility, on the other hand, can be presented through comparatively small body size, childlike schemas, and behavioral forms of cuteness. The contingency of machine behavior is made manageable by presentations of servility in the sense that the human counterpart is placed in a position of directive authority: Whatever the machine does, the supremacy over essential processes is assigned to the human. Juvenility presentations, on the other hand, have a function that exonerates the machine. They excuse deficiencies and inabilities a priori and characterize the respective machine as an unfinished, immature machine being.

What unites all these types of technical, i.e. non-human social actors are two aspects: First, machines of this kind deal with control systems that are opaque in everyday practice. These systems are secondly complemented by practically motivated presentations in the sense of Harold Garfinkel: by the presentation of individually accountable and socially predictable properties (like those of responsiveness, servility, etc.). By assigning and combining such properties, it becomes vivid and explainable how to deal with the contingency of machine behavior ideally: by cooperative communication in the case of responsive machines, by instructions in the case of servile machines, by caution in the case of solipsistic machines, and by forbearance and partial responsibility-taking in the case

of juvenile machines. The machines are thus committed in their design to a principle (a "characteristic") that makes the behavior of the machines by no means entirely transparent and predictable, but nevertheless socially processable.

4 Conclusion

Now, we finally return to the question posed at the beginning: Is it adequate to describe machines (of the kind mentioned) analytically as social actors? It should have become clear that, in our view, the question can be answered in the affirmative. But this necessarily requires specifying the conditions under which this description can be adequate.

First of all, there is the condition that such a terminology does not implicitly equate man and machine. In order to prevent a theoretical anthropomorphization, a decidedly sociological definition of the term is required. Following Goffman's "frame analysis" [17] it could read like this: *A social actor is the product of a presentation (a designation, design, or ritual action) by which a non-trivially behaving corporeal or intelligible entity (e.g., a machine, a network) is ascribed properties toward which a counterpart can orient itself and which can be used as a standard of social judgment of the actor thus designated.*

Second, such a use of the term would be adequate only if the ontological differences between, say, human and technical actors were explicitly taken into account: The behavioral basis of a machine is quite different from that of a human. We still have pointed out the interplay of machine learning, internet of things, and remote control.

Third, the heuristic character of such a conceptualization should be emphasized: It serves the reconstruction of concrete concepts of actors and not the blanket description of machines as actors. Thus, the question is where and how such concepts are developed and applied; to what extent they are adopted in everyday practice; and above all: how they function as coping strategies for dealing with those artificial uncertainties that result from technical innovations.

We conclude our reflections at this point with a quotation from Hans Blumenberg: "The *homo pictor* is not only the producer of cave images for magical hunting practices, but the being that overplays the lack of reliability of its world with the projection of images" [25, p. 14 (translation M.R.M & A.S.]. Religion designs—along these lines—divine beings to counter the unreliability of existence. Law and morality design the person to come to grips with the unreliability of the other. Sociology designs–as a scientific theorem–the rational agent in order to be able to describe the unreliability of its object at least as a deviation. Robotics, finally, design juvenile, servile, responsive or autonomous machine beings. All of these concepts of actors (the divine being, the person, the rational agent, the juvenile, servile, responsive or autonomous machine) serve a comparable function: they are used to gain reliability where things and occurrences would otherwise remain uncertain or insecure.

Acknowledgements Funding Funded by the Deutsche Forschungs-gemeinschaft (DFG, German Research Foundation)—Project-ID 416228727-SFB 1410.

Conflicts of Interest The authors declare no conflict of interest.

References

1. Alač, M. (2016). Social robots: Things or agents? *AI & Society, 31*(4), 519–535. https://doi.org/10.1007/s00146-015-0631-6
2. Muhle, F. (2018). Sozialität von und mit Robotern? Drei soziologische Antworten und eine kommunikationstheoretische Alternative: Sociality of and with Robots? Three Sociological Answers and a Communication-Theoretical Alternative. *Zeitschrift für Soziologie, 47*(3), 147–163. https://doi.org/10.1515/zfsoz-2018-1010
3. Lindemann, G. (2016). Social interaction with robots: Three questions. *AI & Society, 31*(4), 573–575. https://doi.org/10.1007/s00146-015-0633-4
4. Pfadenhauer, M. (2013). On the sociality of social robots. A sociology-of-knowledge perspective. *Science, Technology & Innovation Studies, 10*(1), 135–153.
5. Rammert, W., & Schulz-Schaeffer, I. (Eds.). (2002). Können Maschinen handeln? Soziologische Beiträge zum Verhältnis von Mensch und Technik. Campus.
6. Nass, C., Steuer, J., & Tauber, E. R. (1994). Computers are social actors. In *Conference Companion on Human Factors in Computing Systems-CHI'94*, 204. https://doi.org/10.1145/259963.260288
7. Elias, N. (2000). *Was ist Soziologie?* (11. Aufl). Juventa-Verl.
8. Horstmann A. C., & Krämer, N. C. (2020). Expectations versus actual behavior of a social robot: An experimental investigation of the effects of a social robot's interaction skill level and its expected future role on people's evaluations. *Plos One, 15*(8). https://doi.org/10.1371/journal.pone.0238133
9. Komatsu, T., Kurosawa, R., & Yamada, S. (2012). How does the difference between users' expectations and perceptions about a robotic agent affect their behavior?: An adaptation gap concept for determining whether interactions between users and agents are going well or not. *International Journal of Social Robotics, 4*(2), 109–116. https://doi.org/10.1007/s12369-011-0122-y
10. Simmel, G. (1984). *Grundfragen der Soziologie: Individuum und Gesellschaft* (4., unveränderte Aufl). W. de Gruyter.
11. Häußling, R. (2010). Formale Soziologie. In C. Stegbauer & R. Häußling (eds.), *Handbuch Netzwerkforschung* (241–254). VS Verlag für Sozialwissenschaften. https://doi.org/10.1007/978-3-531-92575-2_22
12. Schutz, A. (1962). *Collected papers*. Martinus Nijhoff.
13. Muhle, F. (2019). Humanoide Roboter als, technische Adressen'. *Sozialer Sinn, 20*(1), 85–128. https://doi.org/10.1515/sosi-2019-0004
14. Lüdtke, N., & Matsuzaki, H. (2011). Einleitung. In N. Lüdtke & H. Matsuzaki (eds.), *Akteur-Individuum-Subjekt* (11–20). VS Verlag für Sozialwissenschaften. https://doi.org/10.1007/978-3-531-934 63-1_1
15. Luckmann, T. (1970). On the boundaries of the social world. In *Phenomenology and Social Reality* (pp. 73–100). Springer, Dordrecht. https://doi.org/10.1007/978-94-011-7523-4_5
16. Lindemann, G. (2012). Die Kontingenz der Grenzen des Sozialen und die Notwendigkeit eines triadischen Kommunikationsbegriffs. *Berliner Journal für Soziologie, 22*(3), 317–340. https://doi.org/10.1007/s11609-012-0192-1
17. Goffman, E. (1986). *Frame Analysis: An essay on the organization of experience*. Northeastern University Press.
18. Müller, M. (2023). Social Displays. Creating Accountability in Robotics. *Österreichische Zeitschrift für Soziologie/Austrian Journal of Sociology, 48*, 469–487. https://doi.org/10.1007/s11614-023-00534-2
19. van Kemenade, M., Konijn, E. A., & Hoorn, J. (2015). Robots humanize care—Moral concerns versus witnessed benefits for the elderly. In *Proceedings of the International Conference on Health Informatics* (pp. 648–653). https://doi.org/10.5220/0005287706480653
20. Topitsch, E. (1960). *Probleme der Wissenschaftstheorie*. Springer.
21. Schulz-Schaeffer, I. (2007). Zugeschriebene Handlungen: Ein Beitrag zur Theorie sozialen Handelns (1. Aufl). Velbrück Wissenschaft.
22. Seibt, J., Vestergaard, C., & Damholdt, M. F. (2020). Sociomorphing, not anthropomorphizing: Towards a typology of experienced sociality. In M. Nørskov, J. Seibt, & O. S. Quick (Eds.), *Frontiers in Artificial Intelligence and Applications* (pp. 51–67). IOS Press.
23. von Foerster, H. (2019). *Wissen und Gewissen: Versuch einer Brücke* (1. Aufl., [Nachdr.]). Suhrkamp.
24. Chesher, C., & Andreallo, F. (2021). Robotic faciality: The philosophy, science and art of robot faces. *International Journal of Social Robotics, 13*(1), 83–96. https://doi.org/10.1007/s12369-020-00623-2
25. Blumenberg, H. (1979). *Arbeit am Mythos* (1. Aufl). Suhrkamp.
26. Glaser, B. G., & Strauss, A. L. (2010). *The discovery of grounded theory: Strategies for qualitative research* (5. paperback print). Aldine Transaction.

Towards Hybrid Personae?

Stefanie Meyer⬝, Michael R. Müller⬝, Anne Sonnenmoser⬝,
Sarah Mandl⬝, Anja Strobel⬝, and Dagmar Gesmann-Nuissl⬝

Abstract

When we think of future hybrid societies, this goes far beyond conventional scenarios. We have long since moved away from humans working only by hand or with the help of machines, and we already find technologized humans, humanized technology, and hybrid entities within a society, and even more so in the future. In this paper, we explore how all these entities fit into existing legal, psychological, and sociological constructions. We shed light on the attributions to an acting entity from the respective disciplines and explore whether hybrid entities can possess person-ality, paving the way for a hybrid personae. It becomes clear that at the core of conventional considerations is the biological person. Based on common definitions of person and personality, psychological considerations do not, up to now, grant personality to artificial actors. This explicitly excludes all kinds of technology where a human being is involved, irrespective of the level of technicity. Sociology can also designate machine entities as actors. The concept of person is always connected with the human being. In legal science, it is conceivable and not excluded in advance, to extend the concept of personhood to hybrid entities as well, at least under certain conditions and perspectives.

The original manuscript was submitted in November 2022; it was revised in December 2022. The conference and the corresponding talk took place in March 2023. The paper has not been revised in terms of content since then.

S. Meyer (✉) · D. Gesmann-Nuissl
Private Law and Intellectual Property Rights, Faculty of Economics and Business Administration, Chemnitz University of Technology, Chemnitz, Germany
e-mail: stefanie.meyer@wiwi.tu-chemnitz.de

D. Gesmann-Nuissl
e-mail: dagmar.gesmann@wiwi.tu-chemnitz.de

M. R. Müller
Visual Communication and Media Sociology Institute for Media Research, Chemnitz University of Technology, Chemnitz, Germany
e-mail: michael-rudolf.mueller@phil.tu-chemnitz.de

A. Sonnenmoser
Institute for Media Research, Chemnitz University of Technology, Chemnitz, Germany
e-mail: anne.sonnenmoser@phil.tu-chemnitz.de

S. Mandl · A. Strobel
Personality Psychology and Assessment Faculty of Psychology, Chemnitz University of Technology, Chemnitz, Germany
e-mail: sarah.mandl@psychologie.tu-chemnitz.de

A. Strobel
e-mail: anja.strobel@psychologie.tu-chemnitz.de

Keywords

Hybrid societies · Person · Humanized technology · Technologized human · Hybrid entities · Hybrid personae

1 Introduction

With increasing digitalization and new technologies, societies are expected to include not only human but also artificial actors. This kind of future of societies raises new questions about coexistence, tasks, and the significant characteristics of the actors. The principles of today's society are—legally, psychologically and sociologically—closely related to human beings. The coexistence of the actors of societies is oriented to the abilities and capacities for action that human beings exhibit as well as the positions they take up within the social world. This may change in the foreseeable future, as shown by the application of artificial autonomous agents in everyday life (e.g., in production, but also in care) [1, 2]. A variety of disciplines and perspectives are involved in designing these future societies—so-called hybrid societies. This requires a common understanding of the actors of hybrid societies and thus of a possible status of artificial actors as (hybrid) personae. This paper aims at clari-

B. Meyer et al. (eds.), *Hybrid Societies*, Advances in Science, Technology & Innovation,
https://doi.org/10.1007/978-3-032-03488-5_22

fying the mentioned concepts from a legal, psychological, and sociological perspective.

2 Object of Consideration

When we talk about the different technologized actors of a hybrid society, we can divide these actors into a total of three groups:

- humanized technology
- technologized human
- hybrid entities.

Of course, these groups can be further divided into definable categories [3]. The first of the aforementioned groups consists essentially of technology (machines, software, etc.) that has been assigned certain characteristics that make it appear to be human (anthropomorphism) to reduce any existing fear of contact. This refers not only to the exterior, i.e., the appearance of an android [4, 5]. Social robots, for example, are also considered to possess the behavior of a human being; they often speak, gesticulate and act in a human-like manner [6].

The second group essentially represents a human being, but with certain components that have been technologized. This may be (medically) indicated, such as a brain or heart pacemaker [7, 8], or it may be done without necessity (as in the case of Neil Harbisson) [9]. Here the subject "human being" is still in the first place.

In the third group—hybrid entities—it is increasingly unclear whether or to what extent the entity is a human or a technology (transition and interface between hybrid and cyborg to clarify the shape of the existing terminologies) [3] due to the increasing intermingling of human and technological aspects. Such hybrid entities are encountered, for example, in the form of robots or vehicles that are controlled or piloted, temporarily or proportionally, by human beings. The difference to the first group of humanized technology is precisely the opacity and thus the lack of explainability of the acting part within the actor (rather the human component—i.e., the human brain—or the technical one) and the question of which motivation the entity has to act (an emotional one or a rational one). Insofar, the dividing line between human and technology is blurred, or rather, the human being and the technology merge into a single entity, so that the latter cannot be clearly categorized as either human or technology in one direction or the other.

3 Hybrid Personality

In order to consider in more detail what it would mean to accept a hybrid personae in our society, a decided look at this entity from different disciplines is necessary. For this reason, we take a look at the preconditions that an actor already occupies in a society so that we can think further and, in a second step, consider whether the hybrid personae could exist in the perspective of legal science, psychology and sociology and what preconditions would be required for this existence. Thereby, the question of a status can be clarified.

3.1 Aspects of Personality

In a first step, we examine the capabilities that the respective discipline—legal science, psychology, and sociology—associates with a human's personality.

3.1.1 Legal Science

As far as the legal perspective on hybrid personality is concerned, we will restrict this contribution to a consideration of the constitutional prerequisites and principles, with a special focus on the general freedom of action protected by fundamental rights. However, due to the indirect third-party effect of this fundamental right [10–12], the following also applies to the other areas of law. However, their consideration (especially the special aspects of civil and criminal law) is covered in a subsequent article. The considerations presented here therefore focus on the preconditions of legal personality or the person in law from the perspective of the German Basic Law (GG) and the European Charter of Human Rights (ECHR). However, looking at the preconditions that the entire legal system attaches to the application of legal concepts, it is noticeable that the addressee and beneficiary of legal norms is always the human being. Actually, this is not really surprising when we consider the meaning and purpose of law: Law is created to regulate the coexistence of society (which basically consists of a multitude of human beings) in a legally binding way. In particular, it is intended to secure the society's conditions of life [13]. Some philosophers of law take a more differentiated perspective on the reference to this purpose, such as Kant, who considers the content of law to be determined by its purpose, but the form to be determined by justice [14, 15]. The concept of person or legal subject can only refer to a responsible human being, which in Kant's words means that the ability to act according to the categorical imperative has to be given [15]. Although there is no specific definition of justice in the German Basic Law (GG) or in comparable European legal acts such as the ECHR, it is the primary objective of every legal expert. The principle of justice is characterized by the fact that the governmental legal system does not prescribe to human beings the objectives of their actions, that it basically refrains from patronizing specifications of purpose, and that it restricts its scope to a binding framework of freedom. In short, the fundamental right is characterized in particular by rights of freedom and equality. Rights of freedom represent possibilities which the entitled person can make use of or also reject [16]. Following the intention of the legislator, the person entitled to freedom [16–18] by law is, in principle, the human being. It is only the human being as a natural person who is the bearer of all the fundamental rights that ensure their freedom. Only the natural person is basically able to produce a legal effect with their own actions (i.e., they have legal capacity), which is based in particular on their capacity for insight. The fact that legal persons can also apply certain fundamental rights, at least insofar as they are applicable by their nature (cf. Article 19 (3) GG), is a mere auxiliary construction for associated persons entitled to freedom. An expansion of the scope of those entitled to freedom does not occur here. The situation is different, however, in the case of the "hybrid entity" described at the beginning—here it is a case of an organism acting independently, whose characteristic as a person entitled to freedom still has to be classified.

These aspects cannot be considered in separation from the legal consequences accorded to or imposed on a person entitled to freedom. The granting of human dignity (cf. Article 1 (1) of the GG and Article 1 of the Charter of Fundamental Rights of the European Union) is conceptually linked to the essential nature of the human being. In addition, the general right of personality (Art. 2 (1) in conjunction with Article 1 (1) of the GG or Article 8 of the ECHR and Article 7 of the Charter of Fundamental Rights) is protected. This fundamental rights protection is directed at aspects which are not the subject of special guarantees of freedom, but which are of equal importance regarding their constitutive significance for the personality of the individual [19, 20]. The protected legal interest is the human being's claim to validity in the social world which characterizes them and which they, for their part, design by trying to be recognized through their actions [19]. The scope of protection includes a wide range of areas: the protection of private and intimate sphere, the right to self-expression or any other protection of personal autonomy. It is primarily natural persons, i.e., human beings, who can invoke this protection [19]. However, the German Constitutional Court (BVerfG) emphasizes the importance of the effectiveness of dynamically adaptable protection of fundamental rights, which is particularly relevant when human personality is exposed to actual or perceived new threats due to social or technological developments [21].

3.1.2 Psychology

From a psychological viewpoint, if we talk about hybrid personalities, the distinction between *person* and *personality*

needs to be established. Stern [22] stated that persons as such are, "despite the manifoldness of the parts, one real, peculiar, and intrinsic unity which accomplishes a unified single-minded wholeness." Hence, "personality refers to the person as long as its inner determination shines through." We infer from this that embodied artificial actors could probably be perceived as persons in Stern's sense due to the physical presence in the physical world, but not as having personality. As *persons*, that is, as actors in hybrid societies, human beings are still unique within the group of actors of hybrid societies: They are the only full ethical actors and as such, responsible for their actions [3]. Since it is not possible to locate human beings on a continuum, as long as a human being—regardless of the level of technicity—is involved, the actor counts as such. Full ethical actors are required to possess moral agency. Moral agents are able to discern morally relevant information, make moral judgments based on this information, and initiate action based on the judgment [23]. Therefore, moral agency should be considered a prime, but not the only capacity for moral responsibility [3]. As Meyer et al. [3] pointed out, a total of seven capacities (i.e., capacity to act, legal capacity, autonomy, liability, explainability, moral agency, and trust) needs to be met for an actor to be responsible actor within hybrid societies. Artificial actors are currently not in possession of these capacities, hence, there is a clear distinction between human and artificial actors in terms of their actual capabilities. However, the perception of the actors might be changing in terms of which attributes actors (seem to) possess. Personality, on the other hand, refers to "the enduring configuration of characteristics and behavior that comprises an individual's unique adjustment to life, including major traits, interests, drives, values, self-concept, abilities, and emotional patterns [...] various theories explain the structure and development of personality in different ways, but all agree that personality helps determine behavior." [24]. Another definition refers to personality as the entirety of biologically resp. genetically (temperament) and culturally (character) influenced characteristics [25], both of which an artificial agent clearly cannot possess. Thus, artificial actors are currently not in possession of personality in a narrow sense.

3.1.3 Sociology

From a sociological perspective, the concept of the person captures the social existence of an individual: what he or she is as a person always results from his or her status, position, prestige and affiliation. At the same time, the concept of the person captures the human being's endowment of reason as well as the rights of freedom resulting from this endowment [26]. Becoming a person thus is the result of socialization as well as of being at a reflective distance from oneself and the (social) world. "An individual," Helmuth Plessner states, who is not only a living body, but who is able to observe and form his body and behavior from the outside, so to speak "is called a person" [27, p. 11].

Plessner's concise definition is instructive for our discussion, because it points out that artificiality does not stand in opposition to human nature, but is rather characteristic of becoming a person in the sense of man outgrowing nature. The ability of the natural organism to observe itself, to shape its behavior, to use tools and to produce artifacts is what makes the living body a person. From such a sociological perspective, therefore, there can be no doubt that the extension of the human body by robotics or its endowment with robotics (see above "technologized human") is covered by the concept of person. It does not make sense to speak of a cyborg mixture of the natural and the artificial here, because artificiality is not opposed to the nature of man, but is what makes him or her a person.

However, it is more difficult to answer the reverse question whether the concept of the person can be transferred to technical devices. Obviously, the concept of the person—in the sense understood here—is related only to biological bodies and living beings who transcend themselves within the framework of the societal knowledge horizon, and not to purely artificial embodiments of human abilities. We will come back to this later.

3.2 Focus: Acting Entity (Actor)

In a second step, we examine the capabilities that the respective discipline—legal science, psychology, and sociology—associates with the status of the actor.

3.2.1 Legal Science

If, for example, the constitutionally protected legal interest and the regulation of the general right of personality are considered, the protected legal interest is the human being's claim to validity in the social world that characterizes them and that they, for their part, form by endeavoring to be recognized through their actions [19]. It is noticeable that, according to the wording of the definition, these requirements are only applicable to human beings, but an extension beyond the human being is not excluded in principle by the wording of the law. On the contrary, case law requires dynamic protection of fundamental rights [21, 28]. The intention of this requirement is first of all to protect the human being intrinsically and to protect the human being from dangers to their personality. However, this idea of protection can also be extended. This requirement initially aims to protect human beings intrinsically and to protect them from dangers to their personality. However, this concept of protection can also be extended. A "hybrid entity," as defined by Meyer et al. for example, presupposes that the personality of the human being inherent

in this entity has been further developed by technology, so that the possible characteristics can no longer be attributed to technology or to the human being [3]. However, if this "hybrid entity" is not to be denied a certain personality, the protection of fundamental rights should also be designed dynamically in this regard. The decisive factor will be whether the "hybrid entity" is capable of participating in structuring the social world and whether its actions can be regarded as equivalent to those of human beings. Indeed, the mere wording of the law does not exclude the protection of a hybrid personae in advance-on the contrary, case law requires a dynamic protection of fundamental rights. This claim can also be asserted against the "hybrid entity" if they are appropriately integrated into the structures of society.

3.2.2 Psychology

As described above, artificial actors are not in possession of the same capacities as human beings, but they might be perceived similarly to human beings: Nass et al. [29] described that computers are social actors (CASA; Computers Are Social Actors), that is, that in human-computer interaction, social rules are applied unconsciously. Furthermore, the phenomenon of anthropomorphism has been discussed in various disciplines. It includes the attribution of human characteristics to non-human objects. Psychology characterizes anthropomorphism as "a basic process of induction" that "works through a similar process of starting with highly accessible knowledge structures as an anchor or inductive base that may be subsequently corrected and applied to a non-human target" [30, 30, 31]. Research has shown that robots, as artificial actors, are attributed traditional human capacities included in the Stereotype Content Model (SCM) [32], that is, warmth (i.e., the individuals' intentions, from good to bad) and competence (i.e., the individuals' ability to act on those intentions) [33–35]. Furthermore, artificial actors are seen as trustworthy [36], as gendered [37], or as deserving of moral care (i.e., moral patiency) [38].

While artificial entities may of course be actors, they are not fully responsible/ethical actors in a human sense. As full ethical actors, human beings are required to possess moral agency. Moral agency includes different capacities [39], which refer to rationality (e.g., reasoning), autonomy (e.g., critical reflection of values), intentionality (e.g., believing), or sentience (e.g., consciousness), among others [3]. These aspects, some of which are uniquely human, whereas others might be attributed to non-human entities, require a closer look for the definition of hybrid personalities. Sentience, for example, cannot be installed in artificial actors, but they might *seem* to possess them. For instance, Scholtz mentions how a robotic dog might be perceived as a being with emotions which, vice versa, evokes emotional responses from the users [40]. By doing so, the artificial actor is attributed capacities and subsequently instills expectations it cannot fulfill [41]. If the trust placed in the artificial actor is violated in such a way, it could have tremendous effects on the actual use of the technology: Users could abandon the technology completely or use it only in a very limited way [42].

3.2.3 Sociology

In sociology, the concept of the actor has different ontological connotations. At one extreme are theoretical positions that attribute the ability to act exclusively to humans, since humans have consciousness. Another extreme is represented by the Actor Network Theory (ANT), which affirms the question whether the concept of action is also applicable to technical artifacts or things: Artifacts as well as natural things and humans are referred to as "actants" in ANT [43]. While the first, essentialist position reserves the term exclusively for humans, according to the second position almost anything can be an actor.

More instructive for our problem, however, is the approach of Erving Goffman, which is empirically focused. Goffman assumes that agency is a social attribution that occurs when the behavioral contingency of an entity or system can no longer be technically controlled [44]. The attribution of agency allows us to impose certain behavioral expectations on the respective entity or system and to sanction them if necessary. Influence on entities or systems is then no longer exerted by technical interventions, but by socio-normative behavioral expectations and regulations. Actors in this sense can also be animals or institutions or any entity that behaves non-trivially and thus challenges our efforts to control the world around us.

Based on such an interactionist understanding, it becomes possible to speak of artificial actors without equating them ontologically with human actors. Rather, the term now acquires a heuristic character [45]: it concerns the reconstruction of different concepts of artificial actors and not about a blanket description of machines as actors. Thus, it is about the question where and how concepts of artificial actors are developed and applied and how they function as coping strategies for dealing with those uncertainties and insecurities that result from technical innovations.

Empirically, at least four different types of technologically realized artificial actors can already be distinguished in the current state of technical development: These are *firstly* responsive machines that are receptive to human communication and can adapt to communicatively presented behavioral expectations; *secondly*, so-called autonomous machines that dissimulate the necessity or possibility of spontaneous human control; *thirdly* servile machines, which indicate by their appearance that they subordinate their self-activity serving humans; and *fourthly* juvenile machines, which already clarify their limited self-activity by their design and excuse it a priori [45, 46].

4 Hybrid Personae?

4.1 Legal Science

To the extent that a hybrid entity is to be granted a legally protected status that goes beyond the protection of a mere "thing" or object within the meaning of Section 90 of the German Civil Code (BGB) (cf. Article 14 of the GG), it is first necessary to integrate the hybrid entity into a social framework and to classify its appearance in governmental society. In this context, the question does not arise—as is the case with legal persons—as to whether the fundamental rights are applicable by their nature. In contrast to legal persons, hybrid entities do not require an auxiliary construction, as is necessary in the case of corporations. Rather, social clarification is required as to whether hybrids can act like human beings, i.e., whether they act in and with the social world and perceive and are perceived by it. Insofar as they can develop a certain claim to validity in society, the wording of the law does not fundamentally exclude legal personality for hybrid personae.

Nevertheless, Kant's fundamental legal philosophical interpretations and considerations always referred to the specific situation he was faced with during his lifetime. In other words, he argued on the basis of the thoughts that were predominant in the time of the Enlightenment or the Age of Reason and focused on human being as the center of action. It can be assumed that Kant, if he were aware of today's developments, would take these into account in his considerations and possibly move away from the human being as the center of action. In view of hybrid societies, this would have the consequence that fundamental principles of the philosophy of law, such as Kant's [15] view that purpose determines the content of law and justice determines the form, would have to be reconsidered. After all, justice is always about justice in the coexistence of people and the functioning of society. Consequently, the question arises whether such a way of thinking is still appropriate with regard to technologized hybrid societies. Or whether this central focus on the human being actually needs to be fundamentally reconsidered and thus also law—derived from Kant, among others—redefined and interpreted.

Here, however, the question arises as to how legal personality with allocated rights and obligations can be established. If a hybrid entity can be described as being able to recognize its entitlement to freedom and to act accordingly, this also constitutes a type of legal capacity. A hybrid entity can participate in a legal transaction, dispose of its assets and be the addressee of legal obligations. By virtue of its legal capacity, it would be the bearer of rights and obligations under civil law. In addition, they would be able to exercise fundamental rights from a constitutional perspective and commit criminal acts [47]. This is already controversial in relation to humanized technology [48–51]. However, it is easier to justify in the Anglo-American legal sphere, where legal capacity is not necessary for a "person" who is not human to sue and be sued (in the case of an action against a company, for example, the addressee is the company as such without being represented by its managing director, see [52]); this is different in other cultures, such as Japan, where even technical entities can be socially recognized as "persons" [53]. Thus, if an affirmative answer is to be given to the question of the personhood of the hybrid entity in the legal sense, a reorientation and redescription of the characteristics of legal personhood is required.

4.2 Psychology

Artificial actors cannot be seen as comparable to human beings in the narrow sense, therefore, irrelevant of any technical details, as long as a human being is involved, the actor will count as a human being. However, an artificial actor might be perceived as possessing human-like capacities and therefore evoke reactions comparable to a human social partner. If this happens, it should be considered as a double-edged sword: It might on one hand facilitate a more natural, easier communication between human and artificial actors. On the other hand, it could elicit exaggerated trust being placed in the artificial actor. This could, if the trust is violated, lead to negative reactions up to the abandonment of the artificial actor.

4.3 Sociology

The sociological finding is, as seen, a double. On the one hand, machines today are increasingly developed and circulated as artificial actors: they are staged as interaction partners (as social robots, as automotive vehicles, as smart assistants), which trigger certain behavioral expectations due to their external appearance and which can fulfill or disappoint precisely these behavioral expectations. It is precisely this relationship, which is no longer merely instrumental but also socionormative, that characterizes them as (artificial) actors. On the other hand, a transfer of the concept of persona to machines seems not only inappropriate, but also wrong, if one assumes with good reasons that the status of person presupposes the existence of a biological body.

From a sociological point of view, the concept of the hybrid persona can therefore only be appropriate if it refers to those human parts in the development, in the intervention, the control and the explanation of such machines, without which such machines would not be possible [54]. For also in this, the human being grows out of himself as a person, that he invents machines, which are self-acting and to which he gives an agency. Both moments of agency, that of the human being and that of the machine, merge, as Werner Rammert notes, into a "distributed agency" [55]. The structural sociomoral problem thus consists in the fact that human agency manifests itself in concrete machine behavior, but that the latter (the concrete machine behavior) does not result from human agency alone and can no longer be causally related to concrete natural persons.

The concept of the *hybrid persona* could thus be discussed—from a sociological point of view—as a potential social construct that maps and weights the institutional interaction of several humans and the technical interaction of humans and machines in such a way that relevant human shares of action become sociomorally visible and manageable. In this sense, the *hybrid persona* is thus not a basic sociological concept, but a concept that refers to a socially virulent problem of robotics.

5 Conclusion

As the preceding investigation shows, each of the disciplines focuses on the human being in terms of a conventional approach. Nevertheless, each discipline has a different perspective on the possibility of establishing a hybrid personae.

From the perspective of legal sciences, it can be stated that the scope of law initially includes humans as norm addressees and norm recipients. However, this circumstance is based on the fact that the law was developed for the society in order to regulate the coexistence of its actors. When the law was developed, no consideration of beings other than human beings as part of society occurred at all. In this respect, this assumption is based on the (psychologically justified) fact that only human beings can act completely independently. However, the wording of the laws does not exclude hybrid beings from the scope of application. Rather, such an enhancement of the law is intended by the courts; a dynamic protection of fundamental rights is desired. The legal question of whether a hybrid being can act as part of a society and is also accepted as such consequently has to be preceded by the social question. Insofar as this is the case, the law cannot exclude the attribution of a personality.

From a psychological point of view, we conclude that in the case of hybrid personalities, transfer processes of human attributions to non-human entities, in this case artificial actors, play a relevant role. However, in the deployment of artificial actors it is of great importance to stress that even if an artificial actor might be attributed human capacities, it is still in its core not human. With regard to the current state of technological development, it can therefore not be seen as a fully responsible actor.

From a sociological point of view, there is little to be said against referring to machines as artificial actors, if appropriate, and asking what the social taxonomy of artificial actors is at a given point in social and technological development. As artificial actors, machines also acquire a socially significant accountability to which they are obliged in the sense of granting functional freedoms [55–57]. The status of the person, instead, refers to humans as living beings and to their responsibility for their actions. Nevertheless, the construct of *hybrid personae* could be suitable for sociomoral processing of complex, i.e., institutionally and technically distributed agency between humans and machines. Whether such a construct will develop in future societies, is an empirical question. The problem at hand, however, is the responsibility of man that goes beyond the accountability of machines.

In summary, it can be concluded that in the controversial topic of granting hybrid personae, the disciplines examined here interact closely. While psychology recognizes the human being as the actor in every case, i.e., regardless of the degree of technologization, and only this individual can possess a personality of their own, both sociology and legal science permit an attribution under certain conditions. Either way, this perspective also make sense in terms of legal history, since human beings alone have always been at the core of legal activity and only they can act in a legally (and morally) relevant way. Society, however, as described at the beginning, is constantly developing and technologizing. A hybrid personae may therefore be appropriate from a sociological perspective if it refers to human parts of an entity. However, the actions of the human parts and the technical parts are merging. This development also has to be addressed by law, which in its wording does not exclude a further development of norm addressee and norm recipient and thus provides an open path to the establishment of a hybrid personae.

Acknowledgements Funded by the Deutsche Forschungsgemeinschaft (DFG, German Research Foundation)—Project-ID 416228727-SFB 1410.

Conflicts of Interest The authors declare no conflict of interest.

References

1. Bendel, O. (2018). Roboter im Gesundheitsbereich. In O. Bendel (Ed.), *Pflegeroboter* (pp. 195–212). Springer Gabler, Wiesbaden. https://doi.org/10.1007/978-3-658-22698-5_11
2. Wein, S., Dassen, Y., Pallasch, C., Miny, T., Storms, S., & Brecher, C. (2021). Konzept und Anwendung Autonomer Industrie 4.0-Komponenten auf Basis Agenten-basierter Ansätze. *At-Automatisierungstechnik, 69*(6), 430–441. https://doi.org/10.1515/auto-2020-0117
3. Meyer, S., Mandl, S., Gesmann-Nuissl, D., & Strobel, A. (2023). Responsibility in hybrid societies: concepts and terms. *AI and Ethics 3*, 25–48. https://doi.org/10.1007/s43681-022-00184-2
4. Decker, M. (2010). Ein Abbild des Menschen: Humanoide Roboter. In M. Bölker, M. Guthmann, & W. Hesse (Eds.), *Information und Menschenbild* (pp. 41–62). Springer.
5. Beck, S. (2014). Brauchen wir ein Roboterrecht? Ausgewählte juristische Fragen zum Zusammenleben von Menschen und Robotern. In Japanisch-Deutsches Zentrum (Ed.), *Mensch-Roboter-Interaktionen aus interkultureller Perspektive* (pp. 124–146). Japan und Deutschland im Vergleich.
6. Micklitz, H.-W., Namyslowska, M., Jablonowska, A. (2020). §6 KI und Verbraucherrecht. In M. Ebers, C. Heinze, T. Krügel & B. Steinrötter (Eds.), *Künstliche Intelligenz und Robotik* (1st ed., pp. 202–241). C.H. Beck.
7. Brunhöber, B. (2012). Individuelle Autonomie und Technik im Körper. In S. Beck (Ed.), *Jenseits von Mensch und Maschine* (pp. 77–104). Nomos, Baden-Baden.
8. Müller, M. F. (2014). Roboter und Recht. Eine Einführung. AJP/PJA 5/2014 (pp. 595–608).
9. Forbes. (2021). Herr der Sinne. Retrieved August 24, 2022, from https://www.forbes.at/artikel/herr-der-sinne.html
10. Hahn, L., Petras, M., Valentiner, D., & Wienfort, N. (2022). § 9 Mittelbare Drittwirkung der Grundrechte. In L. Hahn, M. Petras, D. Valentiner & N. Wienfort (Eds.), Grundrechte: Klausur-und Examenswissen (pp. 104–118). De Gruyter. https://doi.org/10.1515/9783110765533-009
11. Engle, E. (2009). Third party effect of fundamental rights (Drittwirkung). *Hanse Law Review, 5*(2), 165–173. https://ssrn.com/abstract=1481552
12. Frantziou, E. (2019). The horizontal effect of the EU charter of fundamental rights: A constitutional analysis. Oxford Studies in European Law.
13. von Jhering, R. (1877). *Der Zweck im Recht*. Breitkopf & Härtel.
14. Kant, I. (1797). Grundlegung zur Metaphysik der Sitten, 1st ed. Riga (Hartknoch).
15. Penski, U. (2004). Der Zweck des Rechts ist das Recht – Zur Teleologie und Selbstbezüglichkeit des Rechts. *Archiv für Rechts-und Sozialphilosophie, 90*(3), 406–418.
16. Kirchhof, P. (2005). Was erwarten wir vom Recht?, Studentische Zeitschrift für Rechtswissenschaft (StudZR) (pp. 3–16).
17. Kirchhof, P. (2003). Der Vertrag als Ausdruck grundrechtlicher Freiheit. In M. Habersack, P. Hommelhoff, U. Hüffner, & K. Schmidt (Eds.), Festschrift für Peter Ulmer zum 70. Geburtstag am 2. Januar 2003. https://doi.org/10.1515/9783110877038.1211
18. Morgenthaler, G. (1999). Freiheit durch Gesetz. Mohr Siebeck.
19. Di Fabio, U. (2021). Art. 2 Abs. 1 GG. In G. Dürig, & R. Herzog (Eds.), Grundgesetz-Kommentar. C.H. Beck.
20. *Decisions of the German Federal Constitutional Court, BVerfGE, 54*, 148 (153).

21. Albers, M. (2005). Informationelle Selbstbestimmung. Nomos.

22. Stern, W. (1923). Person und Sache. System des kritischen Personalismus, Band 2: Die menschliche Persönlichkeit. J. A. Barth.

23. Gray, H. M., Gray, K., & Wegner, D. M. (2007). Dimensions of mind perception. *Science, 315*(5812), 619. https://doi.org/10.1126/science.1134475

24. American Psychological Association. (n.d.), Personality. In *APA dictionary of psychology*. Retrieved September 15, 2022 from https://www.apa.org/topics/personality

25. Cloninger, C. R., Svrakic, D. M., & Przybeck, T. R. (1993). A psychological model of temperament and character. *Archives of General Psychiatry, 50*(12), 975–990. https://doi.org/10.1001/archpsych.1993.01820240059008

26. Bohn, C. (2006). Individuen und Personen. In C. Bohn (Ed.), Inklusion, Exklusion und die Person (pp. 49–70). UVK Verlagsgesellschaft.

27. Plessner, H. (2015). Mit anderen Augen: Aspekte einer philosophischen Anthropologie. Reclam.

28. *Decisions of the German Federal Constitutional Court, BVerfGE, 65*, 1 (41).

29. Nass, C., Steuer, J., & Tauber, E. R. (1994). Computers are social actors. In *Proceedings of the SIGCHI Conference on Human Factors in Computing Systems* (pp. 72–78).

30. Epley, N., Waytz, A., & Cacioppo, J. T. (2007). On seeing human: A three-factor theory of anthropomorphism. *Psychological Review, 114*(4), 864–886. https://doi.org/10.1037/0033-295X.114.4.864

31. Waytz, A., Cacioppo, J., & Epley, N. (2010). Who sees human? The stability and importance of individual differences in anthropomorphism. *Perspectives on Psychological Science, 5*(3), 219–232. https://doi.org/10.1177/1745691610369336

32. Fiske, S. T., Cuddy, A. J. C., Glick, P., & Xu, J. (2002). A model of (often mixed) stereotype content: Competence and warmth respectively follow from perceived status and competition. *Journal of Personality and Social Psychology, 82*(6), 878–902. https://doi.org/10.1037/0022-3514.82.6.878

33. Bretschneider, M., Mandl, S., Asbrock, F., Strobel, A., & Meyer, B. (2022). Social Perception of embodied digital technologies - a closer look at bionics and social robotics. Gruppe. Interaktion. Organisation. Zeitschrift für Angewandte Organisationspsychologie (GIO), 53, 343–358. https://doi.org/10.1007/s11612-022-00644-7

34. Carpinella, C. M. Wyman, A. B., Perez, M. A., & Stroessner, S. J. (2017). The robotic social attributes scale (RoSAS): development and validation. In *Proceedings of the 2017 ACM/IEEE International Conference on Human-Robot Interaction* (pp. 254–262). https://doi.org/10.1145/2909824.3020208

35. Mandl, S., Bretschneider, M., Meyer, S., Gesmann-Nuissl, D., Asbrock, F., Meyer, B., & Strobel, A. (2022). Embodied digital technologies: First insights in the social and legal perception of robots and users of prostheses. *Fronties in Robotics and AI, 9*, Article 787970. https://doi.org/10.3389/frobt.2022.787970

36. Hancock, P. A., Billings, D. R., & Schaefer, K. E. (2011). Can you trust your robot? Ergonomics in design. *The Quarterly of Human Factor Applications, 19*(3), 24–29. https://doi.org/10.1177/1064804611415045

37. Mandl, S., Laß, J. S., & Strobel, A. (2024). Associations Between Gender Attributions and Social Perception of Humanoid Robots. In L. M. Camarinha-Matos, A. Ortiz, X. Boucher, A. M. Barthe-Delanoë (Eds.): Navigating Unpredictability: Collaborative Networks in Non-linear Worlds. PRO-VE 2024. IFIP Advances in Information Communication Technology, vol 726. Springer, Cham. https://doi.org/10.1007/978-3-031-71739-0_6

38. Nijssen, S. R. R., Müller, B. C. N., van Baaren, R. B., & Paulus, M. (2019). Saving the robot or the human? Robots who feel deserve moral care. *Social Cognition, 37*(1), 41–52. https://doi.org/10.1521/soco.2019.37.1.41

39. Hakli, R., & Mäkelä, P. (2016). Robots, autonomy, and responsibility. *Frontiers in Artificial Intelligence and Applications, 290*, 145–154. https://doi.org/10.3233/978-1-61499-708-5-145

40. Scholtz, C. (2008). Und täglich grüßt der Roboter. Volkskunde in Rheinland Pfalz. Informationen der Gesellschaft für Volkskunde in Rheinland-Pfalz e.V., 23, 139–154.

41. Malle, B. F., Fischer, K., Young, J. E., Moon, A., & Collins, E. (2020). Trust and the discrepancy between expectations and actual capabilities of social robots. In D. Zhang & B. Wei (Eds.), *Human-robot interaction: Control, analysis, and design* (p. 24). Cambridge Scholars Publishing.

42. Hancock, P. A., Billings, D. R., Schaefer, K. E., Chen, J. Y. C., de Visser, E. J., & Parasuraman, R. (2011). A meta-analysis of factors affecting trust in human-robot interaction. *Human Factors: The Journal of the Human Factors and Ergonomics Society, 53*(5), 517–527. https://doi.org/10.1177/0018720811417254

43. Latour, B. (1990). Technology is society made durable. *The Sociological Review, 38*(1_suppl), 103–131. https://doi.org/10.1111/j.1467-954X.1990.tb03350.x

44. Goffman, E. (1986). *Frame analysis: An essay on the organization of experience*. Northeastern University Press.

45. Müller, M., & Sonnenmoser, A. (2022). Sociomorphic technologies. In *CRC Hybrid Societies Conference, 15-17 March 2023*, Chemnitz, Germany.

46. Müller, M. (2023). Social displays. Fabricating accountability in robotics. *Austrian Journal of Sociology* 48(4). https://doi.org/10.1007/s11614-023-00534-2

47. Robotics-open letter. (2021). Open letter to the european commission artificial intelligence and robotics. Retrieved December 12, 2022, from http://www.roboticsopenletter.eu/

48. Bryson, J. J. (2010). Robots should be slaves. In Y. Wilks (Ed.), *close engagements with artificial companions: Key social, psychological, ethical and design issues* (pp. 63–74). John Benjamins. https://doi.org/10.1075/nlp.8.11bry

49. Bertolini, A. (2013). Robots as products: The case for a realistic analysis of robotic applications and liability rules. *L. Innovation Technology, 5*(2), 214–247. https://doi.org/10.5235/17579961.5.2.214

50. Darling, K. (2016). Extending legal protection to social robots: The effects of anthropomorphism, empathy, and violent behavior toward robotic objects. In R. Calo, A. M. Froomkin & I. Kerr (Eds.), *Robot law* (pp. 213–231). Edward Elgar Publishing.

51. Gunkel, D. J. (2018). The other question: Can and should robots have rights? *Ethics and Information Technology, 20*(2), 87–99. https://doi.org/10.1007/s10676-017-9442-4

52. Kraakman, R., Armour, J., Davies, P., Enriques, L., Hansmann, H., Hertig, G., et al. (2017). *The anatomy of corporate law: A comparative and functional approach*. Oxford University Press.

53. Gesmann-Nuissl, D., & Meyer, S. (2023). Robot Land – ロボットランド. *InTeR – Journal of Innovation and Technology Law, 3*, 110–119.

54. Schulz-Schaeffer, I., Meister, M., Wiggert, K., & Clausnitzer, T. (2020). The social construction of human-robot co-work by means of prototype work settings. Bde. 2–2020, Technische Universität Berlin, Fak. VI Planen, Bauen, Umwelt, Institut für Soziologie Fachgebiet Techniksoziologie.

55. Rammert, W. (2008). Where the action is: Distributed agency between humans, machines, and programs. In U. Seifert, J. H. Kim & A. Moore (Eds.), Kultur- und Medientheorie (1st ed., pp. 62–91). transcript Verlag. https://doi.org/10.14361/978383940 8421-004

56. Suchman, L. A. (2007). *Human-machine reconfigurations plans and situated actions*. Cambridge University.

57. Pfadenhauer, M. (2013). On the sociality of social robots. A sociology-of-knowledge perspective. *Science, Technology & Innovation Studies, 10*(1), 135–153.

58. Meister, M., & Schulz-Schaeffer, I. (2016). Investigating and designing social robots from a role-theoretical perspective: Response to "Social interaction with robots—Three questions." *AI & Society, 31*, 581–585. https://doi.org/10.1007/s00146-015-0635-2

Make It More Human! A Systematic Literature Review of the Anthropomorphic Processes on Empathy

Sebastian Jansen, Oliver Rehren, Katharina Jahn, Peter Ohler, and Günter Daniel Rey

Abstract

Today, interactions with Embodiment Digital Technologies (EDT) are becoming quite common. The concept of robots interacting with humans sounds simple. But there is an obstacle: while the robot has no problem interacting with humans, humans may not like the interaction. To improve acceptance and thus interaction, the EDT is often anthropomorphized. In addition to physical features, mental states such as empathy can also be attributed to an EDT and influence the interaction. However, it is important to understand the direction of empathy. Does the human show empathy toward the EDT or is empathy used as a feature to anthropomorphize the EDT? This systematic literature review aims to examine and compare these findings to identify and evaluate the literature from recent years on the influence of empathy and anthropomorphism in interactions with EDTs, and to draw conclusions on how consistent the findings are on anthropomorphic processes on empathy. This is an initial review of the literature to build upon and prepare for a meta-analysis. It was found that both empathy and anthropomorphism are mainly self-reported and that a mental attribution of human abilities seems to be more significant than the visual appearance of an EDT.

S. Jansen (✉)
University of Zurich, Zurich, Switzerland
e-mail: sebastian.jansen@ife.uzh.ch

O. Rehren · K. Jahn · P. Ohler · G. D. Rey
Chemnitz University of Technology, Chemnitz, Germany
e-mail: oliver.rehren@phil.tu-chemnitz.de

K. Jahn
e-mail: katharina.jahn@phil.tu-chemnitz.de

P. Ohler
e-mail: peter.ohler@phil.tu-chemnitz.de

G. D. Rey
e-mail: guenter-daniel.rey@phil.tu-chemnitz.de

Keywords

Anthropomorphism · Empathy · EDT · Uncanny valley

1 Introduction

Nowadays, interaction with robots is widespread around the world. With the increasing use of advanced technologies and the influence of a helping hand in form of a hominid and android assistant in social interactions, it is of great interest to understand the user experience and psychological constructs behind the corresponding human-robot interaction (HRI). Nowadays, robots already work together with humans—they are used in nursing homes, serve as waiters in restaurants, become mobile wayfinders in warehouses or exhibitions, as well as being used in schools. The concept of robots interacting with a human actor sounds simple. But there is an obstacle to this idea: While it is no problem for the robot to interact with humans, we know that we are still dealing with a machine and have to rely on behavioral patterns in the way we interact. Because of this problem, we try to make HRI as natural and human-like as possible to increase the acceptance of EDTs [1]. One way to address this is to adopt a more human-like appearance for the robot [2]. The attribution of human characteristics to non-human entities can be described as anthropomorphism [3]. This rather vague construct describes how a human-like entity is perceived or how much human-likeness individuals attribute to it [4]. This anthropomorphic process can therefore be done in different ways [3]—we can modify the appearance and make the entity look more like ourselves, or make it imitate us through movement and speech. In addition to these more observable human attributes, we also have the possibility of attributing mental abilities to the EDT. This can involve cognitive abilities such as intention or goals [5, 6] but also affective abilities such as expressing emotions or displaying empathic behavior [7, 8]. The latest one is of particular interest here. In general,

B. Meyer et al. (eds.), *Hybrid Societies*, Advances in Science, Technology & Innovation,
https://doi.org/10.1007/978-3-032-03488-5_23

empathy is believed to be of great importance, as it gives us the ability to better understand our counterparts and to comprehend and understand behaviors and emotions [9–11]. From a conceptual perspective, it is interesting to explore how we can design an EDT that we can empathize with, on the one hand, and how we can make an EDT appear emphatic, on the other. This should increase the acceptance of humans toward an artificial interaction partner [5, 8]. But this is also where problems can arise when we try to make our artificial interaction partner as human-like as possible since in both dimensions, appearance, and mental abilities, it is not yet technically possible to fully *recreate* a human. Here, Mori et al. [12] postulated the theory of the uncanny valley, which states that an EDT appears eerie to us when it reaches a certain level of realism and we find it difficult to categorize it as human or artificial. Since we interact with other people daily, we are accordingly extremely well trained in recognizing human expressions, behaviors, and also mental abilities in our counterparts. This makes it challenging to model human attributes realistically in the development of robots. Nevertheless, recent research in HRI has shown that straight mental abilities exert a strong influence on anthropomorphic attribution [8, 13] and lead to an entity being perceived as more human-like [7]. In recent years, there have been many technical improvements that make EDTs appear nearly human. This concerns the complete spectrum of anthropomorphic processes. For example, the highly sophisticated robots from Bosten Dynamics [14] can now hold entire soccer matches and are capable of running upstairs and even jumping over obstacles. Furthermore, new algorithms and neural networks can simulate complex behaviors. We can therefore expect these artificial entities to have an even greater impact on various aspects of our lives. However, a technologically advanced robot is of less use to us if it is not accepted or if it causes discomfort.

Due to the strong influence of empathy on the interaction between actors and the ever-increasing influence of EDTs in our everyday lives and the simultaneous rapid technological development, this literature review addresses the research question of how this influence has been investigated in recent years, what insights have been gained, and whether a coherent investigation of this research topic has been established. It will initially describe the theoretical constructs of anthropomorphism and empathy. The search procedure is then explained, and the studies included are analyzed and their results are presented. Once the results have been presented, they are classified in the final section, and a conclusion and possible research directions are described.

2 Theoretical Basis for the Systematic Review

Previous studies have investigated a lot of different aspects of empathy in human-robot interactions. The focus has been on various aspects of human communication, such as speech or linguistics, touch, empathy, or interpretation of the robot's behavior [1]. Most experiments have been conducted in the area of language and communication, with a focus on human likeness and understanding [15]. Niculescu et al. [16] investigated how verbal features of spoken language, e.g., voice characteristics (pitch) and speech cues (empathy/humor expression) affect the quality of interaction with a social robot as a receptionist. Voice pitch was found to have a strong influence on how users rated the robot's perceived acceptability, quality of interaction, and overall enjoyment. Furthermore, while empathy influenced the way users rated the robot's behavior and the interaction, users' perceptions of enjoyment in the speaking style and personality task seemed to improve. Empathy therefore plays a key role in HRI. Further evidence of the importance of empathy can be found in the work of Leite et al. [17], who were able to show that users toward whom the robot behaved empathically perceived the robot as friendlier. Furthermore, autonomous robots with empathic behavior are not only perceived as friendlier, but they can also stimulate discussions in groups [18].

Another important factor for empathy is the robot's appearance. The degree of human likeness that robots should have when interacting with humans has long been a research question. Riek et al. [8] find that humans show more empathy toward human-like robots and less empathy toward mechanical-looking robots. They explain these results using simulation theory [19]. This states that people mentally *simulate* the situation of others to understand their mental and emotional state. The more similar the other actor is, the stronger the empathy. These results are consistent with those of Krach et al. [20]. The authors found that the higher the degree of anthropomorphisation, the more neurologically similar people perceived robots to be to humans. This is consistent with the findings of Riek et al. [8]. However, no predictive value was found for a person's overall ability to show empathy toward robots. But humans do not just show more empathy toward visually anthropomorphic robots. In addition to the robot's design, its appearance can also be influenced by its movements and gestures. Kontogiorgos et al. [21] pointed out that an anthropomorphic social robot that speaks and demonstrates social gaze behavior through head movements is perceived by users as more social than smart speakers or other less anthropomorphic robots. Salem et al. [22, 23] demonstrated that people anthropomorphize a robot more when it uses gestures during an interaction. Participants reported a greater connection with the robot and showed more intention for future contact than when the robot spoke without gestures. These results show that anthropomorphic perception, as well as the mental models that people build of a robot during an interaction, are influenced by the communicative nonverbal behaviors of a robot. Examining the effects of these aspects in isolation is an important contribution to further understanding the importance and weight of each aspect and direction of action of anthropomorphism. However, as there is an interaction between empathy and anthropomorphism, recently confirmed by a meta-analysis on HCI, modeling a combination is closer to reality and should not be ignored [1].

This paper aims to assess the effect and interaction of empathy and anthropomorphism and the significance of a combination of these effects in recent literature. The process of anthropomorphism through the emotional dimension of empathy is the focus of the literature reviewed. In the following theoretical overview, anthropomorphism and empathy will be introduced.

2.1 Empathy

Empathy can be described as a process that can be divided into three subprocesses [9]. The first of these subprocesses are motor empathy. It is based on innate structures in the motor areas of the brain. This includes neural mirror mechanisms that enable motor imitation of what is perceived [9, 24]. The second subprocess is affective empathy. This provides an imitation of emotions when they are visible in the object through motor cues [9, 25]. Both processes provide a rudimentary empathic response to the environment and thus have the advantage that we are still capable of empathic action even when our cognitive capacity is depleted. The third subprocess is cognitive empathy [9, 26]. It extends the empathic process to include the possibility of evaluation. In this process, we create a mental representation in which the perceived state of the object is reproduced as accurately as possible. The representation created corresponds to an attempted simulation of the situation as perceived by the person being observed. The simulation is then related to the person's own perception of the situation. On this basis, a final evaluation is made [9]. Depending on the outcome, cognitive empathy influences both affective and motor empathy. As a result, various functions related to empathic responses, e.g., emotions and facial expressions, can be adjusted accordingly [9]. Various factors (e.g., physiological responses) provide information about the extent to which the other subprocesses have adapted to the situation and the other persons. Dissonance can thus be revealed by the cognitive process and reconfigured if necessary. New information also constantly changes the simulation. In this way, the system remains responsive even as the situation or interaction changes. In this three-subprocess model, empathy does not necessarily require that all be fully effective to complete an empathic process. The extent of empathy perceived depends on the interaction of the various components. It is important to point out that the process need not result in an emotional response that is consistent with the situation or the emotions displayed by the other person; it may even result in no emotional response at all [9]. The process may be only partially completed or may be only a cognitive appraisal that allows for emotions based on thought processes that do not match the expected emotional response. In this regard, there is an area of research in HRI which particularly often approaches empathy through physiological measures. This involves so-called *robot abuse studies* [27–29], which are particularly well suited for objective measurement. Here, robots are mistreated and an inference of empathy is drawn via the affective and physiological response of the person observing. The purpose is to push the boundaries of normal HRI [30]. In more recent studies, however, the focus shifts to the perception of the human interaction partner. For example, Rosenthal-von der Pütten et al. [31] compared the emotional response to video recordings of robots. They found that participants responded with increased arousal to the torture video compared to the friendly video and expressed *empathic concern* toward the robot. These results are consistent with the findings of Bartneck et al. [31] and lend credence to the predictions of *media equation theory* [31] and provide additional data for the assumption that robots can be perceived as partners in social contexts and support that assumed emotional attachments to robots can form [31, 32]. Another study that attempted to measure empathic response is from Suzuki et al. [33] which used electroencephalography (EEG) to examine the response of human subjects to images showing injured humans or robots. Painful events, such as a knife cutting into a finger, were presented. The results showed that the participants did not feel the robot's pain as strongly, but reported similar levels of empathy. While these abuse studies, despite their rather dark side, have helped HRI provide insights into this area of research. Bartneck et al. [30] conducted Milgram-like experiments and thereby found that human subjects are less inhibited when instructed to harm robots, as in the original experiment [34], compared to humans.

2.2 Anthropomorphism

Anthropomorphism is a rather vague construct that describes how human-like an entity is perceived or how much human-likeness we attribute to it [3]. Thus, human-likeness is not an objectively measurable property, but rather a subjective perception of various attributes that an entity may possess. As mentioned above previous studies have categorized these attributes into traits such as appearance, thoughts, and emotions [4]. A key difference in many of these studies, however, is that anthropomorphism is considered a human disposition—in other words, an aspect of personality that makes some people more likely than others to attribute human-likeliness to a non-human being [3].

However, anthropomorphism can also be applied to the entity itself, i.e., a robot is perceived as more human-like if it behaves and looks more like a human being and can be attributed to mental abilities. Thus, it can be measured how this anthropomorphism affects the person observing and interacting with the EDT. To achieve such human likeness, studies have already shown that features such as voice [3, 35] and movement [36, 37] are considered human-like features that evoke anthropomorphism in, a robot. Non-verbal expressions such as gestures and posture also contribute to making an entity appear more human-like [38]. These features are observable attributes that come from the robot itself and are then categorized as human-like by the observer. Considering only these characteristics would not take into account the interaction between humans and robots and the thoughts and feelings that the observer develops toward the robot. As mentioned, empathy plays a crucial role in any interaction with an actor because regardless of how human-like a robot looks and moves, thoughts and feelings are separately attributed to the robot [6]. This has been shown in studies on voice assistants [38]. Nevertheless, these studies do not lead to conclusive evidence that empathy or the resulting sympathy is a driving factor for anthropomorphism in HRI, as the effect could be triggered exclusively by a human-like voice. In this context, the achieved anthropomorphism is often associated with a positive influence on other factors such as trustworthiness [39], perceived competence [40], and authenticity [41]. However, in this context, the development of more human-like entities can also lead to a negative effect. For example, Mori [12, 42] postulated at an early stage that robots that are too human can have an uncanny effect. However, MacDorman et al. [43, 44] were able to show with photorealistic avatars that they did not trigger discomfort when presented in a highly realistic way.

3 Method

3.1 Search Strategy

This systematic literature review aims to cover the recent work that investigated artificial intelligence (AI), robots, and empathy. The search was guided using the PRISMA method [45] and intended an initial overview of the current literature to further elaborate on this research. To ensure that a wide range of studies related to this topic is found, the key terms used in

Figure 1 are given for searching the different databases. Using synonyms of the main terms *anthropomorphism, empathy,* and *robot,* to search through the English language literature in the databases Scopus (S) and Web of Science (WoS). In addition, a search in Google Scholar (GS) was performed manually with the main terms, since this database otherwise records any articles in which these words occur in the entire text. Here, this search includes all scientific articles published in English on the mentioned databases between 2019 and 2022, and the search was completed on 03.04.2022. Searches using the 13 operators resulted in 105 articles. During the review of the literature, 90 studies were excluded.

In the context of this systematic literature review, we will focus on EDT presented and displayed anthropomorphically in experiments. In addition, empathy had to be determined as a variable in the experiments. Here, we were not limited to how this was measured and if measured by questionnaires, whether it was measured by an overall score or specifically by one of the sub- constructs. Accordingly, articles were considered relevant if the entities were designed to be human-like in some way, and empathy was captured in an interaction, or if the EDT had a mental attribution of empathy to be more anthropomorphic. As the keywords captured some literature that addressed the topic from more diverse disciplines or research approaches, the following are the exclusion criteria used to remove publications that were not in line with the research objectives:

- Did not contribute to the further elucidation of anthropomorphism and the emergence of empathy.
- Covered only one aspect of the study.
- Design recommendations for robots.
- Technical Studies such as robot programming.
- Focused on movies.
- Research papers that did not involve manipulation.
- Measured related constructs and inferred possible effects about them.

From the captured articles and the mentioned exclusion criteria, 13 relevant publications were identified. While the articles were being captured, another backward search [46] was conducted and two additional articles were captured (see Fig. 1). In the initial top-down process, the relevant literature was identified using the keywords and the categories of empathy, anthropomorphism, and EDT. This was done by differentiating within the psychological constructs of empathy and anthropomorphism, according to the understanding of these constructs explained in the introduction. In addition, it was recorded which robots were used and which design the studies employed. After reviewing the relevant publications, a bottom-up approach was applied to refine and specify the categorization. Further differentiation was made as to the exact manipulation involved and whether an anthro-

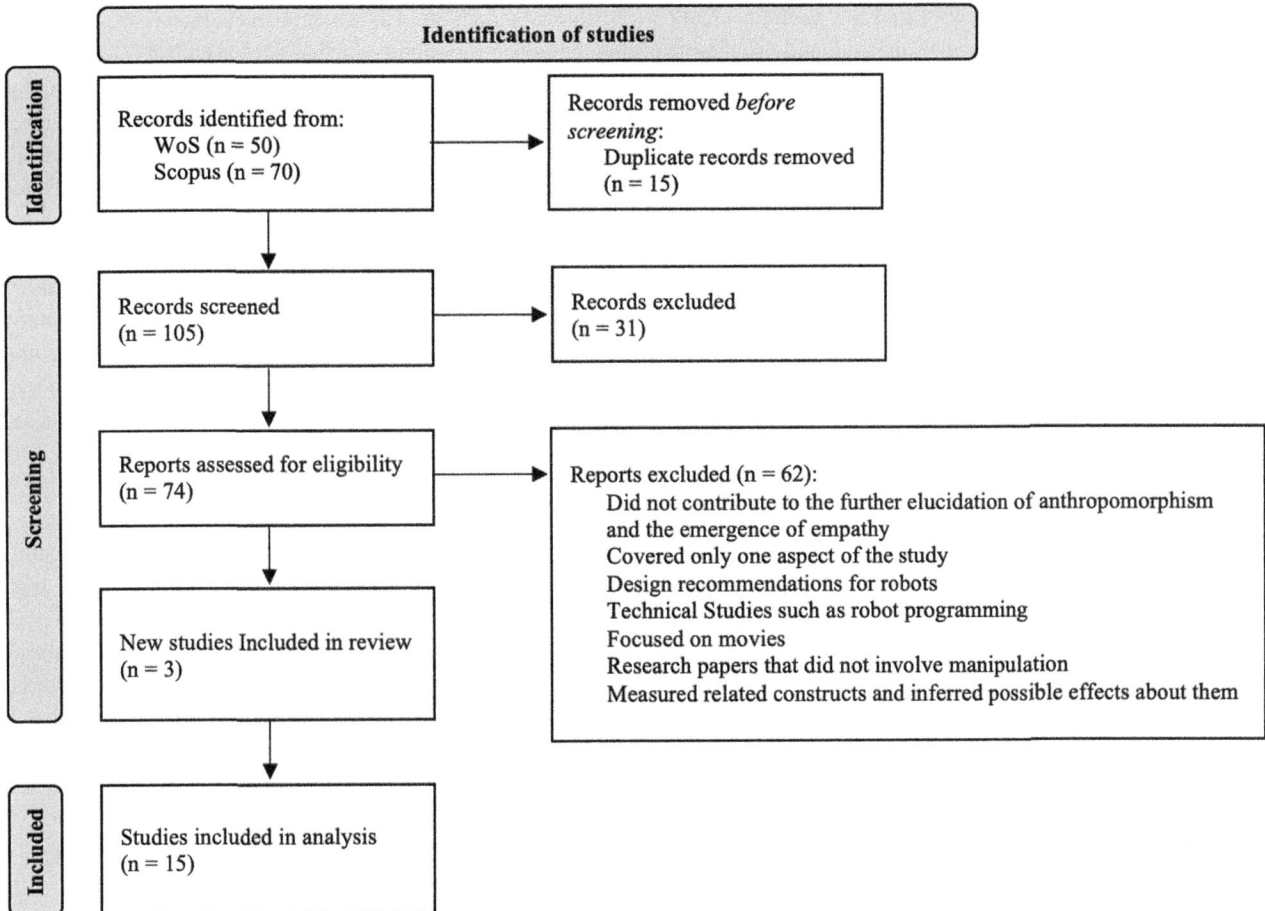

Fig. 1 PRISMA flow-chart. WoS = Web of Science

pomorphic design was used or anthropomorphic attribution occurred. Further differentiation was also made in the report on the empathy condition. Here it was differentiated whether a global value of empathy was measured or whether within the sub-areas motor, affective and cognitive it was or could be differentiated.

4 Results

The literature reviewed deals with different works on anthropomorphism and empathy and mostly does not differentiate between their sub-areas. It is noticeable that the works that focus on experienced empathy toward entities still like to fall back on the mistreatment of robots [47, 48]. In the dimension of anthropomorphism, it comes down to both anthropomorphic design [47, 49, 50] and the attribution of mental abilities to the entity [51, 52]. In the research papers that refer to an anthropomorphic design, the comparison is predominantly between scrambled entities, or a comparison is made with animals or humans [49–52]. The design of the studies is predominantly between-subjects

(10), with fewer within-subjects (4). The average number of subjects in this literature is 105 ($SD = 75$). The studies were conducted as laboratory experiments as well as online experiments.

4.1 Main Results

The studies included in the analysis mostly measure empathy and anthropomorphism through questions (see Table 1). It is interesting to note that anthropomorphic differences were implemented in quite different ways in the experimental studies. However, it is clear that the researchers largely relied on an anthropomorphic design for their entities. Mattiassi et al. [55] thus distinguished between humans, animals, human robots, animal robots, and a vacuum cleaner robot, and used attached violence (pushing by a human interaction partner) to try to draw inferences about subjects' cognitive empathy. Surprisingly, their results attribute the highest empathic response to the animal in this experiment—a cat. In their anthropomorphisation condition, they were also able to show that more empathy is attributed to the human-like robot, followed by the animal-like robot and the vacuum cleaner robot. It should be noted, however, that the authors hypothesize that empathy is moderated by the attributed experiential values and the moral- decision-making ability of the interaction partner. In this context, a human is attributed correspondingly more than a robot. However, they were able to show that more is attributed to a humanized robot than to a neutral robot or even an object. In addition to these results, the studies of Küster et al. [58] and Küster and Swiderska [55] must be mentioned. These also deal with measured empathy toward EDTs when they are inflicted with suffering. They found that when robots are given mental abilities, they received more empathy from humans. Riddoch and Cross [48] are the latest authors found via this research to address *robot abuse*. In doing so, they examined the *Hesitance to Hit* paradigm and attempted to draw an inference about empathy toward the robot via the subjects' delayed response. They found that negative attitudes toward robots influenced empathy and the tendency to anthropomorphise. Pozharliev et al. [50] found conflicting results on the effect of anthropomorphism and empathy in their experiment. In their study, empathy was a minor consideration, as it actually measured satisfaction with the interaction partner, and the anthropomorphic design of the robot was only realized through gender. In doing so, they attempted to replicate the findings that a female robot is perceived as more human [53]. One study examines interaction with a chatbot and only passes the criteria of this literature review because it also presented an anthropomorphic avatar. In this study, the authors compared the empathic response behavior of this bot with a neutral one and were able to show that this could increase satisfaction with the interaction [54]. Interestingly, the authors found that female subjects responded to this more strongly than males. Thus, their results line up with earlier ones, which also found gender differences in chatbot interaction [55]. Herdel et al. [56] showed that facial expressions on drones led to increased acceptance of the drones and that they were more anthropomorphised by the attribution of mental abilities. Their experiment shows that facial expressions can be used to humanize a very rudimentary EDT and increase acceptance of the technology. In the area of empathy, they found that emotional facial expressions elicited varying degrees of empathic response in the participants. Strong emotions such as sadness, fear, and happiness were particularly important. When it comes to showing emotional states, the experiment by Hickton et al. [51] is in line with this. They were also able to elicit an empathic response from their participants in a robot that was not designed to be anthropomorphic, using only its movements and behaviors. They conducted this exploratory experiment with undergraduate students for validation and were able to corroborate their findings. Pelau et al. [13], showed that, in addition to the mental anthropomorphic

attributions mentioned above, human characteristics alone did not influence the quality of interaction with EDTs, and this could be increased by showing mental abilities like empathy. On the contrary, they found that anthropomorphic features alone could reduce interaction quality. Furthermore, the findings of Erel et al. [57] add to the evidence for the importance of emotions in social interaction. In their online experiment, they were able to show that an empathic reaction from an interaction partner, compared to a neutral one, led to a more human-like chatbot, as human characteristics were attributed to it via empathy. Finally, two studies approach the topic and the measurement of empathy from a physiological and more objective direction. For example, Ionta et al. [58] conducted an fMRI study. While this was again about *robot abuse*, they were able to show that when a robot hand is injured, similar areas are activated as in a human being. Ceh and Vanman [59] measured the musculature of the face in their measurement and tried to conclude an empathic response from this. The robots elicited more empathic and threatening responses in the study. The objective measurements of facial features were consistent with the subjective ratings of the subjects. Finally, the study by Lo et al. [60] investigated whether a robot could contribute to the adaptation of its environmental behavior. They found that an anthropomorphically designed robot could induce this better and influence the behavior of the subjects more strongly if the subjects reported empathy toward the robot.

5 Discussion

When it comes to face-to-face interaction, empathy is mostly measured as a side effect, as in previous experiments [39–41]. For example, in the experiment by Mattiassi et al. [49] the attributed experience and moral decision-making ability or interaction satisfaction [13, 44, 50]. Nevertheless, when it comes to measuring empathic behavior toward an EDT, *robot abuse studies* are still used [47, 48, 59]. However, fMRI [59] or facial electromyography [58] studies are also used here to provide a more objective measure. These studies have shown that the empathic response to a robot is not as strong as to a human being. The authors mention that this could be because we know we are looking at a human-made device. Pelau et al. [13] add to existing research that anthropomorphic design features alone have no major impact on interaction quality unless the EDT can demonstrate empathy. This was also the only study included in this review that attributed empathic abilities to the EDT itself. These effects are further supported by the fact that even empathic gestures from a robotic arm can contribute to enhancing the quality of an interaction, and that anthropomorphisation could only be established through presented gestures [57]. Recent research also shows how anthropomorphic robots can change our behavior.

Table 1 Studies reviewed

Author, year of study	Study design & sample size	EDT used	Implementation of anthropomorphism	Empathy measurement
Mattiassi et al. (2021)	BS & 163	Atlas	Design	Questionnaire
Küster et al. (2021)	BS & 172	Atlas	Design	Questionnaire
Küster and Swiderska (2020)	BS & 253	Virtual robot	Design	Questionnaire
Herdel et al. (2021)	WS & 98	Drone	Faicial expression	Questionnaire
Pozharliev et al. (2021)	BS & 225	Peper	Design & voice	Questionnaire
Diederich et al. (2019)	BS & 112	Chatbot	Mental capability	Questionnaire
Ionta et al. (2020)	BS & 55	Robot-hand images	Design	fMRT
Bagherie et al. (2020)	WS & 40	Pepper	Movement, Mimik	Questionnaire
Hicktonet et al. (2020)	Qualitativ & 9	Hexapod roboter	Mental capability	Observation
Pelau et al. (2021)	BS & 188	Images of service robots	Mental capability	Questionnaire
Erel et al. (2022)	BS & 64	Non-human robot arm	Mental capability	Questionnaire
Lo et al. (2022)	WS & 77	Zenbo	Facial expressions	Questionnaire
Riddoch and Cross (2021)	BS & 84	Pepper	Design	Questionnaire
Hickton et al. (2020)	WS & 180	Hexapot robot	Movement	Questionnaire
Ceh and Vanman (2018)	BS & 210	Images of androids, humanoid & mechanical robots	Design	Questionnaire

Note WS = Within-Subject Design, BS = Between-Subject Design

Lo et al. [60] showed in their study that we improved our environmental behavior when we had an anthropomorphic robot in front of us compared to a less anthropomorphic tablet. Except for one study by Pozharliev et al. [50], all of the other studies reviewed were able to show a significant impact of empathy on anthropomorphism and how attributing empathic behavior to an EDT can make a significant contribution to it being perceived as more human-like. Thus, these results support the existing literature that it is the attribution of mental abilities that plays a critical role in the anthropomorphic process and can help improve HRI. However, it should be noted that among the experiments, many were conducted via online surveys. Videos of the robots were shown [47, 49, 52] followed by a subjective questionnaire survey. This might bring the reliability of the experiments into question, as there was no real interaction between the EDTs and human beings. This should also be considered in the context of the pandemic situation, as the researchers may not have had the opportunity to collect data in another way. The research question addressed was to determine whether the literature is coherent and what insights are added by current research. The influence of empathy on anthropomorphism has been investigated from many different perspectives and a wide variety of methods have been used. However, it should be noted that the results of the studies are predominantly positive in terms of the relationship between anthropomorphisation and empathy, and also in terms of

objective measures. That is, the studies show that EDTs who are given the ability to show empathy are perceived as more human and, conversely, EDTs who are anthropomorphised are perceived as social actors and were shown empathy toward.

5.1 Empathy

As described in the part above, empathy has been treated differently in the research. In summary, however, the examined studies show that an empathic reaction toward the EDT is possible, but this is not as pronounced as with living beings [49]. This may be because we can create anthropomorphic EDTs, but they can hardly achieve the full degree of human-likeness, but it does support the *media equation theory*. Apart from this, however, the reviewed literature shows that empathy can be associated with a variety of other useful constructs [40, 57, 60]. Among them, anthropomorphism is one of the most significant. Thus, the literature indicates that empathy can be attributed to an EDT as a mental attribute, and thus the robot can be perceived as more human. If a more human-like perception can be created, this may lead us to simplify our interactions with EDT because it is easier for us to accept them [13].

5.2 Anthropomorphism

Pelau et al. [13] were able to show that anthropomorphism alone cannot increase acceptance toward EDT. Only when mental attributions are also made in the process and cognitive and affective abilities can be attributed to the EDT we begin to develop higher acceptance and even trust [39]. A particularly important aspect of the anthropomorphic process is therefore whether we attribute to a robot a degree of human qualities that are not directly observable. Here, the exploratory research, later validated by another experiment, by Hickton et al. [51] offers interesting results. They were able to show that despite their very mechanical robot, which simulated various states of arousal via its behavior, thereby mimicking human behavior was able to elicit an empathic response from the subjects and made the robot appear more human-like.

5.3 Limitations

Only a small number of studies found a place in this work, due to the now very large existing literature. Due to the limited scope of the initial literature review, a large number of studies with the selected keywords were included. However, this limitation was deliberately chosen to be able to capture the specific constructs of empathy and anthropomorphism within the last four years. It should be noted that the short time period was chosen to ensure that the EDTs used was, as far as possible, at a similar level of technical development, as technical development in this area is making great progress. As this is an initial literature review, it is intended to extend the time to look more closely at the development of the technical dimension.

5.4 Implications and Future Research

The review of the current work should summarize the results of the HRI to conclude possible implications. For example, it would be of great interest to examine the various results of long-term studies. As has been shown in part, the attribution of mental attributes is of great importance within anthropomorphism. This could lead to changes in attitude or acceptance over a longer time. When recommending the design of EDTs, it can be assumed that it may be advantageous to avoid overly realistic human replicas and to focus on the simulation of mental abilities, so that we do not develop any discomfort EDTs. However, this also raises a question for the future. For example, mental processes could at some point be implemented in a similar way to the observable anthropomorphic designs to a very realistic degree and possibly serve the uncanny valley. In this respect, research should continue to explore the processes that facilitate human interaction with machines.

Acknowledgements This publication has been partially funded by the Deutsche Forschungsgemeinschaft (DFG, German Research Foundation) Project-ID 416228727-SFB 1410, the German Ministry of Culture and Science of the Land of North Rhine-Westphalia (reference number: 005-1706-0006).

References

1. Roesler, E., Manzey, D., & Onnasch, L. (2021). A meta-analysis on the effectiveness of anthropomorphism in human-robot interaction. *Science Robotics, 6*(58). https://doi.org/10.1126/scirobotics.abj5425
2. Fink, J. (2012). Anthropomorphism and human likeness in the design of robots and human-robot interaction. In *International Conference on Social Robotics* (pp. 199–208).
3. Epley, N., Waytz, A., & Cacioppo, J. T. (2007). On seeing human: A three-factor theory of anthropomorphism. *Psychological Review, 114*(4), 864–886. https://doi.org/10.1037/0033-295X.114.4.864
4. Ruijten, P. A. M., Haans, A., Ham, J., & Midden, C. J. H. (2019). Perceived human-likeness of social robots: Testing the Rasch model as a method for measuring anthropomorphism. *International Journal of Social Robotics, 11*(3), 477–494. https://doi.org/10.1007/s12369-019-00516-z
5. Gruber, D., Aune, A., & Koutstaal, W. (2018). Can semi-anthropomorphism influence trust and compliance?: Exploring image use in app interfaces, In *Proceedings of the Technology, Mind, and Society on ZZZ-TechMindSociety'18* (pp. 1–6). https://doi.org/10.1145/3183654.3183700
6. Kim, Y., & Sundar, S. S. (2012). Anthropomorphism of computers: Is it mindful or mindless? *Computers in Human Behavior, 28*(1), 241–250. https://doi.org/10.1016/j.chb.2011.09.006
7. Airenti, G. (2015). The cognitive bases of anthropomorphism: From relatedness to empathy. *International Journal of Social Robotics, 7*(1), 117–127. https://doi.org/10.1007/s12369-014-0263-x
8. Riek, L. D., Rabinowitch, T.-C., Chakrabarti, B., & Robinson, P. (2009). How anthropomorphism affects empathy toward robots.
9. Leiberg, S., & Anders, S. (2006). The multiple facets of empathy: A survey of theory and evidence. *Progress in brain research, 156,* 419–440.
10. Preston, S. D., & de Waal, F. B. M. (2002). Empathy: Its ultimate and proximate bases. *Behavioral and Brain Sciences, 25*(1), 1–20. https://doi.org/10.1017/S0140525X02000018
11. Blanke, E. S., Rauers, A., & Riediger, M. (2016). Does being empathic pay off?-Associations between performance-based measures of empathy and social adjustment in younger and older women. *Emotion, 16*(5), 671–683. https://doi.org/10.1037/emo0000166
12. Mori. (1970). The uncanny valley. *Energy, 7*(4), 33–35.
13. Pelau, C., Dabija, D.-C., & Ene, I. (2021). What makes an AI device human-like? The role of interaction quality, empathy and perceived psychological anthropomorphic characteristics in the acceptance of artificial intelligence in the service industry. *Computers in Human Behavior, 122,* Article 106855. https://doi.org/10.1016/j.chb.2021.106855
14. Atlas™ | Boston Dynamics. Retrieved April 07, 2022, from https://www.bostondynamics.com/atlas
15. James, J., Balamurali, B. T., Watson, C. I., & MacDonald, B. (2020). Empathetic speech synthesis and testing for healthcare robots. *International Journal of Social Robotics, 1–19.*
16. Niculescu, A., van Dijk, B., Nijholt, A., Li, H., & See, S. L. (2013). Making social robots more attractive: The effects of voice pitch, humor and empathy. *International Journal of Social Robotics, 5*(2), 171–191.

17. Leite, I., Pereira, A., Mascarenhas, S., Martinho, C., Prada, R., & Paiva, A. (2013). The influence of empathy in human-robot relations. *International Journal of Human-Computer Studies, 71*(3), 250–260.

18. Alves-Oliveira, P., Sequeira, P., Melo, F. S., Castellano, G., & Paiva, A. (2019). Empathic robot for group learning: A field study. *ACM Transactions on Human-Robot Interaction (THRI), 8*(1), 1–34.

19. Goldman, A. I. (2006). *Simulating minds: The philosophy, psychology, and neuroscience of mindreading* (Oxford University Press on Demand).

20. Krach, S., Hegel, F., Wrede, B., Sagerer, G., Binkofski, F., & Kircher, T. (2008). Can machines think? Interaction and perspective taking with robots investigated via fMRI. *Plos One, 3*(7), Article e2597.

21. Kontogiorgos, D., Pereira, A., Andersson, O., Koivisto, M., Gonzalez Rabal, E., Vartiainen, V., & Gustafson, J. (Eds.). (2019). The effects of anthropomorphism and non-verbal social behaviour in virtual assistants. In *Proceedings of the 19th ACM International Conference on Intelligent Virtual Agents.*

22. Salem, M., Eyssel, F., Rohlfing, K., Kopp, S., & Joublin, F. (2011). Effects of gesture on the perception of psychological anthropomorphism: A case study with a humanoid robot. In *International Conference on Social Robotics* (pp. 31–41).

23. Salem, M., Eyssel, F., Rohlfing, K., Kopp, S., & Joublin, F. (2013). To err is human(-like): Effects of robot gesture on perceived anthropomorphism and likability. *International Journal of Social Robotics, 5*(3), 313–323.

24. Gallese, V. (2001). The 'shared manifold' hypothesis. From mirror neurons to empathy. *Journal of Consciousness Studies, 8*(5–6), 33–50.

25. Wondra, J. D., & Ellsworth, P. C. (2015). An appraisal theory of empathy and other vicarious emotional experiences. *Psychological Review, 122*(3), 411.

26. Früh, W., & Wünsch, C. (2009). Empathie und Medienempathie. *Publizistik, 54*(2), 191–215.

27. Bartneck, C., Rosalia, C., Menges, R., & Deckers, I. (2005). Robot abuse-a limitation of the media equation, Proceedings of the Interact 2005 Workshop on Abuse, September 12, Rome

28. Brahnam, S., & De Angeli, A. (2008). Special issue on the abuse and misuse of social agents (Vol. 20, No. 3, pp. 287–291). Oxford University Press.

29. De Angeli, A., & Brahnam, S. (2008). I hate you! Disinhibition with virtual partners. *Interacting with Computers, 20*(3), 302–310.

30. Bartneck, C., & Hu, J. (2008). Exploring the abuse of robots. *Interaction Studies, 9*(3), 415–433.

31. Rosenthal-von der Pütten, A. M., Krämer, N. C., Hoffmann, L., Sobieraj, S., & Eimler, S. C. (2013). An experimental study on emotional reactions towards a robot. *International Journal of Social Robotics, 5*(1), 17–34.

32. Nomura, T., Kanda, T., Suzuki, T., & Kato, K. (2008). Prediction of human behavior in human-robot interaction using psychological scales for anxiety and negative attitudes toward robots. *IEEE Transactions on Robotics, 24*(2), 442–451.

33. Suzuki, Y., Galli, L., Ikeda, A., Itakura, S., & Kitazaki, M. (2015). Measuring empathy for human and robot hand pain using electroencephalography. *Scientific Reports, 5*(1), 1–9.

34. Milgram, S. (2020). *Das Milgram-Experiment: Zur Gehorsamsbereitschaft gegenüber Autorität* (21st ed.). Rowohlt.

35. Schroeder, J., & Epley, N. (2016). Mistaking minds and machines: How speech affects dehumanization and anthropomorphism. *Journal of Experimental Psychology: General, 145*(11), 1427–1437. https://doi.org/10.1037/xge0000214

36. Bartneck, C., Kulić, D., Croft, E., & Zoghbi, S. (2009). Measurement instruments for the anthropomorphism, animacy, likeability, perceived intelligence, and perceived safety of robots. *International*

Journal of Social Robotics, 1(1), 71–81. https://doi.org/10.1007/s12369-008-0001-3

37. Thompson, J. C., Trafton, J. G., & McKnight, P. (2011). The perception of humanness from the movements of synthetic agents. *Perception, 40*(6), 695–704. https://doi.org/10.1068/p6900

38. Cowell, A. J., & Stanney, K. M. (2005). Manipulation of non-verbal interaction style and demographic embodiment to increase anthropomorphic computer character credibility. *International Journal of Human-Computer Studies, 62*(2), 281–306.

39. Schroeder, J., & Schroeder, M. (2018). Trusting in machines: How mode of interaction affects willingness to share personal information with machines. In *Proceedings of the 51st Hawaii International Conference on System Sciences.*

40. Araujo, T. (2018). Living up to the chatbot hype: The influence of anthropomorphic design cues and communicative agency framing on conversational agent and company perceptions. *Computers in Human Behavior, 85*, 183–189.

41. Wuenderlich, N. V., & Paluch, S. (2017). A nice and friendly chat with a bot: User perceptions of AI-based service agents, In Proceedings of the 38th International Conference on Information Systems (ICIS 2017), Seoul, Korea.

42. Mori, M., MacDorman, K. F., & Kageki, N. (2012). The uncanny valley [from the field]. *IEEE Robotics & Automation Magazine, 19*(2), 98–100.

43. MacDorman, K. F., Green, R. D., Ho, C.-C., & Koch, C. T. (2009). Too real for comfort? Uncanny responses to computer generated faces. *Computers in Human Behavior, 25*(3), 695–710. https://doi.org/10.1016/j.chb.2008.12.026

44. MacDorman, K. F., & Chattopadhyay, D. (2016). Reducing consistency in human realism increases the uncanny valley effect; increasing category uncertainty does not. *Cognition, 146*, 190–205. https://doi.org/10.1016/j.cognition.2015.09.019

45. Page, M. J., et al. (2021). The PRISMA 2020 statement: An updated guideline for reporting systematic reviews. *International Journal of Surgery, 88*, Article 105906.

46. Gavetti, G., & Levinthal, D. (2000). Looking forward and looking backward: Cognitive and experiential search. *Administrative Science Quarterly, 45*(1), 113–137.

47. Küster, D., & Swiderska, A. (2021). Seeing the mind of robots: Harm augments mind perception but benevolent intentions reduce dehumanisation of artificial entities in visual vignettes. *International Journal of Psychology, 56*(3), 454–465. https://doi.org/10.1002/ijop.12715

48. Riddoch, K. A., & Cross, E. S. (2021). 'Hit the Robot on the Head With This Mallet'—Making a case for including more open questions in HRI research. *Frontiers in Robotics and AI, 8*. Retrieved April 08, 2022, from https://www.frontiersin.org/article/https://doi.org/10.3389/frobt.2021.603510

49. Mattiassi, A. D. A., Sarrica, M., Cavallo, F., & Fortunati, L. (2021). What do humans feel with mistreated humans, animals, robots, and objects? Exploring the role of cognitive empathy. *Motivation and Emotion, 45*(4), 543–555. https://doi.org/10.1007/s11031-021-09886-2

50. Pozharliev, R., De Angelis, M., Rossi, D., Romani, S., Verbeke, W., & Cherubino, P. (2021). Attachment styles moderate customer responses to frontline service robots: Evidence from affective, attitudinal, and behavioral measures. *Psychology & Marketing, 38*(5), 881–895. https://doi.org/10.1002/mar.21475

51. Hickton, L., et al. (2020). Does Expression of Grounded Affect in a Hexapod Robot Elicit More Prosocial Responses?. In *UKRAS20 Conference: "Robots Into the Real World,"* (pp. 40–42). https://doi.org/10.31256/Hz3Ww4T

52. Küster, D., Swiderska, A., & Gunkel, D. (2021). I saw it on YouTube! How online videos shape perceptions of mind, morality, and fears about robots. *New Media & Society, 23*(11), 3312–3331. https://doi.org/10.1177/1461444820954199

53. van Doorn, J., et al. (2017). Domo Arigato Mr. Roboto: Emergence of automated social presence in organizational frontlines and customers' service experiences. *Journal of Service Research, 20*(1), 43–58. https://doi.org/10.1177/1094670516679272

54. Diederich, S., Janßen-Müller, M., Brendel, A., & Morana, S. (2019). Emulating empathetic behavior in online service encounters with sentiment-adaptive responses: Insights from an experiment with a conversational agent. Proceedings of the 40th International Conference on Information Systems (ICIS 2019)

55. Chiu, H.-C., & Wu, H.-C. (2002). Exploring the cognitive and affective roles of service quality attitude across gender. *Service Industries Journal, 22*(3), 63–76.

56. Herdel, V., Kuzminykh, A., Hildebrandt, A., & Cauchard, J. R. (2021). Drone in love: Emotional perception of facial expressions on flying robots. In *Proceedings of the 2021 CHI Conference on Human Factors in Computing Systems* (pp. 1–20). https://doi.org/10.1145/3411764.3445495

57. Erel, H., Trayman, D., Levy, C., Manor, A., Mikulincer, M., & Zuckerman, O. (2022). Enhancing emotional support: The effect of a robotic object on human-human support quality. *International Journal of Social Robotics, 14*(1), 257–276. https://doi.org/10.1007/s12369-021-00779-5

58. Ceh, S., & Vanman, E. (2018). The robots are coming! The robots are coming! Fear and empathy for human-like entities. https://doi.org/10.31234/osf.io/4cr2u

59. Ionta, S., Costantini, M., Ferretti, A., Galati, G., Romani, G. L., & Aglioti, S. M. (2020). Visual similarity and psychological closeness are neurally dissociable in the brain response to vicarious pain. *Cortex, 133*, 295–308. https://doi.org/10.1016/j.cortex.2020.09.028

60. Lo, S.-Y., Lai, Y.-Y., Liu, J.-C., & Yeh, S.-L. (2022). Robots and sustainability: Robots as persuaders to promote recycling. *International Journal of Social Robotics*. https://doi.org/10.1007/s12369-021-00828-z

Toward TechnoSapiens: Experiencing Embodied Technologies in Augmented Reality

C. Rudolph, S.-A. Dadgar, M. Bretschneider, B. Meyer, F. Asbrock, and G. Brunnett

Abstract

The number of individuals using some form of bionic technology that is merged to their bodies (e.g., prostheses, exoskeletons) is likely to increase in the future. Such Embodied Digital Technologies (EDTs) will affect the psychological processes underlying social interaction, perception, and stereotyping, especially when users and non-users of these technologies meet and need to coordinate in public. Therefore, understanding the psychological processes underlying the use and the perceptions of embodied technologies is important for designing such devices for smooth coordination. However, today's limited availability of both, technology and their users limits the possibilities for conducting psychological studies in this area. We thus suggest employing Mixed Reality (MR) to simulate wearing those technologies and to simulate encounters between users for studying associated processes. We call our system *TechnoSapiens* to emphasize the merging of human bodies with technology. We describe major technological requirements, propose a psychological framework to assess the system's capabilities and lay out the current state-of-the-art in the respective fields.

C. Rudolph (✉) · S.-A. Dadgar · M. Bretschneider · B. Meyer · F. Asbrock · G. Brunnett
Chemnitz University of Technology, Chemnitz, Germany
e-mail: carsten.rudolph@informatik.tu-chemnitz.de

S.-A. Dadgar
e-mail: seyed-amin.dadgar@informatik.tu-chemnitz.de

M. Bretschneider
e-mail: maximilian.bretschneider@psychologie.tu-chemnitz.de

B. Meyer
e-mail: bertolt.meyer@psychologie.tu-chemnitz.de

F. Asbrock
e-mail: frank.asbrock@psychologie.tu-chemnitz.de

G. Brunnett
e-mail: guido.brunnett@informatik.tu-chemnitz.de

Keywords

Mixed reality · Computer vision · Human-computer interaction · Social perception · Stereotyping

1 Introduction

On their way toward hybrid societies, contemporary societies are characterized by ongoing digitalization and automatization processes, leading to ubiquitous digital technologies affecting nearly every aspect of our daily lives.

This becomes not only visible in latest information— and communication technologies or autonomous vehicles, but also in the area of technology for restoring human physical and cognitive abilities. New developments at the intersection of computer science, engineering, robotics, and medicine include exoskeletons for people with paraplegia, powered and computer-controlled leg prostheses, fully articulate bionic hands, cochlear implants for people who are deaf, and eye-tracking gaze-to-speech synthesizers for people who cannot communicate verbally. Such bionic EDTs merge the human body with technology, often quite visibly. While this technology is typically designed for therapeutic functions (increasing diminished capabilities), its future application in the area of enhancing function (increasing normal capabilities) is also likely (e.g., exoskeletons for soldiers or workers). These bionic EDTs promise to enhance the abilities of able-bodied individuals and to restore abilities for people with physical disabilities.

Recent research has shown that using bionic EDTs affects how others perceive users in multiple ways. Apart from changing the abilities of the users, bionic EDTs are likely to affect non-users perceptions of users and the self-perception of the users, their motivation, and their (inter-)actions. According to the stereotype content model (SCM) [1], the two main dimensions of social perception and stereotyping are attributed warmth (ranging from warm = good intentions to cold = bad intentions) and competence (ranging from incompetent to competent). Groups and individuals are perceived in terms of warmth and competence, which elicit specific emotional and behavioral reactions [2] and self-construal [3]. According to an extension of the Stereotype Content Model, the BIAS Map [2], groups perceived as warm elicit active facilitation (help) or active harm (attack), while groups perceived as competent elicit passive facilitation (cooperate) or passive harm (ignore). These effects are mediated by specific emotions, namely admiration, envy, contempt, and pity [2, 4]. Thus, attributions of warmth and competence are likely to affect coordination between users and non-users of bionic EDTs in public.

Individuals with physical disabilities are typically perceived as warm and incompetent [5]. However, wearing a bionic prosthesis changes attributions of competence (but not of warmth) by others: Users of bionic prostheses are perceived as almost as competent as able-bodied individuals [5–7]. Focusing on user's self- and meta-perceptions, on the other hand, new findings indicate that users of bionic prostheses perceive themselves as competent and experience being seen as competent by others [8].

In a nutshell, previous research has shown that using bionic EDTs affects how users perceive themselves and how others perceive them. These changes may also affect attitudes toward users and culminate in specific behavior toward them. So the ongoing introduction of such restorative or augmenting devices might lead to unintended social consequences (e.g., avoidance, alleging certain intentions, or discrimination of users). Therefore, it is a pressing concern to examine the effects that come with the usage of bionic EDTs to mitigate their possible unwanted side effects on social interactions and coordination.

However, these current findings and initial insights remain mixed, and research on the self-and other perception of bionics users still remains scarce. Despite the ongoing proliferation of bionic EDTs, their availability is still limited, so the number of participants for potential studies remains small. This scarcity of potential study participants inevitably affects the corresponding research bionic EDTS such that empiric studies in real-world settings are currently not possible.

To account for this limitation, we suggest using a Mixed Reality system, which we call *TechnoSapiens* to emphasize the merging of human bodies with digital technology. The system enables participants to experience body alteration or modification so that their body is virtually extended with bionic EDTs. This approach allows many individuals to experiment with artificial body parts, opening this important area of research and societal debates for many participants. We describe the development and prerequisites of such a system in the following before evaluating it and discussing

its future research potential. We have earlier developed and first version of our system and validated it in a user study [9]. In the following, we describe the fundamental technological requirements and discuss research questions for further research.

2   Mixed Reality

The modalities of virtual content presentation strongly affect the extent to which users experience immersion 1.4). The use of stereo head-mounted displays (HMDs) makes it possible to provide a sense of depth and scale to the user, which significantly increases the sense of immersion and is generally understood as the main factor for separate virtual reality (VR) from virtual environments (VE) [10]. Further aspects that are unique to virtual reality systems are interactivity and user engagement, among other. All of them are aiming to provide a more *immersive* experience than other digital systems.

Applications that use the virtual reality paradigm can be classified similarly as a subset of virtual reality systems. In this paper, we use the following terms to describe different degrees of "virtuality."

- **Virtual Reality** (VR) describes applications that simulate the entire experience of a digital immersive environment. This includes aspects, such as virtually rendered three-dimensional scenes, spatial audio, and haptic feedback for virtual objects. The user experiences an environment, that is completely artificial and decoupled from the real-world surrounding him.
- **Mixed Reality** (MR) is used to describe applications, that blend the virtual experience with the real world, for example by blending virtual objects into the user's vision. This way it can be seen as a superset of the following categories.
- **Augmented Reality** (AR) applications are used to add virtual objects (i.e., *augmentations*) into the real world. Such virtual content can help making information more accessible by framing it with a spatial context.
- **Diminished Reality** (DR) is similar to AR, but aims to virtually remove and/or alter real-world objects from the user's vision.

Applications that blend real and virtual environments have higher technological demands, than the purely simulated ones (see Sect. 2.1). However, mixed reality technologies promise to improve immersion, especially with regard to body ownership since users does not experience an avatar, but rather their own body in a virtually altered environment (see Sect. 4). Since body ownership is the most important measure for studying subjective psychological phenomena in altered

Fig. 1 Example of a diminished reality application for interacting with a virtual arm prosthesis. Left: user wears a green glove, which is used for chroma-keying as a cheap way to segment the arm within his vision. Right: the arm is removed from his vision and a virtual prosthesis model is shown instead

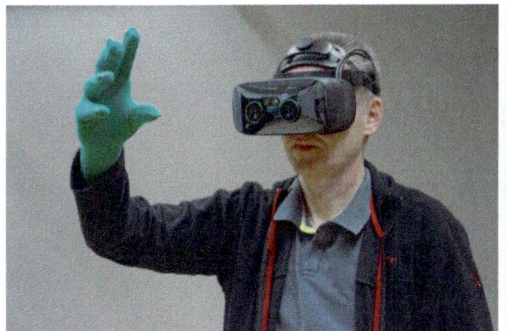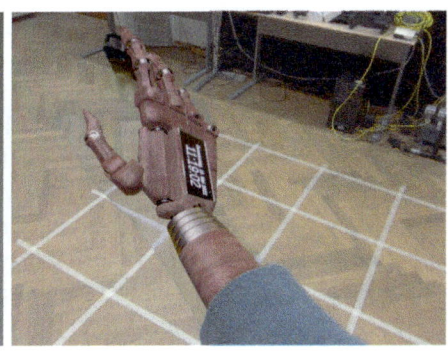

environments, mixed reality applications should perform better, despite the higher technological demands.

2.1 Mixed Reality Hardware

To create a mixed reality, specialized hardware is required, similar to how a stereo HMD is used to create a virtual reality experience. There are two approaches to blend the real world with virtual content: optical and visual see-through [11]. Optical see-through systems (OST) describes systems that project virtual imagery on semi-transparent mirrors. Examples of such "holo-graphic displays" are *Microsoft HoloLens*, *Google Glass,* or *Epson Moverio*. Besides their relative compactness compared to traditional stereo HMDs, such systems often integrate hand tracking for intuitive interaction. However, they are currently less suited for experiencing altered realities for several reasons. Despite technological advancements, the image quality of OST systems does not yet match traditional stereo HMDs in terms of brightness and resolution. Compared to visual see-through (VST) approaches, they do not perform as well with regards to depth perception [11].

It is challenging to create diminished reality experiences with OST systems, since holograms are projected on top of the user's natural vision. Being able to manipulate the image before presenting it to the user is advantageous for such applications. Visual see-through (VST) systems use camera systems to capture the real-world world. For proper depth perception, it is important to capture a separate image for each eye (stereo VST). First stereo VST systems have been proposed in 2000 [12], however, commercial systems were not available until 2016, when the *HTC Vive Pro* first released. Comparable systems are the *Valve Index* and the *HP Reverb*, which feature higher resolution cameras. The *Oculus Quest 2* has several low-resolution cameras that are blended together for a wider field of view. Additionally, the *Stereolabs ZED-mini* is a camera system that can be attached to different VR HMDs to provide high resolution stereo VST. The *Varjo XR-3* is another VST HMD that also integrates hardware-based hand tracking and depth sensing.

Generally, all of the systems presented above are viable to create the envisioned experience. During our research, we build three different prototypes, using the *HTC Vive Pro*, a combination of the *Oculus Quest 2* and the *Stereolabs ZED-mini* camera system, as well as the *Varjo XR-3*. However, for the system used in our study we found the *XR-3* to be much better suited, due to the integrated systems (such as hand tracking and VST cameras) and the higher display resolution (see Fig. 1).

2.2 Diminished Reality

As described earlier, diminished reality refers to the removal or replacement of real-world objects from the vision of the user of a mixed reality system. For our research, diminished reality is used to visually remove the user's arm or parts of it from their vision and replace it with a virtual prosthesis model or other body augmentations.

The technological requirements for a diminished reality system are higher than for augmented and virtual reality. First, it needs to detect and track regions or objects in the image that should be altered (*Segmentation*). Second, the detected regions need to be overdrawn with virtual content in a way that matches the surrounding visual context closely (*Inpainting*). This way, the tracked regions are removed from the image. Third, a virtual rendering of a replacement object can be drawn in place of the removed object, similar to traditional augmented reality applications.

2.2.1 Segmentation

Image segmentation is an active field of research in computer vision. For diminished reality, an approach should be fast, accurate and robust. Segmentation needs to be performed on every image shown to the user, so real-time performance is key for fast, low-latency systems. Furthermore, the approach should be accurate and robust, i.e., it should find as many pixels that belong to the targeted objects as possible, while minimizing the amount of false-positives. This prevents inpainting in regions that do not need to be overdrawn, while not showing parts that should have been removed. Segmentation of body parts, especially hands and arms, remains a challenging problem with regards to the aforementioned criteria. *Bandini et al.* give a good overview over hand tracking from ego-centric ("first person") vision [13]. Those approaches are effective when designing experiments that focus on self-perspective. For experiments on other-perspective, however, it may be required to use different approaches or to train on different datasets [14]. Since many people tend to have their hands in front of their body for many interactions, it can also be viable to rely on depth data for improved accuracy [15].

To minimize performance cost, we found that using a colored glove and segment the image using chroma-keying instead of using algorithmic approaches can be a good compromise (see Fig. 1, left). The glove does not interfere much with the user's movement and chroma-keying can be implemented efficiently. However, this might not be a viable solution in other scenarios, for example if the pose tracking system (see Sect. 3) does not support it.

2.2.2 Inpainting

Image inpainting refers to the problem of removing objects from or restoring missing regions in an image. For diminished reality, inpainting refers to the removal of a real-world object from the user's vision. Mori et al. give an overview over different approaches with regards to diminished reality [16]. The problem is challenging, because it requires a coherent restoration of occluded background information. It can be tackled using hardware or algorithmic approaches.

Hardware-based approaches rely on one or multiple cameras that capture the regions that are occluded from the user's point of view. The camera image is then re-projected to align with the user's vision [17, 18]. Such approaches produce good quality results; however, they require low-latency synchronization and precise calibration.

Algorithmic approaches traditionally operate on the 2D image by extrapolating the boundary of the removed regions into its interior. This is typically an expensive task, but there are variants that work in real time [19, 20]. Additionally, stereoscopic vision can improve inpainting by using information that is only visible for one camera [21–23].

Instead of estimating the inpainting contents from the image itself, it is possible to first reconstruct the background from pre-recorded imagery that provides an unoccluded view. This has been shown to work on 2D images [24] and has also been extended to 3D scenes [25]. Such approaches work well on static scenery; however, recently advancements have been made to combine reconstruction and inpainting into a continuous process to work in dynamic contexts [26, 27]. In the self-perspective provided in our current system, dynamic objects (e.g., other persons, etc.) can be avoided in the study design, so we used a reconstruction approach. However, for future versions, we think it is a necessity to further research real-time algorithmic inpainting.

3 Human Motion Tracking

For a convincing virtual body augmentation experience it is important provide an intuitive user interaction. Approaches that rely on body attachments, such as tracking markers or data gloves are sub-optimal, since they limit the user's freedom of movement and make the alteration of their vision harder (see Sect. 2.2). For those reasons, this section focuses on optical tracking approaches.

Most body restoration targets limbs (i.e., arms and legs), since their impairment is most limiting for an affected person on how they interact with their environment. Especially arms and hands play significant roles in our daily lives. Examples of such roles are (1) Pointing to a person or an object, (2) Conveying information about space, shapes, the objects' number, or the temporal characteristics of motions, (3) Interacting relentlessly with objects in operating rooms, airplanes, laboratories, and factories [28], (4) Carrying out unconscious gesticulation to express ideas, and (5) Conducting conscious communication with sign language [29]. On the other side, Humans can perceive and interpret detailed information in the gesticulations of others. Thus recreation of believable hand motions for a virtual environment is an important task, which calls, before anything, for an accurate estimation of such movements. Due to the anatomical similarities, we think that a generalization of hand tracking approaches to legs is viable.

The realm of three-dimensional hand pose estimation research aims at designing such systems to detect the joint configuration of human hands in 3D space. For that, there are two main classes of technologies. One is to recognize hand gestures is *marker-based*, in which the subject wears data gloves or puts on sensors (optical, mechanical, ultra-sound, or magnetic sensors). These devices are accurate and easy to implement. However, their high cost, the difficulty of wearing its tools, and sometimes their health-hazard limit its applicability in many real-life scenarios, substantially [28, 30]. The other class of technology is *vision-based*, which provides a noninvasive, easy, and natural environment. Here the motions and gestures are analyzed and interpreted using captured video sequences by one or several cameras [30]. However, there are an ample amount of challenges the design must solve before we employ it.

To overcome those vision-based technical difficulties researchers have systematically considered three broad approaches [28, 34]: First, the *appearance- based* [35] (bottom-up or discriminative) approaches, which compare the parameters "directly" extracted from input images or videos (color, texture, shape, depth, and motion) with the parameters of the training data (set of learnt images or videos). Second, the *model-based* [35] (top-down or generative) approaches, which estimate the hand parameters by comparing the input images to the possible 2D appearance, projected from the

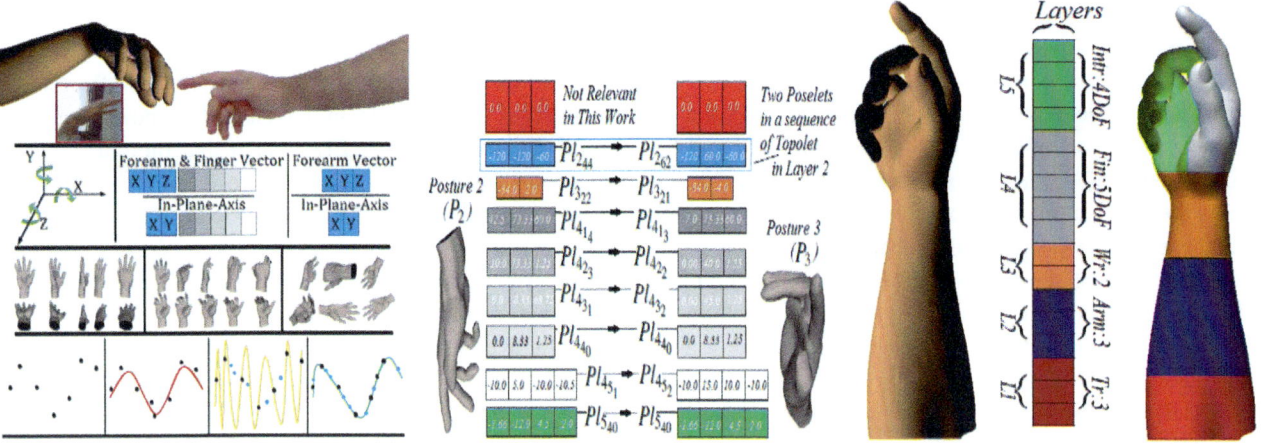

Fig. 2 General trend of Modules considered in the analysis-by-synthesis approach. From left to right: **a** Bottom-up part (e.g., Tracking [31]), **b** Top-down modules (e.g., Pose Estimation [32]), and **c** Classification and Pose Definition module [33])

3D kinematic hand "model." Finally, *analysis-by-synthesis* approach [34], which combines the low-level cues by using spatial grouping rules, obtained from the bottom-up proposal, with the top-down process. That is, on one side, the bottom-up proposal activates some hypotheses, known as *candidate gestures*, about the hand and the scene structure. On the other side, the top-down process *accepts or rejects* those hypotheses by "direct" comparison with the image features. In that context, the winning hypothesis (accepted gesture) will be the one whose projected model is closest to the low-level cues (exhibits the highest probability) [28].

To scratch the surface of the literature in the direction of pose estimation we consider the following definition of the pose: According to the Oxford English Dictionary, a pose is "An attitude or posture of the body, or of a part of the body, especially one deliberately assumed, or in which a figure is placed for effect, or artistic purposes." This definition points to two aspects of a pose [36]: (1). It is a configuration that body parts arranged in 3D space. (2) The resulting captured scene has a 2D appearance. According to these two aspects, considering the vision-based class of technology and, in the scope of the analysis-by-synthesis approach, would be a beneficial choice.

However, still the design calls for addressing many unforeseen challenges. The setting of the project's design would demand the specificity of those challenges to be redefined anew. However, there is a general trend of modules that can attend to several significant obstacles. Those modules include but are not limited to the following: Bottom-up modules Fig. 2a) such as (I) tracking and (II) segmentation to locate and extract human hands within 2D image scenes. Top-down modules (Fig. 2b) such as (III) spatio-temporal models to effectively relate hand's postures with mathematical representations, and (IV) optimization techniques to efficiently search through the high dimensional search space. Finally,

(V) Classification frameworks (Fig. 2c) for converting the estimated postures/gestures to the semantically meaningful commands.

One of the earliest works in finding some global features on fingers, based on which they estimated the poses, proposed by [37]. This and many similar works focused on the estimation of the hand poses on a single image and a limited number of pre-defined poses. Therefore, improvements in the area weighed on improving the accuracy for estimation of an unlimited number of poses, enhancing the time-complexity for pose estimation of videos, and stability of the system. To touch on the challenges of the matter, as an example, stability of a tracking method means [37] the measured position does not change, as long as the tracked object does not move. However, there are several possible sources of instability, such as changing light conditions, the motion of distracting objects, and acquisition noise even if the object does not move.

Since then, many techniques, from statistical Bayesian networks [38], to depth technologies [39] and multi-view setup [40], and from temporal models [41], to convolutional neural network [42, 43], are proposed to tackle the above issues, and to enhance the performance consistently.

4 Psychological Framework and System Evaluation

A significant goal of prosthetic treatment is the perceptual integration of a prosthesis into an amputee's body representation, i.e., the experience that the device is an actual part of the body. In the context of our system, we consider embodiment a prerequisite for obtaining significant effects in upcoming research on bionic EDTs based on the developed system [9]).

The term embodiment is somewhat controversial and ambiguous [44]. Embodiment is occasionally associated with the sub-concepts of self-location, agency, and body owner-ship [45]; ownership/integrity, agency, and anatomical plau-sibility [46]; or sense of ownership and agency [47]. However, re- searchers agree that the sense of body ownership, that is, the sense that a part of the body belongs to oneself and the perception of the body as the source of an experienced sensa-tion, is the key feature of embodiment [48–50]. Research on body ownership dates back to the rubber hand illusion exper-iment [51], where a sense of ownership over a rubber hand is induced by applying synchronous visual-tactile stimulation to a seen rubber hand and the unseen real hand. Until today, most investigations of body ownership employ the rubber hand illu-sion (RHI) paradigm [52] and apply adaptions of the RHI embodiment scale [9].

To evaluate the suitability of our system, we conducted a first evaluation study on self-presence, immersion, and sense of body ownership experienced by users of the system. While the first two qualities are of general importance for studies of virtual technologies, the significance of the latter results from our particular application.

As described in [9], we conducted a quasi-experimental study with 27 participants assigned to the same condition and rated their experiences afterward by filling in a questionnaire. We assessed the following study variables:

- Sense of Ownership: We measured the degree to which participants felt that the virtual prosthesis was actually part of their body (i.e., the sense of body ownership conveyed by the virtual prosthesis) with the embodiment scale for the rubber hand illusion [50].
- Self-presence: We measured self-presence in the VR envi-ronment with the corresponding sub-scale of the Multi-modal Presence Scale for virtual reality environments [53].
- Immersion: We measured immersion with the physical presence sub-scale of the Multimodal Presence Scale for virtual reality environments [53].
- User Experience: To determine the quality of the overall user experience, we measured it with eleven items from the Evaluation of Virtual Reality Games questionnaire [54] and six system-specific self-developed items.

We showed that participants experience a significant sense of body ownership with regard to the virtual prosthesis. Partic-ipants also experienced profound levels of self-presence and immersion and reported a favorable user experience (see [9] for detailed results). Overall, these results indicate that the system works as intended, conveys a sense of owner-ship, and is suitable for further research on the influence of bionic EDTs on social perception and stereotypes toward users.

5 Conclusion

Suppose we understand how bionic EDTs will affect social interactions in the future. In that case, this knowledge could be crucial for designing EDTs such that they do not lead to unforeseen adverse side effects, for example, by evoking negative stereotypes that ultimately lead to less instead of more inclusion. Answering these questions and understanding the underlying processes will also help to gauge the social cost of merging humans with technology.

As the availability of EDTs is limited, our approach empha-sized the use of VR technologies as a platform for studies aiming at these questions. Consequently, we developed a simulation environment based on DR/AR technologies that allows to virtually blend the human body with artificial parts and experience bionic EDTs in real time. The assessment of the systems capabilities as well as the evaluation of the user experience showed promising results. It indicated that our system is suitable for future research and facilitates the intended experimental manipulation. Accordingly, our system will enable researchers to conduct large-scale controlled experimental studies on the effects of merging humans with technology. So it allows future research to gain more profound insights and successively answer those pressing research questions that come along with these developments.

In sum, we believe that answering these questions by using our virtual experimental platform can help society in its transformation into a digital future, where the lines between humans and technology start to blur. Also, the insights gained by analyzing the impact of EDTs on social perception, poten-tial fears, and the potential for excluding EDT users can help guide societal and legal regulations governing the use of bionic EDTs.

Acknowledgements This work was funded by the Deutsche Forschungsgemeinschaft (DFG, German Research Foundation) project number 491193532 as a part of the collaborative research centre 1410 "Hybrid Societies" and Chemnitz University of Technology. The funding organization had no influence on the design, results and decisions made during the creation of this work. The authors have no competing interests to declare that are relevant to the content of this article.

References

1. Fiske, S. T., Cuddy, A. J., Glick, P., & Xu, J. (2002). A model of (often mixed) stereo-type content: Competence and warmth respec-tively follow from perceived status and competition. *Journal of Personality and Social Psychology, 82*(6).
2. Cuddy, A. J., Fiske, S. T., & Glick, P. (2008). Warmth and compe-tence as universal dimensions of social perception: The stereotype content model and the bias map. *Advances in Experimental Social Psychology, 40*, 61–149.
3. Guimond, S., Chatard, A., Martinot, D., Crisp, R. J., & Reders-dorff, S. (2006). Social comparison, self-stereotyping, and gender

differences in self-construals. *Journal of Personality and Social Psychology, 90*(2), 221.

4. Becker, J. C., & Asbrock, F. (2012). What triggers helping versus harming of ambivalent groups? Effects of the relative salience of warmth versus competence. *Journal of Experimental Social Psychology, 48*(1), 19–27.

5. Meyer, B., & Asbrock, F. (2018). Disabled or cyborg? How bionics affect stereotypes toward people with physical disabilities. *Frontiers in psychology*, 2251.

6. Bretschneider, M., Mandl, S., Strobel, A., Asbrock, F., & Meyer, B. (2022). Social perception of embodied digital technologies—A closer look at bionics and social robotics. *Gruppe. Interaktion. Organisation. (GIO)*.

7. Mandl, S., Bretschneider, M., Meyer, S., Gesmann-Nuissl, D., Asbrock, F., Meyer, B., & Strobel, A. (2022). Embodied digital technologies: First insights in the social and legal perception of robots and users of prostheses. *Frontiers in Robotics and AI*, 68.

8. Bretschneider, M., Meyer, B., & Asbrock, F. (2023). The impact of bionic prostheses on users' self-perceptions: A qualitative study. Acta Psychologica, 241, 104085. https://doi.org/10.1016/j.actpsy.2023.104085

9. Rudolph, C., Brunnett, G., Bretschneider, M., Meyer, B., & Asbrock, F. (2024). TechnoSapiens: Merging humans with technology in augmented reality. In *The Visual Computer* (pp. 1021–1036).

10. Wohlgenannt, I., Simons, A., & Stieglitz, S. (2020). Virtual reality. *Business & Information Systems Engineering, 62*(5), 455–461.

11. Medeiros, D., Sousa, M., Mendes, D., Raposo, A., & Jorge, J. (2016). Perceiving depth: Optical versus video see-through. In *Proceedings of the 22nd ACM Conference on Virtual Reality Software and Technology* (pp. 237–240).

12. Kanbara, M., Okuma, T., Takemura, H., & Yokoya, N. (2000). A stereoscopic video see-through augmented reality system based on real-time vision-based registration. In *Proceedings IEEE Virtual Reality 2000 (Cat. No. 00CB37048)* (pp. 255–262). IEEE.

13. Bandini, A., & Zariffa, J. (2020). Analysis of the hands in egocentric vision: A survey. *IEEE Transactions on Pattern Analysis and Machine Intelligence*.

14. Urooj, A., & Borji, A. (2018). Analysis of hand segmentation in the wild. In *Proceedings of the IEEE Conference on Computer Vision and Pattern Recognition* (pp. 4710–4719).

15. Augustauskas, R., & Lipnickas, A. (2017). Robust hand detection using arm segmentation from depth data and static palm gesture recognition. In *2017 9th IEEE International Conference on Intelligent Data Acquisition and Advanced Computing Systems: Technology and Applications (IDAACS)* (Vol. 2, pp. 664–667). IEEE.

16. Mori, S., Ikeda, S., & Saito, H. (2017). A survey of diminished reality: Techniques for visually concealing, eliminating, and seeing through real objects. *IPSJ Transactions on Computer Vision and Applications, 9*(1), 1–14.

17. Meerits, S., & Saito, H. (2015). Real-time diminished reality for dynamic scenes, in *2015 IEEE International Symposium on Mixed and Augmented Reality Workshops* (pp. 53–59). IEEE

18. Mori, S., Maezawa, M., Ienaga, N., & Saito, H. (2016). Detour light field rendering for diminished reality using unstructured multiple views. In *2016 IEEE International Symposium on Mixed and Augmented Reality (ISMAR-Adjunct)* (pp. 292–293). IEEE.

19. Herling, J., & Broll, W. (2010). Advanced self-contained object removal for realizing real-time diminished reality in unconstrained environments. In *2010 IEEE International Symposium on Mixed and Augmented Reality* (pp. 207–212). IEEE.

20. Herling, J., & Broll, W. (2014). High-quality real-time video inpainting with PixMix. *IEEE Transactions on Visualization and Computer Graphics, 20*(6), 866–879, IEEE.

21. Hervieu, A., Papadakis, N., Bugeau, A., Gargallo, P., & Caselles, V. (2010). Stereoscopic image inpainting: Distinct depth maps and images inpainting. In *2010 20th International Conference on Pattern Recognition* (pp. 4101–4104). IEEE.

22. Hervieux, A., Papadakis, N., Bugeau, A., Gargallo, P., & Caselles, V. (2011). Stereoscopic image inpainting using scene geometry. In *2011 IEEE International Conference on Multimedia and Expo* (pp. 1–6). IEEE.

23. Wang, L., Jin, H., Yang, R., & Gong, M. (2008). Stereoscopic inpainting: Joint color and depth completion from stereo images. In *2008 IEEE Conference on Computer Vision and Pattern Recognition* (pp. 1–8). IEEE.

24. Cosco, F. I., Garre, C., Bruno, F., Muzzupappa, M., & Otaduy, M. A. (2009). Augmented touch without visual obtrusion. In *2009 8th IEEE International Symposium on Mixed and Augmented Reality* (pp. 99–102). IEEE.

25. Mori, S., Shibata, F., Kimura, A., & Tamura, H. (2015). Efficient use of textured 3D model for pre-observation-based diminished reality. In *2015 IEEE International Symposium on Mixed and Augmented Reality Workshops* (pp. 32–39). IEEE.

26. Bescos, B., Fácil, J. M., Civera, J., & Neira, J. (2018). DynaSLAM: Tracking, mapping, and inpainting in dynamic scenes. *IEEE Robotics and Automation Letters, 3*(4), 4076–4083.

27. Bescos, B., Campos, C., Tardós, J. D., & Neira, J. (2021). DynaSLAM II: Tightly-coupled multi-object tracking and slam. *IEEE Robotics and Automation Letters, 6*(3), 5191–5198.

28. Rautaray, S. S., & Agrawal, A. (2015). Vision based hand gesture recognition for human computer interaction: A survey. *AI Review, 43*(1), 1–54.

29. Starner, T. E., & Pentland, A. (1995). Visual recognition of American sign language using Hidden Markov models. In *Media* (pp. 189–194). http://www.eecs.ucf.edu/~gitars/cap6938/starner95visual.pdf

30. Garg, P., Aggarwal, N., & Sofat, S. (2009). Vision based hand gesture recognition. *World Academy of Science, Engineering and Technology, 49*, 972–977. http://pdf.aminer.org/000/238/273/visionbasedgesturerecognitionareview.pdf

31. Dadgar, A., & Brunnett, G. (2020). SaneNet: Training a fully convolutional neural network using synthetic data for hand detection. In *SAMI 2020-IEEE 18th World Symposium on Applied Machine Intelligence and Informatics, Proceedings* (pp. 251–256).

32. Dadgar, A., & Brunnett, G. (2019). Topolet: From atomic hand posture structures to a comprehensive gesture set. Proceedings of the 14th International Joint Conference on Computer Vision, Imaging and Computer Graphics Theory and Applications, Volume 5: VISAPP (pp. 157-164). SciTePress.

33. Dadgar, A., & Brunnett, G. (2018). Multi-forest classification and layered exhaustive search using a fully hierarchical hand posture/gesture database. Proceedings of the 13th International Joint Conference on Computer Vision, Imaging and Computer Graphics Theory and Applications, Volume 5: VISAPP. Funchal, Madeira, Portugal: SciTePress.

34. Yuille, A., & Kersten, D. (2006). Vision as Bayesian inference: Analysis by synthesis? *Trends in Cognitive Sciences, 10*(7), 301–308.

35. Fang, Y., Wang, K., Cheng, J., & Lu, H. (2007). A real-time hand gesture recognition method. In *2007 IEEE International Conference on Multimedia and Expo* (pp. 995–998).

36. Bourdev, L., & Malik, J. (2009). Poselets: Body part detectors trained using 3D Human Pose annotations. In *IEEE International Conference on Computer Vision* (pp. 1365–1372). http://ieeexplore.ieee.org/document/5459303/

37. Hardenberg, C. V. (2001). Bare-hand human-computer interaction. Proceedings of the 2001 Workshop on Perceptive User Interfaces (pp. 1-8). ACM Press.

38. Suk, H.-I., Sin, B.-K., & Lee, S.-W. (2010). Hand gesture recognition based on dynamic Bayesian network framework. *Pattern Recognition, 43*(9), 3059–3072. http://www.sciencedirect.com/science/article/pii/S0031320310001366

39. Keskin, C., Kırac, F., Kara, Y. E., & Akarun, L. (2011). Real time hand pose estimation using depth sensors. In *IEEE International Conference on Computer Vision Workshop.*

40. Jojic, N., Brumitt, B., Meyers, B., Harris, S., & Huang, T. (2000). Detection and estimation of pointing gestures in dense disparity maps. In *Proceedings—4th IEEE International Conference on Automatic Face and Gesture Recognition, FG 2000* (pp. 468–475).

41. Dittmar, T., Krull, C., & Horton, G. (2015). A new approach for touch gesture recognition: Conversive Hidden non-Markovian models. *Journal of Computational Science, 10*, 66–76. https://doi.org/10.1016/j.jocs.2015.03.002

42. Wei, S.-E., Ramakrishna, V., Kanade, T., & Sheikh, Y. (2016) Convolutional pose machines. 2016 IEEE Conference on Computer Vision and Pattern Recognition (CVPR), Las Vegas, NV, USA, 2016, pp. 4724-4732. https://doi.org/10.1109/CVPR.2016.511.

43. Zimmermann, C., & Brox, T. (2017). Learning to estimate 3D hand pose from single RGB images. In *Proceedings of the IEEE International Conference on Computer Vision* (Vol. 2017-Octob, pp. 4913–4921).

44. Pielli, L., & Zlatev, J. (2020) The cyborg body: Potentials and limits of a body with prosthetic limbs. *Cognitive Semiotics, 13*(2).

45. Kilteni, K., Groten, R., & Slater, M. (2012). The sense of embodiment in virtual reality. *Presence: Teleoperators and Virtual Environments, 21*(4), 373–387.

46. Foell, J., Bekrater-Bodmann, R., Flor, H., & Cole, J. (2011). Phantom limb pain after lower limb trauma: Origins and treatments. *The International Journal of Lower Extremity Wounds, 10*(4), 224–235.

47. Zbinden, J., Lendaro, E., & Ortiz-Catalan, M. (2022). Prosthetic embodiment: Systematic review on definitions, measures, and experimental paradigms. *Journal of NeuroEngineering and Rehabilitation, 19*(1), 1–16.

48. Bekrater-Bodmann, R. (2020). Perceptual correlates of successful body–prosthesis interaction in lower limb amputees: Psychometric characterisation and development of the prosthesis embodiment scale. *Scientific Reports, 10*(1), 1–13.

49. Piryankova, I. V., Wong, H. Y., Linkenauger, S. A., Stinson, C., Longo, M. R., Bu"lthoff, H. H., & Mohler, B. J. (2014). Owning an overweight or underweight body: Distinguishing the physical, experienced and virtual body. *PloS One, 9*(8), e103428.

50. Romano, D., Maravita, A., & Perugini, M. (2021). Psychometric properties of the embodiment scale for the rubber hand illusion and its relation with individual differences. *Scientific Reports, 11*(1), 5029.

51. Botvinick, M., & Cohen, J. (1998). Rubber hands 'feel' touch that eyes see. *Nature, 391*(6669), 756.

52. Tsakiris, M., & Haggard, P. (2005). The rubber hand illusion revisited: visuotactile integration and self-attribution. *Journal of Experimental Psychology. Human Perception and Performance, 31*(1), 80–91.

53. Makransky, G., Lilleholt, L., & Aaby, A. (2017). Development and validation of the Multi-modal Presence Scale for virtual reality environments: A confirmatory factor analysis and item response theory approach. *Computers in Human Behavior, 72*(7), 276–285.

54. Norman, K.L. (2018). Evaluation of virtual reality games: Simulator sickness and human factors. In *GHItaly18: 2nd Workshop on Games-Human Interaction.*

Human–Robot Interaction in Telemanipulation—An Overview

Stephan Andreas Schwarz and Ulrike Thomas

Abstract

Teleoperation and haptic telemanipulation is a common solution to perform tasks from a remote distance. It is very suitable and can be used in dangerous or unreachable environments, such as nuclear power plants, space missions, or under water. In recent years, especially due to the COVID-19 pandemic, telemanipulation is increasingly used to perform tasks involving other human participants. This paper gives an overview of the state of the art regarding control concepts to improve human–robot interactions on the follower side of a telemanipulation system. In this context, system architectures and shared control approaches are considered. We also present the work done in the Collaborative Research Center 1410 regarding telemanipulation including a safety mechanism as well as two shared control concepts to improve human-likeness, safety, and mobility of the follower motion.

Keywords

Telemanipulation · Human–robot interaction · Control · Shared control · Safety

1 Introduction

Teleoperation in general allows the operator to perform a task from a remote place. The term telemanipulation specifies this process by defining the task as a manipulation task performed by a manipulator, e.g., a robotic arm or a gripper. The operator controls an input device, so-called leader, whose motion commands are transferred to the follower [11]. In the past, these systems were and still are used in hazardous environments, such as nuclear power plants [29], space missions [2], and under water exploration [5], but also in medical applications when a highly specialized surgeon is unable to travel to the location of the surgery [6].

In recent years, the scope of telemanipulation broadens toward interactive applications. Integrating these systems more into the community and the daily life has huge potential to improve life quality, reduce stress, and save resources. Especially the COVID-19 pandemic shows that remote working is an indispensable requirement and has to be enabled for more people. In combination with telepresence, first applications of telemanipulation are already used in healthcare [30], but there is still a massive, unused potential in social and interactive areas.

On the downside, interactions with other humans increase the requirements on several aspects of the telemanipulation system, such as safety, predictability, and human-likeness, which have to be considered before implementing a new application. Despite the extensive research, aspects of human–robot interactions on the follower side are rather unexplored. Therefore, this paper gives an overview about the current state of the art of telemanipulation systems considering human–robot interactions and the resulting requirements. The focus lies on existing system architectures and shared control approaches in combination with safety aspects. Further, we present the work done in our lab as well as our ongoing research. These depictions lead to open research questions and unused potential of telemanipulation in human–robot environments that might be worth exploring in the future.

2 State of the Art

In this section, we give an overview about the current research in the field of telemanipulation. We focus on possibilities and approaches regarding desired and undesired human–robot

S. A. Schwarz (✉) · U. Thomas
Robotics and Human-Machine Interaction Laboratory at Chemnitz University of Technology, Chemnitz, Germany
e-mail: stephan-andreas.schwarz@etit.tu-chemnitz.de

U. Thomas
e-mail: ulrike.thomas@etit.tu-chemnitz.de

B. Meyer et al. (eds.), *Hybrid Societies*, Advances in Science, Technology & Innovation,
https://doi.org/10.1007/978-3-032-03488-5_25

interactions and look into system architectures and shared control concepts.

2.1 System Architectures for Telemanipulation

As already explained, a telemanipulation system consists of a leader and a follower device. The operator moves the leader device, and the follower receives control commands to follow the predetermined movement depending on the active control approach. Further, feedback information are sent back to the leader to give the operator insights about the follower environment. This information mostly consist of visual and force data but can also contain tactile feedback to provide the operator a sense of touch. In most cases, a communication channel lies between the leader and the follower which inserts a time delay. The general system structure is shown in Fig. 1.

In a simple position-position controller, the pose of the leader is sent as a command to the follower and the position of the follower is fed back to the leader [12].

In a position-force architecture, the leader's pose is again the control command for the follower, but now the follower returns the interaction force [9]. These forces are measured by a force-torque sensor mounted on the end effector or joint-torque sensors in the joints. Since both of these static approaches do not consider the dynamics of the individual systems, instability, and undesired dynamic behavior may occur [21]. Especially time delays resulting from the communication channel between the leader and follower increase these risks. This problem received special attention since it is a common issue in telemanipulation. Using wave variables [23] or time-domain passivity approaches [3, 28], can ensure stability during constant and varying time delays.

These static approaches are mostly to restrictive to be applied in scenarios where a human interacts with the robot. Limited forces and a compliant behavior are required during such interactions. Therefore, more complex system architectures are used to control the dynamic behavior of the overall system.

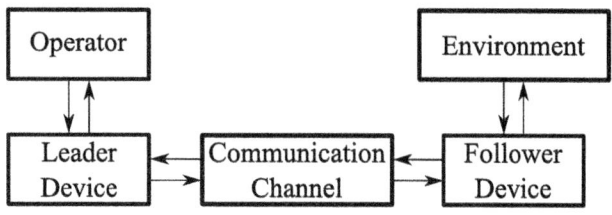

Fig. 1 General data flow of a telemanipulation system consisting of an operator, a leader device, a communication channel, a follower device and the environment on the follower side

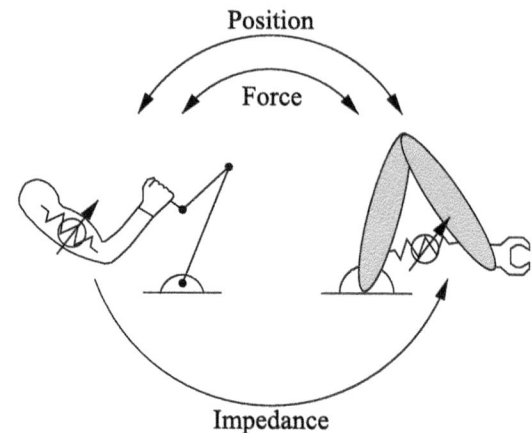

Fig. 2 User-controlled variable impedance telemanipulation. The measured grip force of the operator adjusts the impedance of the follower robot [37]

One approach is the use of impedance control, first introduced in [15]. Compensating the dynamics of the robot and simulating a desired dynamic behavior enables the system to safely interact with the environment. The commonly chosen mass-spring-damper dynamic allows the follower to move compliant. A displacement of the leader results in a command force for the follower. The compliance of the system can be adjusted by adapting the control parameters as presented in [8]. Walker et al. [37] present a user-controlled variable impedance controller that adjusts the impedance of the follower robot depending on the applied grip force of the operator. The mechanism is shown in Fig. 2. This approach allows a natural and intuitive adaptation of impedance, since the grip force correlates with the human impedance.

Furthermore, it is possible to invert the approach of the impedance control which results in an admittance controller. In this case, a force applied on the leader results in a position displacement of the follower. The dynamic behavior is again simulated as a mass-spring-damper system. Landi et al. [20] present a parameter adaptation algorithm for admittance control to reduce the effort for the user during physical human–robot interactions.

The decision between these two approaches depends on the control capabilities of the systems as well as the integrated sensors, e.g., existing force-torque or joint-torque sensors.

Furthermore, it is possible to scale the commands of the leader depending on the current requirements. Krishnan et al. [19] present a study to investigate the usability of velocity scaling in telemanipulation. The result is that an environment-based scaling performs best compared to a scaling relative to the control speed. Reducing the speed close to the interaction point can improve human–robot interactions on the follower side and thus offers potential for improvement.

2.2 Shared Control

The goal of shared control is to reduce the physical and mental workload of the operator while controlling the telemanipulation system during complex tasks [7]. It also can improve the completion time and accuracy of performed tasks. There already exist many approaches to improve the task performance and usability of telemanipulation, such as presented in [1, 16, 22, 25, 32, 33].

On the other side, there is less research done concerning shared control algorithms that improve human–robot interactions on the follower side, which will be presented in this section. Most of the work rises from telesurgery and telehealthcare since these fields have unavoidable contacts with other humans.

In general, shared control approaches can be divided in two categories, haptic guidance and shared autonomy.

During haptic guidance, the control algorithm limits the workspace or guides the movement of the operator by applying virtual forces, so-called virtual fixtures, to the system [27]. The authors in [4] present a telemanipulation system to perform oropharyngeal swab samplings. A forbidden-region virtual fixtures algorithm is used to improve safety. Virtual forces drag the operator away from undesired contacts with the oral cavity. The forbidden regions are marked by the operator during the process. An alternative mechanism is presented in [14]. Here, an impulsive force is computed to reduce the velocity toward the forbidden regions quicker than in the classical approach. This renders the feeling of a fully plastic collision before touching the vulnerable areas. These concepts can also be used for training purposes.

Nudehi et al. [24] introduce an algorithm for minimal invasive telesurgical training. Two leader devices are coupled with one follower and the control authority is split between both of them. This allows the trainee to continuously increase his amount of impact during surgery.

On the other hand, shared autonomy denotes all algorithms where the robot performs some parts of the tasks fully autonomously while the operator focuses on the remaining components. Vogel et al. [36] proposes a multi-priority Cartesian impedance controller which contains collision detection and reflex reaction. This approach is built for people with physical disabilities. Based on measured neural signals and a decision-and-control strategy, the system provides support to perform the desired task. A constraint-based strategy to define skills for shared control approaches is described in [26]. The authors in [34] use the redundancy of a seven-degree-of-freedom manipulator to provide compliant behavior. The compliance allows interactions between the robot and other medical staff while maintaining end effector accuracy. Furthermore, they integrate an adaptive fuzzy compensator to guarantee accuracy of the task during spontaneous human–robot interactions. In [35], the authors improved this approach

and used the null space to autonomously guarantee a remote center of motion constraint which is a key requirement during minimally invasive surgery.

Besides the safety aspects, the predictability of the follower is a key component for smooth human–robot interaction. Zhou et al. [38] tackle this issue using a dual-arm cooperative robot. The motion of the leader is tracked by wearable motion capture devices and mapped to the follower. Virtual objects are created to constrain the robot motion to improve its human-likeness.

3 Research of the Robotics and Human–Machine Interaction Laboratory

In this section, we present our published and ongoing research in the field of human-centered telerobotics. Therefore, our telemanipulation setup with the current system architecture is described. Furthermore, we introduce a safety mechanism to increase safety during undesired contacts in telemanipulation. Finally, a brief overview of our ongoing research regarding shared control and the aimed goals is given.

3.1 System Architecture

Our current telemanipulation system, as shown in Fig. 3, consists of a Franka Emika Panda robot with an attached 6d force-torque sensor on its end effector as follower system and a Haption Virtuose 6D input device as the leader. We use a F6D100-50 sensor from ME-Meßsysteme GmbH. Figure 4 shows the bilateral impedance-force controller. The position, velocity and acceleration of the leader are transferred to the follower and used to generate the desired dynamic behavior. The measured forces and torques acting on the end effector

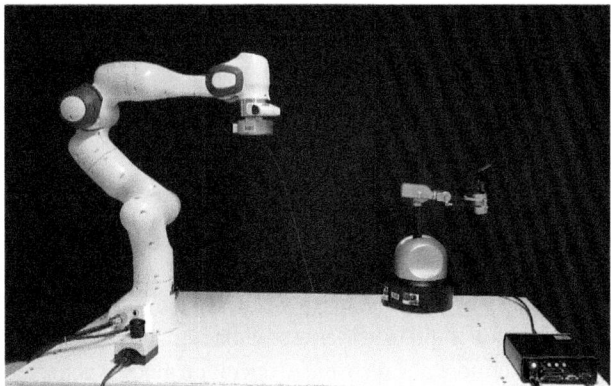

Fig. 3 Telemanipulation setup consisting of a Franka Emika Panda robot with a 6d force-torque sensor as follower (left) and a Haption Virtuose 6D input device as leader (right) [31]

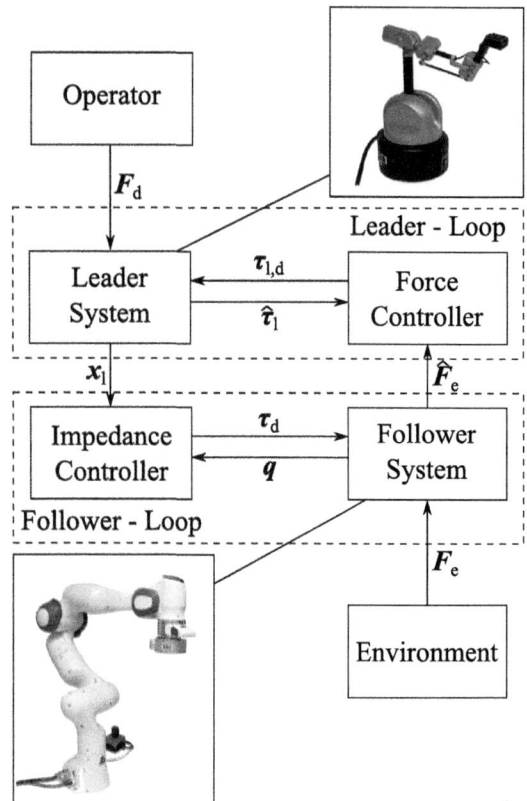

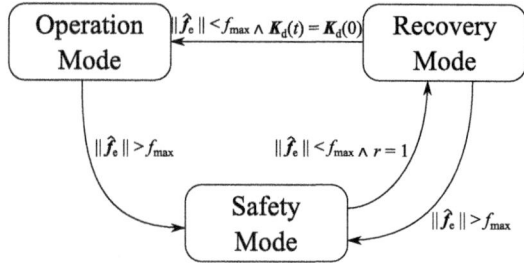

Fig. 5 State machine of the safety mechanism. The system changes into the compliant safety mode, when external forces $\hat{f}_e$ violate a force threshold $f_{\max}$. A user input r can start the recovery process. Reaching the initial stiffness $K_d(0)$ returns the system to operation mode [31]

Fig. 4 Impedance-force control architecture of the presented telemanipulation system. The operator applies a desired force F_d on the leader system. The resulting pose x_l and its derivations are sent to the impedance controller of the follower. The desired torques τ_d are computed by measuring the current joint angles q. Interaction forces F_e are measured and returned to the leader $\widehat{F}_e$. The force controller of the leader applies the joint torques $\tau_{l,d}$ on the device using the currently acting torques $\hat{\tau}_l$ [31]

Fig. 5 is introduced that ensures stability during recovery by using an energy tank. This approach not only acts as emergency mechanism but also increases the system's usability. It enables participants on the follower side to decouple the follower from the leader and to move it to a desired position. This allows for altering the workspace setup without restarting the system. Figure 6 shows a full cycle of the state machine. It can be seen, that a force violation switches the system into the safety mode and the leader and follower are decoupled. The user input starts the recovery process which slowly increases the stiffness back to its nominal value. The generated motion of the follower reduces the level of the energy tank but it stays in its valid range and thus stability is guaranteed.

3.3 Ongoing Research

3.3.1 Vison-Based Shared Control

As already mentioned, human–robot interactions often occur in telemedicine and telehealthcare. Due to the ongoing COVID-19 pandemic, the urge to improve telemanipulation in healthcare is present more than ever. We aim at implementing a vision-based virtual fixture approach to improve the ease of telemanipulated nasopharyngeal swab samplings for COVID testing. Based on a RGBD camera, we use the MediaPipe software to estimate the pose of the nostrils as it is shown in Fig. 7 [13]. Virtual forces drag the operator toward the correct orientation of the swab. With this, the operator only has to perform the radial movement into the nostril. The goal is to reduce the mental workload of the operator and to decrease the execution time for a single procedure. Furthermore, it increases the safety of the patient since minor head movements are compensated automatically by the controller. The safety can be further improved by adding the already presented safety mechanism. Also, this approach can be easily adjusted for other scenarios by simply adapting the desired target point. The overall setup is shown in Fig. 8.

are fed back to the leader and applied to the operator by a force controller. This architecture allows for telemanipulation with a compliant follower which increases safety during human–robot interactions.

3.2 Safety Mechanism

In [31], we introduce a mechanism that increases safety and usability of the telemanipulation system in case of interactions with humans or the environment. The soft stiffness of the human combined with the varying stiffness of other surroundings demands for a rapid switch from hard to soft stiffness of the system. We use an adaptive stiffness approach to drastically increase compliance in case of violating a force threshold $f_{\max}$. Furthermore, the recovery to operation mode is achieved by increasing the stiffness back to its nominal value. Since this increase might cause instabilities, a state machine as shown in

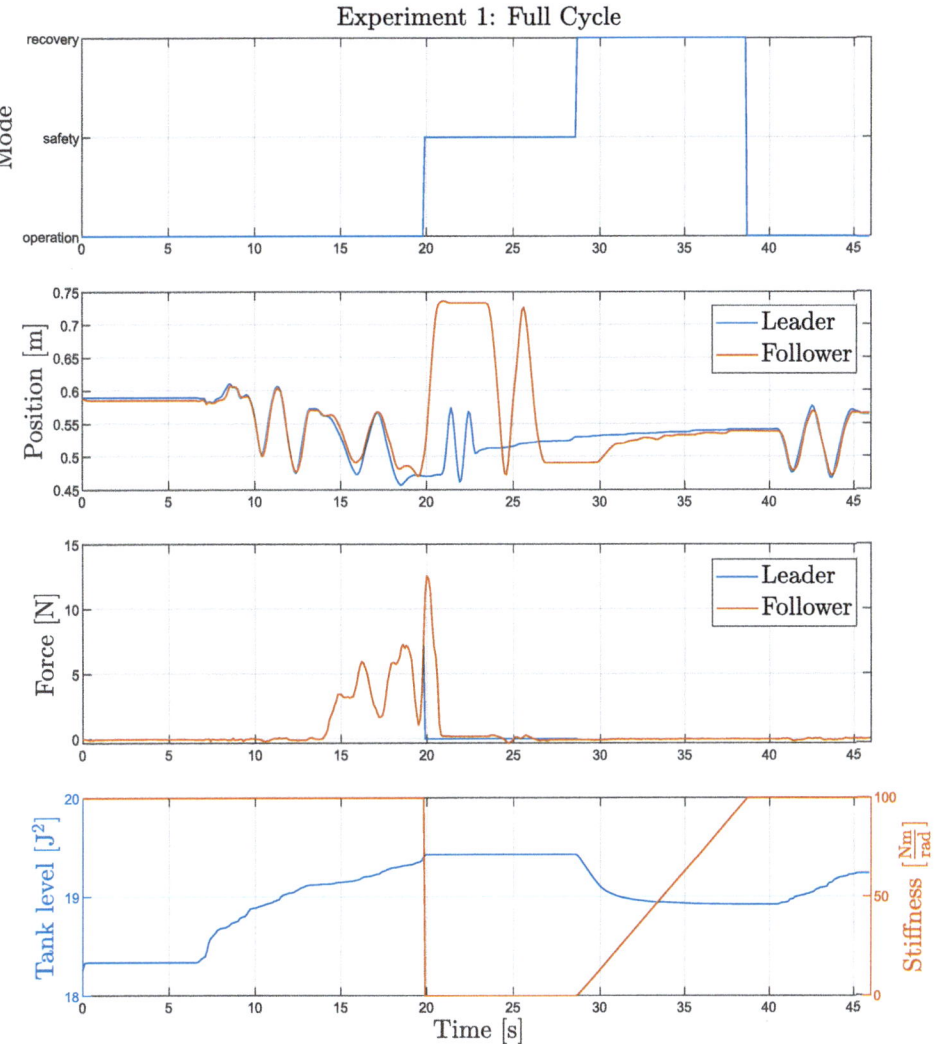

Fig. 6 Full cycle of the safety mechanism. The active mode as well as the z-component of the position and force of the leader and follower together with the tank level and the stiffness are considered [31]

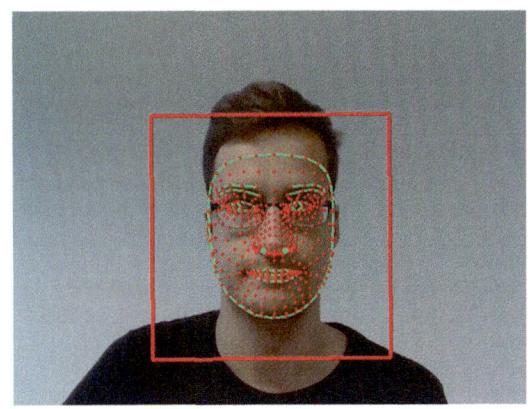

Fig. 7 Face mesh with nostril detection for the nasopharyngeal swab sampling using MediaPipe

3.3.2 Improving Adaptability and Human-Likeness During Task Execution

Safe human–robot interactions demand several requirements to the motion of the follower. The robot has to be able to adapt to unexpected events such as collisions or changing environments as well as performing its movements in a predictable way. These requirements have to be fulfilled during the execution of the desired task. Therefore, we aim at implementing a shared control approach that improves both of them. The system detects the intention of the operator either by combing a VR system with a gaze detector or by using a vision-based algorithm to detect the target object depending on the motion of the follower. With this intention, it is possible to plan a

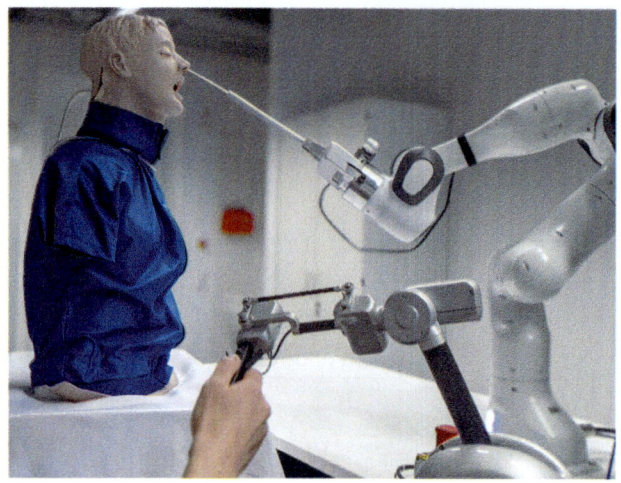

Fig. 8 Setup of the nasopharyngeal swab sampling scenario using a dummy and the presented telemanipulation system

trajectory toward the desired object. Using virtual fixtures allows the system to guide the user along this path. In [17, 18], we already presented a novel path planning algorithm to maximize mobility, which consists of manipulability and joint limit costs during collision-free motion. Combining this with a compliant control architecture, e.g., an impedance controller, the follower is highly capable to evade unexpected collisions during the motion. In a next step, we aim at improving the human-likeness of the robot's motion by integrating biophysical characteristics of human arm-motions into the planning process. We already introduced an approach to generate human-like motions in [10]. We assume that this combination increases the predictability for other participants and the adaptability of the robot in case of a collision.

4 Conclusion

Telemanipulation received a lot of attention during the last decades. Many control strategies are developed to improve the transparency and to add further feedback mechanisms. This aims at providing a realistic impression of the follower's environment to the operator, which increases the usability and also the safety from his point of view. On the other side, the broadening field of application increasingly includes interactions between the follower and other human participants. Especially healthcare and telemedicine demand for flexible human–robot interactions.

This paper gives an overview about recent approaches to improve human–robot interactions on the follower side. It reveals that there is limited work done so far and shows the research gap regarding safety and predictability for desired and undesired human–robot interactions. Especially the aspect of human-likeness is barely considered in the context of telemanipulation.

This paper further presents the published and ongoing work of the Robotics and Human–Machine Interaction Laboratory in conjunction with the Collaborative Research Center Hybrid Societies in this area. As presented, shared control algorithms and compliant control strategies offer huge potential to improve the usability of telemanipulation in human–robot scenarios. We show that there are many uncovered possibilities to improve human–robot interactions and recommend further research in this area. With the collaboration of the colleagues in the CRC 1410, we are able to evaluate our system from a physiological point of view which allows us to further improve the system. Moreover, we believe that combining human-likeness and improved safety concepts with a telemanipulation system would be a huge step towards the integration of these systems into the daily life and making them accessible to a wider field of application.

Acknowledgements This project was funded by the Deutsche Forschungsgemeinschaft (DFG, German ResearchFoundation)—Project-ID 416228727—SFB 1410.

References

1. Abi-Farraj, F., Pedemonte, N., & Robuffo Giordano, P. (2016). A visual-based shared control architecture for remote telemanipulation. In *IEEE/RSJ International Conference on Intelligent Robots and Systems (IROS)* (pp. 4266–4273).
2. Artigas, J., Balachandran, R., Riecke, C., Stelzer, M., Weber, B., Ryu, J. H., & Albu-Schaeffer, A. (2016). Kontur-2: Force-feedback teleoperation from the international space station. In *IEEE International Conference on Robotics and Automation (ICRA)* (pp. 1166–1173).
3. Artigas, J., Ryu, J.-H., Preusche, C., & Hirzinger, G. (2011). Network representation and passivity of delayed teleoperation systems. In IEEE/RSJ International Conference on Intelligent Robots and Systems (pp. 177–183). IEEE. https://doi.org/10.1109/IROS.2011.6094919
4. Chen, W., Zhou, J., Cheng, S. S., Lu, Y., Zhong, F., Gao, Y., Wang, Y., Xue, L., Tong, M. C. F., & Liu, Y.-H. (2021). Tele-operated oropharyngeal swab (TOOS) robot enabled by TSS soft hand for safe and effective sampling. *IEEE Transactions on Medical Robotics and Bionics, 3*(4), 1040–1053.
5. Cho, G. R., Ki, G., Lee, M. J., Kang, H., Kim, M. G., & Li, J. H. (2021). Experimental study on tele-manipulation assistance technique using a touch screen for underwater cable maintenance tasks. *Journal of Marine Science and Engineering, 9*(5).
6. Douissard, J., Hagen, M. E., & Morel, P. (2019). *The da Vinci surgical system* (pp. 13–27). Springer International Publishing.
7. Dragan, A. D., & Srinivasa, S. S. (2013). A policy-blending formalism for shared control. *The International Journal of Robotics Research, 32*(7), 790–805.
8. Ferraguti, F., Secchi, C., & Fantuzzi, C. (2013). A tank-based approach to impedance control with variable stiffness. In *IEEE International Conference on Robotics and Automation* (pp. 4948–4953).
9. Flatau, C. (1977). Sm-229: A new compact servo master-slave manipulator. *Transactions of the American Nuclear Society, 27.*
10. Gäbert, C., Kaden, S., & Thomas, U. (2021). Generation of human-like arm motions using sampling-based motion planning. In *IEEE/*

RSJ International Conference on Intelligent Robots and Systems (IROS) (pp. 2534–2541).

11. Goertz, R. C. (1952). Fundamentals of general-purpose remote manipulators. *Nucleonics (U.S.) Ceased publication, 10*(11).

12. Goertz, R. C., Blomgren, R. A., Grimson, J. H., Forster, G. A., Thompson, W. M., & Kline, W. H. (1961). The ANL model 3 master-slave electric manipulator–its design and use in a cave. *Transactions of the American Nuclear Society, 4*(2).

13. Google. (2022). Mediapipe. Retrieved August 16, 2022, from https://github.com/google/mediapipe

14. Hennekens, D., Constantinescu, D., & Steinbuch, M. (2008). Continuous impulsive force controller for forbidden-region virtual fixtures. In *IEEE International Conference on Robotics and Automation* (pp. 2890–2895).

15. Hogan, N. (1984). Impedance control: An approach to manipulation. In American Control Conference (pp. 304–313). IEEE. https://doi.org/10.23919/ACC.1984.4788393

16. Javdani, S., Admoni, H., Pellegrinelli, S., Srinivasa, S. S., & Bagnell, J. A. (2018). Shared autonomy via hindsight optimization for teleoperation and teaming. *The International Journal of Robotics Research, 37*(7), 717–742.

17. Kaden, S., & Thomas, U. (2019). Maximizing robot manipulability along paths in collision-free motion planning. In *19th International Conference on Advanced Robotics (ICAR)* (pp. 105–110).

18. Kaden, S., & Thomas, U. (2021). Optimizing mobility of robotic arms in collision-free motion planning. *Journal of Intelligent & Robotic Systems, 102*.

19. Krishnan, A. U., Lin, T.-C., & Li, Z. (2022). Design interface mapping for efficient free-form tele-manipulation. In *IEEE/RSJ International Conference on Intelligent Robots and Systems (IROS)*.

20. Landi, C. T., Ferraguti, F., Sabattini, L., Secchi, C., & Fantuzzi, C. (2017). Admittance control parameter adaptation for physical human-robot interaction. In *IEEE International Conference on Robotics and Automation (ICRA)* (pp. 2911–2916).

21. Lawrence, D. (1993). Stability and transparency in bilateral tele-operation. *IEEE Transactions on Robotics and Automation, 9*(5), 624–637.

22. Nicolis, D., Palumbo, M., Zanchettin, A. M., & Rocco, P. (2018). Occlusion-free visual servoing for the shared autonomy teleoperation of dual-arm robots. *IEEE Robotics and Automation Letters, 3*(2), 796–803.

23. Niemeyer, G., & Slotine, J.-J. (1991). Stable adaptive teleoperation. *IEEE Journal of Oceanic Engineering, 16*(1), 152–162.

24. Nudehi, S., Mukherjee, R., & Ghodoussi, M. (2005). A shared-control approach to haptic interface design for minimally invasive telesurgical training. *IEEE Transactions on Control Systems Technology, 13*(4), 588–592.

25. Panzirsch, M., Artigas, J., Tobergte, A., Kotyczka, P., Preusche, C., Albu-Schaeffer, A., & Hirzinger, G. (2012). A peer-to-peer trilateral passivity control for delayed collaborative teleoperation. In P. Isokoski, & J. Springare (eds.), *Haptics: Perception, devices, mobility, and communication* (pp. 395–406). Springer, Berlin, Heidelberg.

26. Quere, G., Hagengruber, A., Iskandar, M., Bustamante, S., Leidner, D., Stulp, F., & Vogel, J. (2020). Shared control templates for assistive robotics. In *IEEE International Conference on Robotics and Automation (ICRA)* (pp. 1956–1962).

27. Rosenberg, L. (1993). Virtual fixtures: Perceptual tools for telerobotic manipulation. In *Proceedings of IEEE Virtual Reality Annual International Symposium* (pp. 76–82).

28. Ryu, J. H., Artigas, J., & Preusche, C. (2010). A passive bilateral control scheme for a teleoperator with time-varying communication delay. *Mechatronics, 20*(7), 812–823. Special Issue on Design and Control Methodologies in Telerobotics.

29. Schilling, R. (1992). Telerobots in the nuclear industry: A manufacturer's view. *Industrial Robots, 19*(2), 3–4.

30. Schwarz, M., Lenz, C., Rochow, A., Schreiber, M., & Behnke, S. (2021). Nimbro avatar: Interactive immersive telepresence with force-feedback telemanipulation. In *IEEE/RSJ International Conference on Intelligent Robots and Systems (IROS)* (pp. 5312–5319).

31. Schwarz, S. A., & Thomas, U. (2022). Variable impedance control for safety and usability in telemanipulation. In *IEEE/RSJ International Conference on Intelligent Robots and Systems (IROS)*.

32. Selvaggio, M., E, A. M. G., Moccia, R., Ficuciello, F., & Siciliano, B. (2019a). Haptic-guided shared control for needle grasping optimization in minimally invasive robotic surgery. In *IEEE/RSJ International Conference on Intelligent Robots and Systems (IROS)* (pp. 3617–3623).

33. Selvaggio, M., Robuffo Giordano, P., Ficuciello, F., & Siciliano, B. (2019b). Passive task-prioritized shared-control teleoperation with haptic guidance. In *International Conference on Robotics and Automation (ICRA)* (pp. 430–436).

34. Su, H., Sandoval, J., Makhdoomi, M., Ferrigno, G., & De Momi, E. (2018). Safety-enhanced human-robot interaction control of redundant robot for teleoperated minimally invasive surgery. In *IEEE International Conference on Robotics and Automation (ICRA)* (pp. 6611–6616).

35. Su, H., Yang, C., Ferrigno, G., & De Momi, E. (2019). Improved human–robot collaborative control of redundant robot for teleoperated minimally invasive surgery. *IEEE Robotics and Automation Letters, 4*(2), 1447–1453.

36. Vogel, J., Haddadin, S., Jarosiewicz, B., Simeral, J., Bacher, D., Hochberg, L., Donoghue, J., & van der Smagt, P. (2015). An assistive decision-and-control architecture for force-sensitive hand–arm systems driven by human–machine interfaces. *The International Journal of Robotics Research, 34*(6), 763–780.

37. Walker, D. S., Wilson, R. P., & Niemeyer, G. (2010). User-controlled variable impedance teleoperation. In *IEEE International Conference on Robotics and Automation* (pp. 5352–5357).

38. Zhou, H., Yang, G., Lv, H., Huang, X., Yang, H., & Pang, Z. (2020). Iot-enabled dual-arm motion capture and mapping for telerobotics in home care. *IEEE Journal of Biomedical and Health Informatics, 24*(6), 1541–1549.

Grasp Pose Generation for Human-to-Robot Handovers Using Simulation-to-Reality Transfer

Carl Gaebert, Chaitanya Bandi, and Ulrike Thomas

Abstract

Human-to-robot handovers play an important role in collaborative tasks in industry or household assistance. Due to the vast amount of possible unknown objects, learning-based approaches gained interest for robust and general grasp synthesis. However, obtaining real training data for such methods requires expensive human demonstrations. Simulated data, on the other hand, is easy to generate and can be randomized to cover the distribution of real-world data. The first contribution of this work is a dataset for human grasps generated in simulation. For this, we use a simulated hand and models of 10 objects from the YCB dataset [1]. It can also be easily extended to include new objects. The method thus allows for generating an arbitrary amount of training data without human interactions. Secondly, we combine a generative neural grasp generator with an evaluator model for grasp pose generation. In contrast to previous works, we obtain grasp poses from simulated RGB images which allows for reducing the negative effects of depth sensor noise. To this end, our generator model is provided with a cropped image of the human hand and learns the distribution of grasps in the wrist system. The evaluator then narrows down the list of grasps to the most promising ones. The presented approach requires the model to extract relevant features from images instead of point clouds. A cost-efficient method for generating large amounts of training data is therefore needed. We test our approach in simulation and transfer it to a real robot
system. We use the same objects as in the training dataset but also test the generalization capabilities toward new objects. The presented dataset is available for download: https://tuc.cloud/index.php/s/g3noZD7oCqbQR9d.

Keywords

Handover · Grasp generation · Human-robot interaction

1 Introduction

Handing over objects is a basic building block for various complex tasks and interactions between people. With the increasing appearance of collaborative robots in agile production or elderly assistance, the interest in solving the robot-to-human (R2H) and human-to-robot (H2R) handover problems has increased drastically.

In this work, we focus on grasp generation for the H2R handover problem and its main challenges. The first one is the number of possible objects involved. Even with a clear focus on a specific application domain like household assistance, the number of possible objects is incalculable due to the vast number of variations of the same object type. Hence, a generalizable approach is needed that does not require objects to be known beforehand. Second, the H2R problem naturally bears the challenge of mutual occlusion between object and hand. In consequence, the simultaneous estimation of object and hand pose for traditional grasp planning can be impractical. Considering the first two challenges, it is desirable to rely on learning-based approaches for robust and general grasp pose generation. The nature of the problem, however, renders the generation of training data for such approaches difficult because it involves human demonstrations. Even though such datasets exist [2, 3], they can only cover a small set of objects and are at the same time expensive to generate.

In this work, we address this challenge by proposing a pipeline for the automatic generation of a handover training

C. Gaebert (✉) · C. Bandi · U. Thomas
Robotics and Human-Machine Interaction Lab, Chemnitz University of Technology, Chemnitz, Germany
e-mail: carl.gaebert@etit.tu-chemnitz.de

C. Bandi
e-mail: chaitanya.bandi@etit.tu-chemnitz.de

U. Thomas
e-mail: ulrike.thomas@etit.tu-chemnitz.de

B. Meyer et al. (eds.), *Hybrid Societies*, Advances in Science, Technology & Innovation,
https://doi.org/10.1007/978-3-032-03488-5_26

dataset in simulation. This significantly reduces the cost of extending existing datasets to arbitrary new objects since no human is involved in the process. Second, we present a novel grasp planner based on a generative neural network model and a grasp pose evaluator. In contrast to other works, we only train our model on RGB images. By doing so, we mitigate the challenges that arise from relying on depth data in real-world domains that have been reported in previous works [3]. Such setups are subject to sensor noise due to lighting conditions and object properties. In contrast, our approach takes a cropped image of the hand instead and learns a distribution of grasp poses in the wrist coordinate system. We thus reduce the dependency on depth data to estimating the wrist pose in 3D. Finally, we evaluate our approach on a simulated test set and transfer it to a real robot. Hence, we test whether the system generalized across the simulated training domain.

2 Related Work

Several recent works have addressed the H2R and R2H problems. In [17], the authors provide a thorough analysis of object handovers. It includes several aspects such as human perception, intention of handover, motion planning, and error handling. The analysis shows that a significant amount of related work focuses on R2H handovers and few works concern the H2R problem. In this section, we focus on the latter problem. A straightforward yet useful methodology for handover of objects is to directly place an object in the robot's gripper. While this approach can work under certain circumstances, it is not appropriate in collaborative applications where the human movement may be limited. The earlier works [12, 13, 23] rely on sensing devices or custom-built controllers for handover of objects to robots. Pan et al. [18] investigated the issue of recognizing the handover intent based on human skeleton tracking. The works mentioned above focus on individual aspects of the handovers. In contrast, Rosenberger et al. [21] developed a full pipeline for human-to-robot handover including human and robot perspective modules. The overall pipeline consists of different modules, where the RGB-D stream is passed to a YOLO V3 based object detector [20] which is trained on 80 object categories from the well-known coco dataset [11] for hand and body segmentation. Based on this information, grasps are generated using a modified GG-CNN [15]. In order to reduce the complexity of sim-to-real gap of hand recognition, Yang et al. [24] explored the idea of grasp classification for the handover of objects. The approach considers 5 different grasp categories with objects and two categories without objects. The segmented point cloud of the hand is used as an input to the PointNet++ model [19]. Based on the classification result, a motion is triggered for taking

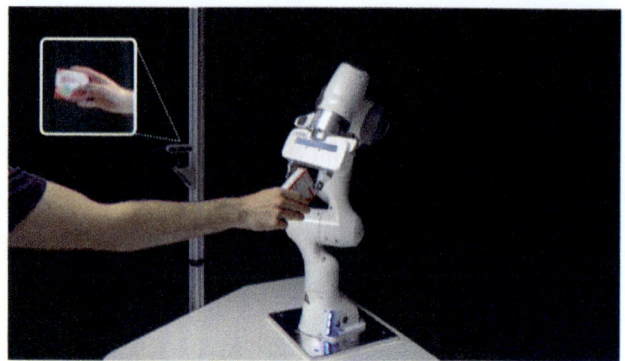

Fig. 1 A human-to-robot handover action can be seen. The setup involves a calibrated RGB-D camera and an industrial manipulator. The input to our model is an RGB image of the human hand holding the object (white box)

the object out of the human hand. Yang et al. [25] further extended this idea to grasp arbitrary objects using a 6-DOF GraspNet [16] on the segmented point cloud of objects and hands. Of relevance for our approach are also the works of Chao et al. in which the authors provide a large dataset of real-world handover data for YCB objects [3] and transfer it to a simulated benchmark environment [2]. In contrast to this work, our method uses simulated training data only (Fig. 1).

3 Methodology

In this section, we present the pipeline for solving the H2R handover problem. The overall structure can be seen in Fig. 2. The architecture consists of a grasp pose generator and a grasp pose evaluator. The former generates a distribution of grasps for a given RGB image. In a subsequent step, the grasps are filtered using the evaluator network. The grasps with a quality score above a certain threshold are then considered during the motion planning stage. The backbone of the grasp pose generator is a conditional variational autoencoder (CVAE).

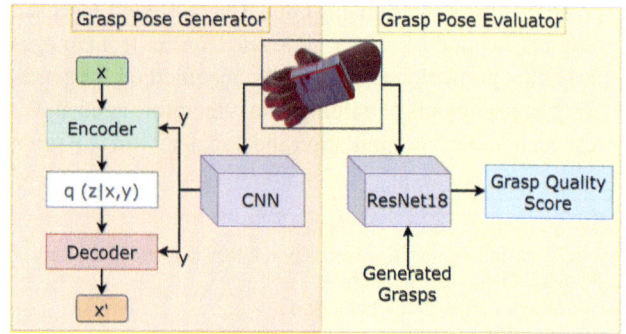

Fig. 2 Overview of the pipeline

3.1 Conditional Variational Autoencoders

A CVAE [22] is a deep conditional generative model architecture that is widely used in robotics-related fields such as motion planning [4, 9]. In the context of our problem, the model is provided with a grasp pose x in the human's wrist system and a conditional variable y. For the latter, we use a learned encoding of images showing the human hand and the object to grasp. The CVAE consists of an encoder network that approximates $q_\phi(z|x, y)$ with z being a latent distribution. The latter is typically chosen to be a unit-variance Gaussian. The second part is a decoder network that approximates $p_\theta(x|z, y)$. Provided with the image encoding y and a sample from the latent distribution, one can generate a grasp pose from the learned distribution $p_\theta(x|z, y)$ using the decoder. The encoder weights φ and decoder weights θ are learned using the loss function shown in 1. Minimizing this loss maximizes the evidence lower bound (ELBO) between the estimated posterior and true posterior distribution.

$$\mathcal{L}(\theta, \phi) = -\mathbb{E}_{z \sim q_\phi(z|x,y)}\big[\log p_\theta(x|z, y)\big] \\ + \beta D_{KL}\big(q_\phi(z|x, y) \| p_\theta(z|y)\big) \quad (1)$$

It consists of a L_2 reconstruction loss representing the distance between the provided grasp x and the reconstruction x' generated by the decoder. The second component in 1 is the Kullback-Leibler divergence between the learned latent distribution and the unit-variance Gaussian. It is used as a regularizer to keep the variational posterior close to the prior distribution over latent variables. The two loss components can be weighted using the parameter β that balances the capacity in the latent variables with the reconstruction error [8].

3.2 Grasp Pose Generator

The encoder is provided with the ground truth grasp and the image embedding. The latter is obtained by applying three layers of convolution on the input RGB images. The resulting features are then further compressed using two linear layers. The decoder network itself consists of three linear layers and outputs the parameters of the latent Gaussian distribution. The decoder network is the provided with samples from the latent distribution as well as the image embedding. It uses four linear layers to reconstruct the input grasp. The complete generator process can be observed in Fig. 3.

3.3 Grasp Pose Evaluator

The aim of the grasp pose evaluator is to select the most promising grasps among the generated ones. It thus serves as an additional stage that reduces the influence of domain shift or inaccurate hand tracking. The model takes RGB image as an input and generates image embeddings using ResNet [7] with 18 layers. The image embeddings are then concatenated with grasps from the generator and forwarded to a set of fully connected layers to obtain the grasp quality score. The architecture is shown in Fig. 4.

3.4 Input Representation and Model Parameters

The input is a grasp pose x which consists of rotation R and translation t. There are different types of representation for rotations such as Euler angles, quaternions, and rotation

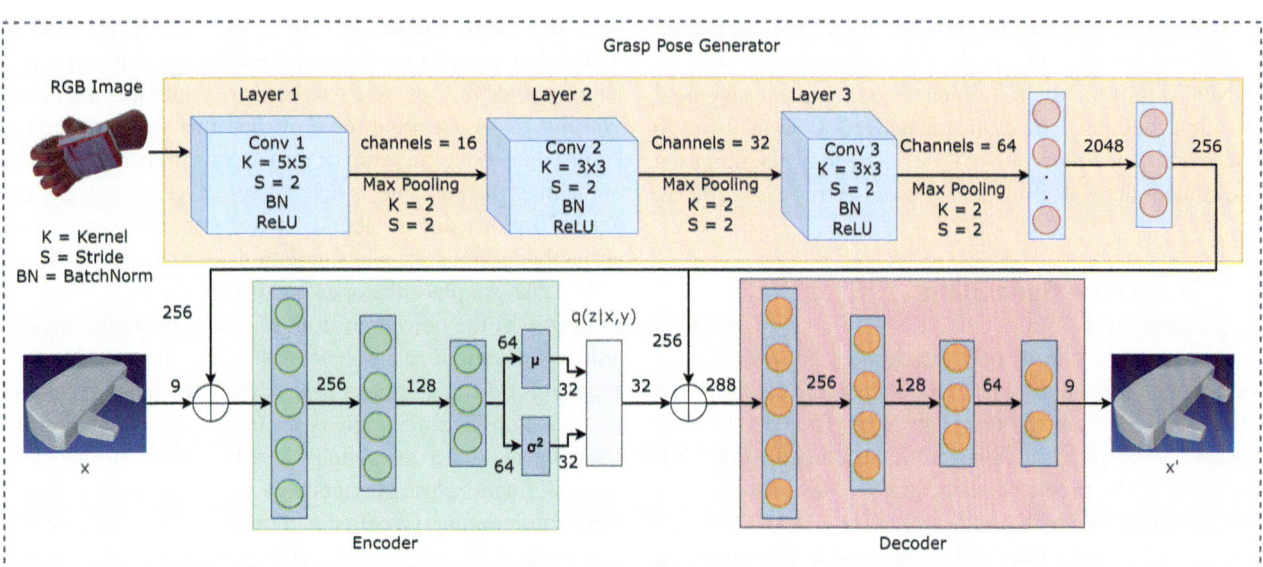

Fig. 3 Overall architecture of the Grasp pose generator network which consists of the convolutional layers, the encoder and the decoder

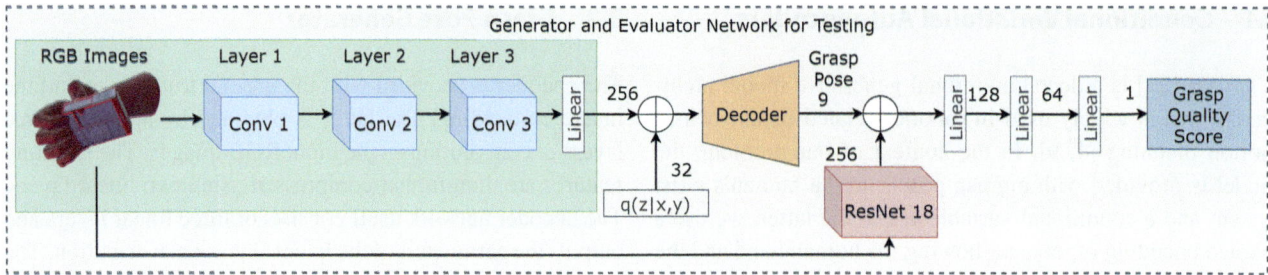

Fig. 4 Grasp pose regression and evaluation network for testing the simulated and real data

matrices. Euler angles and quaternions have limitations like singular-ities and antipodal issue for regression. Furthermore, it has been demonstrated in [27] that any rotation representation in 3D with fewer than five dimensions is discontinuous and harder to learn. We therefore exploit the orthogonal properties of a rotation matrix using the Gram-Schmidt process and construct an orthonormal basis from two vectors (see 2). According to this process, the parameters third column are redundant. Multiplying the first second column vector e_1 with the second column vector e_2 results in the complete rotation matrix R. So we use first two columns from the rotation matrix as the 6D representation for learning.

$$R = \begin{pmatrix} r_{11} & r_{12} & r_{13} \\ r_{21} & r_{22} & r_{23} \\ r_{31} & r_{32} & r_{33} \end{pmatrix} = \begin{pmatrix} \vec{e_1} & \vec{e_2} & \vec{e_1} \times \vec{e_2} \end{pmatrix} \quad (2)$$

$$x = \begin{pmatrix} t_1 & t_2 & t_3 & r_{11} & r_{21} & r_{31} & r_{12} & r_{22} & r_{32} \end{pmatrix} \quad (3)$$

The architectures of the grasp pose generator and evaluator are trained independently. For the generator, 30 epochs, a learning rate of 0.001 and a weight decay of 1^{-6} is used. The generator loss is the ELBO described in 1. We use a "warm-up" strategy [8] to gradually increase the KL weight β from 0.0001 to 0.1 with each epoch. The evaluator is a 18 layer ResNet [7] followed by fully connected layers and a sigmoid activation function. The evaluator network is trained for 20 epochs with a learning rate of 0.001. The loss function used for evaluator network is the binary cross-entropy loss (BCE).

3.5 Grasp Pose Generation

When generating grasps, only the decoder network is used (see Fig. 4). The batched image embeddings are concatenated with n samples from the Gaussian prior. The decoder then generates n grasps by sampling from the learned conditional distribution. The predicted set of grasp poses and the input image is then passed to the evaluator network. Next, the grasps with a score below the threshold t are removed. The remaining grasp poses and their corresponding approach poses are then passed to an inverse kinematics solver. Among all solutions, the closest grasp in Cartesian space is selected for execution.

4 Dataset Generation

Generating real-world training data for the handover task is expensive as it always involves a human holding the object. In recent work [3] for example, 1000 trials were performed to collect real-world data for the YCB object dataset [1]. This approach, however, quickly becomes intractable when the number of objects increases. To mitigate this problem, we propose generating training data for handover tasks in simulation. Our approach allows for creating large datasets for an arbitrary amount of objects and does not require human subjects. For this, we use the Unity3D game engine together with textured models for the hand and objects. For the scope of this work, we consider objects from the YCB dataset that can be grasped with a parallel gripper (see Fig. 5). The process of human grasp generation is complex and takes prior knowledge of the objects as well as social norms and the next intended actions into account. The range of possible grasps is thus intractable and highly dependent on the context and the possibly unknown object. In consequence, modeling accurate human-like grasps in simulation for arbitrary objects is challenging since the concrete real handover setting is unknown. We thus rely on a brute force method that does not require knowledge about the setup or the object's properties. Hence, our aim is to generate a diverse set of grasps to achieve generalization using the neural network rather than generating perfectly natural grasps in simulation.

For this, we place the open hand at the origin of the scene and position the objects in front of it. For obtaining diverse grasps, we randomize the object pose within the hand. We thus account for different object poses during the pickup phase in the real setup. The grasp generation then avoids complex grasp planning and relies on a simple heuristic instead. In the beginning, all fingers close continuously at a fixed speed. A finger then stops moving if a collision with the object is detected. A grasp is considered successful if the thumb and at least three other fingers collide with the object. This reduces the amount

Fig. 5 **a–f** Example grasps for the considered objects **f** unrealistic grasp resulting from the grasp heuristic

(a) marker   (b) small jello   (c) mustard   (d) tuna can   (e) big jello   (f) unnatural grasp

Fig. 6 Subset of valid grasps (blue) for the object to grasp (red) can be seen. Invalid grasps are removed by collision checking with the exported hand mesh (green)

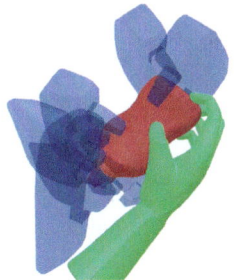 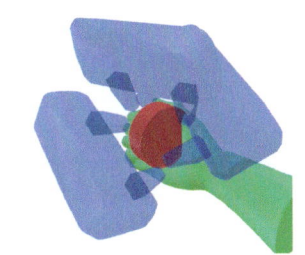 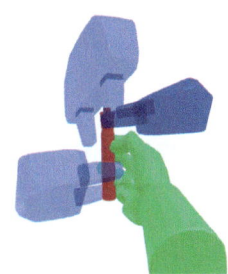

of unnatural grasps while avoiding explicit grasp planning. Given a successful grasp, we generate ten images showing the object and hand models. For each image, we randomize the camera pose, lighting and hand texture. Examples can be seen in Fig. 5. Due to the heuristic used, some grasps in the dataset may not reflect natural human grasping (see Fig. 5f). After closing the hand, we export the object and hand meshes and save their relative poses together with the corresponding images. In a subsequent step, we consider 500 valid parallel gripper grasps for each exported object. They were generated using the dataset provided in [5]. Alternatively, one could use available grasp planners such as GraspIt! [14] for novel objects. Next, we add the hand mesh and remove all the grasps that collide with it. A few examples can be seen in Fig. 6. Finally, we obtain the grasp poses in the hand's wrist system which can be tracked easily. In consequence, our approach does not require estimating the hand or object poses.

5 Experiments

5.1 Evaluation in Simulation

To evaluate our model, we generated a simulated test set using the same objects as described above. The test set contains ten simulated grasps for each of the five objects. For each grasp, ten pictures were taken. We evaluate the accuracy in simulation by generating grasps using the generator. A model of the open gripper is then placed at each generated grasp pose. If the gripper collides with the object or hand, the grasp is considered not successful. If not, we close the gripper and check if a collision with the object occurs. In case it does, the grasp was successful. This method is a pessimistic estimate on the grasp accuracy since even minor collisions are considered a failure. Such small deviations, however, could be easily compensated

Fig. 7 Effect of using the evaluator network can be seen. Light blue grasps are discarded by the evaluator and not considered during grasping. Dark blue grasps are considered. It can be seen on the left how the evaluator filters out unsuccessful grasps

by the human hand which is typically not perfectly steady. For our experiment, we generated 25 grasps for each image which leads to 250 generated grasps per object. In a further step, we introduce the evaluator network. It is used to filter out grasps by using a threshold value of $t = 0.9$. It thus reduces the number of considered grasps but increases the quality of the selected ones (see Fig. 7). We then test the remaining grasps only. The results for each object can be seen in Table 1. It can be seen that using the evaluator network increased the grasp success for each object. However, the overall grasp success clearly depends on the object. While the small jello and the tuna can be grasped relatively reliably, the marker was grasped in only 44% of the trials.

Table 1 Accuracy for simulated test set

Objects	Without evaluator (%)	With evaluator (%)
Marker	35.1	44.8
Mustard	77.3	81.04
Small jello	54.8	72.01
Tuna can	79.3	88.03
Big jello	58.6	61.7

Fig. 8 **a** Corresponding real counterparts of the simulated objects used during training. From left to right: marker, mustard, jello small, tuna, jello big. **b** Hold- out test objects not used during training. From left to right: tape measure, glue stick, clamp, scissors, can

(a) Training objects   (b) Test objects

5.2 Simulation-to-Reality Transfer

In the second experiment, we test our approach including the evaluator using the setup shown in Fig. 1. It involves a calibrated RGB-D camera and an industrial manipulator. In line with our training data, the camera faces the object and the inside of the right hand. Grasping the object is done in several steps. First, we track the human hand using mediapipe hands [26]. Due to occlusions, the estimated finger poses vary a lot in between frames which is why we do not explicitly reconstruct the hand's model. Instead, we consider the not occluded wrist landmark and obtain its relative pose to the robot. In a subsequent step, we crop the hand and object from the input image and pass it to the grasp generator. The resulting 30 grasps are then rated using the evaluator using $t = 0.9$. After confirming the reachability using inverse kinematics, the robot moves to the closest grasp pose in Cartesian space. For this, we use our in-house motion planning library and the RRT*-Connect algorithm [10]. Static obstacles as well as the human wrist are considered as obstacles to ensure a collision-free grasp.

We first use the same objects as in the training dataset (see Fig. 8a). We present each object ten times to the robot while keeping the hand as steady as possible. A grasp is considered successful if the gripper closes at the object and the object does not slip out of the gripper. After that, we also evaluate the model's performance on five unseen objects (see Fig. 8b). Each of them was again presented ten times. The results are shown in Table 2. It can be seen that the grasp success rate is lower than in the simulated experiments. This is caused by the domain shift as well as inaccuracies in hand tracking. Future work in this direction is discussed below.

Table 2 Accuracy for real grasps

Train objects	Success (%)	Test objects	Success (%)
Marker	50	Tape measure	50
Mustard	40	Glue stick	50
Small jello	60	Clamp	30
Tuna can	60	Scissors	60
Big jello	40	Can	50

In addition, the grasp success rate is even lower for some hold-out objects such as the clamp. Hence, the model does not transfer well to objects with more complex geometries than the ones in the training set.

6 Discussion and Conclusion

The H2R handover approach presented in this paper generates grasp poses from RGB images of the human hand. Compared to other methods, it reduces the use of depth information and is thus less prone to sensor noise. On the other hand, it does not consider relevant 3D information and has to extract meaningful features from images only. This requires a vast amount of image data from various angles. The presented generative model is thus trained on simulated data. For this, a brute force grasp generator was used in combination with a simple heuristic. In the future, we would like to improve the quality of the simulated grasps while maintaining our method's generalizability. Utilizing a grasp taxonomy [6] could be a promising next step in this direction. However, using the presented approach already results in cost-efficient dataset generation and enables fine-tuning the model to new objects in the future. The results of our first experiment show that the grasp success rate is object- dependent. While objects like the tuna can be grasped in about 79%, the success rate for the marker is only 35%. This indicates that the model does not generalize well to arbitrarily complex objects. One reason for that is the lack of 3D information. Extracting relevant featured from RGB images can partially compensate this but does not work reliably enough for objects like the marker. Changing the setup to include a top-view camera or even multiple angles at once could be a possible future step to resolve this. Moreover, the CVAE uses a Gaussian latent distribution and only maximized the evidence lower bound. The estimated posterior may thus not fit the true posterior well enough. Hence, a more flexible latent distribution could lead to a more accurate estimation. Therefore, other generative models should be considered in the future. The first experiment also shows that using the evaluator model consistently improves the grasp success rate for all objects. It is thus recommended to always utilize a binary classifier to narrow down the list of considered grasps. Future work in this direction could also include application-specific constraints such as allowing grasps from a specific side only.

In our second experiment, the model was tested in a real setup. It can be seen that the success rates are slightly lower than in the simulated experiments. The reasons for grasps failures in the real setup are manifold. They include blurry images due to small hand movements and inaccurate estimation of the hand pose. Also, the hand dimensions of the human do not necessarily align with the simulated hand which can lead to failures as well. Optimizing the camera position and image filtering could thus improve hand tracking and the overall grasp success rate.

In the future, we would like to improve the general performance of our model in reality by evaluating other generative models. Moreover, a combination of images from different points of view could be a future step, too.

Acknowledgments Funded by the Deutsche Forschungsgemeinschaft (DFG, German Research Foundation)—Project-ID 416228727—SFB 1410.

References

1. Calli, B., Singh, A., Walsman, A., Srinivasa, S., Abbeel, P., & Dollar, A. M. (2015). The ycb object and model set: Towards common benchmarks for manipulation research. In *2015 International Conference on Advanced Robotics (ICAR)* (pp. 510–517). IEEE.
2. Chao, Y. W., Paxton, C., Xiang, Y., Yang, W., Sundaralingam, B., Chen, T., Murali, A., Cakmak, M., & Fox, D. (2022). HandoverSim: A simulation framework and benchmark for human-to-robot object handovers. In *IEEE International Conference on Robotics and Automation (ICRA)*.
3. Chao, Y. W., Yang, W., Xiang, Y., Molchanov, P., Handa, A., Tremblay, J., Narang, Y. S., Van Wyk, K., Iqbal, U., Birchfield, S., Kautz, J., & Fox, D. (2021). DexYCB: A benchmark for capturing hand grasping of objects. In *IEEE/CVF Conference on Computer Vision and Pattern Recognition (CVPR)*.
4. Cheng, R., Shankar, K., & Burdick, J. W. (2020). Learning an optimal sampling distribution for efficient motion planning. In *IEEE International Conference on Intelligent Robots and Systems* (pp. 7485–7492). IEEE.
5. Eppner, C., Mousavian, A., & Fox, D. (2019). A billion ways to grasps—an evaluation of grasp sampling schemes on a dense, physics-based grasp data set. In *Proceedings of the International Symposium on Robotics Research (ISRR)*, Hanoi, Vietnam.
6. Feix, T., Romero, J., Schmiedmayer, H.-B., Dollar, A. M., & Kragic, D. (2016). The grasp taxonomy of human grasp types. *IEEE Transactions on Human-Machine Systems, 46*(1), 66–77.
7. He, K., Zhang, X., Ren, S., & Sun, J. (2016). Deep residual learning for image recognition. In *Proceedings of the IEEE Conference on Computer Vision and Pattern Recognition* (pp. 770–778).
8. Higgins, I., Matthey, L., Pal, A., Burgess, C., Glorot, X., Botvinick, M., Mohamed, S., & Lerchner, A. (2017). B-VAE: Learning basic visual concepts with a constrained variational framework. In *5th International Conference on Learning Representations, ICLR 2017—Conference Track Proceedings*.
9. Ichter, B., Harrison, J., & Pavone, M. (2018). Learning sampling distributions for robot motion planning. In *Proceedings of the IEEE International Conference on Robotics and Automation* (pp. 7087–7094). IEEE.
10. Klemm, S., Oberländer, J., Hermann, A., Roennau, A., Schamm, T., Zollner, J. M., & Dillmann, R. (2015). RRT*-connect: Faster, asymptotically optimal motion planning. In *2015 IEEE International Conference on Robotics and Biomimetics (ROBIO)* (pp. 1670–1677). IEEE.
11. Lin, T.-Y., Maire, M., Belongie, S., Hays, J., Perona, P., Ramanan, D., Dollár, P., & Zitnick, C. L. (2014). Microsoft coco: Common objects in context. In D. Fleet, T. Pajdla, B. Schiele, & T. Tuytelaars (Eds.), *Computer vision—ECCV 2014* (pp. 740–755). Springer International Publishing.
12. Medina, J. R., Duvallet, F., Karnam, M., & Billard, A. (2016). A human-inspired controller for fluid human-robot handovers. In *2016 IEEE-RAS 16th International Conference on Humanoid Robots (Humanoids)* (pp. 324–331).
13. Meyer zu Borgsen, S., Bernotat, J., & Wachsmuth, S. (2017). Hand in hand with robots: Differences between experienced and naive users in human-robot handover scenarios. In A. Kheddar, E. Yoshida, S. S. Ge, K. Suzuki, J. J. Cabibi-han, F. Eyssel, & H. He (Eds.), *Social robotics* (pp. 587–596). Springer International Publishing.
14. Miller, A. T., & Allen, P. K. (2004). Graspit! a versatile simulator for robotic grasping. *IEEE Robotics & Automation Magazine, 11*(4), 110–122.
15. Morrison, D., Leitner, J., & Corke, P. (2018). Closing the loop for robotic grasping: A real-time, generative grasp synthesis approach. In *Proceedings of Robotics: Science and Systems*, Pittsburgh, Pennsylvania.
16. Mousavian, A., Eppner, C., & Fox, D. (2019). 6-dof graspnet: Variational grasp generation for object manipulation. *2019 IEEE/CVF International Conference on Computer Vision (ICCV)* (pp. 2901–2910).
17. Ortenzi, V., Cosgun, A., Pardi, T., Chan, W. P., Croft, E. A., & Kulić, D. (2021). Object handovers: A review for robotics. *IEEE Transactions on Robotics, 37*, 1855–1873.
18. Pan, M. K., Knoop, E., Bächer, M., & Niemeyer, G. (2019). Fast handovers with a robot character: Small sensorimotor delays improve perceived qualities. In *2019 IEEE/RSJ International Conference on Intelligent Robots and Systems (IROS)* (pp. 6735–6741).
19. Qi, C. R., Yi, L., Su, H., & Guibas, L. J. (2017). Pointnet++: Deep hierarchical feature learning on point sets in a metric space. In *Proceedings of the 31st International Conference on Neural Information Processing Systems*, NIPS'17 (pp. 5105–5114), Red Hook, NY, USA: Curran Associates Inc.
20. Redmon, J., & Farhadi, A. (2018). Yolov3: An incremental improvement. Preprint retrieved from arxiv:abs/1804.02767
21. Rosenberger, P., Cosgun, A., Newbury, R., Kwan, J., Ortenzi, V., Corke, P., & Grafinger, M. (2021). Object-independent human-to-robot handovers using real time robotic vision. *IEEE Robotics and Automation Letters, 6*(1), 17–23.
22. Sohn, K., Lee, H., & Yan, X. (2015). Learning structured output representation using deep conditional generative models. In C. Cortes, N. Lawrence, D. Lee, M. Sugiyama, & R. Garnett (Eds.), *Advances in neural information processing systems* (Vol. 28). Curran Associates, Inc.
23. Strabala, K., Lee, M. K., Dragan, A., Forlizzi, J., Srinivasa, S. S., Cakmak, M., & Micelli, V. (2013). Toward seamless human-robot handovers. *Journal of Human-Robot Interaction, 2*(1), 112–132.
24. Yang, W., Paxton, C., Cakmak, M., & Fox, D. (2020). Human grasp classification for reactive human-to-robot handovers. In *IEEE/RSJ International Conference on Intelligent Robots and Systems (IROS)* (pp. 11123–11130).

25. Yang, W., Paxton, C., Mousavian, A., Chao, Y. W., Cakmak, M., & Fox, D. (2021). Reactive human-to-robot handovers of arbitrary objects. In *2021 IEEE International Conference on Robotics and Automation (ICRA)* (pp. 3118–3124).

26. Zhang, F., Bazarevsky, V., Vakunov, A., Tkachenka, A., Sung, G., Chang, C. L., & Grundmann, M. (2020). Mediapipe hands: On-device real-time hand tracking.

27. Zhou, Y., Barnes, C., Lu, J., Yang, J., & Li, H. (2019). On the continuity of rotation representations in neural networks. In *Proceedings of the IEEE/CVF Conference on Computer Vision and Pattern Recognition (CVPR)*.

Flip of a Switch: Designing a Kinetic Dialogue System for Switch Interfaces

Maximilian Kullmann, Jan Ehlers, Eva Hornecker, and Lewis L. Chuang

Abstract

The proliferation of smart technology in everyday objects provide an opportunity for human–machine collaborations, signaling a shift from typical principal-agent relationships. The current work proposes roboticizing familiar user interfaces to allow objects to communicate with their users without the need of an additional communication modality. We implemented a light switch that is capable of kinetic gestures and report an exploratory study that investigated how naïve users (n = 15) would respond to and interact with different gesture designs. A qualitative analysis of recorded interactions and semi-structured interviews reveal that users are surprised by unfamiliar automation and kinetic gestures serve as cues to promote discovery of the system's mechanisms and mitigate assumed helplessness.

Keywords

Automation · Physical prototyping · Interface design · Human–robot interaction

M. Kullmann (✉) · L. L. Chuang
Humans and Technology, Institute for Media Research, Faculty of Humanities, Chemnitz University of Technology, Chemnitz, Germany
e-mail: maximilian.kullmann@phil.tu-chemnitz.de

L. L. Chuang
e-mail: lewis.chuang@phil.tu-chemnitz.de

J. Ehlers
Usability, Faculty of Media, Bauhaus-Universität Weimar, Weimar, Germany
e-mail: jan.ehlers@uni-weimar.de

E. Hornecker
Human-Computer Interaction, Faculty of Media, Bauhaus-Universität Weimar, Weimar, Germany
e-mail: eva.hornecker@uni-weimar.de

1 Introduction

Automated systems typically demonstrate terrible communication skills. This impedes their effectiveness and desirability as partners and co-agents in our everyday activities. To illustrate our point, let us caricaturize how a "bad" driver might use their turn indicator lights, and to contrast this with a "good" driver. A "bad" driver might use the turn indicator light, quite literally, to indicate to the external environment that they intend to change driving lanes. The signal is a feed-forward command, which does not negotiate and is unlikely to provide contextual information (e.g., how soon until the intended lane change). In contrast, a "good" driver verifies that a signal is received before acting upon the indicated intention. A "good" driver is capable of assuming the perspective of other co-agents, to anticipate potential misunderstandings, and to act against one's own intentions pending circumstances. A "good" driver treats the indicator as a cooperative tool to signal intent, by feed-forwarding information [1]. This allows for the coordination and synchronization of swarm traffic behavior. By minimizing surprise, a "good" driver minimizes the untimely likelihood that co-agents lack the mental resources [2] to engage safely with their upcoming actions [3].

With increasing vehicle automation, such negotiations have to be considered—not only between vehicles, but also—between automated processes and their users. It is neither appropriate nor safe to inform a user that vehicle automation is likely to fail, given upcoming circumstances, and requires user intervention. Explicit notifications, even if they are loud and clear, may not always achieve their intended aim of negotiating coordinated actions. In contrast, implicit communications could be more effective, if they are thoughtfully designed to be compatible with the targeted information processing modalities of the user and their circumstances [4]. For instance, takeover requests could consider skeumorphic visual designs that directly cue the desired response

© The Author(s) 2026
B. Meyer et al. (eds.), *Hybrid Societies*, Advances in Science, Technology & Innovation,
https://doi.org/10.1007/978-3-032-03488-5_27

modalities (e.g., skeumorphic icons) [5], tailor auditory alarms to match the appropriate cognitive mechanisms for interpreting traffic situations [6, 7], introducing ambient peripheral indicators to report the ongoing reliability of vehicle automation [8, 9], and provide contextual prompts to clarify the reason for a takeover request [10, 11]. The premise of this work is that user interfaces can be effectively designed to enable machines—not only notify, but also—to communicate with their users.

Automated and smart systems are prevalent in our everyday lives. Let us consider a commonplace example, whereby the lack of bidirectional communication is, while unlikely to be a safety-critical issue, highly likely to cause immense frustration. Many of us have experienced being plunged in sudden darkness and waving frantically to catch the "attention" of the motion detector of the room's lighting system, which has decided from our lack of movement that we no longer exist. The lighting system might even deign to precede this action with a series of taunting beeps, which we are unable to readily interpret. This sudden loss of a functionality that is "ready-to-hand" results in a "break" in our interactions [12]. We are compelled to disengage from being present (i.e., *Dasein*) and to analyze the situation that is "present-at-hand." From a psychological perspective, such repeated acts of disengagement and re-engagement can result in mental fatigue [13]. To return to our original analogy, it is the equivalent of having to deal with unpredictable drivers who swerve into and out of our driving lanes at the moment's notice, at the prompt flick of their turn indicator lights. Such a situation should be avoided in our everyday experiences. The objective of this work is to explore the design space of a roboticized light switch that is able to articulate the basis of its actions as well as demonstrate an understanding of the user's potential needs. In other words, we strive to demonstrate a system that enables its user to "understand" its behavior as well as indicate "understanding" of its user.

This work reports the subjective experiences of participants who were tasked with solving a puzzle (i.e., "Devil's Knot") in a room with two sources of light. A ceiling light supported visibility of the task and a UV lamp revealed suggestions for solving the puzzle. A light switch allowed manual toggling between the two lighting conditions and could be manipulated at any time. Moreover, it automatically switched between the two conditions at periodic 3-minute intervals. The interaction experience with the light switch differed between two groups of participants in one vital aspect. Only the test participant group experienced a light switch that would perform autonomous physical switch flipping that generated visible and audible click gestures. These gesticulations were designed to resemble the consequence of an invisible hand that manipulated the switch,

suggestive of the involvement of a collaborative agent that sought to alert participants that changing lighting conditions could assist their attempts at puzzle solving.

2 System Implementation and Description

Most smart systems falsely assume that human users have well-defined goals. The current work challenges this assumption by exploring how a smart lighting system could communicate and collaborate with a human user to solve a puzzle, in an unfamiliar setting where the optimal strategy is not known from the beginning. We sought to develop a bidirectional communication between the lighting system and its user by "roboticizing" the user interface, namely a single-pole single-throw switch that is a common household fixture.

A. Design Motivation

The fundamental idea of a roboticized user interface (RUI) is as follows. Any system can be designed to perform deterministic behavior, pending user instructions. With a RUI, systems will have the physical means to reciprocally communicate with the user. To illustrate this, a set of window blinds could be programmed to be automatically drawn at sunset. However, should the user be immersed in appreciating a beautiful sunset, this automatic and abrupt occurrence would result in an unpleasant "Bruch." A RUI seeks to inform users of its upcoming actions through subtle yet noticeable gestures that minimize abrupt surprises. This is comparable to how a butler might signal to the hosts, engaged in a conversation with their guests, that dinner is ready to be served.

The current work is inspired by two examples, the "Ultimate Machine" and the "Angry Lamp." The "Ultimate Machine" is a well-known example of a useless machine that extends an arm to switch itself off, whenever it is switched on [14]. The "Angry Lamp" is a more recent example of a standing lamp that pulls its own cord switch to turn itself off, when it notices that the room is bright enough or if there is no one in the room. The accompanying concept is that the "Angry Lamp" has the agenda of saving energy [15]. Like these examples, we focused on roboticizing the same interface that users associate with sending instructions to the system, i.e., switch. Unlike these examples, this work considers how a prolonged bidirectional dialogue could be main-tained with a user throughout a continuous activity, namely puzzle solving, with multiple steps. Switch movements serve not only to actuate a consequence, but also to communicate intent.

In Minsky's [14] and Weng's [15] implementation, the intent of the autonomous agent was represented exogenously as an independent actuator that operated the same switch as a human would. In contrast, we adopted an endo-constructive style. This means that the switch appears

Fig. 1 (Left) The mounted prototype, where the cover is detached to show the circuit board. The blue (3D printed) part on the rocker of the switch holds six so-called pogo pins—two for power supply, two for reading the switch's contacts and two for powering the motor which is hidden on the rear of the switch within the flush socket of the wall. (Right) API that allows kinetic gestures to be designed and introduced

as if it is being flicked by an invisible user or that the lighting system is sending messages via the same channel that it receives instructions from. Thus, any existing switch or physical interface element can be modified into a RUI without requiring the introduction of a novel output channel or display interface (see Fig. 1). This accords with the well-known design principle "Most Advanced Yet Acceptable" that strives to introduce advanced technology without compromising on user familiarity [16].

B Implementation

I. The (hidden) Toggle Mechanism

The binary switch changes its state appearance either through user actuation or by a hidden mono-directional rotary motor that drives a forward/backward conveyer. All the electronics are on a printed circuit board (PCB), which is hidden in the cover and thus makes maintenance easy. It carries a microcontroller (ESP8266) that monitors the status of the unit so that it can sense, if it was flicked manually or that it has to fire up the engine; it also contains a network stack for communication with other units. The motor driver (L9110S) and a voltage regulator (NCP1117) as well as a couple of capacitors and resistors are also soldered onto the PCB surface. The device is powered via an adapter by mains supply. If the actual output is a lamp (as with the prototype), a relay must be implemented to interrupt the circuit as the contacts of the actual switch are not connected to mains current (240V AC) anymore but to the microcontroller (3.3V DC). If there is more than one interface element (e.g., several switches in a corridor) it makes sense to have the relay located close to the output (in a separate unit if necessary), whereby cause and effect could be linked together again because the relay also emanates a gentle but perceivable click sound.

II. Software

The current implementation utilizes the Message Queue Telemetry Transport (MQTT) protocol over a wireless network. It is based on the architectural publish-subscribe pattern, which is the standard for many internet-of-things applications. A central control unit (CCU; or, "broker") is part of the output unit (the lamp), providing easy remote access to each separate element (i.e., relays, motors) via any computing device with MQTT support. Locally stored (re-) actions of individual units can also be manipulated during runtime via an API (see Fig. 1 right), which could support Wizard-of-Oz studies. The CCU stores the program that directs the RUI's logic. It distributes messages, so-called topics, across the corresponding units. In addition, each unit runs its own internal program to manage incoming topics and trigger the corresponding actions, or publish own topics back to other units via the CCU.

For example, the control group experiences a CCU that publishes the topic /switch/auto/ every three minutes. All components subscribed to this topic (i.e., lamps) trigger the corresponding action that is stored locally on the unit (i.e., switching the relay, i.e., light) each time it is received. On the other hand, the switch, if manually flipped, publishes the topic /switch/physical/ that switches the light as well as stop the auto mode (according to the logic of the main program). The current work only uses lamps and switches. However, additional sensors and actuators can be introduced to form a more complex system.

III. Kinetic Gestures

Manual activation of the switch will fulfill the user's wish immediately and external inputs from smartphones or programmed behavior can result in sequences of mechanical behavior that are referred to as *kinetic gestures*. Thus, the system can communicate its "intentions" and negotiate what should happen next. Kinetic gestures can accompany as well as precede or proceed from a perceived output effect: For example, a switch could flip back and forth like a countdown timer without affecting the lighting, prior to the lighting system turning itself off. This provides time for the user to intervene.

Kinetic gestures are defined by two commands—switch and hold. Each kinetic gesture is a string pattern of alternating commands, divided by a slash and a number that is understandable and programmable by laypersons. Numbers that follow the switch command determine the amount of switching operations; those that follow the hold command determine the amount of time in milliseconds to pass before the next command is executed. The kinetic gesture switch/1/ constitutes a simple flip of the switch; switch/2/hold/1000/switch/1/ constitutes a double-click, then waits 1 full second, and finishes with flipping the switch to the opposite state. "Excitable" behavior such as switch/17/ would flip the switch 17 times in just one second, which is the maximum speed that the motor and the toggle mechanism of the prototype can support. It is worth noting that the "language of intent is very contextual" [4], and the same kinetic gesture could hold different meanings, depending on the situation. For example, a double-click switch/2/in the dark could communicate "Here I am, use me!", while the same kinetic gesture, when the lights are on, could be communicating "Anyone there? Do you still need me?". Finally, the inclusion of hold/0/ is a non- gesture and disables any reaction to an incoming topic.

3 User Study

What are meaningful kinetic gestures of a RUI that could initiate and facilitate a collaborative dialogue between a human user and the lighting system? In spite of its simplicity, the design space of possible switch behavior is vast and not readily familiar to those accustomed to normal switches. To accelerate the design process for meaningful kinetic gestures, we performed an exploratory study to gain insights on *how people, who have never been in touch with a RUI before, react to and interact with it?*

A. Study Setting

All participants (n = 15; 9 females, 6 males; median age = 36 yrs) were required to solve a wooden puzzle in a room that was fitted with our lighting system. They were naïve to the real purpose of the experiment and informed that they were participating in a cognitive experiment that investigated puzzle solving.

Figure 2 depicts a study schema. Participants experienced a living-room environment with a couch and a neighboring side table. The table held a disassembled wooden puzzle (i.e., Devil's Knot) in a silver box, which participants had to reassemble. The lighting system consisted of three components: A ceiling light, a desk lamp with an ultraviolet (UV) light bulb, and a wall switch that toggles between activating either light source. The UV desk lamp was stationed on a second table that was distanced from the puzzle. Instructions to solve the puzzle were printed on sheets of paper with transparent fluorescent ink that would only be revealed when the UV desk lamp was activated. The wall switch was located next to the entrance door. The study began when the user entered the unlit room, which also initiated the automatic alternation between the ceiling lamp and the UV desk lamp that occurred at regular time

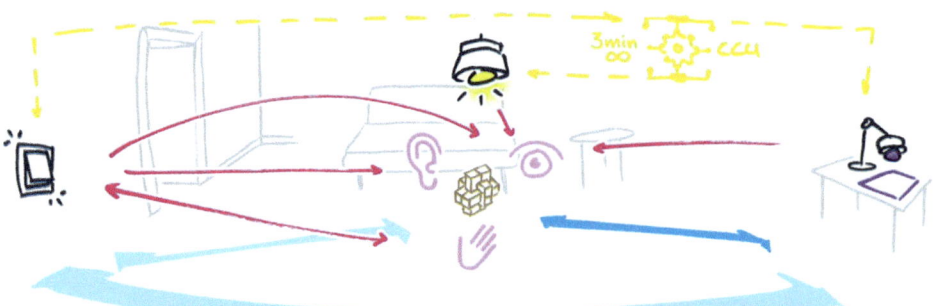

Fig. 2 A RUI in action—bidirectional dialogue (bright pink) between users (light pink) and system components (black); motion profile of the sample (bright blue) and additional paths of the experimental group (light blue); communication between components (yellow). The puzzle, i.e., devil's knot (brown). Invisible instruc- tions (purple). Surrounding furniture (grey)

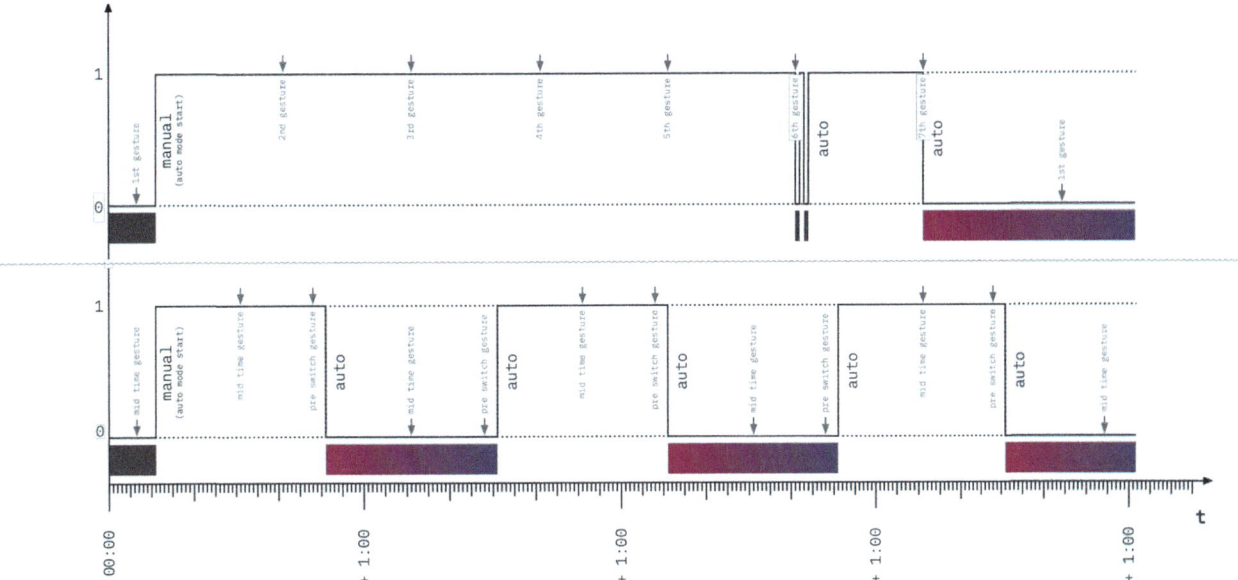

Fig. 3 Diagram depicting the timing algorithms of the experimental group (Phase 1: below; Phase 2: above). Solid lines depict the ceiling light, dotted lines depict the UV light, which alternate (0 is off; 1 is on). The pink-purple (ultraviolet) bars correspond to exactly these alternating light conditions as an additional visual element (black bar ≙ darkness; pink-purple bar ≙ ultraviolet light; no bar ≙ bright light). Grey arrows indicate the triggering of a gesture. Note: The control group runs on the same algorithms that turn on and off the light with the only difference being that the switch does not gesture (not pictured)

intervals regardless of the experimental condition. Thus, all participants were aware of the two lighting modes.

Modifications to the interaction design were introduced halfway through the study, from participant 9 onwards, which are referred to as phase 1 and phase 2. First, phase 1 provided a note, halfway through the hidden instructions, which informed participants that the switch was operable at any time. This was removed in phase 2 to evaluate if RUIs could be implicitly understood. Second, the lighting system alternated between its two modes every 40 seconds in phase 1 and every 3 minutes in phase 2. Third, two different kinetic gestures were employed in phase 1, which were increased in phase 2 to seven gestures, with escalating intensities every 30 secs.

Thus, the critical condition involved the use of the RUI and there were two versions (Compare Fig. 3 below and above, respectively). The baseline condition was an ordinary automated lighting system (OAS) with a non-gesturing switch whereby participant 7 was served as a control for phase 1 and participants 14–15 were controls for phase 2. In phase 1, the RUI performed a mid-time gesture and a pre switch gesture (see grey arrows; Fig. 3), which, respectively, corresponds to gestures #1 and #6 in phase 2. The RUI repertoire was expanded to seven different kinetic gestures. They are described in Table 1 and can be viewed at this video link.

Table 1 Description of RUI gestures with links to a video example—https://www.youtube.com/watch?v=S5IvtWLVn8w—@sec indicates a timestamp

#-TimeStamp	Intended meaning
#1-Gesture @2 s# (mid-time gesture)	Tentatively raising awareness for a novel interface concept ("Excuse me!")
#2-Gesture @32 s	Intended communication: Making acquaintance with the novel feature (of a familiar interface; "This is your switch.")
#3-Gesture @62 s	Reassure that this is an intentional act rather than a defect. ("Yes, I am talking to you.")
#4-Gesture @92 s	Getting a bit excited. ("I want you to flick me.")
#5-Gesture @122 s	Getting very excited. ("Come on, flick me!")
#6-Gesture @148 s (pre-switch)	Explaining the constitutive assignment between switch and light by flicking the switch and the relay analogously each second twice
#7-Gesture @182 s	Ordinary automated behaviour of doing things for the user + maintaining the constitutive assignment between the switches states and the lights states. ("I'm giving up…")

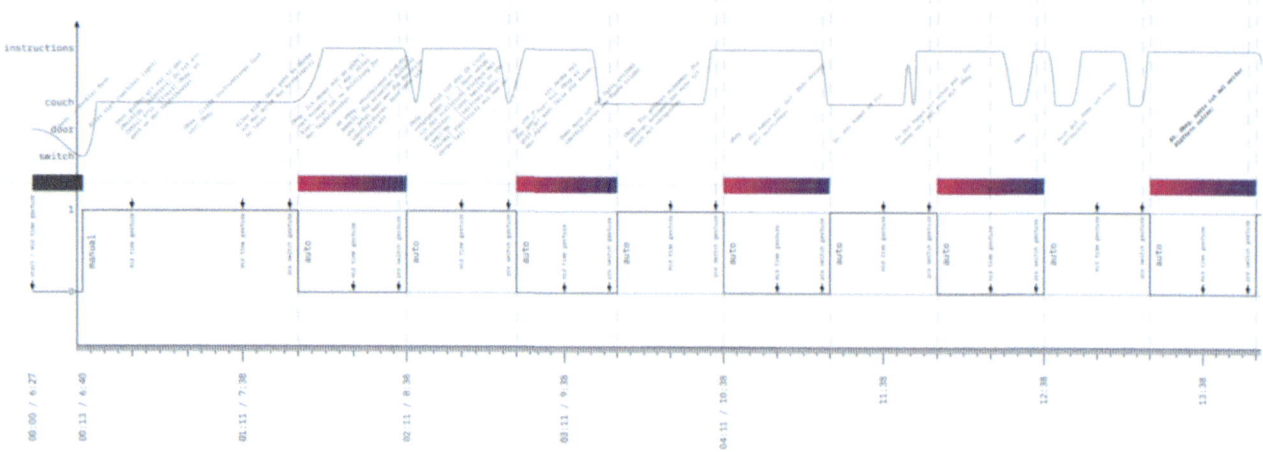

Fig. 4 Section of an interaction plot (cut after 14 min), derived from a video recording of user activity in a RUI condition (Phase 1), which visualizes the flow of (inter-)actions (Y-axis) over time (X-axis). Solid lines depict the ceiling light, dotted lines (Y-axis lower half) depict the UV light (alternating between 0–off – and 1–on) and whether these were triggered automatically or manually (vertical labeling); The pink-purple (ultraviolet) bars (between upper and lower half of the Y-axis) correspond to exactly these alternating light conditions as an additional visual element; black arrows (and corresponding vertical dashed lines) mark the triggering of the different kinetic gestures; dotted lines (Y-axis upper half) depict the subject's movement and interaction with either switch, puzzle, or instructions; think-aloud comments are added in a 45° angle (Y-axis upper half)

B *Procedure*

Participants were welcomed in a separate room and went through a consent form and instructions. Next, they filled a short questionnaire to supply demographic information and to describe their general affinity toward technology and their relation to automatic systems. Participants were introduced to the think-aloud method—where people "voice the words in their minds as they solve a wide variety of problems" [17]—and prompted to apply it during the study. A picture of the room was presented to the participants who were then instructed to sit on the couch upon entering the room and to look into the silver box. The study was typically terminated by the experimenter as soon as the participant has adapted to the automated timing or established a clear strategy of switching between the two lighting conditions to solve the puzzle. This session was video-recorded; the video camera was repositioned from participant 10 onwards. Finally, a semi-structured interview was conducted and audio recorded where the real purpose of the study is also revealed.

Video-recordings of the experiment were summarized as interaction plots—a diagrammatic visualization of time (X-axis) and the flow of (inter-)actions (Y-axis): Information on the subject's movement (interaction with either switch, puzzle, or instructions), changes of the lighting conditions and whether these have been triggered automatically or manually, and the different kinetic gestures. The think-aloud comments are added as well. This allowed relationships between these factors to be visually inferred. The interviews have been transcribed with the help of Rev AI's Speech-to-Text API. Specific passages were assigned to the respective situations of the experiment via time stamps to enrich these with more details.

C *Main Qualitative Findings*

The following findings are based on subjective interpretations of interaction plots (see Fig. 4 as an example), derived from the video-recordings and semi-structured interviews. They are presented here with the objective of informing future development of RUIs as well as to provide insight into user predispositions to machine automation.

All participants (except P6) used the switch when entering the unlit room. This established that most participants assumed the casual relationship between flipping the switch and controlling the room's lighting. All participants expressed surprise when the lighting mode changed automatically and some attributed antagonistic intent to the lighting system. P4 expressed "That is mean!" before realizing the revelation of the hidden instructions. P7 said "Argh, miese Rinde!" [East-German idiom; something like: "Argh! Perfidious!"] when the ceiling light came back on while they were reading the hidden instructions. This highlights the problem that we initially highlighted, namely that smart technology can behave abruptly and be perceived as uncooperative.

P10 and P11 did not realize that activation of the ceiling lamp alternated automatically with the UV desk lamp. Thus, unlike all other participants, they assumed a malfunction that they tried to repair. P11 asked out loud "Is there a motion detector?" and tried waving. Both P10 and P11 eventually reverted to a familiar interaction paradigm to repair the ceiling light by flipping the switch. Ju [4] refers to this repair method as "reiteration." Thus, unexpected automated behavior can either be interpreted as intentional behavior or as a malfunction that will elicit familiar user interactions for repair.

None of the control participants ever touched the switch after their initial interaction. This suggests that they simply accepted that the switch initiated a lighting system that automatically alternated between the ceiling light and UV desk lamp and assumed helplessness. They scheduled their actions to suit the regular alternations of the two light modes.

The kinetic gestures in the RUI condition prompted participants to regain control by experimenting with the switch. ("Uh, does UV turn on when I press here? [flicks the switch] Ah, okay, that's intriguing too, of course! [flicks the switch a couple of times]" P5). In the RUI condition, eight out of twelve participants repeatedly flipped the switch and learned the functional principle of the lighting system. Therefore, RUIs could mitigate assumed helplessness.

For instance, P8 mentioned the clicking sound early during the experiment and correctly inferred that it related to the mechanism of the lighting system and learnt to use the switch as per desire: "[Ceiling light comes on, UV lamp goes off automatically] Okay, now of course I would like to turn the [UV] lamp back on. And over there [points toward the switch] it has been making noises all the time. So I would assume that if I press on it again, intuitively, this light [points to the ceiling light] goes off again and this [points to the UV lamp] goes on again. I'll try it out. [Presses the switch manually] Yes, it works!". Based on anecdotal evidence, the modification in phase 2 that intensified clicking intensity is likely to have accelerated a realization of the system's mechanism.

Hedonistic responses to kinetic gestures were mixed. Our participants indicated consent ("I think it's pleasant" P4), indifference ("I have nothing against it." P13) or aversion ("The click-click is somehow... maybe not a bad signal in the end, but it takes a lot of getting used to at first. It reminds me of flicking the light switch on and off, which I have always done myself up to now. Perhaps a different signal should be found for people like me, a different sound. [...] Voice is too much for me, I would prefer a different signal, because I grew up with this click-click, my brain has to rethink too much. I would take a new signal no matter what it is like, you can get used to it. [...] It's a new message. The functionality is different here and then I have to have a different signal." Others (P6, P9, P10, P11, P12) associated and attributed the audible clicks of switch flipping, not to intentional communication of the lighting system but to more familiar sources, namely a background camera.

We propose that the interpretation of a gesture's meaning is the final of several necessary steps to understanding kinetic gestures: 1. Conscious detection of the kinetic gesture; 2. Localization of the kinetic gesture; 3. Attributing intentionality to the gesture; and 4. assigning meaning to the gesture.

4 Conclusion

This study revealed fundamental challenges to the introduction of smart and unexpected behavior in systems that we are accustomed to being dumb and deterministic. Roboticizing user interfaces can provide a first indication to naïve users that a system is operating as programmed and that unexpected behavior is not a malfunction. A careful design of such an interface's gestures could further promote discovery.

Roboticizing a familiar interface to convert it from being a unidirectional to bidirectional interface can reduce complexity. This is congruent with the "Most Advanced Yet Acceptable" design principle and offers the opportunity to upgrade legacy systems without introducing novel displays and interaction paradigms. Nonetheless, our study reveals that some interaction paradigms are highly learned and may be resistant to being updated. Subsequent research could systematically evaluate and compare the implementational cost of introducing a completely novel interaction system against the psychological cost of modifying one's mental representation of an existing system. The situation presented here of a lighting system is trivial. However, analog switches continue to be widely used in safety-critical scenarios such as flight cockpits, where an unintended reversion to old and outmoded interaction habits could be fatal.

Acknowledgements This work is funded by the Deutsche Forschungsgemeinschaft (DFG, German Research Foundation)—Project-ID 416228727—CRC 1410. Special thanks goes to Sonia Kampel, Julian Bornemeier, and Nico Tauchmann for revision. We thank all participants, without whom this work would still be nothing but wild guesswork. Also gratitude goes to Max Neupert, who catapulted the prototype to the next level of (circuit board) design.

References

1. Norman, D. A. (2013). *The design of everyday things*. Basic Books.
2. Scheer, M., Bülthoff, H. H., & Chuang, L. L. (2016). Steering demands diminish the early-P3, late-P3 and RON components of the event-related potential of task-irrelevant environmental sounds. *Frontiers in Human Neuroscience, 10*, 73.
3. Lee, J. D. (2014). Dynamics of driver distraction: The process of engaging and disengaging. *Annals of Advances in Automotive Medicine, 58*, 24.
4. Ju, W. (2015). *The design of implicit interactions*. Morgan and Claypool.
5. Brandenburg, S., & Chuang, L. (2019). Take-over requests during highly automated driving: How should they be presented and under what conditions? *Transportation Research Part F: Traffic Psychology and Behaviour, 66*, 214–225.
6. Glatz, C., Krupenia, S. S., Bülthoff, H. H., & Chuang, L. L. (2018). Use the right sound for the right job: verbal commands and auditory icons for a task-management system favor different information processes in the brain. In *Proceedings of the 2018 CHI Conference on Human Factors in Computing Systems* (pp. 1–13).

7. Lahmer, M., Glatz, C., Seibold, V. C., & Chuang, L. L. (2018). Looming auditory collision warnings for semi-automated driving: An ERP study. In *Proceedings of the 10th International Conference on Automotive User Interfaces and Interactive Vehicular Applications* (pp. 310–319).

8. Faltaous, S., Baumann, M., Schneegass, S., & Chuang, L. L. (2018). Design guidelines for reliability communication in autonomous vehicles. In *Proceedings of the 10th International Conference on Automotive User Interfaces and Interactive Vehicular Applications* (pp. 258–267).

9. Figalová, N., Chuang, L. L., Pichen, J., Baumann, M., & Pollatos, O. (2022). Ambient light conveying reliability improves drivers 'takeover performance without increasing mental workload. *Multimodal Technologies and Interaction, 6*(9), 73.

10. Borojeni, S. S., Chuang, L., Heuten, W., & Boll, S. (2016). Assisting drivers with ambient take-over requests in highly automated driving. In *Proceedings of the 8th International Conference on Automotive User Interfaces and Interactive Vehicular Applications* (pp. 237–244).

11. Sadeghian Borojeni, S., Boll, S. C., Heuten, W., Bülthoff, H. H., & Chuang, L. (2018). Feel the movement: Real motion influences responses to take-over requests in highly automated vehicles. In *Proceedings of the 2018 CHI Conference on Human Factors in Computing Systems* (pp. 1–13).

12. Heidegger, M. (1962). *Being and time.* Translated by John MacQuarrie & Edward Robinson from the 7th German edition of 1953. Harper & Row Publishers.

13. Hopstaken, J. F., Van Der Linden, D., Bakker, A. B., & Kompier, M. A. (2015). A multifaceted investigation of the link between mental fatigue and task disengagement. *Psychophysiology, 52*(3), 305–315.

14. Minsky, M.. (2011). Making the most useless machine. Retrieved July 29, 2022 from https://youtu.be/C8kU3oZwVJA

15. Weng. X. (2014). *Good medicine tastes bitter.* Final thesis at the Bauhaus-Universität Weimar, Faculty of Arts and Design, Study Program Product Design. https://vimeo.com/yuuedesign/bitterdesign#t=13s

16. Loewy, R. (1951). *Never leave well enough alone.* Simon and Schuster.

17. Charters, E. (2003). The use of think-aloud methods in qualitative research: An introduction to think-aloud methods. *Brock Education: A Journal of Educational Research and Practice., 12*(2), 68–82.

Where Am I? How to Measure and Support Spatial Orientation in Teleoperation

Jennifer Brade⬤, Ning Xie⬤, Sven Winkler⬤, Philipp Klimant⬤, and Georg Jahn⬤

Abstract

A multitude of differently embodied digital technologies, such as autonomous vehicles, delivery robots or telepresence systems and humans will coordinate with each other, interact and work side by side in the near future. The possibilities that this incurs are manifold, but creating a smooth coordination is just as challenging. Because of limits of automation and exceptions to routine operation, there are and will be systems, which can both act autonomously and be controlled remotely by humans using teleoperation. This article pertains to challenges of spatial orientation that teleoperators have to face if they take over and control systems remotely. The ability to orient oneself via the cameras of a teleoperated system in the remote environment plays a crucial role for a smooth and error-free coordination with and in the environment of the teleoperated system. In this article, we will explain why orientation in a remote environment is crucial, which problems exist, how spatial orientation can be measured, and which approaches could be used to provide visual support for operating in a remote environment.

Keywords

Telepresence · Spatial orientation · Human–robot interaction · Teleoperation

1 Introduction

Mobile robots and highly-automated vehicles are Embodied Digital Technologies (EDTs) that can be encountered in public environments. Urban robots such as delivery robots or street-cleaning robots, and automated vehicles on public roads move in public environments among humans and human-driven vehicles. Such EDTs may navigate fully autonomously for some time, but for the foreseeable future will need human assistance or temporary operation by humans to manage exceptional situations and for failure recovery [1]. Teleoperation can improve the performance of EDTs and is the fallback mode for delivery robots and driverless vehicles if they encounter obstacles and exceptional challenges that cannot be resolved by automation [2, 3]. Teleoperators on call for troubleshooting thus often face the challenge of having to quickly orient themselves and to acquire situation awareness in a remote environment that they experience through a more or less immersive interface. They may experience the remote environment in a single, manually controlled egocentric camera perspective on a 2D screen with access to a separate ego-cantered map view, or they may be able to more intuitively explore a stereoscopic view wearing a head-mounted display (HMD) that provides additional information about

Jennifer Brade and Ning Xie are the authors contributed equally.

J. Brade · S. Winkler (✉)
Professorship Production Systems and Processes, Chemnitz University of Technology, Chemnitz, Germany
e-mail: sven.winkler@mb.tu-chemnitz.de

J. Brade
e-mail: jennifer.brade@mb.tu-chemnitz.de

N. Xie · G. Jahn
Department of Psychology, Chemnitz University of Technology, Chemnitz, Germany
e-mail: ning.xie@psychologie.tu-chemnitz.de

G. Jahn
e-mail: georg.jahn@psychologie.tu-chemnitz.de

P. Klimant
Professorship Virtual Technologies, Hochschule Mittweida - University of Applied Sciences, Mittweida, Germany

Fraunhofer Institute for Machine Tools and Forming Technology IWU, Chemnitz, Germany

P. Klimant
e-mail: philipp.klimant@hs-mittweida.de

© The Author(s) 2026
B. Meyer et al. (eds.), *Hybrid Societies*, Advances in Science, Technology & Innovation,
https://doi.org/10.1007/978-3-032-03488-5_28

the environment and leaves the hands free to directly control the remote EDT perhaps with multimodal feedback available. Depending on the task to be performed or problems to be solved, a more immersive interface that increases the experience of telepresence may provide advantages [4]. But an immersive interface could also impede, because, for instance, quickly switching perspectives or accessing data displays and various physical controls is more difficult than just looking and grasping [4, 5]. In general, less immersive interfaces seem more appropriate for remote assistance in which teleoperators, for instance, provide assistance in object recognition, designate paths to follow or choose among alternative manoeuvres. More immersive interfaces may provide advantages in remote driving when teleoperators control EDT movements directly.

Quickly gaining and retaining spatial orientation and situation awareness despite limited sensory information and a certain decoupling of self-motion and 'ego-' motion in the remote environment is a challenge that is typical of teleoperation [6]. It has been described and alleviating measures have been sought for teleoperation in space, underwater, for operating unmanned aerial and ground vehicles, in teleoperating forklifts and in teleoperating mobile robots for search and rescue, maintenance or in dangerous and toxic environments [7]. The difficulties and proven support of spatial orientation and situation awareness also apply to teleoperation of EDTs in public environments, however, it seems worthwhile to consider certain specific conditions of urban environments and public roads as well as of navigating among and coordinating with other human and artificial agents. Public environments are usually well-mapped in contrast to, for example, disaster sites. Possible routes, distances, estimated travel times, navigation goals, landmarks and other points of interest are usually known and can be visualized or their location can be indicated. Connections to infrastructure or other EDTs may be available as sources of information and for coordination. Infrastructure may provide additional camera perspectives. For smoothly coordinating with human agents, mutual observability, predictability and directability [8] should be ensured in appropriate ways. Human agents should be able to assess the operation mode of an EDT that they encounter. The teleoperation mode should be apparent and it may help for certain challenges of coordination for individuals on site to be able to perceive the current attention allocation of a teleoperator and to be able to communicate verbally or via an audio-visual interface. Furthermore, human agents in the vicinity or on board can be a valuable resource in teleoperation. They can be addressed and asked for help and assistance to overcome the teleoperators' limited possibilities for acting in the remote environment. Human helpers may also contribute to improving spatial orientation and situation awareness. They may, for instance, operate mobile cameras to provide additional viewpoints. For such support, appropriate ways of directing human helpers need to be in place.

Spatial orientation and spatial situation awareness encompass knowledge about enduring spatial features and spatial representations in working memory modulated by attention. Working memory representations can quickly adapt to changes. For example, a teleoperator may be familiar with the route of a driverless bus line but needs to be aware also of the current location and orientation of the vehicle as well as of temporary static obstacles and moving agents in the vicinity. In addition, teleoperators must be able to correctly estimate distances to other road users and people in order to enable smooth coordination. For characterizing requirements of a specific task, for assessing spatial orientation and for designing support, it is useful to also discern allocentric and egocentric spatial representations [9]. For directing attention and for navigating and acting in an environment, egocentric representations are necessary that are tied to the ego-perspective from the current location with the current orientation. Self-localization is possible based on egocentric relations to elements of an environment. Distances and directions to landmarks, objects and agents can be represented egocentrically (see Fig. 1 with different viewpoints of different participants). Environmental features, locations and spatial relations can also be represented allocentrically independent from the own current location and orientation, for example, map-like knowledge about a route in a city or the layout of an environment. Allocentric representations as knowledge background are helpful for self-localization and for (re-)constructing egocentric spatial representations based on perceptual input. For instance, imagine that you as a teleoperator log into a camera feed on a traffic scene. If the locations of visible landmarks are familiar to you, you know what permanent features of the environment are behind before changing the camera perspective even if you have never been there yourself. A camera view from a drone on the traffic could also provide information on what is behind, but from an allocentric perspective if the egocentric perspective is the one of the camera on the vehicle. Allocentric perspectives can support in instantaneous self-localization and in acquiring and refining allocentric as well as egocentric representations. (allocentric locations, spatial relations, extensions, egocentric directions and distances).

Self-motion in an environment incurs the requirement to update egocentric spatial representations to maintain spatial orientation. Spatial updating is possible with similar processes as right after logging into a remote system (instantaneous spatial updating). But when moving with the own body, humans as animals experience effortless continuous spatial updating of the egolocation and egocentric relations based on sensory information (visual, vestibular, proprioceptive, auditory) and motor efference. In teleoperation, the input for continuous spatial updating typically is impoverished because the perception of the remote environment is compromised due to technical limitations or missing sensory information, which

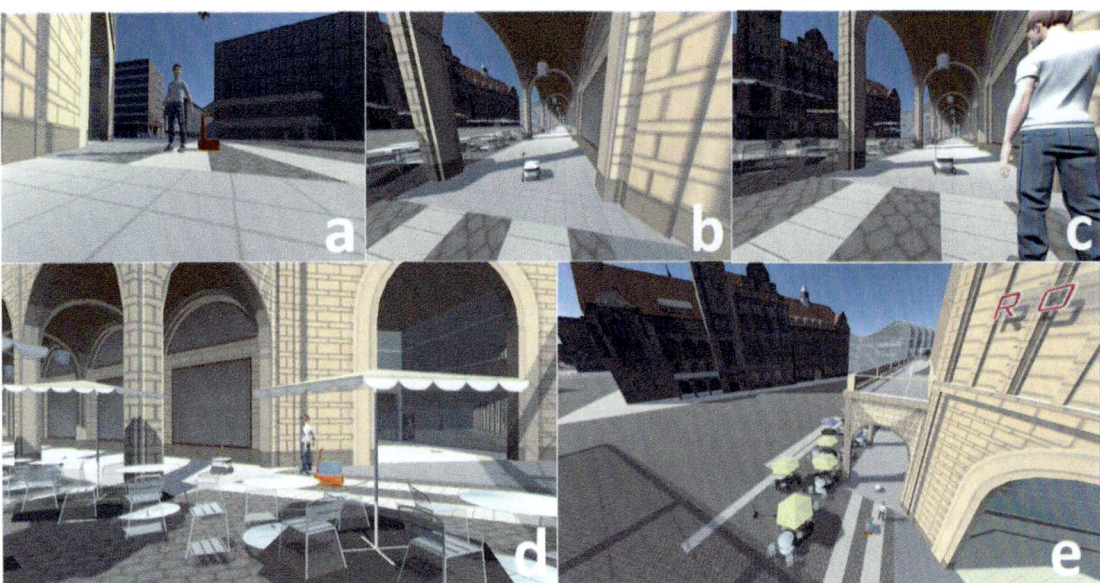

Fig. 1 Exemplary presentation of a futuristic space sharing conflict with different participants and systems and their viewpoints (**a** View teleoperator via delivery robot; **b** view human, **c** view transport robot **d** teleoperator via telepresence system, **e** view flying drone)

can have detrimental effects on situation awareness and thus on task completion [10]. Hence, technical support for maintaining spatial orientation may improve task performance. Assessing spatial orientation is important both for identifying the need for technical support as well as for evaluating the effectiveness of technical support. This article provides a brief overview of assessment tasks with some examples from our own research and results. Furthermore, we present which types of technical support offer further research potential.

2 Experiments and Methods for Measuring Spatial Orientation

An indirect way of assessing spatial orientation is to measure task performance in a task that requires spatial orientation, for example, how quickly a teleoperator of a delivery robot manages to distribute three meals to customers at three different addresses. Of course, the time this takes is influenced by many other factors than just the spatial orientation of the teleoperator, particularly, if this task is performed in the real world and not just in a simulation. More direct ways of assessing spatial orientation that are established methods in research on spatial cognition encompass probing individuals' spatial representations by asking them to indicate directions and locations, for instance, by pointing, turning or navigating to a location. How well spatial orientation is maintained by spatial updating is assessed by probing after self-motion. An example of such a task is triangle completion that requires moving to a location, turning and moving to a second location, and then to move or point to the starting

location. The travelled outward path does not have to be two sides of a triangle but can be any path. Veridical representation of distances and extensions (perceived, travelled or inferred) can be assessed by asking individuals to reproduce or compare spatial intervals. Methods particularly suitable for assessing allocentric long-term memory representations typically involve imagery. Egocentric representations from a certain perspective have to be constructed or reconstructed in working memory to, for instance, draw a map or to indicate directions from imagined locations and orientations (e.g., 'Imagine standing at the courthouse facing the belltower, please, point to the hospital', a judgement of relative direction). In the following, we provide examples of some of these methods that we have recently applied.

A. *Finding the starting point after forward translation and rotation*

A frequently used task to test the updating of position and orientation in an environment is the already mentioned triangle completion task (TCT). During the tests, participants move along two sides of a triangle and are prompted at the end of the second outward leg to indicate (point or move to) the starting position. The participant must integrate the distance travelled on the first two legs with the turn angle to produce an accurate homebound trajectory [11]. The type of movement along the sides can be simulated or self-controlled and the rotations at the triangle points can be simulated or real. We used the TCT to examine how well participants perform a path integration task under different visual cue conditions while using a remote-controlled telepresence robot [12]. To do this, the participants

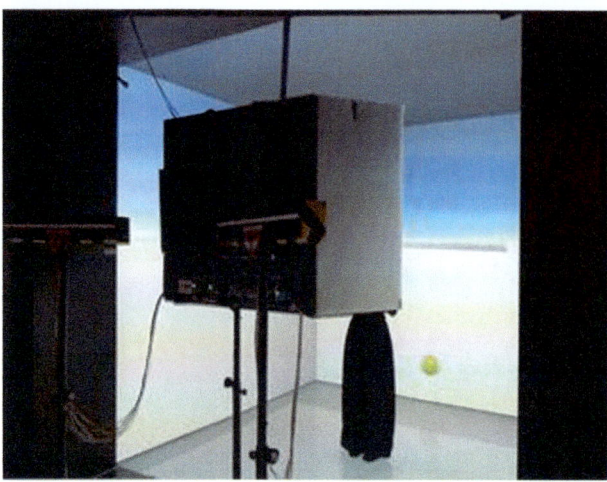

Fig. 2 View in the 5-sided CAVE incl. sound projector, participant and appearance of the virtual environment [8]

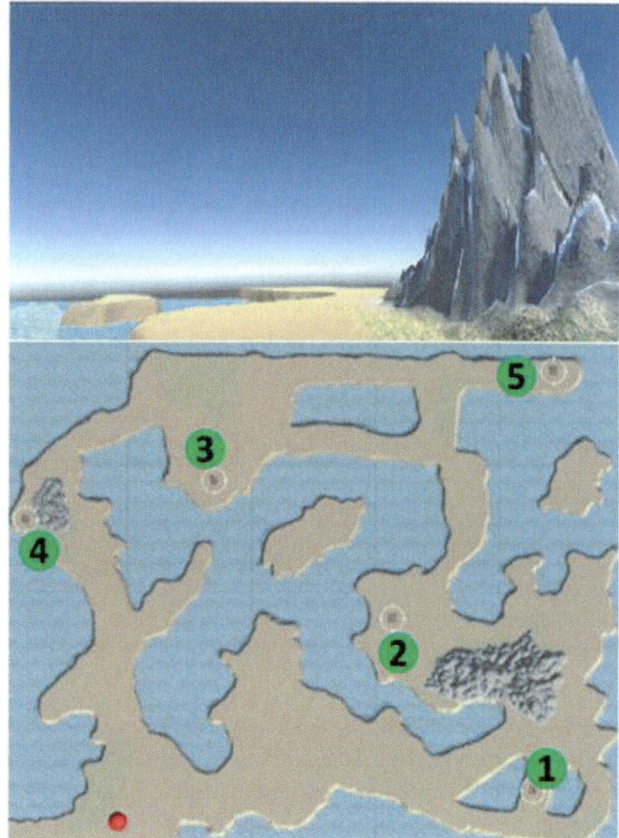

Fig. 3 View in the virtual environment (above) and map with targets (below)

first steered the telepresence robot to two spotlights projected on the floor that appeared and then disappeared one after the other and then steered to the presumed starting point. We explored the influence of the number of visual cues available in the field of view on the performance in the TCT and showed a reliably better performance with rich compared to sparse visual cues. In another study [13] with the TCT, which took place in a virtual environment—the 5-sided CAVE—we examined whether the presence of a continuous auditory cue can support spatial updating in VR similar to visual landmarks when body-based cues are sparse (see Fig. 2). Our results indicate that the auditory cue could improve performance for a subset of participants to a similar degree as visual cues compared to the complete absence of cues.

B. *Memorizing and visiting locations in a specific sequence*

If navigation performance rather than spatial updating is of interest, goal locations in navigation tasks do not match the starting locations as in the TCT. Previous studies often asked participants to find a special location [14] or a target [15], and their performance was quantified by travel times or distances they travelled with the number of errors they made. Sometimes, participants were asked to report the strategies they applied [16] or to draw a map [17] after navigation. We have assessed navigation performance for comparing different map systems. Following the Sea Hero Quest game by Hyde et al. [18], a treasure-hunting game was created in a virtual reality environment. Participants, wearing an HMD, were instructed to find five (see Fig. 3) treasure boxes (targets) in a particular sequence after memorizing the whole map of an island for about 15 s. After the map had disappeared, participants used a controller to translate forward in the virtual environment and body rotation for turning. They could push a button to access a map system to help them locate the targets during

the whole time, but could only navigate in first-person view on the ground in between using the map system. We explore whether a continuous perspective change into the map system helps participants more compared to a switch to a traditional static map system. The continuous change into the map view provides them with the whole process of moving their first-person view on the ground to a bird's-eye view in the sky. Data collection for this study is still ongoing.

C. *Pointing to remembered object locations after forward self-motion*

Memory for exact locations in the current close environment can be tested by asking participants to move to these locations (navigation as in A and B), but it can also be assessed by pointing. To study spatial memory updating in VR, we have explored the effects of spatial reference (landmarks, boundaries) and sensory feedback about self-motion (increased optical flow, real walking) in a sparse virtual scene presented in an HMD. Participants were asked to encode the locations of two objects presented to the left and to the right of the forward axis. Then, the objects disappeared and participants experienced forward self-motion, which changed the egocentric relations to the object locations, however, both object locations remained in vista space. Immediately after self-motion stopped, participants were probed to point to either the left or the right object location. The pointing performance reflected the known underestimation of distances in virtual reality (e.g., [19, 20]) and demonstrated supporting effects of close landmarks and boundaries. We found that spatial updating of object locations across forward self-motion that remain in vista space seems to be performed based on visual sensory feedback and is not improved by real walking. Primacy of visual cues in spatial updating across forward translation is consistent with previous results, additional sensory feedback and motor efference do, however, improve spatial updating if self-motion includes extended orientation changes (rotation) (e.g., [21]).

D. *Estimate and reproduce distances*

Pointing to remembered locations after self-motion involves distance estimation, but it may be of interest how well encoded distances that are not changed by subsequent self-motion are reproduced. To study distance reproduction in typical conditions of teleoperation, we asked teleoperators to estimate egocentric distances using a commercially available telepresence system in two perceptual matching tasks [22]. To do this, the participants had to reproduce distances that they either had travelled with the telepresence robot themselves beforehand or that were presented to them visually for a certain duration. Our results were in line with the common underestimation of distances in mediated environments. But we also showed that a time-based estimation, which was possible in the task where participant drove the distance with the telepresence robot before reproducing it, resulted in a slightly better performance than the purely visual experience of distance. In a second study, we used a self-developed automated guided vehicle (AGV) with a 360° camera on top as telepresence robot. Participants saw the live stream of the robot camera in a HMD and operated the robot via a controller (see Fig. 4) [23]. We used different distance estimation tasks, in which participants had to either drive the AGV over a certain distance to an object or estimate a given distance verbally. Participants could verbally estimate distances in the range of 0.7 m to 3.7 m very precisely and reproduced distances with an accuracy of approx. 20 cm.

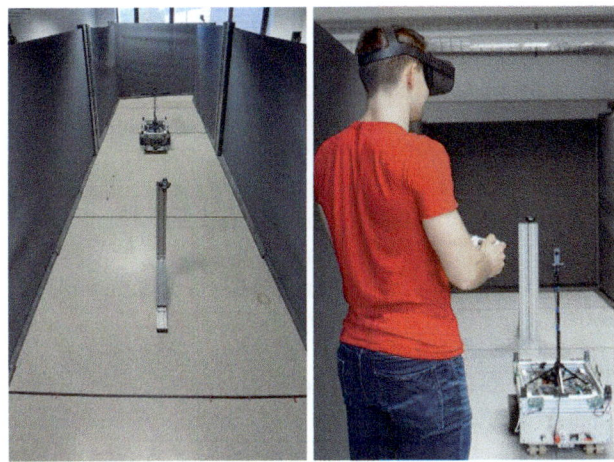

Fig. 4 Sparse remote environment with AGV (left) participants with HMD and controller (right)

3 Visual Support for Remote Perception

To increase the user's situational awareness as much sensory information as possible should be transmitted to the operator [24]. However, it is technically problematic to record all sensory information correctly and present it to the teleoperator in an easy and understandable way. Thus, most studies and developments focus on the visual sense. Different factors can contribute to improving visual remote perception, for example, the field of view or the viewpoint in the remote environment, the type of interaction and thus the amount of body-based cues, as well as the output system itself [10]. In the following, we would like to go into more detail about the type of output system and the viewpoint in the scene and how they can support remote perception.

A. *With what and how is the remote environment displayed?*

Various output systems exist to display the view of the remote environment and some are also typical virtual reality output systems. The range extends from simple 2D desktop screens, to 2D 360° videos, to recording the video stream via stereoscopic cameras and displaying it on head-mounted displays or projective VR systems. Of course, the characteristics of each output system influence how the remote environment is perceived. Stereoscopic visualization has been advocated because this is closer to the way humans naturally see the world, but not all studies show advantages of 3D over 2D output systems: Georg et al. [25] compared a conventional monitor and a head-mounted display (HMD) in a driving study concerning the factors immersion, driving behaviour and workload. They showed that a head-mounted-display increases immersion, but does not necessarily improve the driving performance and controllability of the vehicle. Chellali and Baizid [26] compared regular monitors and HMDs in a localisation task and showed that participants performed better with regular monitors. In contrast, Kratz and Ferriera [27] showed in a collaborative assembly task that the use of an HMD reduces task error rates and improves collaborative success and quality of visualization compared to the monitor condition. Another study by Forster et al. [28] compared 2D and stereoscopic 3D simulations in a driving simulator study investigating distance estimation. No differences between 2 and 3D in a verbal distance estimation task were found, but in a car-following task, participants were able to adjust distances more accurately in the 3D condition compared to 2D. Boustila et al. [29] also showed that stereoscopic vision led to a more accurate distance estimation in a virtual house visit task than in the monoscopic vision condition. Although there is no uniform picture with regard to the comparison of 2D and 3D output systems, it is assumed that stereoscopic output systems provide more depth information, which means that spatial perception with these systems tends to be better [30]. In addition to the type of visualization, display properties of the output system also play an important role in influencing remote perception. One factor that plays a special role here is the field of view: Compared to the field of view that is available to people in everyday life, typical output systems for synthetic environments usually have a smaller field of view. Practitioners complain about limited angular view [31] and more effort is needed to gain situation awareness comparable to direct viewing [10]. In addition to this effect, known as the keyhole effect, a narrow view limitation is also associated with what is known as cognitive tunnelling—the individual failing to recognize land- marks, even those which are in view [32], errors in distance and depth judgments [33] and an impaired ability to detect targets or obstacles [34]. A wider field of view can be achieved using wide-angle cameras or with additional cameras that show other viewpoints in the scene. Studies show that an extended FOV goes hand in hand with a higher cognitive load on the part of the user and that, especially with wide-angle cameras, the user's sense of speed changes when moving forward [10]. This raises the question of which FOV is best for interaction via telepresence. Johnson et al. [35] compared a narrow (45°), a wide-angle (180°) and a panoramic (360°) horizontal field-of-view in a between-subjects experiment. In each condition, participants controlled a telepresence robot remotely and collaborated with an on-site user in a room-redecorating task. Their results showed that participants using the wide-angle and panoramic views completed the task more quickly and caused fewer collisions compared to the narrow view users. Compared to the narrow and wide-angle view, the panoramic view improved participants ability to form accurate mental maps of the remote environments, but was also perceived to be more difficult to use. Other studies varying the horizontal field-of-view [36, 37] showed the typical underestimation in synthetic environments [38] also for wider FOVs, which indicated, that a wider FOV does not lead to accurate estimations. According to the currently available research, both a FOV that is too small and a FOV that is too wide have disadvantages. Direct recommendations for action do not exist, however, providing several viewpoints on the remote environment can alleviate disadvantages of a limited FOV.

B. *From which perspective is the remote environment perceived?*

A small FOV may cost an operator more effort to finish a task or solve a problem [39]. To improve that, additional cameras on different EDTs or in the environment can help operators receive information from different viewpoints on a remote area [40]. There are many examples of this in our everyday life, such as a bird's-eye view generated by a camera when parking or camera flights from station cameras at sporting events. Keyes et al. [40] compared cameras from different viewpoints and multiple cam- eras to a single camera. Results suggested that two cameras from different directions increase situation awareness during a maze travel. Another way to improve operators' performance is to change the viewpoints in the environment [39, 41]. Chadwick et al. [39] conducted an experiment which asked people to navigate an environment by controlling a virtual robot. They manipulated the field of view and camera viewpoints separately and found that operators performed better if they were provided with a third-person view instead of a first-person view especially when the displayed field of view was small. If two views are provided effectively in parallel, getting lost and keyhole effects disappear [42]. Morison et al. [42] presented a situation that required to judge the passability for a robot. When presenting a third-person view from another device, people could easily see how high the robot was relative to the door, and when presenting a bird's-eye view, people could see the whole environment to choose alternative routes for the robot to pass if necessary. However, it is not generally better to provide more camera views in teleoperation. Because more information may create heavier cognitive load for people, it is very important to select and integrate information from different cameras. Several factors should be considered including task type, time limitation, location, price, integrating camera views, switching and coupling issues [43, 44]. In the study of Hughes and Lewis [43], a single camera or multiple cameras were provided for participants in a navigation task. Their results showed that multiple cameras prove beneficial for route-drawing systems and may allow for a more cooperative collaboration. However, they also showed that a single camera independently controllable from the orientation of the vehicle may also yield significant benefits in some situations. Besides, they found evidence to support the use of separate cameras for different navigational subtasks. Users should consider their task and be able to decide whether they need multiple cameras or which camera they should rely on. It is also important to consider realistic problems during real world operation. For example, it is hard to use large equipment with multiple cameras if people need to perform a rescue task in a dangerous area [45]. Because it is not easy for people to choose cameras, some researchers start to develop dynamic camera systems or integrated systems for operators [44, 46, 47]. Rakita et al. [46] for example, featured a novel viewpoint adjustment algorithm to provide operators with an enhanced view of the task environment. However, the problem of lacking depth perception still remains, which may cause problems in tasks requiring veridical depth perception.

4 Summary and Future Steps

The possibility of controlling EDTs from remote provides the possibility to act in distant or dangerous environments and is a fallback mode for automation. Despite the further development of robots, teleoperation remains a difficult task [5]. The reasons for this are manifold and one of them is the perception of the remote environment mediated by technical systems compared to natural perception. Even though spatial orientation can be improved by integrating various sensory modalities such as visual and body-based cues by real walking and turning, it can be assumed that the control of teleoperated EDTs will be carried out using classic locomotion methods (e.g., controller) for virtual environments in the near future. The reasons for this are the lower space requirement and the available workstations, which are typically based on classic office workstations. Thus, in order to enable a more coordinated and error-free interaction with the remote environment and humans on site, other methods to support the teleoperators must be found. Visual support in the form of different viewpoints on the scene, stereoscopic displays with a sufficient FOV, and additional information provided via augmented reality support are promising approaches [48]. Nevertheless, no approach has yet been found that corresponds to the performance of normal vision in a scene, because there are still limitations with regard to spatial perception including the assessment of distances and extensions, and updating spatial representations across self-motion. There is therefore still a need for research into which type of output enables the teleoperator to optimally apprehend the circumstances and offers all the necessary information about the environment and the participants in the environment without overtaxing the teleoperator. In future, research must therefore be carried out into how the teleoperator receives both a suitable initial orientation and how he/she can constantly capture his/her surroundings spatially. Possible approaches for this are, for example,

the continuous 'flying' into a first-person perspective in the remote environment. Cameras from drones or surveillance technology could provide the teleoperator with a first impression of the remote environment and the system to be teleoperated. It would then be possible to 'fly into' the EDT perspective. Changes of perspective would be possible by 'flying out' from the robot to other viewpoints. In addition, the possibility of using on-site humans as helpers should be further investigated. People could volunteer to support EDTs in urban areas and be certified. Recognition of these people using, for instance, an RFID tag would be conceivable, so that teleoperators can contact the relevant people. These human helpers can then either remove obstacles directly or give the teleoperator an external view of the EDT using their own mobile devices, for example. In summary, a wide variety of EDT systems can be expected in urban areas in the future, but also in the work context. How smoothly the coordination and collaboration with these systems runs depends, among other things, on how well their teleoperators can orient themselves in remote environments for direct and indirect control, and how they can implicitly and explicitly communicate with on-site humans.

Acknowledgements This work was funded by the Deutsche Forschungsgemeinschaft (DFG, German Research Foundation)—Project-ID 416228727—CRC 1410.

References

1. Moniruzzaman, M., Rassau, A., Chai, D., & Islam, S. M. S. (2021). Teleoperation methods and enhancement techniques for mobile robots: A comprehensive survey. *Robotics and Autonomous Systems, 103973.*
2. Bogdoll, D., Orf, S., Töttel, L., & Zöllner, J. M. (2022). Taxonomy and survey on remote human input systems for driving automation systems. In *Future of Information and Communication Conference* (pp. 94–108).
3. Mutzenich, C., Durant, S., Helman, S., & Dalton, P. (2021). Updating our understanding of situation awareness in relation to remote operators of autonomous vehicles. *Cognitive Research: Principles and Implications, 6*(1), 1–17.
4. Almeida, L., Menezes, P., & Dias, J. (2020). Interface transparency issues in teleoperation. *Applied Sciences, 10*(18), 6232.
5. Rea, D. J., & Seo, S. H. (2022). Still not solved: A call for renewed focus on user-centered teleoperation interfaces. *Frontiers in Robotics and AI, 9.*
6. Opiyo, S., Zhou, J., Mwangi, E., Kai, W., & Sunusi, I. (2021). A review on teleoperation of mobile ground robots: Architecture and situation awareness. *International Journal of Control, Automation and Systems, 19*(3), 1384–1407.
7. Cui, J., Tosunoglu, S., Roberts, R., Moore, C., & Repperger, D. W. (2003). A review of teleoperation system control. In *Proceedings of the Florida Conference on Recent Advances in Robotics* (pp. 1–12).
8. Johnson, M., Bradshaw, J. M., Feltovich, P. J., Jonker, C. M., Van Riemsdijk, M. B., & Sierhuis, M. (2014). Coactive design: Designing support for interdependence in joint activity. *Journal of Human-Robot Interaction, 3*(1), 43–69.

9. Ekstrom, A. D., Spiers, H. J., Bohbot, V. D., & Rosenbaum, R. S. (2018). *Human spatial navigation.* Princeton University Press.

10. Chen, J. Y., Haas, E. C., & Barnes, M. J. (2007). Human performance issues and user interface design for teleoperated robots. *IEEE Transactions on Systems, Man, and Cybernetics Part C (Applications and Reviews), 37*(6), 1231–1245.

11. Chrastil, E., & Warren, W. (2008). Testing models of path integration in a triangle completion task. *Journal of Vision, 8*(6), 1153–1153.

12. Brade, J., Hoppe, T., Winkler, S., Jahn, G., & Klimant, P. (2023). Visual cues improve spatial orientation in telepresence as in VR. In *6th International Conference on Intelligent Human Systems Integration: Integrating People and Intelligent Systems (IHSI 2023).*

13. Breitkreutz, C., Brade, J., Winkler, S., Bendixen, A., Klimant, P., & Jahn, G. (2022). Spatial updating in virtual reality–auditory and visual cues in a cave automatic virtual environment. In *IEEE Conference on Virtual Reality and 3D User Interfaces (VR)* (pp. 719–727).

14. Meilinger, T., & Knauff, M. (2008). Ask for directions or use a map: A field experiment on spatial orientation and wayfinding in an urban environment. *Journal of Spatial Science, 53*(2), 13–23.

15. Li, R., & Klippel, A. (2012). Wayfinding in libraries: Can problems be predicted? *Journal of Map & Geography Libraries, 8*(1), 21–38.

16. Lawton, C. A. (1996). Strategies for indoor wayfinding: The role of orientation. *Journal of Environmental Psychology, 16*(2), 137–145.

17. Muffato, V., Meneghetti, C., & De Beni, R. (2019). Spatial mental representations: The influence of age on route learning from maps and navigation. *Psychological Research Psychologische Forschung, 83*(8), 1836–1850.

18. Hyde, M., Scott-Slade, M., Scott-Slade, H., Hornberger, M., Spiers, H., Dalton, R., Hoelscher, C., Wiener, J., & Bohbot, V. (2016). Sea hero quest: The world's first mobile game where anyone can help scientists fight dementia.

19. Witmer, B. G., & Kline, P. B. (1998). Judging perceived and traversed distance in virtual environments. *Presence, 7*(2), 144–167.

20. Lampton, D. R., McDonald, D. P., Singer, M., & Bliss, J. P. (1995). Distance estimation in virtual environments. In *Proceedings of the Human Factors and Ergonomics Society Annual Meeting* (pp. 1268–1272).

21. Chance, S. S., Gaunet, F., Beall, A. C., & Loomis, J. M. (1998). Locomotion mode affects the updating of objects encountered during travel: The contribution of vestibular and proprioceptive inputs to path integration. *Presence, 7*(2), 168–178.

22. Brade, J., Winkler, S., Beisig, L. S., Flucke, J., Jasniewcz, L., Laaser, J., Seehars, A., Lorenz, M., Jahn, G., & Klimant, P. (2022). How far is it? Distance estimation and reproduction through a double 3 telepresence robot. In *IEEE International Symposium on Mixed and Augmented Reality (ISMAR)*. Singapore.

23. Winkler, S., Weidensager, N., Brade, J., Knopp, S., Jahn, G., & Klimant, P. (2022). Use of an automated guided vehicle as a telepresence system with measurement support. In: *CIVEMSA 2022.* Chemnitz, Germany.

24. Lichiardopol, S. (2007). A survey on teleoperation. Technische Universitat Eindhoven, DCT report 20, 40-60

25. Georg, J. M., Feiler, J., Diermeyer, F., & Lienkamp, M. (2018). Teleoperated driving, a key technology for automated driving? comparison of actual test drives with a head mounted display and conventional monitors. In *21st International Conference on Intelligent Transportation Systems (ITSC)* (pp. 3403–3408).

26. Chellali, R., & Baizid, K. (2011). What maps and what displays for remote situation awareness and ROV localization? In *Symposium on Human Interface* (pp. 364–372).

27. Kratz, S., & Ferriera, F. R. (2016). Immersed remotely: Evaluating the use of head mounted devices for remote collaboration in robotic telepresence. In *25th IEEE International Symposium on Robot and Human Interactive Communication (RO-MAN)* (pp. 638–645).

28. Forster, Y., Paradies, S., Bee, N., Bülthoff, H., Kemeny, A., & Pretto, P. (2015). The third dimension: Stereoscopic displaying in a fully immersive driving simulator. In *Proceedings of DSC 2015 Europe Driving Simulation Conference & Exhibition* (pp. 25–32).

29. Boustila, S., Bechmann, D., & Capobianco, A. (2017). Effects of adding visual cues on distance estimation, presence and simulator sickness during virtual visits using wall screen. In *Proceedings of the Computer Graphics International Conference* (pp. 1–6).

30. Livatino, S., Muscato, G., & Privitera, F. (2009). Stereo viewing and virtual reality technologies in mobile robot teleguide. *IEEE Transactions on Robotics, 25*(6), 1343–1355.

31. Casper, J., & Murphy, R. R. (2003). Human-robot interactions during the robot-assisted urban search and rescue response at the world trade center. *IEEE Transactions on Systems, Man, and Cybernetics Part B (Cybernetics), 33*(3), 367–385.

32. Thomas, L. C., & Wickens, C. D. (2000). *Effects of display frames of reference on spatial judgments and change detection,* Unpublished Master Thesis.

33. Witmer, B. G., & Sadowski, W. J. (1998). Nonvisually guided locomotion to a previously viewed target in real and virtual environments. *Human Factors, 40*(3), 478–488.

34. Van Erp, J. B., & Padmos, P. (2003). Image parameters for driving with indirect viewing systems. *Ergonomics, 46*(15), 1471–1499.

35. Johnson, S., Rae, I., Mutlu, B., & Takayama, L. (2015). Can you see me now? How field of view affects collaboration in robotic telepresence. In *Proceedings of the 33rd Annual ACM Conference on Human Factors in Computing Systems* (pp. 2397–2406).

36. Alexandrova, I. V., Teneva, P. T., De La Rosa, S., Kloos, U., Bülthoff, H. H., & Mohler, B. J. (2010). Egocentric distance judgments in a large screen display immersive virtual environment. In *Proceedings of the 7th Symposium on Applied Perception in Graphics and Visualization* (pp. 57–60).

37. Grechkin, T. Y., Nguyen, T. D., Plumert, J. M., Cremer, J. F., & Kearney, J. K. (2010). How does presentation method and measurement protocol affect distance estimation in real and virtual environments? *ACM Transactions on Applied Perception (TAP), 7*(4), 1–18.

38. El Jamiy, F., & Marsh, R. (2019). Distance estimation in virtual reality and augmented reality: A survey. In *IEEE International Conference on Electro Information Technology (EIT)* (pp. 63–68).

39. Chadwick, R. A., Pazuchanics, S. L., & Gillan, D. J. (2006). What the robot's camera tells the operator's brain. In *Human factors of remotely operated vehicles* (pp. 373–384). Emerald Group Publishing Limited.

40. Keyes, B., Casey, R., Yanco, H. A., Maxwell, B. A., & Georgiev, Y. (2006). Camera placement and multi-camera fusion for remote robot operation. In *Proceedings of the IEEE International Workshop on Safety, Security and Rescue Robotics* (pp. 22–24).

41. Cmentowski, S., Krekhov, A., & Krueger, J. (2019). Outstanding: a perspective-switching technique for covering large distances in VR games. In *Extended Abstracts of the 2019 CHI Conference on Human Factors in Computing Systems* (pp. 1–6).

42. Morison, A. M., Woods, D. D., & Murphy, T. (2015). 42 Human-robot interaction as extending human perception to new scales. In The Cambridge Handbook of Applied Perception Research, edited by Robert R. Hoffman, Peter A. Hancock, Mark W. Scerbo, Raja Parasuraman, and James L. Szalma. Cambridge: Cambridge University Press, 848–868.

43. Hughes, S. B., & Lewis, M. (2005). Task-driven camera operations for robotic exploration. *IEEE Transactions on Systems, Man, and Cybernetics-Part A: Systems and Humans, 35*(4), 513–522.

44. Praveena, P., Molina, L., Wang, Y., Senft, E., Mutlu, B., & Gleicher, M. (2022). Understanding control frames in multi-camera robot telemanipulation. In *Proceedings of the 2022 ACM/IEEE International Conference on Human-Robot Interaction* (pp. 432–440).

45. Burke, J. L., Murphy, R. R., Coovert, M. D., & Riddle, D. L. (2004). Moonlight in Miami: Field study of human-robot interaction in the context of an urban search and rescue disaster response training exercise. *Human-Computer Interaction, 19*(1–2), 85–116.

46. Rakita, D., Mutlu, B., & Gleicher, M. (2019). Remote telemanipulation with adapting viewpoints in visually complex environments. In *Robotics: Science and Systems XV*.

47. Naceri, A., Mazzanti, D., Bimbo, J., Tefera, Y. T., Prattichizzo, D., Caldwell, D. G., Mattos, L. S., & Deshpande, N. (2021). The vicarios virtual reality interface for remote robotic teleoperation. *Journal of Intelligent & Robotic Systems, 101*(4), 1–16.

48. Suzuki, R., Karim, A., Xia, T., Hedayati, H., & Marquardt, N. (2022). Augmented reality and robotics: A survey and taxonomy for AR-enhanced human-robot interaction and robotic interfaces. In *CHI Conference on Human Factors in Computing Systems* (pp. 1–33).

Towards Smooth Human–Robot Interaction Using Potential Gradient-Based Sampling

Sascha Kaden, Carl Gaebert, and Ulrike Thomas

Abstract

Successful human–robot interaction calls for fast generation of collision-free and optimized motions. To this end, sampling-based motion planning algorithms have been widely used. However, they often require long planning times to achieve optimized motions. While not being a critical issue in traditional industrial applications, planning time delays or poorly optimized motions have very negative effects on human–robot cooperation. Including artificial potential fields in the sampling algorithm can drastically improve the quality and planning time of such methods. Previous works in this direction are often tailored towards minimizing distance costs such as path length. In this work, we propose a heuristic based on potential fields that can also be used with a variety of state cost functions. We demonstrate the effectiveness of our approach using two cost functions related to human–robot interaction. We achieve drastically improved results in both scenarios. This allows for reducing total planning time and achieving a smoother interaction between human and robot.

Keywords

Motion planning · RRT* · Optimal path planning · Artificial potential fields · Human–robot interaction

S. Kaden (✉) · C. Gaebert · U. Thomas
Robotics and Human-Machine Interaction Lab at Chemnitz, University of Technlology, Chemnitz, Germany
e-mail: sascha.kaden@informatik.tu-chemnitz.de

C. Gaebert
e-mail: carl.gaebert@etit.tu-chemnitz.de

U. Thomas
e-mail: ulrike.thomas@etit.tu-chemnitz.de

1 Sampling-Based Motion Planning in Human–Robot Interaction

Motion planning in a fundamental skill for many collaborative robotics applications. Planning collision-free motions for manipulators, however, is difficult due to their high-dimensional configuration spaces. To address this problem, Rapidly-exploring random trees (RRT) have been introduced [10]. Instead of considering the whole configuration space, the algorithm relies on random collision-free samples. It attempts to grow a tree with collision-free edges to connect start and goal. It is thus capable of finding feasible paths within a reasonable planning time. The latter can be further reduced using the RRT Connect algorithm [9]. It grows an additional tree from the goal to the start while trying to connect with the other tree. The planner's performance, how- ever, largely depends on the environment and the number of joints of the manipulator. Besides possibly long planning times, the algorithm lacks the capability of optimizing the path according to given criteria. In [1], the authors solved this problem by rewiring the tree locally. The resulting paths thus become optimized with respect to the given cost function (typically the robot's joint space). The algorithm then converges asymptotically to the optimal solution. Obtaining a solution close to the optimum, however, can still require long planning times. This approach was later combined in [8] with the bidirectional approach from the RRT-Connect algorithm. Especially in cases where the goal is located in a narrow passage, this method can drastically reduce planning times. To further reduce planning delays and increase the quality of the paths, one can also modify the randomly generated samples. In contrast to them, our sampling-based motion planning algorithm is extended by different costs. These costs can either be state costs or distance costs. State costs are assigned to a single configuration and then usually summed up along the path. Distance costs are calculated by taking the cost differences between states into account. One example is the Euclidean path length of the robot's tool centre point (TCP).

B. Meyer et al. (eds.), *Hybrid Societies*, Advances in Science, Technology & Innovation,
https://doi.org/10.1007/978-3-032-03488-5_29

In [12, 13], the authors proposed a sampling strategy based on Artificial Potential Field (APF). After generating a new random sample in the configuration space, it is moved along an attractive potential. The latter is typically generated by the goal state while the intensity of attraction is determined by the distance to the goal. In doing so, the random samples are drawn to the goal which leads to more optimized motions. This concept can also be included in bidirectional planners [15, 17]. However, they only con- sider distance costs and are thus tailored towards optimizing for the shortest path. In human-robot interaction (HRI) optimization criteria can be more complex and involve a variety of state costs. This stems from the observation that generating a short motion in joint space is not intuitively understandable and can thus harm the acceptance of the robotic co-worker. A summary on this topic is provided in [3].

In this work, we present an approach for adjusting the random samples using a potential field while considering complex state costs. Our goal is to obtain optimized motions for HRI quickly to decrease reaction times and increase acceptance. For this, we consider two cost functions that are of interest when planning manipulator motions for collaborative scenarios. Both were presented in previous works and are briefly summarized below.

1.1 Mobility of the Robot Arm

In collaborative tasks, the robot has to be able to react to unforeseeable events such as a human entering its workspace. To avoid collisions with the human co-worker, the manipulator thus needs to be flexible enough to react quickly. A measure to quantify this was presented by Kaden et al. [6, 7]. The so-called mobility measure consists of two components: manipulability and the robot's distance to its joint limits. The manipulability $\lambda(\boldsymbol{\theta})$ of a configuration $\boldsymbol{\theta}$ is a scalar measure and describes the robot's ability to translate or rotate in its current configuration [18]. It can be calculated from the robot's Jacobian $J(\boldsymbol{\theta}) \in R^{m \times n}$ as shown in (1). The variables $\sigma_1 \sigma_2 \ldots \sigma_m$ hereby denote the singular values of the Jacobian. If one of the degrees of freedom is lost due to a singular configuration, one of the singular values becomes zero. In consequence, $\lambda(\boldsymbol{\theta})$ becomes zero as well. In order to maximize the mobility one can formulate it as a cost as shown in (2).

$$\lambda(\theta) = \sqrt{\det\left(J J^T\right)} = \sigma_1 \sigma_2 \ldots \sigma_m \qquad (1)$$

$$c_\lambda(\boldsymbol{\theta}) = 1 - \lfloor \lambda(\boldsymbol{\theta}) \rfloor \qquad (2)$$

This measure was applied in various applications [11, 14] as well as in sampling-based motion planning (SBMP) [6, 7].

The second component is the distance to the robot's joint angles. While a high manipulability ensures the capability of moving to any direction in Cartesian space, this measure is important for avoiding dynamic obstacles. In [7], the costs are calculated from the summed distances to the joint range centres. In addition, a scaling function is used for increasing the costs exponentially towards the joint limits. Finally, the mobility costs can then be obtained by simply creating a weighted sum of these costs.

1.2 Human-Inspired Motions

Besides optimizing for mobility, one can also aim at mimicking humans by taking characteristics of human arm motions into account. This can be achieved by recording human activities using motion tracking. The recorded motions can then be mapped to a humanoid robot [4, 16]. Alternatively, one can consider modelling natural human arm motions and resolve the redundancy of a robot accordingly [2]. To the same end, we proposed a cost function in our previous work [5]. Among other components, it considers human joint angles and the so-called swivel angle. The human joint limit costs are state costs that consider the joint limits of a human arm. Minimizing them ensures that the robot stays within natural looking postures. The swivel angle cost takes the hand pose and the elbow elevation into account. It utilizes an existing model that was trained on human motion data. The difference between the current elbow elevation and the one given by the model is used as a cost function. Thus, given the robot's hand pose one can resolve the arm's redundancy in a more natural manner. All details can be found in [5].

2 State Cost-Based Bidirectional Potential Gradient

The proposed method is based on the PRRT*-Connect algorithm [15, 17]. This method combines the RRT*-Connect algorithm [8] with the idea of introducing potential fields to SBMP [12, 13]. The resulting approach can thus find optimized paths quickly by growing two trees from the start and the goal simultaneously. In addition, it adapts the random samples themselves using an APF. The latter moves each randomly generated sample into the direction of the goal which leads to increased convergence. However, the problem of getting stuck in minima does not exist, because the samples are randomly generated. In contrast to this approach, we also consider different distance and state costs of and towards a sample. Our goal is thus to not only minimize distance costs but also optimize for additional state costs. The latter can be arbitrarily complex. Moreover, it is not always possible to obtain a gradient for such a cost function. Our method thus guides the randomly generated sample along the attracting potential field of the goal state θ_{goal} as long as the considered state costs do not increase. In the following, the combined approach is described whereas the main novelty of our work is provided in Algorithm 3.

The PRRT*-Connect algorithm (see Algorithm 1) starts by initializing a start and a goal tree. Next, both trees are extended using the *EXTEND* function given in Algorithm 2. Since we utilize our in-house motion planning library, this is done fully in parallel. During extension, a random sample θ_{rand} is generated by uniformly sampling the robot's configuration space. Next, the closest sample of the existing tree $\theta_{nearest}$ is determined and the new sample θ_{new} is obtained by moving a certain step size s from $\theta_{nearest}$ to θ_{new}. A more detailed description of this can be found in [8]. Next, the sample θ_{new} is moved using our bidirectional potential gradient (BPG) heuristic (see Algorithm 3). The heuristic uses θ_{new} and a target sample θ_{target} as an input. The target configuration is hereby the root of the other tree. This means that the samples of the goal tree are drawn towards θ_{start} and samples of the start tree towards θ_{goal}. After calculating the cost for θ_{new} an attractive force towards θ_{target} is calculated from the distance and a weight parameter λ. Next, the sample is moved along this direction up to k times. If the sample remains collision-free and the state cost decreases, the process continues. In case the state cost does not improve, the new sample is returned. In consequence, the randomly generated sample is drawn towards the goal while reducing the state costs. The *Extend* routine in Algorithm 2 then continues by adding the node to the tree and rewiring the tree to optimize the costs. This is the standard procedure of the RRT*-Connect algorithm as described in [8]. After extending both trees in that manner, the algorithm attempts to connect them. This is done by establishing a collision-free connection between the closest points of both trees. Since this is a fundamental building block of the RRT* Connect algorithm, the reader is referred to [8]. If a connection was found, θ_{start} and θ_{goal} are connected by a collision-free path. This initial solution can then be further optimized by generating new samples and rewiring the tree accordingly.

Algorithm 1 PRRT*-Connect

```
1: function RRT*-Connect(θstart, θgoal, knodes)
2:    Tstart.init(θstart)
3:    Tgoal.init(θgoal)
4:    for i ← 1 to attempts do
5:        extendThread(Tstart, knodes, θgoal)
6:        extendThread(Tgoal, knodes, θstart)
7:        join(Tstart, Tgoal)
8:        if connect(Tstart, Tgoal) then
9:            return path(Tstart)
10:   return failure
```

Algorithm 2 Extend

```
1: function extend(T = (V, E), θtarget)
2:    θrand ← getSample()
3:        θnearest ← nearest(T, θrand)
4:        θnew ← steer(θrand, θnearest)
5:        θnew ← BPG(θnew, θtarget)
6:    if inCollision(θnew) then
7:        return false
8:    // Connect to node with minimum cost.
9:    Θnear ← near(T, θnew)
10:   Θnear.insert(ancestors(Θnear, m))
11:   sortByCosts(Θnear)
12:   for each θnear ∈ Θnear do
13:       if trajectoryColFree(θnew, θnear) then
14:           θnew.cost = calcCost(θnew, θnear)
15:           T.addNode(θnew);
16:           T.addEdge(θnear, θnew);
17:           break
18:   if valid connection found then
19:       rewire(θnew, Θnear \ θnear)
20:   return true
```

Algorithm 3 BPG

```
1: function BPG(θrand, θtarget)
2:    θlast ← θrand
3:    curCost ← calcStateCost(θrand)
4:    Fattr = (θtarget − θrand).normalized() ∗ λ
5:    for i ← 1 to k do
6:        θnew ← θlast + Fattr
7:        if isCollisionFree(θnew) and calcStateCost(θnew) < curCost then
8:            curCost ← calcStateCost(θnew)
9:            θlast ← θnew
10:       else
11:           break
12:   return θlast
```

---

3 Evaluation

To evaluate the proposed BPG heuristic, we conduct experiments in two different environments with the KUKA iiwa manipulator (see Fig. 1). For this, we consider mobility costs [6, 7] and human-likeness costs [5]. Each motion planning scenario involves narrow passages in Cartesian space and joint space. In addition, an intermediate goal position is used in both of them. The planning problem is solved 50 times for each environment. For all evaluations, we use a workstation with two Intel Xeon E5-2670 v3 CPUs, each with 12 processor cores and 64 GB Ram. All tests run with 24 threads and are implemented in C++. For the parameters in Algorithm 3 we use $k = 5$ and $\lambda = 0.3$. In Fig. 2, the results for optimizing mobility in scenario A are shown. It can be seen that using the proposed heuristic increases the manipulability of the manipulator (see Fig. 2a). In consequence, the robot stays away from singular configurations and has more capabilities to react to a dynamic environment. In addition, the joint limit costs decrease, keeping the robot away from its joint boundaries and increasing its reaction capabilities (see Fig. 2b). Since the samples are being moved towards the goal

Fig. 1 **a** In scenario A, the robot moves from the narrow passage inside the shelf (1) to the table (2) and then in the shelf again (3). **b** In scenario B, it starts at in the back (1) and moves from the lower shelf position (2) to the upper one (3)

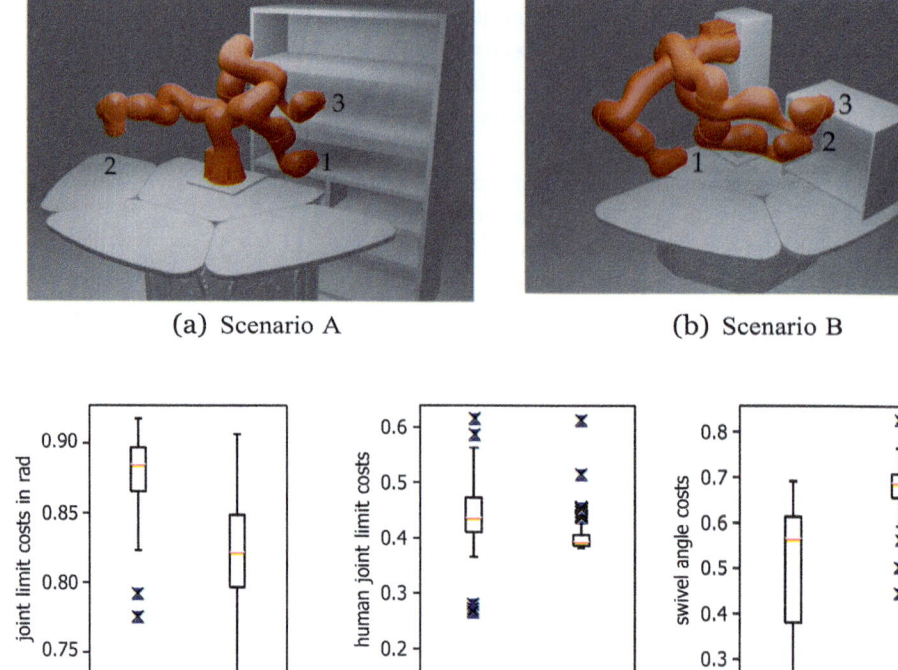

(a) Scenario A   (b) Scenario B

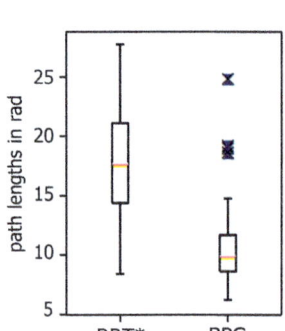

(a) Mean manipulability costs

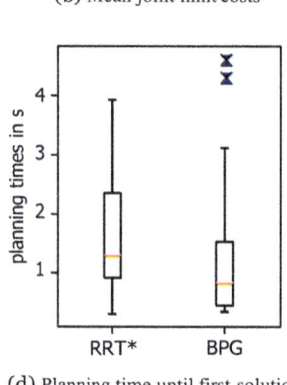

(b) Mean joint limit costs

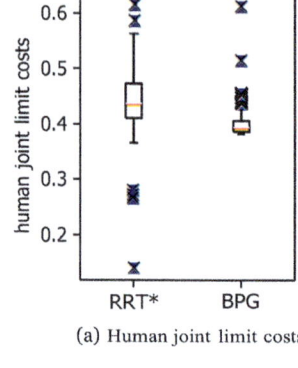

(a) Human joint limit costs

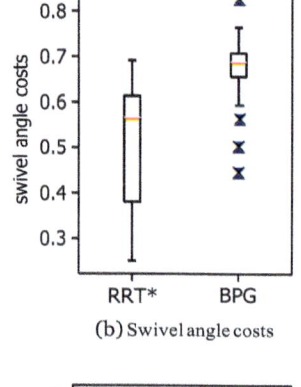

(b) Swivel angle costs

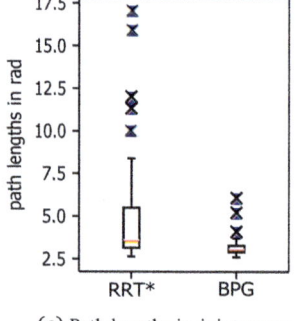

(c) Path lengths in joint space

(d) Planning time until first solution

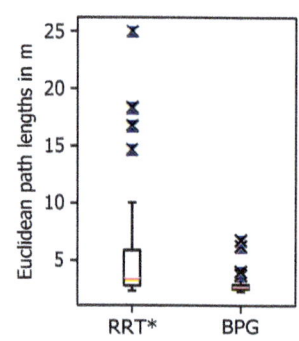

(c) Path lengths in joint space

(d) Path lengths in Cartesian space

Fig. 2 Results for 50 trials of optimizing the mobility in scenario A, with using the BPG heuristic and without it as the default RRT*. **a** Mean manipulability costs per trial **b** mean joint limit costs per trial **c** path lengths in rad **d** planning time in s until first solution found

Fig. 3 Results for 50 trials of optimizing human-likeness costs [5] in scenario B, with using the BPG heuristic and without it as the default RRT*. **a** Mean human joint limit costs per trial **b** mean swivel angle costs per trial **c** path lengths in rad **d** Cartesian path lengths of the end-effector in m

configuration, the path length in joint space decreases as well (see Fig. 2c). A positive side effect of this is that also the planning time to the initial solution decreases (see Fig. 2d).

In the second, scenario we optimize for the human-likeness costs described in [5]. As described earlier, it is a weighted combination of several state and distance costs. Minimizing the total cost can thus lead to an increase in one of the cost components with a lower weight. It can be seen in Fig. 3a that using our heuristic keeps the robot within human-like joint limits. In addition, the Cartesian path lengths (Fig. 3d) and the path length in joint space (Fig. 3c) decrease drastically. On the other hand, the swivel angle costs increase due to its

small influence on the total costs (see Fig. 3b). For both cases, the mean planning time was 0.46 s. This is most likely caused by the fact that calculating the state costs during moving the sample is more expensive than in scenario A.

4 Conclusion

In this work, a sampling method based on potential gradients was introduced. This allows for adapting random samples in SBMP algorithms to minimize state costs while following the attractive potential of the goal configuration. This is of interest for scenarios with optimization criteria such as mobility or human-likeness where the local gradient cannot easily be determined. Using the proposed strategy, one can obtain drastically shorter paths with reduced state costs. This is shown in two simulated experiments for the mobility and human-likeness costs. In the former case, the manipulability cost could be reduced while keeping the robot away from its joint limits. Using the proposed sampling strategy thus increases the robot's capabilities to avoid dynamic obstacles or react to human actions. In case of the human-likeness costs, our method leads to more direct motions while keeping the robot within natural joint limits. Hence, planned motions for humanoid robots appear more goal-directed and unnatural arm configurations can be avoided.

Acknowledgements Funded by the Deutsche Forschungsgemeinschaft (DFG, German Research Foundation)—Project-ID 416228727—SFB 1410

References

1. Akgun, B., & Stilman, M. (2011). Sampling heuristics for optimal motion planning in high dimensions. In *IEEE/RSJ International Conference on Intelligent Robots and Systems* (pp. 2640–2645). IEEE.
2. Asfour, T., & Dillmann, R. (2003). Human-like motion of a humanoid robot arm based on a closed-form solution of the inverse kinematics problem. In *Proceedings 2003 IEEE/RSJ International Conference on Intelligent Robots and Systems (IROS 2003) (Cat. No.03CH37453)*, (Vol. 2, pp. 1407–1412).
3. Chakraborti, T., Kulkarni, A., Sreedharan, S., Smith, D. E., & Kambhampati, S. (2019). Explicability? legibility? predictability? transparency? privacy? security? The emerging landscape of interpretable agent behavior. In *Proceedings of the International Conference on Automated Planning and Scheduling* (Vol. 29, pp. 86–96).
4. Gaertner, S., Do, M., Asfour, T., Dillmann, R., Simonidis, C., & Seemann, W. (2010). Generation of human-like motion for humanoid robots based on marker-based motion capture data. In *ISR 2010 (41st International Symposium on Robotics) and ROBOTIK 2010 (6th German Conference on Robotics)* (pp. 1–8).
5. G¨abert, C., Kaden, S., & Thomas, U. (2021). Generation of human-like arm motions using sampling-based motion planning. In *IEEE/RSJ International Conference on Intelligent Robots and Systems (IROS)* (pp. 2534–2541). IEEE.
6. Kaden, S., & Thomas, U. (2019). Maximizing robot manipulability along paths in collision-free motion planning. In *19th International Conference on Advanced Robotics (ICAR)* (pp. 105–110). IEEE.
7. Kaden, S., & Thomas, U. (2021). Optimizing mobility of robotic arms in collision-free motion planning. *Journal of Intelligent and Robotic Systems, 102*, 1–15.
8. Klemm, S., Oberla¨nder, J., Hermann, A., Roennau, A., Schamm, T., Zollner, J. M., & Dillmann, R. (2015). RRT*-Connect: Faster, asymptotically optimal motion planning. In *IEEE International Conference on Robotics and Biomimetics (ROBIO)* (pp. 1670–1677). IEEE.
9. Kuffner, J., & LaValle, S. (2000). RRT-connect: An efficient approach to single-query path planning. In *Proceedings 2000 ICRA Millennium Conference. IEEE International Conference on Robotics and Automation, Symposia Proceedings (Cat. No.00CH37065)* (Vol. 2, pp. 995–1001).
10. LaValle, S. M. (1998). Rapidly-exploring random trees: A new tool for path planning. Research report 9811.
11. Pardi, T., Ortenzi, V., Fairbairn, C., Pipe, T., Esfahani, A. M. G., & Stolkin, R. (2020). Planning maximum-manipulability cutting paths. *IEEE Robotics and Automation Letters, 5*, 1999–2006.
12. Qureshi, A. H., & Ayaz, Y. (2016). Potential functions based sampling heuristic for optimal path planning. *Autonomous Robots, 40*, 1079–1093.
13. Qureshi, A. H., Iqbal, K. F., Qamar, S. M., Islam, F., Ayaz, Y., & Muhammad, N. (2013). Potential guided directional-RRT* for accelerated motion planning in cluttered environments. In*IEEE International Conference on Mechatronics and Automation (ICMA)* (pp. 519–524). IEEE.
14. Su, H., Li, S., Manivannan, J., Bascetta, L., Ferrigno, G., & De Momi, E. (2019). Manipulability optimization control of a serial redundant robot for robot-assisted minimally invasive surgery. In *International Conference on Robotics and Automation (ICRA)* (pp. 1323–1328). IEEE.
15. Tahir, Z., Qureshi, A. H., Ayaz, Y., & Nawaz, R. (2018). Potentially guided bidirectionalized RRT* for fast optimal path planning in cluttered environments. *Robotics and Autonomous Systems, 108*, 13–27.
16. Terlemez, O¨., Ulbrich, S., Mandery, C., Do, M., Vahrenkamp, N., & Asfour, T. (2014). Master Motor Map (MMM)—framework and toolkit for capturing, representing, and reproducing human motion on humanoid robots. In *IEEE-RAS International Conference on Humanoid Robots* (pp. 894–901).
17. Xinyu, W., Xiaojuan, L., Yong, G., Jiadong, S., & Rui, W. (2019). Bidirectional potential guided RRT* for motion planning. *IEEE Access, 7*, 95046–95057.
18. Yoshikawa, T. (1985). Manipulability of robotic mechanisms. *The International Journal of Robotics Research, 4*, 3–9.

Human–Machine Teaming Agents: A Future Perspective

Michael Teichmann, Marco Ragni, Julien Vitay, Martin Gaedke, and Fred Hamker

Abstract

The rise of artificial agents that excel in complex tasks makes it feasible for humans and machines to form efficient teams. However, the design and requirements of the necessary software agent are still unclear. We identify important factors for human–machine teaming and characterize teaming situations by the presence of shared (sub-) goals, communication, interdependence, and the ability to learn and adapt. Given this, we introduce a frame-work for software teaming agents, which utilizes state-of-the-art deep neural networks and addresses how communication and conceptual information can be incorporated into such a design. Moreover, we suggest information-seeking behavior, based on uncertainty assessment, to deal with the variability of the environment and the agent's imperfectness. Finally, we address some inter-disciplinary research directions in human–machine teaming which arise from our conception.

Keywords

Human-machine teaming · Deep neural networks · Multi-task learning · Multi-channel networks · Information-seeking · Abstraction

M. Teichmann (✉) · M. Ragni · J. Vitay · M. Gaedke · F. Hamker
Chemnitz University of Technology, Chemnitz, Germany
e-mail: michael.teichmann@informatik.tu-chemnitz.de

M. Ragni
e-mail: marco.ragni@hsw.tu-chemnitz.de

J. Vitay
e-mail: julien.vitay@informatik.tu-chemnitz.de

M. Gaedke
e-mail: martin.gaedke@informatik.tu-chemnitz.de

F. Hamker
e-mail: fred.hamker@informatik.tu-chemnitz.de

1 Introduction

We here propose a modern framework for future human–machine teaming agents from the viewpoint of modern AI methods, particularly deep neural networks (DNNs). We highlight those tasks they excel at and their current limitations. In our framework, we aim to address challenges to be solved to allow teaming agents to interact with each other at a level of complexity beyond mere joint action. Further, we introduce methods to facilitate the exchange of conceptual information between humans and neural networks. Finally, we discuss actual hurdles and research directions for the realization of such systems.

1.1 Deep Neural Networks as a Method to Empower Teaming Agents on a Cognitive Level

AI systems that can deal with complex problems such as autonomously driving cars, finding hypotheses in big data, beating world champions in chess and Go, and proving mathematical theorems. Due to this success, the time has come to ask if such artificial systems could also team with humans on a cognitive level. While most systems excel in one specific area, researchers wondered if they can develop systems that demonstrate a form of Artificial General Intelligence (AGI [10]). Concepts of human intelligence research (identified as the so-called factor "g" [29]) that emphasize certain cognitive abilities or mental skills can be generalized to all forms of cognitive agents, including artificial ones. Research in AI [24] has developed a classification of agents from so-called reflexive agents to learning agents, which refers to the cognitive abilities behind their autonomous behavior. However, the minimal required abilities for teaming agents are less clear. Present systems likely lack the cognitive capacity for mutual understanding, i.e., the full understanding of the present situation including the context, the

B. Meyer et al. (eds.), *Hybrid Societies*, Advances in Science, Technology & Innovation,
https://doi.org/10.1007/978-3-032-03488-5_30

recognition of potential problems, and the proposition of solutions.

Artificial neural networks, particularly when combined with deep learning methods, have become the most powerful and ubiquitous tool in artificial intelligence and machine learning within a decade. After the initial breakthroughs in computer vision [17], they spread to reinforcement learning [28], natural language processing [2] and robotics [18]. Deep learning methods became the go-to method for feature extraction and generative modeling on almost any modality (image, video, sound, text, graphs), allowing a rapid increase in the applicability of AI methods in the industry. Recent progress in self-supervised learning, through the introduction of the Transformer architecture [32] and contrastive learning [13], further removed the necessity of expensively labeled data.

However, deep learning methods are mostly limited to single domains and applications: their end-to-end training procedure, where a single loss function is minimized by the neural network, the well-documented catastrophic forgetting property of neural networks—the impossibility of learning two tasks one after another—and the inductive bias required by various modalities (convolutional layers, bottlenecks) limit deep neural networks to a single task on a single dataset. Recent attempts at multi-modal, multi-task [6, 36], multi-embodiment generalist agents such as Deepmind's GATO [23] are able to learn an impressive amount of tasks, but require the simultaneous learning of all tasks and can only extrapolate to closely related tasks. This is the opposite of lifelong/continuous learning, where tasks can be learned one after another and even build upon each other, a hallmark of human intelligence.

Another major drawback of deep learning approaches to cognition and behavior is their acceptance and trustworthiness: neural networks are black boxes that often make correct predictions but are hard to interpret and are not able to explain why they arrived at that conclusion. In the context of teaming, this may limit the trust of the human in the system and impair their cooperation. Neural networks are statistical learners, and they sometimes fail inexplicably in situations that seem obvious to humans, as they learn solely from data and do not have any common sense [19]. Moreover, they are prone to adversarial perturbations [11] that allow attackers to purposefully fool them. Many methods have been proposed to perform posthoc analyses of decisions (e.g., layer-wise relevance propagation [3]), but the field of explainable AI (XAI), where the system should be able to explain its decisions while making them (or running them through their human partners before acting), is still in its infancy.

The goal is clear: Artificial systems need to be empowered when they should become human companions [4] and teaming partners on a cognitive level.

1.2 What Can Constitute Teaming with an Artificial Agent?

As human beings, we have all experienced some form of teaming in activities such as sports or research. Whenever there are shared goals with a partner, some complementary capabilities, and shared resources (such as time or space), we may form a team, intentionally or unintentionally, with another human in order to overcome the current situation and reach the goal together. What makes it possible is that humans—in contrast to systems—are equipped with similar physical and cognitive capabilities. Humans can infer, by using a "theory of mind" [7], what the other can or cannot do, what they can learn, what they may or may not know, and so forth. The question is, however, if and when we can call a joint action with an artificial system "teaming" in the same sense as we would call it by observing some specific joint actions among humans. How can we classify in general common actions that several agents perform as a team?

Several factors have been identified that we will briefly introduce based on common grounds in the literature [14, 15]: In contrast to joint action, teaming requires first to share (sub-)goals. A goal is a specific state that both agents want to reach by the means of their actions. This in turn requires a shared "representation of one's own and a partner's action" [8]. The second prerequisite is some form of explicit or implicit communication to synchronize the goals and coordinate actions. Hence, communication is necessary to share subgoals, acquire knowledge/learn, coordinate actions (including changing plans), and share mental models for a first skeletal plan (see below) to reach the shared subgoals. Note that there are different requirements to communicate between humans and systems compared to machine-machine communication relevant in pure artificial multi-agent systems. This is due to the different nature of the neuro-cognitive system in humans and artificial agents by this date. So there needs to be some common base for both that we will propose and outline in the next section.

A third key aspect is learning or the adaptability of teaming partners. This can be realized by the active acquisition of knowledge (which we call information seeking below) and even understanding rewards. This is especially relevant in the fourth factor—the interdependence, i.e., that two agents do depend, either for reasons of efficiency or necessity, on the other agent to reach a goal. For instance, a system can perform complex calculations much quicker than humans, but humans can direct the computations. Interdependence is "the necessary junctures in task-oriented communication and actions among people and machines that make the joint activity in which they are engaged productive" [15]. In the interaction between team players, a so-called shared skeletal plan [14] is formed that in turn depends on the shared goals above and is a form to reach the goal. This capability to design and execute a plan is critical for any teaming interaction together with "observability," "predictability," and "directability" as the three cornerstones that "facilitate" teamwork.

2 Future Teaming System

2.1 Conception of the Framework for Teaming Agents

Since no universal AI system is on the horizon, we have to focus our conception on existing technologies which can be enhanced with modern concepts from machine learning or neuroscience in terms of network design and learning concepts, as well as from social science in terms of concepts for information representation and communication, and novel ways to represent human cognition. Indisputably, an AI agent has to cover three main abilities: sensing, cognition, and acting. Modern AI systems, in terms of end-to-end deep neural networks, excel in many disciplines on specific tasks. However, this comes with an enormous effort in terms of data acquisition and training. Thus, for a composed system, the effort for training the agent has to be reduced and its specialization has to be widened from single tasks to operation domains. A viable path is the use of pre-trained components, which provide a convincing set of domain-relevant abilities. Since a single agent will not cover all possible types of use, its software components should be exchangeable with components suitable for other domains, or suitable for other physical implementations of the agent.

However, a classical modular conception is not sought, where each ability is computed in a dedicated software component. Because this often comes with a quasi-symbolic representation within the single components, which is known to make the processing of complex problems difficult or impossible. Therefore, we argue to aim for a modern end-to-end trainable deep neural network architecture (Fig. 1), where internal representations are not reduced to a quasi-symbolic level. This in turn comes with the difficulty to access the internal representations of such a system, which would be high-dimensional and distributed across the entire network. Such a system cannot easily be analyzed by observing its internal state from the outside. Rather, the system has to be established with communication channels to convey its internal processing in some form of language that can be understood by humans. Hence, we propose to modularize the system in a way that allows the transmission of any required information and at the same time has areas where specific information is represented. The separation into different information channels will ease the access of a specific communication component to its required information and will allow it to provide information on a more abstract—conceptual—level to the system.

2.2 Deep Neural Network Module for Representing Concepts

Particularly for deep neural networks (Fig. 1, green), it is still unclear how concepts can be incorporated and represented within a network. We propose the idea of specialized channels within a deep neural network [9]. The idea is that within these channels high-level representations develop, conveying information from a particular category, such as objects which have been detected and where they have been detected. These channels are intended to be a result of a well-chosen modularized network architecture, which receives its information from multiple sources. Indeed, these channels are just in part a result of the network architecture, they also result from a clever utilization of the network objective function, which is used to enable the learning of suitable representations.

There are different concepts present to implement such an architecture. The first concept is multi-headed neural networks, where multiple heads, i.e., output layers, are used to solve the same or different tasks (Fig. 2). As a consequence, each network segment, having its own output heads, is tuned to solve the related, so-called auxiliary tasks. The use of multiple auxiliary tasks (multi-task learning) allows one to control which particular information should be encoded within a particular segment of the network (channel) by a clever choice of its particular objective. Objectives can be certain questions, such as: which persons are within a scene (face recognition), which postures do they have (posture recognition), where are the tools required for the current task (object detection, with task-specific visual attention, spatial memory), what will be the motor sequence in the next 10 s (planning, motor control, predictions). Another way to foster such specific representations is capsule networks [25]. The theoretical idea of these networks is to route specific information across the hierarchy of a neural network. Therefore, the layers are divided into small groups of neurons, the capsules, and a specialized logic decides, based on the activity of the capsules, where to route its information. As a consequence, such networks develop specific compartments for different properties, for instance, of a visual scene. It is thought that higher-level representations of the resulting network compartments can be more easily linked to abstract representations such as concepts or human-friendly symbolic encoding.

2.3 Concepts Module and Memory of Context

Another important aspect of the network architecture is to advance from reflexive agents to agents with task-spanning memory. While a form of working memory is present in

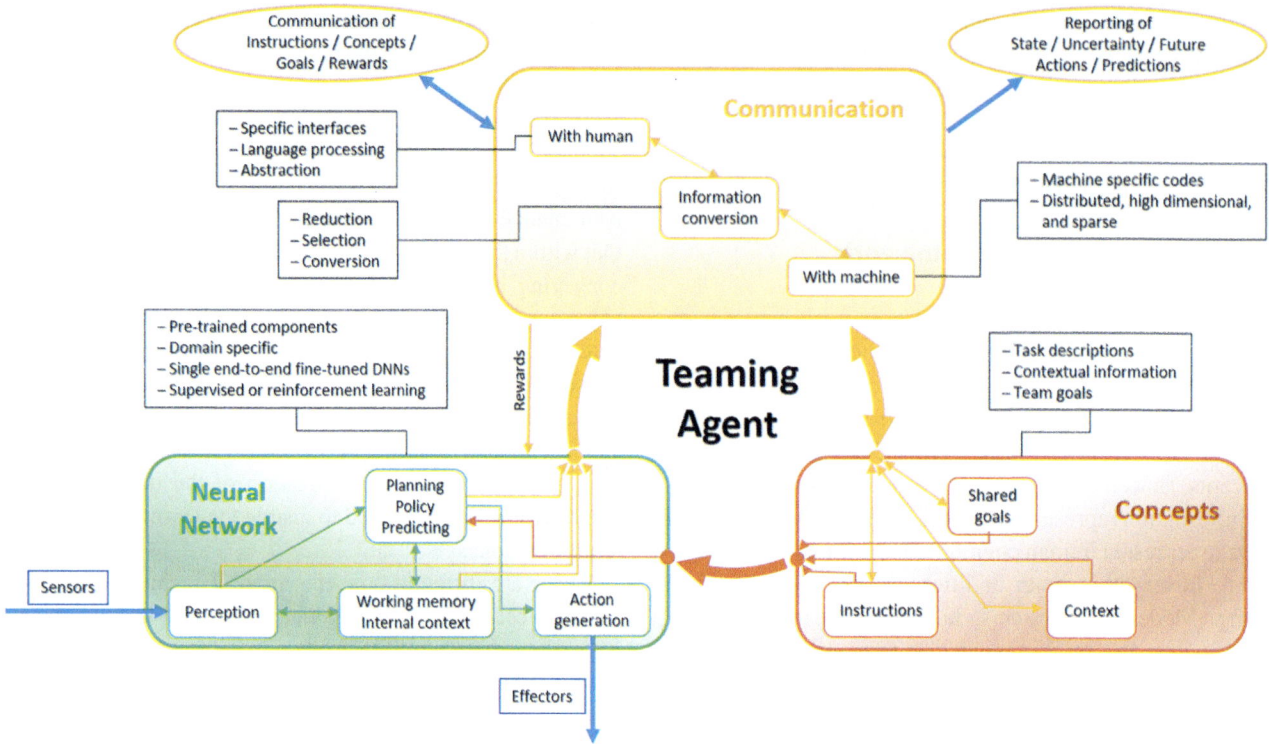

Fig. 1 Framework for a modularized learning agent. The agent is composed of four main structures derived from a cognitive agent: (blue) are the required sensors and effectors to sense and behave in the environment. (green) is a modularized deep neural network providing cognition. (orange) is an abstraction layer, called concepts, containing bottleneck features provided as input to the neural network. (yellow) is the communication interface which converts network states into a human-readable form and converts received additional information into suitable machine representations

modern concepts of recurrent neural networks (e.g., LSTM), which decide by their logic what to store and to forget, this kind of memory is always related to the requirements of a particular task. Consequently, these networks are unable to incorporate long-term knowledge from the previous tasks or other earlier perceived contextual information into their decisions, thus they cannot adapt to the situation and "learn" from previous experiences. Contextual information, such as knowledge of its current teammate or the correctness of the last execution of the task, can be beneficial. This contextual information can also be conceptual knowledge on the task to be performed or (sub-)goals to reach and enable online adaptation by the agent based on previous outcomes or externally communicated feedback. Such information is stored in the concepts component (Fig. 1, orange).

Concepts of tasks or subtasks include task-related instructions, i.e., what should be done and when, which cover fundamental descriptions of new workflows. They are key in so far as new tasks and work steps do not first have to be trained on the basis of a large number of examples. Their execution should be enabled by the internal mapping of their abstract descriptions to known working primitives. Therefore, the agent also requires the ability to receive new concepts.

2.4　Communication Module

To communicate task-related information or inform about the agent's internal processing a communication component is required. This component provides to the agent's decision logic additional external signals in terms of the abovementioned concepts, comprising context, shared goals, and instructions.

One of the two main purposes of this component (Fig. 1, yellow) is to convert communicated information from human teammates (or other agents) into suitable machine representations, which in turn provide an input source for the related channels of the neural network. This information might be a natural language, where it is a challenge on its own to recognize its intended meaning, or it comes from specific interfaces. In both cases, information will be on an abstract level of encoding processable by humans, even if programmed interfaces are used. So the communication component has to be developed to transform such representations into suitable representations to serve as machine input, these are highly distributed neural codes fulfilling certain properties in terms of sparseness or range. Such conversion can result from an end-to-end learning process but is today mainly in the hands of data scientists who have a common knowledge of how to preprocess and shape data

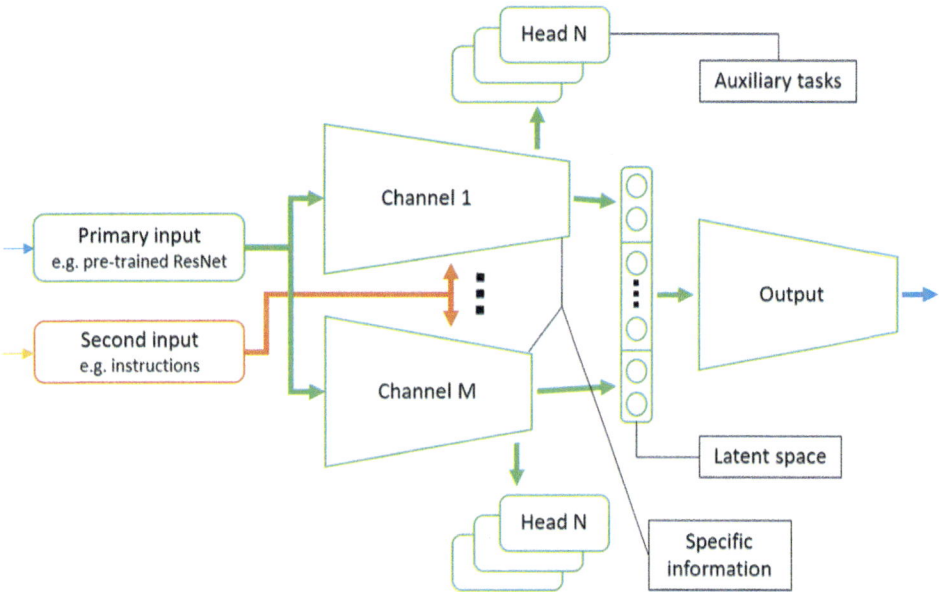

Fig. 2 Multi-task learning within a multi-channel network with multiple input sources and multiple output heads. A head refers to an individual output layer utilized to solve an auxiliary task and a channel is a multi-layer neural network on its own. To foster the development of specific information representations, the processing of the input information is divided into different channels, where each has to solve (multiple) specific auxiliary tasks. The bottleneck features of each channel can be used to access the agent's internal representations. Secondary information sources can be incorporated in deeper processing stages within the channels to allow the utilization of rather abstract information. Finally, all channels contribute to a latent representation that is used to solve the target tasks

to serve as neural network input. Importantly, the communication component has to interpret abstract descriptions and fill in lacking or poorly detailed information with its own knowledge. Thus, it must provide a machine-related form of common ground.

The second purpose is the conversion of the agent's internal representations into human-readable and understandable information. Therefore, an opposed form of conversion has to take place. Complex distributed machine codes have to be reduced, by finding a good level of detail, and the information to communicate has to be selected with a sense of its relevance by internal attentional operations. From the current state of research, it remains unclear how this can be implemented or learned. One of the possible ways would be natural language processing, which has made a lot of progress in recent years. Models for speech recognition and language translation have shown remarkable results [35], but also approaches in fields like automatic image captioning [33], transferring the complexity of a visual scene into a text description, or the automatic generation of abstract programming language codes are available [31].

Once developed, such a conversion can also be provided by a pre-trained module, where the pre-training was performed on tasks that cover the domain well enough. There is, of course, a need for specialized interfaces, which have to be designed task- or domain-specifically, but also for universal components, such as symbolic vectors, natural language, and visualizations of future actions.

2.5 Information-Seeking Behavior

Teaming with a human will lead in several occasions to unpredictable situations for the agent, as it will be difficult to embed the agent with a full theory of mind of its human partner. In situations where the agent is able to fully predict the dynamics of its environment, including the behavior and intentions of its partner, it can rely on its learned behavior and react to its sensory inputs in a straightforward and reactive manner. When the world is uncertain, the agent should rather spend more time looking for information rather than using its present knowledge which may lead to potentially severe errors. This requires the agent to anticipate its performance given its present input by means of internal monitoring of the likelihood of outcomes. As a result, the agent may start interacting with its partner in order to find out his/her intentions or what the current task is about. These two modes of behavior, automatic behavior, and information-seeking behavior have been thoroughly investigated in animals and humans [5] and relate to the well-known dilemma in reinforcement learning (RL): exploration vs. exploitation. When the environment is fully predictable, the RL agent can become greedy and simply maximize the expected return of its actions, but when it is still uncertain about the outcome of its actions, it has to explore its environment and try out different actions that may lead to bigger rewards. Earlier RL algorithms relied on stochastic exploration, either by randomly selecting alternative actions or adding some noise to the preferred action [30]. Such a blind exploration is, however, very inefficient, as the agent might spend a lot of time exploring useless or even dangerous actions, adding to the sample complexity of RL algorithms and making them hard to use in complex online scenarios.

Modern RL approaches try to overcome this issue by developing more focused and task-relevant exploration mechanisms. Maximum Entropy RL [12, 37] embeds the agent with an incentive to behave as randomly as possible (high entropy) while still collecting rewards in the long term. This leads to quite robust policies able to quickly adapt when the environment changes, for example when a preferred path becomes inaccessible: a greedy agent would need to re-learn everything in order to find a new path, while a maximum entropy agent would have implicitly learned all possible paths and could quickly find an alternative solution. Another very important approach to this problem is the use of intrinsic motivation or curiosity to guide exploration [21]. In this paradigm, actions are not only interesting when they lead to an external reward, but also when they lead to a novel situation: the unexpected consequences of an action are an incentive for the agent to further explore this direction, as it is a region of the state space with high uncertainty. Intrinsic curiosity (IC) mechanisms allow RL agents to intelligently explore their environment, spending less time in well-known regions and focusing on uncertain and potentially promising regions [22]. IC modules require the agent to concurrently learn an internal model of the world predicting the sensory consequences of an action, allowing the agent to estimate the novelty and uncertainty of an outcome in order to guide its exploration behavior.

From the recent progress in deep learning-based RL, it becomes clear that the teaming agent should be equipped with an internal predictive model of the world, including the human partner (theory of mind), allowing the agent to switch its behavior from a reflexive mode to an information-seeking mode based on its own uncertainty. Generative models such as variational autoencoders [16] are powerful neural architectures able to learn world models in complex sensory spaces and are currently central to the best performing model-based RL architectures [26, 34].

3 Discussion

We proposed a framework for a modularized deep neural network architecture for future cognitive teaming agents, which connects a primary decision logic with two key components for teaming: communication and concepts. The communication component serves as an interface with the team partners and provides additional conceptual information to the agent as context, instructions, or (sub-)goals. We subsumed this information under the term concepts. This additional information is required to enable situationally adapted behavior and implement the interdependence

between agent and team. It serves as a second input source to different channels of the deep neural network. These channels are designed to represent specific information on important aspects of the internal processing of the agent to facilitate information transmission on an abstract level. This can be achieved by utilizing auxiliary tasks, which ensure that the "right" information is represented by the channels. High-level representations of the different channels, such as bottleneck features, serve as output to the communication module and provide a machine-related form of abstraction, a machine-specific vocabulary. With this available, the agent is enabled to communicate its internal state, predictions, or planned actions, and by receiving conceptual information on this level of abstraction it is enabled to execute multiple tasks by simply receiving the goals or required working steps. Since the agent comes with a domain-specific set of abilities it requires just this abstract description of how to utilize them.

Although there are many challenges and limitations which require future research attempts, several questions arise from our framework. It has to be investigated how the information to communicate can be converted and what representation schemes are appropriate for humans, respectively, machines. This requires technologies to process complex data, such as human language and bottleneck features on the machine site. Related conversion tasks can be found in natural language processing or automatic description generation. Research is also needed into what is important information, for example in a production context, and how is this information perceived. Since the communicated information is often underspecified, a theoretical framework of domain-specific common ground is required. This common ground has to respect the requirements of machines and may therefore differ in the degree of description from human requirements. This is because humans and machines build upon different working primitives, both for their behavior and for their knowledge. But the machine representation itself also has to be investigated and a common language has to be found that allows serving as input to neural networks to advance from narrow special-purpose solutions to domain-covering applications. This language does not only take into account data encoding requirements, but it also has to provide a domain-specific vocabulary that can encode all work and situation-relevant information at a level of abstraction that generalizes beyond individual tasks and specific situations. It is hard to imagine that this can be easily found without experts from the application domain. Since there will be no perfect machine, solutions have to be found to empower the agent to know when it has to initiate communication, for instance, to obtain further instructions or ask for context information. One promising field of research is human-inspired information-seeking behavior. This can be achieved by an assessment of the uncertainty of his actions, either by internal estimations or by the comparison between predictions and outcomes. The internal behavior model can be updated with a few trials and updates to its conceptual knowledge can be triggered, causing adaptation to the context or proactive behavior, such as asking the teammate for further information. Of course, the implementation and training of such an agent come with multiple difficulties. Nowadays, AI agents are tailored for specific tasks. When the complexity of the task advances, the effort to develop and train such an agent also increases. To reduce this effort, the reuse of pre-trained modules (e.g., [27]) and their utilization for new tasks via transfer learning is state-of-the-art. On an advanced level, pre-trained modules might utilize again other pre-trained modules, a common example is the reuse of convolutional neural networks for feature extraction in visual scenes to build upon these universal features with an advanced decision logic [1]. However, combining different neural network modules into one larger network and fine-tuning it for the targeted tasks comes with the issue that each module have had different hyperparameters and should now be trained with another set of hyperparameters. How to train such assembly networks is a question within the deep learning community (e.g., [20]), but becomes critical for complex assemblies. These challenges in computer science and the humanities highlight the need for interdisciplinary research to make such a system feasible and to lift AI from a special-purpose solution to one that covers domains.

Acknowledgements This research has been supported by the German Research Foundation (DFG, 416228727)—SFB 1410 Hybrid Societies and by the Saxony State Ministry of Science and Art (SMWK3-7304/35/3-2021/4819) research initiative "Instant Teaming between Humans and Production Systems" on the basis of the budget passed by the deputies of the Saxony parliament.

References

1. Abramson, J., Ahuja, A., Brussee, A., Carnevale, F., Cassin, M., Fischer, F., Georgiev, P., Goldin, A., Gupta, M., Harley, T., Hill, F., Humphreys, P. C., Hung, A., Landon, J., Lillicrap, T., Merzic, H., Muldal, A., Santoro, A., Scully, G., von Glehn, T., Wayne, G., Wong, N., Yan, C., & Zhu, R. (2022). Creating multimodal interactive agents with imitation and self-supervised learning. Preprint retrieved from arXiv:2112.03763

2. Bahdanau, D., Cho, K., & Bengio, Y. (2016). Neural machine translation by jointly learning to align and translate. Preprint retrieved from arXiv:1409.0473

3. Binder, A., Montavon, G., Lapuschkin, S., Mu¨ller, K. R., & Samek, W. (2016). Layer-wise relevance propagation for neural networks with local renormalization layers. In A. E. Villa, P. Masulli, & A. J. Pons Rivero (Eds.) *Artificial neural networks and machine learning—ICANN 2016* (pp. 63–71). Springer International Publishing.

4. Biundo, S., & Wendemuth, A. (Eds.). (2017). Companion technology. In *Cognitive Technologies*. Springer International Publishing.

5. Bromberg-Martin, E. S., & Monosov, I. E. (2020). Neural circuitry of information seeking. *Current Opinion in Behavioral Sciences, 35,* 62–70.

6. Caruana, R. (1997). Multitask learning. *Machine Learning, 28*(1), 41–75.

7. Devine, R. T. (2016). Theory of mind. In V. Zeigler-Hill, & T. K. Shackelford (Eds.) *Encyclopedia of personality and individual differences* (pp. 1–9). Springer International Publishing, Cham.

8. Farley, J., & Costanza, R. (2002). Envisioning shared goals for humanity: A detailed, shared vision of a sustainable and desirable USA in 2100. *Ecological Economics, 43*(2), 245–259.

9. Gao, J., Li, P., Chen, Z., & Zhang, J. (2020). A survey on deep learning for multimodal data fusion. *Neural Computation, 32*(5), 829–864.

10. Goertzel, B., Pennachin, C., Gabbay, D. M., Siekmann, J., Bundy, A., Carbonell, J. G., Pinkal, M., Uszkoreit, H., Veloso, M., Wahlster, W., & Wooldridge, M. J. (Eds.). (2007). *Artificial general intelligence.* Cognitive Technologies. Springer.

11. Goodfellow, I. J., Shlens, J., & Szegedy, C. (2015). Explaining and harnessing adversarial examples. Preprint retrieved from arXiv: 1412.6572

12. Haarnoja, T., Tang, H., Abbeel, P., & Levine, S. (2017). Reinforcement learning with deep energy-based policies. In *Proceedings of the 34th International Conference on Machine Learning—ICML'17* (Vol. 70, pp. 1352–1361).

13. He, K., Fan, H., Wu, Y., Xie, S., & Girshick, R. (2020). Momentum contrast for unsupervised visual representation learning. In *2020 IEEE/CVF Conference on Computer Vision and Pattern Recognition (CVPR)* (pp. 9726–9735).

14. Johnson, M., & Bradshaw, J. M. (2021). How interdependence explains the world of teamwork. Lecture Notes in Computer ScienceIn W. F. Lawless, J. Llinas, D. A. Sofge, & R. Mittu (Eds.), *Engineering artificially intelligent systems: A systems engineering approach to realizing synergistic capabilities* (pp. 122–146). Springer International Publishing.

15. Johnson, M., Bradshaw, J. M., Feltovich, P. J., Jonker, C. M., van Riemsdijk, M. B., & Sierhuis, M. (2014). Coactive design: Designing support for interdependence in joint activity. *Journal of Human-Robot Interaction, 3*(1), 43–69.

16. Kingma, D. P., & Welling, M. (2014). Auto-encoding variational Bayes. In *2nd International Conference on Learning Representations, ICLR 2014, Banff, AB, Canada, Conference Track Proceedings,* 14–16 April 2014.

17. Krizhevsky, A., Sutskever, I., & Hinton, G. E. (2012). ImageNet classification with deep convolutional neural networks. In *Advances in neural information processing systems (NIPS).*

18. Levine, S., Finn, C., Darrell, T., & Abbeel, P. (2016). End-to-end training of deep visuomotor policies. *Journal of Machine Learning Research, 17*(39), 1–40.

19. Marcus, G. (2018). Deep learning: A critical appraisal. Preprint retrieved from arXiv:1801.00631

20. Mormont, R., Geurts, P., & Mar´ee, R. (2020). Multi-task pretraining of deep neural networks for digital pathology. *IEEE Journal of Biomedical and Health Informatics, 25*(2), 412–421.

21. Oudeyer, P. Y., Gottlieb, J., & Lopes, M. (2016). Chapter 11—intrinsic motivation, curiosity, and learning: Theory and applications in educational technologies. In B. Studer & S. Knecht (Eds.), *Progress in brain research* (Vol. 229 of Motivation, pp. 257–284). Elsevier.

22. Pathak, D., Agrawal, P., Efros, A. A., & Darrell, T. (2017). Curiosity-driven exploration by self-supervised prediction. In *Proceedings of the 34th International Conference on Machine Learning* (Vol. 70, ICML'17, pp. 2778–2787).

23. Reed, S., Zolna, K., Parisotto, E., Colmenarejo, S. G., Novikov, A., Barth-Maron, G., Gimenez, M., Sulsky, Y., Kay, J., Springenberg, J. T., Eccles, T., Bruce, J., Razavi, A., Edwards, A., Heess, N., Chen,

Y., Hadsell, R., Vinyals, O., Bordbar, M., & de Freitas, N. (2022). A generalist agent. Preprint retrieved from arXiv:2205.06175

24. Russell, S., & Norvig, P. (2021). *Artificial intelligence* (Global Edition). Pearson Deutschland.

25. Sabour, S., Frosst, N., & Hinton, G. (2017). Dynamic routing between capsules. In *Advances in neural information processing systems* (pp. 3856–3866).

26. Schrittwieser, J., Antonoglou, I., Hubert, T., Simonyan, K., Sifre, L., Schmitt, S., Guez, A., Lockhart, E., Hassabis, D., Graepel, T., Lillicrap, T. P., & Silver, D. (2020). Mastering atari, go, chess and shogi by planning with a learned model. *Nature, 588*(7839), 604–609.

27. Schro¨der, E., Braun, S., M¨ahlisch, M., Vitay, J., & Hamker, F. (2019). Feature map transformation for multi-sensor fusion in object detection networks for autonomous driving. In K. Arai, & S. Kapoor (Eds.), *Advances in computer vision.* Advances in Intelligent Systems and Computing (pp. 118–131). Springer International Publishing.

28. Silver, D., Huang, A., Maddison, C. J., Guez, A., Sifre, L., van den Driessche, G., Schrittwieser, J., Antonoglou, I., Panneershelvam, V., Lanc- tot, M., Dieleman, S., Grewe, D., Nham, J., Kalchbrenner, N., Sutskever, I., Lillicrap, T., Leach, M., Kavukcuoglu, K., Graepel, T., & Hassabis, D. (2016). Mastering the game of go with deep neural networks and tree search. *Nature, 529*(7587), 484–489.

29. Sternberg, R. J., Conway, B. E., Ketron, J. L., & Bernstein, M. (1981). People's conceptions of intelligence. *Journal of Personality and Social Psychology, 41*, 37–55.

30. Sutton, R. S., & Barto, A. G. (2017). *Reinforcement learning: An introduction* (2nd ed). MIT Press.

31. Svyatkovskiy, A., Deng, S. K., Fu, S., & Sundaresan, N. (2020). Intellicode compose: Code generation using transformer. In *Proceedings of the 28th ACM Joint Meeting on European Software Engineering Conference and Symposium on the Foundations of Software Engineering, ESEC/FSE 2020* (pp. 1433–1443). New York, NY, USA: Association for Computing Machinery.

32. Vaswani, A., Shazeer, N., Parmar, N., Uszkoreit, J., Jones, L., Gomez, A. N., Kaiser, L., & Polosukhin, I. (2017). Attention is all you need. In *Proceedings of the 31st International Conference on Neural Information Processing Systems, NIPS'17* (pp. 6000–6010). Red Hook, NY, USA: Curran Associates Inc.

33. Vinyals, O., Toshev, A., Bengio, S., & Erhan, D. (2015). Show and tell: A neural image caption generator. In *2015 IEEE Conference on Computer Vision and Pattern Recognition (CVPR)* (pp. 3156–3164). Los Alamitos, CA, USA: IEEE Computer Society.

34. Wu, P., Escontrela, A., Hafner, D., Goldberg, K., & Abbeel, P. (2022). DayDreamer: world models for physical robot learning. Preprint retrieved from arXiv:2206.14176

35. Young, T., Hazarika, D., Poria, S., & Cambria, E. (2018). Recent trends in deep learning based natural language processing. *IEEE Computational Intelligence Magazine, 13*(3), 55–75.

36. Zhang, Y., & Yang, Q. (2022). A survey on multi-task learning. *IEEE Transactions on Knowledge and Data Engineering, 34*(12), 5586–5609.

37. Ziebart, B. D., Maas, A., Bagnell, J. A., & Dey, A. K. (2008). Maximum entropy inverse reinforcement learning. In *AAAI Conference on Artificial Intelligence*, p. 6.

Proposals for Communicative and Cooperative AI to Promote Synergies in Hybrid AI-Augmented Socio-technical Arrangements—A Humanities Perspective

Arne Sonar and Christian Herzog

Abstract

We propose to explicitly adopt the requirements of cooperation and communication as constitutive features that can govern the design of hybrid artificial intelligence (AI)-augmented socio-technical arrangements. We believe that truly harnessing the synergies within concrete human–machine interaction scenarios calls for a fundamental reevaluation of the normative ethical groundwork based on which the hybridization of actions, processes, and structures is advanced. For this purpose, we conduct a qualitative-explorative literature analysis as well as review and evaluate corresponding design implications of the constituents of cooperation and communication. We use the AI-augmented diagnosis of deep-vein-thromboses at the point-of-care as an illustrative example.

Keywords

Cooperative AI · Communicative AI · Hybridity · Socio-technical arrangements

1 Introduction

Technical applications and components are fundamental elements in the immediate actions, processes, and organizational structures of a wide variety of social areas. Especially with the social diffusion of technological concepts such as artificial intelligence (AI) it is to be expected that the hybrid coexistence of and interaction between humans and technology will significantly increase in both scope and depth. In this sense and for the purposes of this paper, we see hybridity, beneath its incorporated dimension of meaning, as an intensified (externalized) co-operation of human and technical components in the most diverse contexts of action, process and structure. The hybridization of actions, tasks, processes, and organizational structures, which goes hand in hand with continuously evolving technical capabilities, reveals an increased need for tapping the synergistic potentials of human and technological counterparts. Conceptual approaches for an application-oriented, coordinative design of the hybrid co-operative interaction between humans and technology are to be regarded as increasingly important in this respect. We recognize the relevance of power (asymmetries) within the organizational structures as well. However, for the purposes of this paper, we will assume no oppressive or exploitative intentions on behalf of persons and institutions of power that commission robotic or AI systems meant to achieve some specific task in human–machine cooperation. While we acknowledge that this is a limiting factor, we still believe the goal of substantiating constitutive features of cooperative and communicating AI a meaningful endeavor. For instance, we hope that our work can alleviate issues such as errors resulting from automation bias—to name just a single example—which can result even in settings where automation is deployed with only the best of intentions.

Particularly in the context of AI, the specific perception and attribution toward technical applications is changing in the associated discourses. In this sense, innovative technological applications are no longer exclusively understood as useful tools and functional instruments for specific tasks, but are increasingly associated as proactive interaction partners and communicators themselves [1–3]. Accordingly, with growing technical (adaptive) and communicative capabilities, the aspect of cooperation becomes increasingly important in view of an AI-enabled hybridization [4, 5]. Thus, a perspective on technical applications as "conceptual entities capable of reactively and/or proactively influencing their dynamic

A. Sonar · C. Herzog (✉)
Ethical Innovation Hub (EIH), Universität zu Lübeck, Lübeck, Schleswig Holstein, Germany
e-mail: christian.herzog@uni-luebeck.de

A. Sonar
e-mail: arn.so@web.de

B. Meyer et al. (eds.), *Hybrid Societies*, Advances in Science, Technology & Innovation,
https://doi.org/10.1007/978-3-032-03488-5_31

environment" [6], p. 227] seems to be a fruitful one. Two further quotations are intended to support this assumption:

"In other words, the instrumental definition of technology, which had effectively tethered machine action to human agency, no longer adequately applies to mechanisms that have been deliberately designed to operate and exhibit some form, no matter how rudimentary, of independent action or autonomous decision making. (...) The point is that the instrumentalist perspective, no matter how useful and seemingly correct in some circumstances for explaining some technological devices, does not exhaust all possibilities for all kinds of devices" [2], p. 11].

"Algorithms are social agents. Their presence and role are now central and indispensable in many sectors of society, both as tools to do things (such as machines) and as communicative partners" [1], p. 249].

To a certain extent, such pragmatic perspectives will be connected to in the following article. The paper will focus on the two terms of *cooperation* and *communication*. For the purposes of this paper, *cooperation* is to be understood as working together to achieve a (specific) common goal [7, 8]. *Communication* will refer to the act of transmitting information between sender and receiver. On the one hand, the receiver is understood to reconstruct the information, while on the other hand—due to its adaptive dimension—the information can also exert influence on the receiver's behavior [9]. In the following, the paper will extract and analyze relevant features for a conceptual approach to *"cooperative and communicative AI."* The general theoretical approach as well as the associated literature analysis and its results presented within the following, were conceived and carried out as part of an applied research project on cooperative and communicating AI methods for medical image-guided diagnostics. It aims at the development of a deep learning-based software interpreting images of a portable ultrasound device for detecting deep (leg) vein thromboses during the examination in real-time. The system is intended to be used even outside specialist practices, i.e., at the point-of-care, where diagnoses might be conducted by, e.g., inexperienced medical personnel. Thus, the application is intended to be used independently of a local consultation by appropriate specialist physicians—e.g., in the context of emergency diagnoses. In view of these objectives, a special characteristic of this application is that it goes beyond augmenting medical image-guided diagnosis and proactively guides the process of examination (itself and conducted by human staff). In ultrasound-based deep vein thrombosis diagnostics, the veins of the legs are systematically examined. At relevant locations, the human physician applies pressure to the vein with the ultrasound device, ideally compressing the vein's volume. Successful compression indicates the absence of a thrombosis, while compression fails, if the vein is obstructed. The deep learning-based systems guides the human in systematically assessing all relevant veins and additionally supports

the compression-based diagnosis process itself. Hence, the approach goes beyond simply augmenting the ultrasound imagery, but involves active guidance.

This is why we propose a conceptual, visionary approach of a cooperating and communicating AI (in medicine/medical imaging). In order to approach this question, in the following we present significant results of a qualitative-explorative literature analysis. Specifically, it should be noted that the presentations and explanations done here are mainly conceptual in nature. Starting with some brief remarks on methodology (II-A), core results of the literature review are presented (II-B) and discussed further within the context of the application of our research project (III). This work will conclude with critical remarks on possible limitations (IV) and an outlook (V).

2 Methodology and Results of Literature Analysis

2.1 Methodology

Within our systematic literature review a two-step approach was used for identifying relevant work: On the one hand, we conducted a randomized search for relevant publications and other contributions within Google Scholar based on the general nouns "cooperation," "communication," "technology," and "artificial Intelligence." Especially in the case of the term set "communication technology," we quickly noticed that primarily texts from the field of information and communication technology (ICT) were tapped, which we classified as less suitable for our purpose. Therefore, we switched here from our originally intend of using only the terms "cooperation," "communication" and "technology," and changed it through "artificial intelligence" for both, the term of "cooperation" and "communication." The aim of such an open and at the same time more general search was to obtain a multidisciplinary and context-independent perspective on the individual terms and the characteristic connections associated with them. The papers selected were those whose titles and abstracts dealt with aspects of cooperation and communication—favored, but not necessarily as an exclusive criterion in connection with technology. An equally relevant selection criterion was the accessibility to the corresponding papers and works—accordingly, papers, although supposedly relevant to our analysis, could not be accessed and thus, were not considered in our analysis.

Following this, we undertook a further, but this time systematic literature search within the "Web of Science"—database. Here, we used the following sets of search terms: "cooperating—artificial intelligence," "cooperative—artificial intelligence," "communicating—artificial intelligence," and "communicative—artificial intelligence." Papers were selected according to the relevance of their titles as well as an in-depth analysis of the respective abstracts. Due to the exploratory nature of our analysis work, there was no limitation or restriction on the year of publication, which is why papers from the 1980s onwards were also included. As described before, the access factor was an exclusion criterion for potentially relevant paper. In total, 79 paper identified by the keyword sets mentioned above (randomized and systematically) were considered relevant for our analysis. Its key results will be presented in the following.

2.2 Results/Findings

In the following, we will present the main findings of our literature analysis. Based on the results generated through our qualitative-analytical literature study, in a first step we were able to identify and define five general categories for each of the two sets of keyword links used. For the keyword range 'cooperation—cooperative—cooperating' these are (1.) the cooperative motives and modes/arrangements, (2.) the cooperation- and task fields as well

as associable (decision) rules, (3.) shared/common goals, understandings and approaches, (4.) interactivity and (5.) authority (see Table 1). For the keyword range 'communication—communicative—communicating', in contrast, the items (1.) functional- and perception variation, (2.) system and prose capabilities communication, (3.) (communicative) grounding, (4.) information exchange and (5.) subject design of a communicating "other"/communicator have been identified (Table 2). During further clustering steps, these superordinate clusters were assigned further aspects, which were likewise treated within the analyzed literature, as specifically associable sub-features. Tables 1 and 2 show the corresponding higher-level clusters and various associated sub-characteristics, which are both associable discussed in more detail and in relation to our application example below.

A fundamental point to note here is that the general categories presented in Tables 1 and 2, are not hierarchically related to each other. Rather, it needs to be emphasized that they are in an equal relationship with each other. Furthermore—and although we considered cooperation and communication within our study as separate dimensions—connections between the individual points (presented in the

Table 1 Overview of results/findings for terms "cooperation," "cooperating," and "cooperative"

Results/Findings for terms "cooperation," "cooperating," "cooperative"	
General category	Specifications
Motive, modes/arrangements	– Classification and goal of cooperation [10], Cooperative framework conditions [5], scenario and problem analysis of cooperative interaction [11] – Distribution of functions [5, 12], risk analysis [13], balance in perspectives (developers/user) [14] – Design according to needs and expectations [15], relevance of acceptance, perception of usefulness and trust [15, 16], – Adaptation management [16], requirement analysis [17], diverging adaptation habits [16], influences of affective reactions (evaluations, judgments, decisions) [18]
Cooperation-/task fields and (decision) rules	– Identifying activities cooperative potentials [8], delimitation of cooperation fields, tasks and rules (decision relevance) [11, 19], descriptive clarification of situational conditions [10] – Coordination of activities/tasks [8], content-/task-related and technical-/media-related coordination [10], arbitrate responsibility distribution [19] – Integrative network view (mutuality/interdependence) [5], novel system dynamics and complexities [12], variable degrees of interaction (environmental conditions [19], agreement versus guardianship versus competition versus cooperation [19]
Common goal, understanding and approach	– Synergy of specific skills, capabilities. features and characteristics [20–22], mutual mediation/transfer of perspectives [14], through group model/team cognition (common basis) [21], specific roles (entities) and procedures (task execution [23] – Confidence in and familiarity with application [23], (objective and subjective) perceived safety and acceptance [5, 13], trust calibration/adjustment [24], – knowledge acquisition and application [14], substantial preservation and expansion potentials of specialized knowledge and objects of work[14]
Interactivity	– Explicit extension of interaction concepts (e.g., multi agent perspective) [19, 25], interaction management (adjustment related to condition, difficulties encountered and specific needs) [5], – Reliable recognition of availability and intention (successful feedbacks) [11, 19], positive perception of interaction processes [7], negative correlation perceived user-friendliness and technology Awareness [15] – Informational interface interactivity [20], focus on task and advantageousness [26], ergonomic interface design [15], Interaction parameters (strategy, style, model), modes (channel), interfaces (realization) [19] – dynamic degrees of interaction [19], static/dynamic components (e.g., agents, skills, tasks, resources [11, 27]
Authority	– Assignment of roles and associated (social) identities [10], equitable and non-hierarchical, partnership and equal co-operation [28, 29], antagonistic (e.g., competitive/manipulative) [25] – distributed control, reactive and proactive influences [6], situational cooperation circumstances (e.g., task relevance, mutual dependence, degree of intensity [11, 19]

Table 2 Overview of results/findings for terms "communication," "communicating," and "communicative"

Results/findings for terms "communication," "communicating," "communicative"	
General category	Specifications
Functional- and perceptual variability	– Reactively input respond (communication partner) and proactively generating meanings (content producer) [3], functional dimensions, relationship dynamics and metaphysical implications [3], alternative ways of design, function, perception, and interaction [3] – Dynamic-interactive feedback (mutual communicative exchange) [30], avoiding deterministic approaches (simple extension/multiplication of human capabilities) [30] – Trust and perception of attributes (dominance, friendliness) [31], adaptive adjustment to users, messages and contexts [3], Interaction changes under changing conditions (e. g., stressful situations, different lengths of service) [32]
Capability communication (System/Prose)	– Purposes/functions of applications, related changes in workflows (prose) [14], – Identification of work activities and organizations, mutual communication of points of view (system) [14] – Purpose (e.g., information type), means (methods of transmission) and direction (human to technology) of communication [4], calibrated uncertainty models (communication of uncertainty levels [33] – Understanding of objectives and target functions (transparency) [34], exploiting system capabilities (natural/accustomed communication modes [35]
Grounding	– H-T communication is language-based, relational, and normatively regulated [36], degrees of human characteristics/traits [3, 37], individual characteristics of language approaches and selective characteristics (style, structure, content) [38]
	– Shared/common meanings and codes [30, 39], common (linguistic) basis of understanding, shared content and cultural backgrounds/foundation [10] – Common basis on important functions and features [34]
Information exchange	– Immediate interface exchange and discursive communication [1, 36], immediate influence (production, provision, dissemination and transmission of information) [9], communication method and interface selection [35] – Adaptation of and adjustment to applications [39], mutual learning processes (expansion of skills and knowledge) [14], – Interaction and communication concepts (multimodality and productivity premise) [35, 38], intuitive, ergonomic and user-friendly interface designs [11, 20, 37], different levels of interfaces (high-level, low-level) [4, 35, 38], multi user interfaces [38] – Multimodality in/at display channels versus simplicity (necessary of displayed content [8, 40], relevance, adequate quantity and quality of information (premises information presentation) [35], ensure systems reliability, effectiveness and functionality [38]
Subject design of the "other"/communicator	– Instrumentalist design perspective versus expanded role understandings [2, 3, 32], overcoming anthropocentric (human) singularity [1–3, 36] – Contingent, confronting, challenging and demanding appropriate responses counterpart/protagonist [1], active influence on messages/recommendations (omitting and inserting information) [32], influence on opinion and behavior [3] – Static and dynamic degrees of interaction [3], dependencies on specific environmental conditions [41], intentional (communicative) paternalism [38]

respective circles, see Fig. 1) were revealed during the analysis. As can be seen in Fig. 1, for example, the cooperative aspect of authority connects interdependently with the communicative aspect of shaping/designing a communicating "other" or the cooperative aspect of interactivity with the communicative momentum of information exchange. These relationships, which we have assumed, are intended to be taken up in the further explanations.

3 Discussions/Reflections on Proposed Features

As described above, it should be emphasized that the communicative dimension as well as the cooperative dimension of AI technology are indeed distinct areas, but they are nevertheless interrelated and interdependent. Instead of (re-) presenting these individual areas and their sub-points individu-

ally, they will be addressed below in precisely those contextual connections and tensions. This will serve to clarify the connections between communicative and cooperative dimensions for hybrid human-AI constellations by especially addressing our application example. With recourse to the contents presented in Tables 1 and 2, a central thesis or assumption within our considerations should therefore be the following: The correlative relationship between the communicative and the cooperative dimension in human-technology interaction, and in particular in human-AI interaction, is specifically explicated in the tension (and its resolution) between the requirements for a general, i.e., fundamental design of a system/application (and its interfaces) and the necessity of addressing specific, among other things, user-individual moments of interaction. For our application example of an application for real-time examination and diagnosis of deep vein thrombosis, six focal points can be identified, which can and will be used to illustrate this in more detail:

Fig. 1 Schematic presentation of the interrelationships of the relevant factors

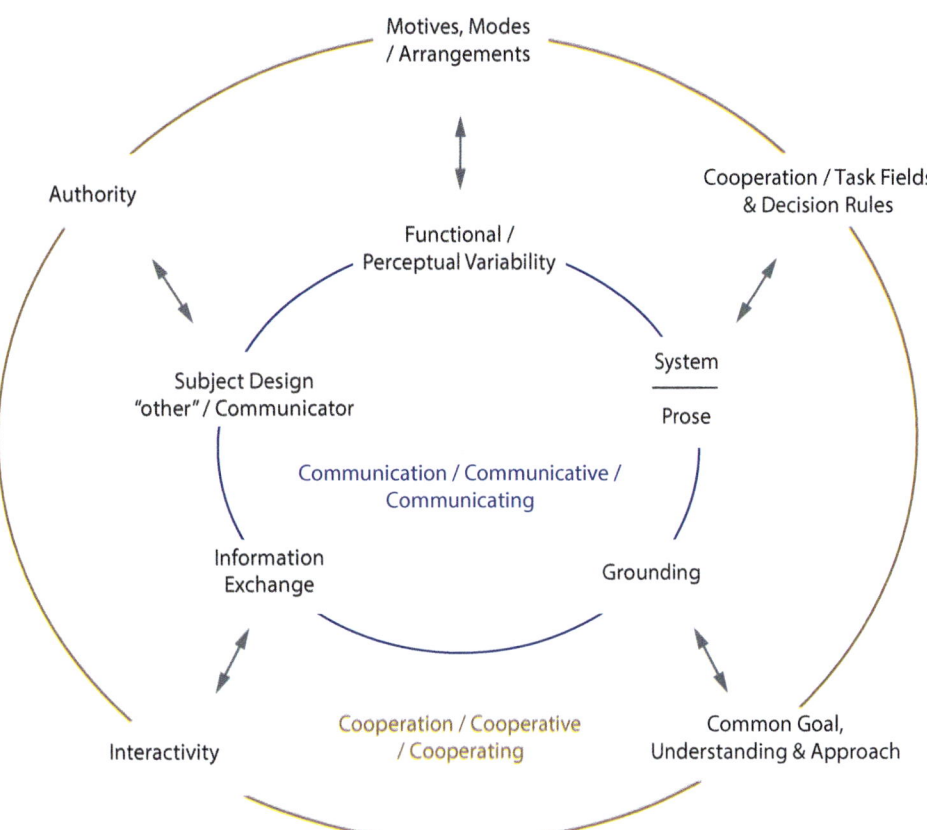

- the determination of specific requirements, aims and needs,
- customizability,
- coordination and synergy,
- authority levels,
- capability understanding/awareness and
- trustworthiness and -adjustment (see Fig. 2).

In order to precisely analyze and *determine the general requirements and needs* of the application area, as well as *to define concrete aims* of the application development as close to practice as possible, the fundamental inclusion of stakeholders is (also) essential for our development project. Involving stakeholders directly in the process of development proves to be important not only to define requirements-specific objectives for the human-technology cooperation itself. It is also applicable, e.g., to get an experience-based insight into the cooperative relevance of situational, environmental and context-specific moments (e.g., external stress, the length of a shift in the clinic). Beneath that, gaining (communicative) impressions about the users' evaluation about the dependencies of, e.g., situations, environmental conditions or users' abilities and the types, representations or amounts of information that are required, is an additional benefit for our development process and the concrete design of the application's interfaces. In addition, special procedures such

as focus group workshops or qualitative stakeholder/expert interviews can be used to elicit the necessity, permissibility and practicability of specific levels and types of interaction—which in turn is of particular relevance when it comes to their empirically-based normative justification. This indeed becomes relevant, when it comes to considering and ensuring the user-specific moments of communicative and collaborative human-AI interaction. Specifically for our application example, for considering the immediate interaction processes, we have to assume that different user groups with, in turn, individual user types, experiences and competencies will be taken into account. Accordingly, it is important for us to include within our considerations and reflections not only competence- and skill-related cooperation specifics, but also the specific communication needs of divergent user groups that can be associated with them.

Customizability implies a need for establishing a common shared basis between users and the applications, both on the communicative as well as the cooperative level of interaction. Likewise, specific adaptation intentions, abilities, habits, and attributes of different users could be taken on and addressed by this. It can be assumed, for example, that inexperienced users have a greater need for a more system-led interaction and greater quantitates of information than experienced physicians, who may, in turn, perceive too much guidance and infor-

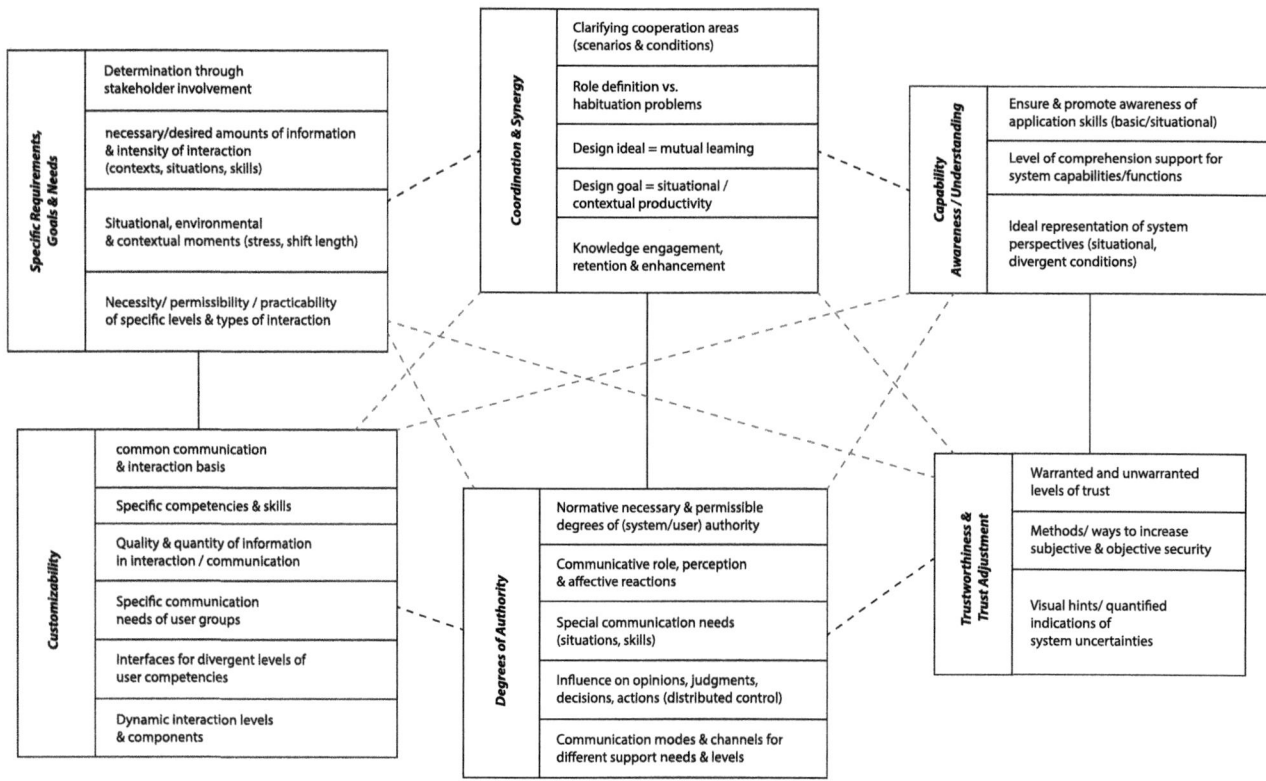

Fig. 2 Visualization of the content transfers to the application example and their likability

mation as a hindrance. However, it is also possible that people who are inexperienced both in the examination itself and in the use of the application are, on the one hand, quantitatively overloaded when searching for and evaluating relevant information, and, on the other hand, there may be too little information available for a qualitative assessment of the examination and diagnosis itself for low levels of medical competence. Therefore, the inclusion of dynamic interaction levels as well as considerations of static or dynamic interaction components are also relevant for our application. This indeed, e.g., could enable variable multi-user interfaces to fundamentally and simultaneously address different levels of user competencies. Thus, a human-centered design perspective could be achieved, which may improve interaction processes in its entirety, e.g., by a mutual unleashing or (targeted) taming of potentials [17]. To tap and bring together the potentials of human users and a technical application in communicative and cooperative interaction networks to an optimal degree, i.e., in ways that are *coordinative and designed for synergy*, for the example of our application it is also crucial to clarify precisely the areas of cooperation between user and application, taking into account a wide range of scenarios and conditions. While achieving this, we also have to define the respective roles of users and the application in the course of interactive cooperation. This as well, has to contribute to a critical reflection toward a direct habituation of these roles' understandings.

Varying situational contexts and a potential inability to adapt to these might be accompanied by rigid, inflexible habituation tendencies. In particular, it is important to ask how possible tendencies of omission and commission [42] can be countered in stressful situations, after long and strenuous shifts, or as a result of the everyday use of the application. From our perspective, the concrete design measures for the communicative and cooperative interface design of our application example are therefore to be oriented primarily toward two aspects: On the one hand, the design ideal of (mutual) learning must be pursued. On the other hand, the design goal is to achieve the highest possible level of situational and contextual productivity, i.e., efficiency and effectiveness of interaction processes—again in both, the communicative as well as the cooperative dimension of interaction. Here as well, a certain amount of attention needs to be paid to the tension and possible divergences between a general version of system, application, and interface design and the more specific, individual user needs. For instance, highly capable diagnosticians may not wait for system prompts, which could then be dynamically faded out based on user feedback (buttons, voice commands, speed of handling the equipment), or faded in, should users wait for system responses. However, the effectiveness of such means needs to be carefully studied, of course. In addition, with regard to the diversity of intended user groups (e. g. paramedics, general practitioners, non-specialist physicians, medical-technical personnel), we must also consider how necessary knowledge bases of the examination and diagnosis process itself, as well as specific forms of knowledge of the respective specific user groups can be integrated into the interaction process—and, respectively, be maintained and expanded.

Especially in view of the possibility of differently qualified and experienced user groups in our AI application example, along with the synergetic coordination of technical and human capabilities and competencies, the question of different *degrees of authority* between the application and its users, taking into account the specific and individual needs, requirements, and necessities, arises. What needs to be questioned in terms of our application example is not only the normative necessary or appropriate authority relations, e.g., to what extent inexperienced users need a greater degree of guidance and/or support from the system during the examination than it would be necessary for more experienced users. Rather, it is also relevant to question how to deal with the tension between specific degrees of authority and the immediate design of the communicating counterpart, i.e., how and in what way the device communicates when there are different needs and degrees of support—e.g., using hypothetical or categorical formulations for recommendations and advice). So, e.g., certain situations imply a different communication requirement than originally intended, while specific user groups, e.g., inexperienced users, require more intensive cooperative and communicative support or accompaniment than experienced or stress-resistant users. Since the technical application in our case not only accompanies the diagnostic process, but also may guide the examination process itself proactively, it also is necessary to elicit, in which situations the application can or should act in, e.g., accompanying or guiding matters. This should be accompanied by considerations about by which criteria this, in turn, can be justified normatively (and legally). Thus, a highest possible (communicative and co-operative) individuality of the application may need to be ensured. Corresponding considerations have to aim at situation- and ability-specific communication needs of respective user groups, Additionally, their connection with communicative roles, perceptions and affective reactions on the one hand, as well as possible influences on opinions, judgments, decisions, or actions and such cooperative relationships on the other hand, need to be regarded.

A further "focal point," ties in with the omission and commission issue already alluded to above: it is the aspect of *awareness and understanding of the capabilities* of the application. A specific value here is to be placed on how to promote the user's knowledge about the system capabilities and functions, both prior to use and within immediate interaction processes. In particular, it is necessary to investigate whether and which information (qualitative and quantitative) from the system can be incorporated into the communicative

interaction, e.g., the communication of uncertainties during examinations or diagnoses, and via which specific communication channels this should be achieved. Accordingly, on the one hand, it requires an evaluation of what the respective users perceive as ideal representations of system perspectives, during interaction in different situations and under divergent conditions. On the other hand, however, it also requires eliciting the extent to which a clear understanding of system capabilities and functions can be supported on the user side. This indeed is necessary, to ensure users' awareness both from the ground up and situationally.

Those, in turn, are closely linked to factors such as expectations, needs, acceptance and usefulness as well as establishing appropriate relationships of trust on the side of users toward the application itself. Considering the tension between the *trustworthiness* of an application and the *trust adjustment* on the part of the users toward it, creating and ensuring adequate, that means warranted and unwarranted levels of trust in relation to the application is another factor of essential relevance. For instance, visual cues or explicitly quantified displays indicating the level of the AI algorithms' uncertainty imbued in the separate outputs (such as vein compression or the amount of veins accounted for during diagnosis) can be given to mitigate user trust that is unwarranted for. It is therefore essential to examine how users' subjective as well as objective sense of security can be increased, and to what extent specific communication features, e.g., in terms of specific language approaches (e.g., formulations) or in- formational characteristics (e.g., specific amounts of available information) can contribute to this.

We are convinced that an orientation toward the factors of communicative and cooperative AI, which we have elaborated on above, not only contributes to improving human-AI interaction per se, but that an appropriately designed system, application, and interface can also address normatively relevant aspects and normative premises. Due to the brevity of this paper, this aspect could only be touched upon in a rather abstract sense. Corresponding descriptions of these considerations and interrelationships in greater depth are therefore not possible here and must necessarily be given elsewhere.

4 Future Work

The discussion here represents only an excerpt of the potentially available literature—it includes, in particular, those papers that were accessible to us (keyword: Open Access). In addition, only papers written in German or English were included. It should be noted, however, that some evidence during our research also points to the existence of relevant texts in other languages. It should also be pointed out that a new search with identical search terms within the previously used database now achieves a larger number of basic hits. We further recognize that we have only been able to provide a cursory sketch of the implications for our particular application example—a sketch that often remains on the conceptual level, which we believe needs to be brought down to concrete design ideas and evaluation strategies as part of a highly interdisciplinary development exercise. This work is forthcoming.

5   Conclusion

The main aim of this paper was to present key findings of a qualitative and at the same time exploratory literature study on possible constitutive factors of a conceptual approach of "cooperative and communicative" AI—a presentation geared toward sparking discourse aimed at practical implementation ideas and evaluative studies. The features presented and discussed here are purely exploratory in nature and have a primarily theoretical-conceptual character. They serve as initial proposals to fundamentally open up the field for a conceptual approach to *cooperative and communicative AI*. The theoretical explorations made here still need to be substantiated in the course of further, especially empirically oriented investigations (e.g., interviews, design and user studies), in order to create an extended basis for a concept of cooperative as well as communicative and thus partnership-based AI, originating in the idea of empirical aesthetics as well as empirical ethics [17]. This, in turn, could provide a broad foundation for embedding the utility of AI-based technical applications in the context of ethically grounded development, design, and implementation.

Acknowledgements Partially funded by the German Federal Ministry of Science and Education, reference 01GP1908, project title "CoCoAI—Kooperative und kommunizierende KI-Methoden für die medizinische bildgeführte Diagnostik."

References

1. Esposito, E. (2017). Artificial communication? The production of contingency by algorithms. *Zeitschrift für Soziologie., 46*(4), 249–265.
2. Gunkel, A. J. (2017). Communication technology and perception. In P. Plaisance (Ed.), *Handbook of communication and media ethics* (pp. 453–470) Berlin/Munich/Boston: de Gruyter
3. Guzman, A. L., & Lewis, S. C. (2020). Artificial intelligence and communication: A human-machine communication research agenda. *New Media Society, 22*(1), 70–86.
4. Klingspor, V., Demiris, J., & Kaiser, M. (1997) Human-robot-communication and machine learning. *Applied Artificial Intelligence, 11*(7).
5. Bellet, T., Hoc, J. M., Boverie, S., & Boy, G. A. (2011). From human-machine interaction to cooperation—towards the integrated co-pilot. In C. Kolski (Ed.), *Human-Computer Interactions in Transport* (pp. 129–155). Wiley-ISTE.
6. Lanzola, G., Gatti, L., Falasconi, S., & Stefanelli, M. (1999). A framework for building cooperative software agents in medical applications. *Artificial Intelligence in Medicine, 16*(3), 223–249.
7. Johnson, D. W., & Johnson, R. T. (2004). Cooperation and the use of technology. In D. H. Jonassen (Ed.) *Handbook of research on educational communications and technology (2nd ed.)* (pp. 785–811) New York: Routledge
8. Schröter, D. (2018) Entwicklung einer Methodik zur Planung von Arbeitssystemen in Mensch-Roboter-Kooperation. In T. Bauernhansl, Verl & E. Westkämper (Eds.) *Stuttgarter Beiträge zur Produktionsforschung—Band 81*, Stuttgart: Fraunhofer Verlag
9. Tjøstheim, T. A., Stephens, A., Anikin, A., & Schwaninger, A. (2020). The cognitive philosophy of communication. *Philosophies, 5*(4), 39.
10. Reinmann-Rothmeier, G., Mandl, H. (2002) Analyse und Förderung kooperativen Lernens in netzbasierten Umgebungen. *Zeitschrift für Entwicklungspsychologie und Pädagogische Psychologie, 34*(1), 44–57.
11. Zimmermann, K., & Bengler, K. (2012). Bereichsübergreifende Methoden, Techniken und Tools für kooperative Mensch-Maschine-Systeme aus automobiler Multiagentensicht. In Kongress der Gesellschaft für Arbeitswissenschaft, Gesellschaft für Arbeitswissenschaft e.V. (Ed.), *Gestaltung nachhaltiger Arbeitssysteme—58. Conference Paper*, Dortmund, GfA Press, pp. 279–282. https://www.researchgate.net/publication/280153754_Bereichsubergreifende_Methoden_Techniken_und_Tools_fur_koopera-tive_Mensch-Maschine-Systeme_aus_automobiler_Multiagentensicht. Last accessed 01 Sept 2022
12. Hoc, J.-J. (2010). From human-machine interaction to human-machine cooperation. *Ergonomics, 43*(7), 833–843.
13. Elkmann, N. (2013). Sichere Mensch-Roboter-Kooperation—Normenlage, Forschungsfelder und neue Technologien. *Zeitschrift für Arbeits-wissenschaft, 67*, 143–149. https://doi.org/10.1007/BF03374401
14. Mehl, W.-M., & Reisin, F.-M. (1989) Skandinavische Ansätze zur kooperativen Gestaltung computergestützter Systeme. In K.D. Jansen, U. Schwitalla & W. Wicke (Eds.), *Beteiligungsori-entierte Systementwicklung—Sozialverträgliche Technikgestaltung Materialien und Berichte* (pp. 120–132) Wiesbaden: VS Verlag für Sozialwissenschaften
15. Bröhl, C., Nelles, J., Brandl, C., Mertens, A., & Schlick, C. M. (2017) Entwicklung und Analyse eines Akzeptanzmodells für die Mensch-Roboter-Kooperation in der Industrie. In Kongress der Gesellschaft für Arbeitswissenschaft, Gesellschaft für Arbeitswissenschaft e.V. (Ed.), *Soziotechnische Gestaltung des digitalen Wandels—kreativ, innovativ, sinnhaft*, 63. https://www.rese-archgate.net/publication/314238572_Entwicklung_und_Analyse_eines_Akzeptanzmodells_fur_die_Mensch-Roboter-Koopera-tion_in_der_Industrie. Last Accessed 01 Sept 2022.
16. Thomas, D. M., & Bostrom, R. P. (2008). Building trust and cooperation through technology adaptation in virtual teams—empirical field evidence. In *Proceedings of the 41st Hawaii International Conference on System Sciences (HICSS 2008), 423*.
17. Pols, J. (2016). Good relations with technology: Empirical ethics and aesthetics in care. *Nursing Philosophy, 18*(1), 12154.
18. Liu, P., & Du, Y. (2022). Blame attribution asymmetry in human–automation cooperation. *Risk Analysis 42* (8), pp. 1769-1783.

19. Zimmermann, M., Bortot, D., & Bengler, K. (2012). Allgemeine Interaktionsprinzipien für kooperative Mensch-Maschine-Systeme. In Kongress der Gesellschaft für Arbeitswissenschaft, Gesellschaft für Arbeitswissenschaft e.V. (Ed.), *Gestal-tung nachhaltiger Arbeitssysteme—58. Conference Paper,* Dortmund, GfA Press, pp. 469–472. https://www.researchgate.net/publica-tion/280153 754_Bereichsubergreifende_Methoden_Techniken_und_Tools_ fur_kooperative_Mensch-Maschine-Systeme_aus_automo-biler_ Multiagentensicht. Last accessed 01 Sept 2022

20. Murphy, R. R., & Rogers, E. (1996). Cooperative assistance for remote robot supervision. *Presence—Teleoperators and Virtual Environments 5*(2), 224–240.

21. Hakli, R. (2017). Cooperative human-robot planning with team reasoning. *International Journal of Social Robotics, 9,* 643–658.

22. Wanyou, H. Y. (2019). From competition, coexistence to win-win—relationship between intelligent design tools and human designers. *Landscape Architecture Frontiers, 7*(2), 76–83.

23. McCormack, J., Hutchings, P., Gifford, T., Yee-King, M., Llano, M. T., & D'inverno, M. (2020). Design considerations for real-time collaboration with creative artificial intelligence. *Organised Sound 25*(4), 41–52.

24. Okamura, K., & Yamada, S. (2020). Empirical evaluations of frame-work for adaptive trust calibration in human-AI cooperation. *IEEE Access, 8,* 220335–220351.

25. Kirn, S. (2002). Kooperierende intelligente Softwareagenten. *Wirtschaftsinformatik, 44*(1), 53–63.

26. Han, C. (2011). Human-robot cooperation technology—an ideal midway solution heading toward the future of robotics and automation. In *Construction Proceedings of the 28th ISARC,* Seoul, Korea, pp. 13–18.

27. Lüdtke, A., Javaux, D., Tango, F., Heers, R., Benglere, K., & Ronflé-Nadaud, C. (2012). Designing dynamic distributed cooperative Human-machine systems. *Work, 41*(1), 4250–4257.

28. Jennings, N. R., Varga, L. Z., Aarnts, R. P., Fuchs, J., & Skerek, P. (1993). Transforming standalone expert systems into a community of cooperating agents. *Engineering Applications of Artificial Intelligence., 6*(4), 317–331.

29. Leuthner, R., Lexhaller, A., & Steinbrenner, H. J. (1999) Innovationswerkstatt Handwerk—Strategien für technologieorientierte Unternehmen. Bad Wörishofen: Holzmann

30. Hepp, A. (2020). Artificial companions, social bots and work bots—communicative robots as research objects of media and communication studies. *Media, Culture & Society, 42*(7–8), 1410–1426.

31. Cullen, J. (2009). Imitation versus communication—testing for human-like intelligence. *Minds & Machines, 19,* 237–254.

32. Song, Y., & Luximon, Y. (2020). Trust in AI agent: A systematic review of facial anthropomorphic trustworthiness for social robot design. *Sensors, 20*(18), 5087.

33. Gunkel, A. J. (2012). Communication and artificial intelligence—opportunities and challenges for the 21st century. *Communication+1. 1*(1), 1–26.

34. Kompa, B., Snoek, J., & Beam, A. L. (2021). Second opinion needed: communicating uncertainty in medical machine learning. *NPJ Digital Medicine, 4*(4)

35. Huang, S. H., Held, D., Abbeel, P., & Dragan, A. D. (2019). Enabling robots to communicate their objectives. *Autonomous Robots, 43,* 309–326.

36. Karpov, A. A., & Yusupov, R. M. (2018). Multimodal interfaces of human-computer interaction. *Herald of the Russian Academy of Sciences, 88*(1), 67–74.

37. Zhao, Z. (2006). Humanoid social robots as a medium of communication. *New Media & Society, 8*(3), 401–419.

38. Aeschlimann, S., Bleiker, M., Wechner, M., & Gampe, A. (2020). Communicative and social consequences of interactions with voice assistants. *Computers in Human Behavior, 112,* Article 106466.

39. Byrer, J. K., & Jelassi, M. T. (1991). The impact of language theories on DSS dialog. *European Journal of Operational Research, 50*(2), 113–126.

40. Walter, J. (2000). Technological adaptation and learning by cooperation—a case study of a successful onshore technology transfer in Tierra del Fuego. *Journal of Technology Transfer, 25,* 13–22.

41. Schurig, A., & Thomas, C. G. (2017). Designing the next generation of connected devices in the era of artificial intelligence. *The Design Journal, 20*(1), 3801–3810.

42. Inoue, N., & Morita, J. (2021). A behavioral task for exploring dynamics of communication system in dilemma situations. *Artificial Life and Robotics, 26,* 329–337.

43. Neri, E., Coppola, F., Miele, V., Bibbolino, C., & Grassi, R. (2020). Artificial intelligence: Who is responsible for the diagnosis? *La radiologia medica, 125,* 517–521.

Nadaraya–Watson Time Series Early Classification for Gesture Recognition

Florian Kretzschmar, Rim Barioul, Dana Uhlig, Olfa Kanoun, and Alois Pichler

Abstract

The prediction of gestures in shared public spaces is important not only to ensure good functionality of agents but also the safety of humans and their acceptance of the hybrid society concept. In this context, surface electromyography (sEMG) combined with inertial measurement unit (IMU) signals is investigated in this work to predict dynamic hand gestures. Our proposed algorithm is designed to make predictions, while these signals are being observed, combining a Nadaraya–Watson kernel estimator and an entropy-based decision function. We not only achieve a high accuracy, but also a significant earliness of prediction.

Keywords

Gesture recognition · Time series classification · Surface electromyography

1 Introduction

Gestures occur in most cases in a flow, simultaneously with other actions, so that they need to be recognized in a dynamic, varying, and evolving environment. It is important to predict gestures as they build an important mode of communication. Surface electromyography signals (sEMGs) are suitable biological signals for identifying dynamic gestures, but they are ambiguous. Additional input modalities such as inertial measurement Units (IMU) need to be considered as means to improve recognition reliability and to reduce the set of plausible future states, which can be predicted.

F. Kretzschmar (✉) · R. Barioul · D. Uhlig · O. Kanoun · A. Pichler
Chemnitz University of Technology, Chemnitz, Germany
e-mail: florian.kretzschmar@math.tu-chemnitz.de

R. Barioul
e-mail: rim.barioul@etit.tu-chemnitz.de

The sEMG is used for motion intention estimation as it is generated 30–150 ms prior to the corresponding motion [7]. Motion intention recognition based on sEMG is divided into two groups: (1) sEMG-driven musculoskeletal (MS) model-based, where a function between sEMG and joint moment, angular velocity, or angular acceleration is established by bio-mechanical model of the muscles. (2) Machine learning-based, where commonly used methods includes support vector machine (SVM), linear discriminant analysis (LDA), back-propagation neural network (BPNN), and deep learning (DL) [7]. Moreover, EMG can be combined with electroencephalography (EEG) to detect action intention in highly specific scenarios (i.e., vehicle acceleration) mainly to support faster emergency braking in automated vehicles. Recent work suggests that sensorimotor regions of the brain are involved in anticipatory motor recognition and action of sign language [4]. Additionally, deterministic time series analysis has been well-studied in describing physical behavior, e.g., to describe the random behavior. The time series sEMG originating from human movements have some patterns but manifest random behaviors, which are not present in well-studied time series and for this, new mathematical methods need to be developed.

The remainder of this paper is organized as follows: Sect. 2 introduces the mathematical notation and proposed algorithm. Furthermore, we explain the gesture recognition task and the corresponding data set. Section 3 is dedicated to the evaluation of the numerical results and achieved performance of our model. Finally, we conclude the paper and give a preview of possible future work.

2 Methods

A transitory phase could be observed in the time series signals collected from the sEMG and IMU sensors in Figs. 1 and 2, respectively. This transitory phase is indicating the dynamic change in the muscle physiology over time before the gesture manifestation in its final state.

© The Author(s) 2026
B. Meyer et al. (eds.), *Hybrid Societies*, Advances in Science, Technology & Innovation,
https://doi.org/10.1007/978-3-032-03488-5_32

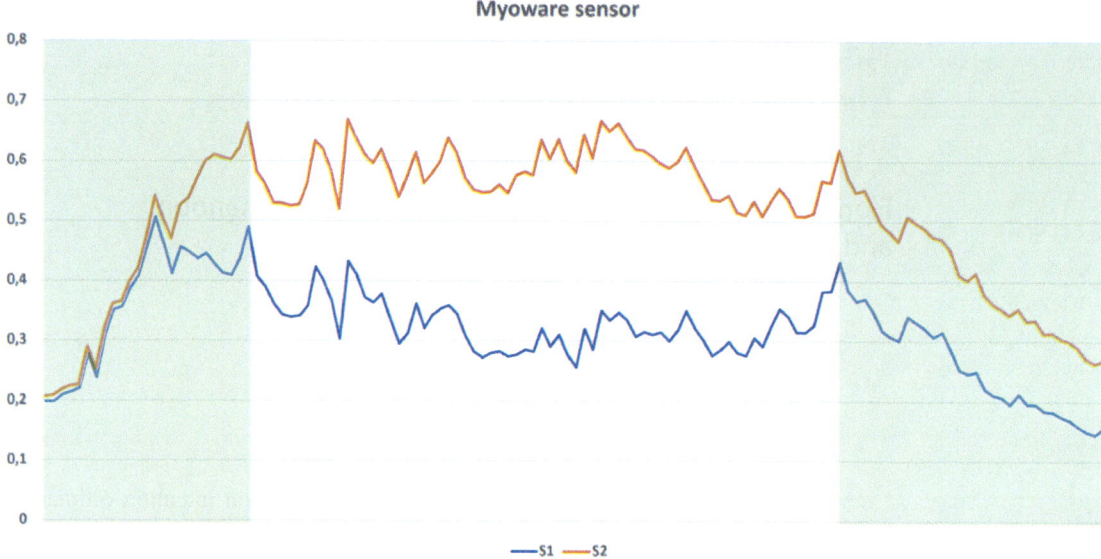

Fig. 1 EsEMG signals of left hand while performing a pointing gesture [2]

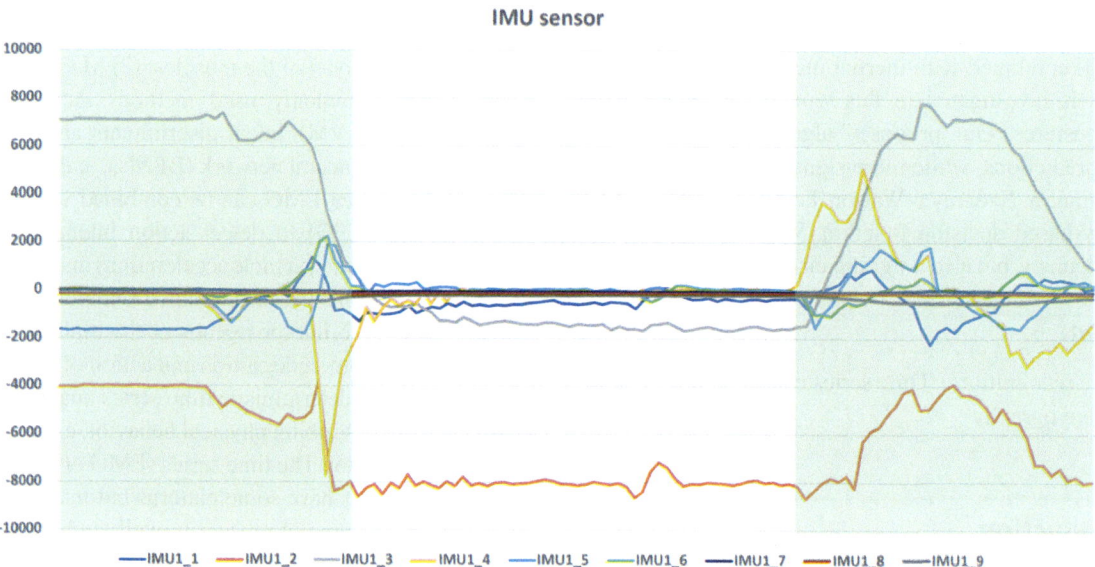

Fig. 2 IMU-9 sensor signals of left hand while performing a pointing gesture [2]

Considering the collected signals as time series this paper aims to investigate the trade-off between the prediction accuracy and the earliness of the gesture recognition.

2.1 Problem Statement and Notation

We will now briefly formulate the objective of early time series classification and clarify the notation used throughout this paper. Let $\mathbf{x}$ be a multivariate.

time series consisting of d components—univariate time series—of length T. Such a time series can be presented as a matrix

$$\mathbf{x} = \begin{bmatrix} x_{1,1} & \cdots & x_{1,d} \\ \vdots & \ddots & \vdots \\ x_{T,1} & \cdots & x_{T,d} \end{bmatrix} \in \mathbb{R}^{T \times d}.$$

Furthermore, we denote by

$$\mathbf{x}_{t:t'} = \begin{bmatrix} x_{t,1} & \cdots & x_{t,d} \\ \vdots & \ddots & \vdots \\ x_{t',1} & \cdots & x_{t',d} \end{bmatrix} \in \mathbb{R}^{(t'-t) \times d}$$

the sub-time series of x from t until t' and in particular $\mathbf{x}_t = \left(x_{t,1}, \ldots, x_{t,d}\right)^T$. Now, let $X \in \mathbb{R}^{N \times T \times d}$ be a data set containing N multivariate time series and $y \in \mathbb{R}^N$ a vector which contains the corresponding class labels $y_i \in \{1, \ldots, C\}$. Given a time series $\hat{\mathbf{x}} \in \mathbb{R}^{T \times d}$, our goal is to not only predict the correct class label but to also make the prediction as early as possible, at some $t' < T$, which we will refer to as earliness from now on. Since we are waiving the information which will be revealed at a later time, the risk of misclassification increases. Hence, there is a trade-off between prediction accuracy and earliness which we ideally want to be able to control by the choice of model parameters.

2.2 Estimation of Class Membership Probabilities

The core of our proposed algorithm will be the Nadaraya–Watson kernel density estimator which was first introduced in [3] and [6] and whose concept we apply to the estimation of a probability mass function, or more precisely class membership probabilities. Given $x, \hat{x} \in \mathbb{R}^d$ we define the multivariate kernel

$$K_H(x, \hat{x}) = \exp\left(-\frac{1}{2}(x - \hat{x})^T H^{-1}(x - \hat{x})\right) \quad (1)$$

where $H \in \mathbb{R}^{d \times d}$ is a symmetric and positive semidefinite matrix. Now, let X be a set of training data which has the structure described above. For each class $k = 1, \ldots, C$, let X_k be the subset containing all $x \in X$ belonging to class k. Given a new time series $x^\wedge$, we can estimate the probability of $x^\wedge$ being a member of class k at time t as

$$p_k(t) = \frac{\sum\limits_{x_t \in X_k} K_H\left(x_t, \hat{x}_t\right)}{\sum\limits_{k=1}^{C} \sum\limits_{x_t \in X_k} K_H\left(x_t, \hat{x}_t\right)} \quad (2)$$

and assign a label according to

$$\hat{y}(t) = \underset{k}{\operatorname{argmax}} \; p_k(t). \quad (3)$$

So far our estimation is based on the observations made at time t, and we will now extend the concept to also account for observations made in the past. For this purpose, we define the kernel

$$K_H^{t,h}(x, \hat{x}) = \exp\left(-\frac{1}{2h} \sum_{l=t-h+1}^{t} \left(x_l - \hat{x}_l\right)^T H_l^{-1}\left(x_l - \hat{x}_l\right)\right), \quad (4)$$

which now uses subseries of length h, i.e., from time $t - h + 1$ to t. The estimation of class membership probabilities and subsequent assignment of a class label can be done analogously to (1.2) and (1.3), respectively. The different possible choices of h and $H_l \in \mathbb{R}^{d \times d}$ will be discussed in the following section.

2.3 Early Classification Framework

This subsection outlines the process of early classification. Given an incomplete time series $\hat{x}_{1:t}$ for some $t \in \{1, \ldots, T\}$, our goal is to obtain a class label using Eqs. (2) and (3) and the training samples $x_{1:t}^{(i)}$, $i = 1, \ldots, N$.

For this purpose, we need to specify the kernel $K_H^{t,h}$, which means choosing a window length h and the matrices Hl. For the latter, we consider

- setting Hl as the sample covariance matrix of $\left\{x_l^{(1)}, x_l^{(2)}, \ldots, x_l^{(N)}\right\}$
- applying Silverman's rule of thumb [5], which suggests using

$$\sqrt{H_{ii}} = \left(\frac{4}{d+2}\right)^{\frac{1}{d+4}} N^{\frac{-1}{d+4}} \sigma_i,$$

where σi is the sample standard deviation of $\left\{x_{l,i}^{(1)}, x_{l,i}^{(2)}, \ldots, x_{l,i}^{(N)}\right\}$.

Regardless of the choice, these matrices will have to be computed in a fitting step before a classification can be made. As for the window length, we choose to do a grid search over different parameter values.

After the kernel is specified, we start the prediction process at time $t = h$. We obtain a class label $y^\wedge(t)$ and a class membership probability vector $p(t)$. Based on the latter, we want to decide whether we trust the current prediction or move forward in time and predict again. For this purpose, we propose a Boolean decision function which will help us choose between the aforementioned options at each time t. First, we compute the entropy

$$H(p(t)) = -\sum_{k=1}^{C} p_k(t) \log_C(p_k(t)),$$

which serves as a tool to assess the uncertainty in our prediction at time t.

For example, assume that our classifier is indecisive, i.e., let $p_i \approx \frac{1}{C}$ for all $k = 1, \ldots, C$. Then

$$H(p(t)) = - \sum_{k=1}^{C} p_k(t) \log_C(p_k(t))$$

$$= - \sum_{k=1}^{C} \frac{1}{C} \log_C\left(\frac{1}{C}\right)$$

$$= 1.$$

On the contrary, we have $\lim_{p \to 0} p \log p = 0$ and $\lim_{p \to 1} p \log p = 0$, meaning that the entropy will shrink if the probability mass becomes more concentrated on a certain entry of the vector p, which we interpret as the classifier becoming more decisive toward the class corresponding to that entry. Therefore, once the entropy falls below a certain threshold, we stop the prediction process and assign the final label. The decision function hence takes the form

$$D(p, \gamma) = \begin{cases} \text{True:} & H(p) < \gamma \\ \text{False} & \text{else} \end{cases}, \qquad (5)$$

where the threshold γ is an additional parameter to be chosen.

If the number of components d of the multivariate time series is large, one might be interested in reducing the dimension by using only a subset of the components. We choose to do this at each time step by computing the ANOVA F-statistic between labels and features and selecting the $d' < d$ components with the highest scores. Finally, the complete procedure is summarized in Algorithm 1.

Algorithm 1: Kernel-based Time Series Early Classification.

Input: time series $\hat{x}$ to be classified, training data set X, window length h and entropy threshold γ, number of components to use d'

Output: predicted label $\hat{y}$ and earliness $t_{\hat{x}}$

Training Step
for For $t = h, \dots, T$ **do**
  Select the d' best components according to the ANOVA F-statistic
  Compute the matrix H_t
end

Prediction Step
Initialize $t = h$
stop = False
while *stop = False* **do**
  Calculate $p(t)$ and $\hat{y}(t)$ using equations (1.2)-(1.4)
  $t_{\hat{x}} = t$
  stop = $D(p(t), \gamma)$
  $t = t + 1$
end
Assign the final class label $\hat{y} = \hat{y}(t_{\hat{x}})$

2.4 The Gesture Dataset

This section briefly describes the data set and the corresponding gesture recognition task we want to address in this paper. Our data originates from an experiment, where

Table 1 Sensors used during experiment and corresponding locations

Sensor type	Location
MyoWare sensor 1	Third part of right forearm
MyoWare sensor 2	Above third part of right forearm
MyoWare sensor 3	Third part of left forearm
MyoWare sensor 4	Above third part of left forearm
IMU-6 sensor 1	Middle forearm of right hand
IMU-6 sensor 2	Middle forearm of left hand
IMU-9 sensor 1	Back of left hand
IMU-9 sensor 2	Back of right hand

a single participant was asked to perform 13 different two-handed gestures shown in Fig. 3 [1]. Dynamic gestures were performed by one right-handed healthy subject aged 28 years old. These gestures are repeated twenty times each with three minutes rest after each gesture.

A total of 20 trials were performed per gesture, each lasting six seconds. The participant has an initial position of the hands on the table; when the data collection starts, the subject starts moving the hand and makes the needed position stable for two seconds and then goes back to the initial position again [2]. Four inertial unit measurement (IMU) and four MyoWare sensors for sEMG collection were attached to the participants hands and arms (cf. Table 1)—operating at a frequency of 20 Hz—resulting in a data set of 260 multivariate time series, each consisting of $d = 34$ components with length $T = 127$. With the help of the algorithm described in the previous section, we aim to correctly recognize the gestures before they have been performed completely.

3 Results

We will now present and discuss numerical results for different choices of the bandwidth matrix, window size h and entropy threshold γ. For each set of parameters, we performed a tenfold cross-validation to obtain averages of prediction accuracy and earliness (presented as a fraction of T) which are.

displayed in Tables 2 and 3, respectively. Our first observation is that, regardless of the bandwidth matrix, there is a positive correlation between the window length h and the two quality measures (cf. Figs. 4 and 5). Regarding the earliness, this is to be expected by design of the algorithm since we start predicting at $t = h$. Increasing the window length is also equivalent to including a larger number of past observations which should intuitively increase the average accuracy of the predictions. We can make a similar observation about the entropy threshold γ: the lower we choose it, the greater

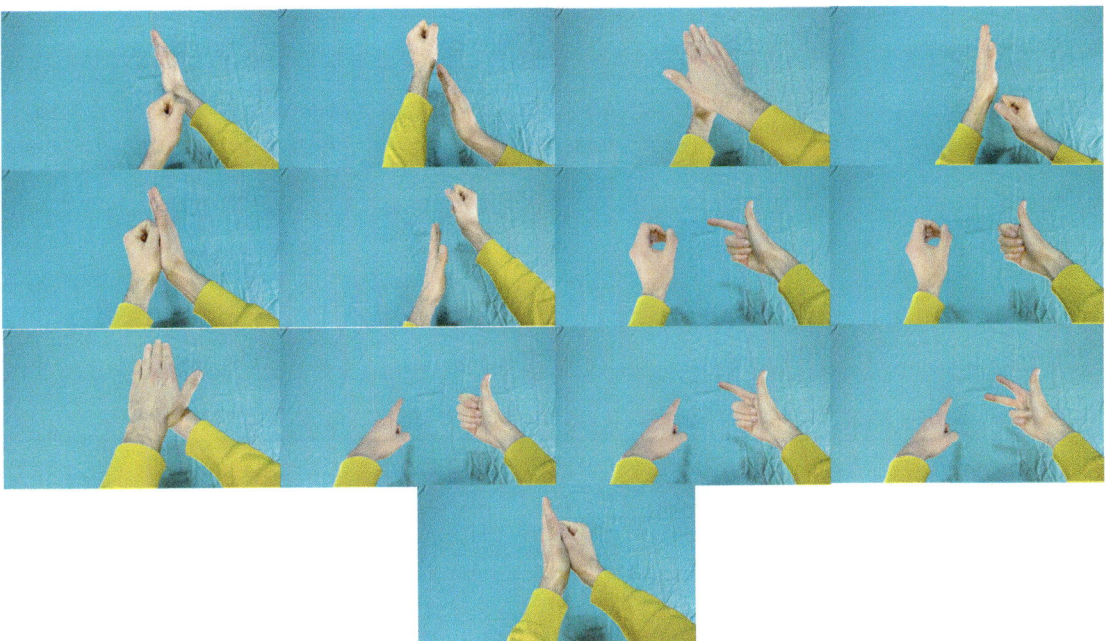

Fig. 3 Collected gestures

Table 2 Average prediction accuracy and earliness using the covariance bandwidth matrix

	Prediction accuracy					
h/γ	0.2	0.3	0.4	0.5	0.6	0.7
5	0.792	0.788	0.788	0.773	0.746	**0.696**
10	0.800	0.812	0.804	0.819	0.762	0.723
25	0.815	0.827	0.835	**0.858**	0.812	0.777
50	0.877	0.873	0.865	0.877	0.862	0.838
75	0.908	0.908	0.904	0.900	0.873	0.877
100	0.904	0.904	0.896	0.896	0.885	0.885
	Earliness					
h/γ	0.2	0.3	0.4	0.5	0.6	0.7
5	0.695	0.599	0.449	0.321	0.197	0.095
10	0.745	0.648	0.515	0.399	0.265	0.147
25	0.831	0.748	0.641	0.526	0.389	0.270
50	0.893	0.841	0.767	0.669	0.560	0.458
75	0.932	0.901	0.848	0.0783	0.695	0.625
100	0.963	0.949	0.923	0.888	0.845	0.804

the prediction accuracy. However, this delays the classification, leading to an increase in earliness. Clearly, the trade-off between accuracy and earliness is apparent and while we can control it by the choice of model parameters, there is no obvious best classifier. Instead, one has to decide how to value the two objectives against each other. One possible way could be to choose weights *wacc* and *wearl* with *wacc* + *wearl* = 1 and compute.

$$h^*, \gamma^* = \underset{h,\gamma}{\mathrm{argmax}}\, w_{acc} A(h, \gamma) + w_{earl}(1 - E(h, \gamma)),$$

where $A(h, \gamma)$ is the prediction accuracy and $E(h, \gamma)$ the earliness of the parameter pair (h, γ) obtained via cross-validation. In case of the covariance bandwidth matrix, for example, choosing $w_{acc} = 0.8$ then yields $(h^*, \gamma^*) = (25, 0.5)$, whereas for $w_{acc} = 0.5$ we get $(h^*, \gamma^*) = (5, 0.7)$. Similarly, for the Silverman bandwidth matrix, the same weights yield the parameter pairs $(h^*, \gamma^*) = (50, 0.2)$ and $(h^*, \gamma^*) = (5, 0.4)$, respectively (cf. Tables 2 and 3 for the corresponding accuracy and earliness).

Table 3 Average prediction accuracy and earliness using the Silverman bandwidth matrix

	Prediction accuracy					
h/γ	0.2	0.3	0.4	0.5	0.6	0.7
5	0.785	0.781	**0.735**	0.650	0.635	0.646
10	0.838	0.788	0.758	0.658	0.654	0.658
25	0.846	0.827	0.758	0.723	0.719	0.719
50	**0.904**	0.858	0.819	0.808	0.800	0.796
75	0.915	0.877	0.865	0.865	0.858	0.858
100	0.908	0.915	0.919	0.912	0.908	0.908
	Earliness					
h/γ	0.2	0.3	0.4	0.5	0.6	0.7
5	0.345	0.205	0.108	0.056	0.038	0.033
10	0.390	0.238	0.132	0.093	0.076	0.072
25	0.457	0.311	0.225	0.198	0.190	0.189
50	0.577	0.460	0.411	0.392	0.388	0.386
75	0.702	0.612	0.592	0.586	0.583	0.583
100	0.835	0.794	0.786	0.782	0.780	0.780

Fig. 4 Relationship of accuracy/earliness and model parameters (covariance bandwidth matrix)

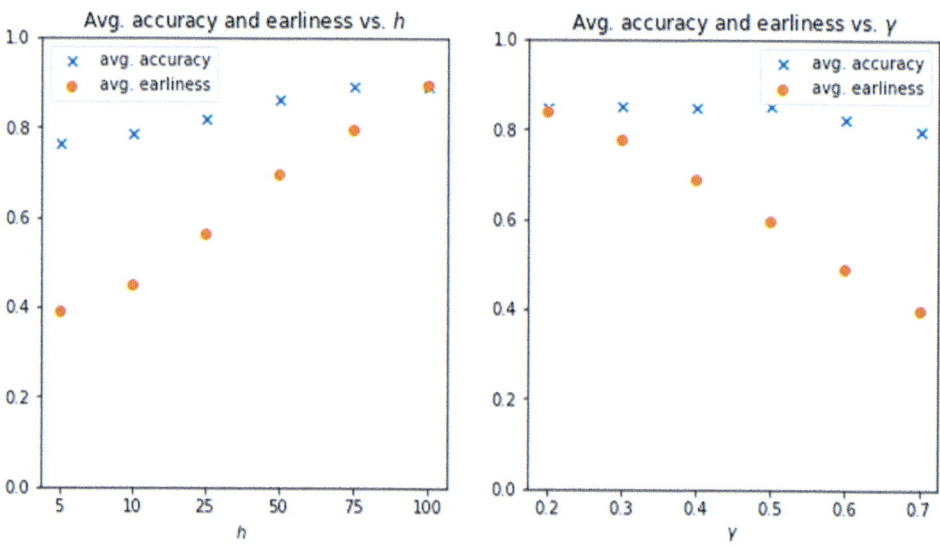

Fig. 5 Relationship of accuracy/earliness and model parameters (Silverman bandwidth matrix)

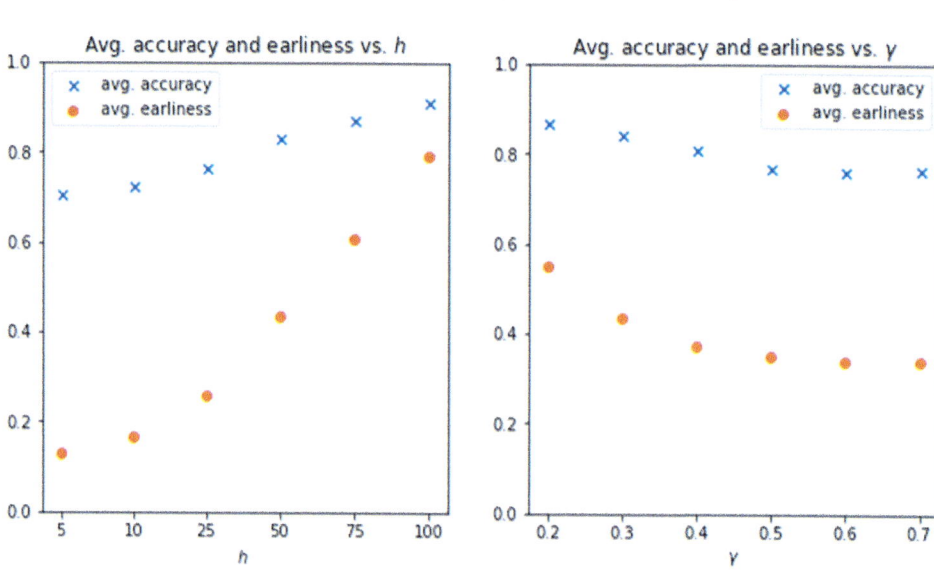

4 Conclusion

In this paper, we proposed a new algorithm for the task of gesture recognition based on sEMG and IMU signals. Using a Nadaraya–Watson kernel estimator, we introduced a classifier which is able to make predictions based on partially observed time series of arbitrary length. We then defined an entropy-based criterion in order to enable early classification. Our model successfully treats the trade-off between the two goals of prediction accuracy and earliness of the prediction. Furthermore, the user is able to control said trade-off by the choice of model hyperparameters. Hence, we conclude that human gestures can be predicted with reasonably high accuracy before the full motion is completed. The subject of future work could be the further development of the proposed algorithm, e.g. regarding the choice of the window length parameter h or the selection of the components of the multivariate time series, as well as the application to other data sets.

Acknowledgements This work is funded by the Deutsche Forschungsgemeinschaft (DFG, German Research Foundation)—Project-ID 416228727—SFB 1410.

References

1. Carl Gabert, Achraf Djemal, Hiba Hellara, Bilel Ben Atitallah, Rajarajan Ramalingame, Rim Barioul, Dennis Salzseiler, Ellen Fricke, Olfa Kanoun, & Ulrike Thomas. Gesture based symbiotic robot programming for agile production. In *2022 IEEE 9th International Conference on Compu- tational Intelligence and Virtual Environments for Measurement Systems and Applications (CIVEMSA)*. IEEE.

2. Hiba Hellara, Achraf Djemal, Rim Barioul, Rajarajan Ramalingame, Bilel Ben Atitallah, Ellen Fricke, & Olfa Kanoun. (2022). Classification of dynamic hand gestures using multi sensors combinations. In *2022 IEEE 9th International Conference on Computational Intelligence and Virtual Environments for Measurement Systems and Applications (CIVEMSA)*. IEEE.

3. Nadaraya, E. A. (1964). On estimating regression. *Theory of Probability and its Applications, 9*(1), 141–142.

4. Lorna, C., Quandt, & Willis, A. S. (2021). Earlier and more robust sensorimotor discrimination of asl signs in deaf signers during imitation. *Language, Cognition and Neuroscience, 36*(10):1281–1297

5. Silverman, B. W. (1998). *Density Estimation for Statistics and Data Analysis*, volume 1. Chapman & Hall/CRC

6. Geoffrey, S., Watson (1964). Smooth regression analysis. *Sankhya: The Indian Journal of Statistics, Series A*, 26(4):359–372.

7. Li Zhang, Geng Liu, Bing Han, Zhe Wang, & Tong Zhang. (2019). sEMG based human motion intention recognition. *Journal of Robotics*, 2019:1–12.

Person Detection and Differentiation in Shopping Scenarios

Divyasha Naik, Martin Reber, Tom Uhlmann,
and Guido Brunnett

Abstract

In the research field of human–robot-interaction, detection of people is of central importance. Moving robotic platforms create additional challenges due to an absence of a defined area of interaction between humans and robots. The use of a mobile robotic platform to provide personalized assistance requires that tracking and identification of the interaction partner in a group of people can be ensured at all times. This work contributes to an application that can detect and distinguish multiple people in public spaces. We use Mask R-CNN followed by DeepSORT for the differentiation and tracking of each individual in a video by assigning a unique ID. We apply our approach to the Multi-Object Tracking and Segmentation (MOTS20) data set and show that our method successfully performs person detection, differentiation, and tracking at 5 frames per second. Using the three evaluation metrics Multiple Object Tracking (MOTA), the Higher Object Tracking (HOTA) and the Identification metric (IDF1), we achieved an accuracy of 49.69%, 48.35% and 56.14%, respectively. Our results show good accuracy, but also possibilities for future improvements. We show how reflections, shadows, human-like clothing, and poor annotations influence the precision negatively.

D. Naik (✉) · M. Reber · T. Uhlmann · G. Brunnett
Computer Graphics and Visualization, Chemnitz, Germany
e-mail: divyasha.naik@informatik.tu-chemnitz.de

M. Reber
e-mail: martin.reber@informatik.tu-chemnitz.de

T. Uhlmann
e-mail: tom.uhlmann@informatik.tu-chemnitz.de

G. Brunnett
e-mail: guido.brunnett@informatik.tu-chemnitz.de

Keywords

Person detection · Person tracking · Mask R-CNN · DeepSORT

1 Introduction

The ubiquity of portable devices with integrated cameras has significantly improved the number of images and videos that are created every day. Apart from the recreational sector, video material is used today for different serious applications as security surveillance and production control. Thus, the field of computer vision has attracted much attention in recent years. One of the most important sub fields of computer vision is object recognition. A machine learning method uses object detection to observe objects in video streams and learn about their environment.

As a prominent example in the research field of human-robot interaction, human recognition is of central importance. In everyday situations, a mobile robotic platform can be used to provide personalized assistance to an individual, which requires that tracking and identification of the interaction partner in a group of people can be ensured at any time. Detecting and tracking multiple people becomes more challenging when the camera is constantly moving. Despite the outstanding performance and advanced level of computer vision systems, machine vision still fails when objects have to be recognized in a dynamic environment. Therefore, a system needs to be developed that provides a sufficiently high success rate even with rapid changes in environment.

In this study, we developed a system for the task of recognizing multiple people and distinguishing them in a crowd. We use two deep learning algorithms Mask R-CNN and DeepSORT to distinguish and track each person in a video by recognizing them at the pixel level and assigning them a unique ID. As test data, we used the publicly available Multiple Object Tracking Benchmark (MOTS20)

B. Meyer et al. (eds.), *Hybrid Societies*, Advances in Science, Technology & Innovation,
https://doi.org/10.1007/978-3-032-03488-5_33

data set. The presented methodology successfully performs person detection, discrimination, and tracking in a real-time application.

This work is divided into 6 sections. Section 2 is intended to provide a detailed study of the state of the art in object detection and differentiation. First, existing machine learning algorithms are examined, then some improvements and drawbacks are discussed. Section 3 deals with the data set used for testing and evaluation. It describes the ground truth data and the annotation techniques used. In Sect. 4, the proposed methodology and data preprocessing techniques are explained in detail. In Sect. 5, various experiments are conducted and their results are evaluated. The performance is visualized and the results are analyzed. The last Sect. 6 summarizes the overall performance of the paper and suggests ways for possible improvements in future experiments.

2 Related Work

Person recognition research began before the 2000s and remains an open and challenging problem. Ross Girshick and collaborators present a simple and scalable person detection algorithm, the Region-based Convolutional Neural Network (R-CNN) approach, which offers a relative improvement of 30 percent over the best previous results on the PASCAL VOC 2012 data set. However, it is still very time-consuming to train the network to classify 2000 region proposals per image. An improved version called Fast R-CNN [2] was developed to overcome the drawbacks of R-CNN, where the image is sent through the CNN only once, but despite the improvement it does not work in real-time scenarios.

Therefore, a new module Faster R-CNN [7] was introduced, which eliminates the need for the selective search algorithm and ensures that the network learns the region proposals itself. This method is currently leading in several benchmarks with bounding boxes. Mask R-CNN [3] provides a pixel mask for each recognized human which adds only a small overhead to Faster R-CNN and runs at 5fps.

In 2016, Alex Bewley used the Kalman filter together with the Hungarian algorithm, in particular to reduce ID switching in person tracking and to track a person even after obstacles appear, called Simple Online and Real-time Tracking (SORT) [1]. The main goal of this algorithm was simply to track multiple objects simultaneously, and it should be sufficient to identify objects for successive images in real-time applications. Nevertheless, the results were not satisfactory because there were many ID changes and the system was not able to recognize the appearance of the same person again in successive images.

In addition to SORT, the deep association DeepSORT [9] was developed to improve accuracy and reduce false detections. This is based not only on the Kalman filter and the Hungarian algorithm, which detects a person based on distance and speed alone, but also on the appearance of an object using CNN.

Tracking of multiple individuals based on Faster R-CNN and Deep Appearance features was developed in 2019 [4] has achieved human tracking limited only to bounding box and fails in dark environmental conditions. The method called Track R-CNN [8] was successfully able to track and segment multiple objects on the data set MOTS at 2 frames per second, which seems insufficient for real-time applications.

3 Data Set

The data set used in this work is the publicly available Multi-Object Tracking and Segmentation (MOTS) [8] which contains pixel level annotations. A detailed description of the data set can be found in Table 1 and sample images in Fig. 1.

Labeling pixel masks for each frame of each person in a video is an extremely time-consuming task. Therefore, the availability of such data is very limited. There are some other data sets with MOT annotations [2], i.e., tracks annotated at the bounding box level. However, for the MOTS task, these

Table 1 Description of the data set

Name	No. of frames	Resolution	Length	FPS
MOTS20-02	600	1920 × 1080	(00:20)	30
MOTS20-05	837	640 × 480	(01:00)	14
MOTS20-09	525	1920 × 1080	(00:18)	30
MOTS20-011	900	1920 × 1080	(00:30)	30

MOTS20-02 MOTS20-05 MOTS20-09 MOTS20-11

Fig. 1 Sample images from the MOTS20 data set

data sets lack segmentation masks. We used the MOTS20 data set for two main reasons.

- The main task of this work was to detect people at the pixel level.
- It contains multiple pedestrians in a shopping scenario with a mobile robotic platform.

Data compression techniques called run-length encoding (RLE) are used to store the pixel masks in the ground truth annotation file.

4 Proposed Method

4.1 Data Preprocessing

The ground truth file of the MOTS data set contains annotations for both people and vehicles. Since the goal of this project is only to detect and track people, the vehicle class was neglected in the annotation file. Second, RLE needs to be converted to a suitable format so that it can be understood by machines. This was done using coco-tools [5]. Pycocotools is a set of Python functions included in cocoapi that can handle different types of annotation data sets. Based on the image height, width, and RLE, the mask is decoded for each individual in the entire video, as shown in Fig. 2.

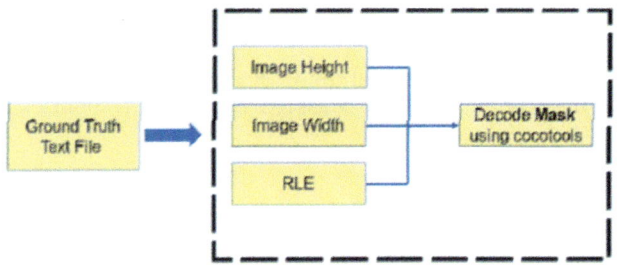

Fig. 2 Block diagram of data preprocessing of ground truth file

4.2 Methodology

There are two main steps in the development of the MOT system: person recognition and association. The quality of recognition has an impact on tracking performance. To detect and distinguish multiple people in the MOTS20 data set, we use the popular Mask R-CNN deep learning module, which adds a mask head to the Faster R-CNN detector. This module uses pre- trained Coco weights with the TensorFlow framework. It is capable of classifying 80 different classes, with the exception of background. The modification of the Mask-R-CNN module was necessary to keep the class ID, the mask and the bounding boxes only for the human class at the output and neglect all other classes.

The other main goal of this project was to track these multiple people detected by Mask R-CNN module in the video stream. Therefore, we used another Deep Learning module called DeepSORT to find moving entities in the video footage. Deep SORT uses the Kalman filter features to predict the position of the person in future frames, the Hungarian algorithm to find the best association, and generates a cost matrix from the Deep Convolutional Neural Network based on motion information and appearance features. The Mask R-CNN module was integrated with the DeepSORT module in two ways (Fig. 3).

- Tracking with Bounding Boxes—The Mask R-CNN module returns class id, which specifies the class of the object, the bounding boxes of the detected object, the confidence value, and the predicted mask as shown in Fig. 4. Of these, class id, bounding boxes, and confidence values were first passed to the DeepSORT module, which tracks the person using only the bounding boxes.
- Tracking with Masks—Detection, discrimination and tracking of multiple pedestrians in successive images beyond 2D bounding boxes is enabled by integrating two modules with "MASK." In this approach, the class of the object, the confidence value, and the mask are fed as input to the DeepSORT module.

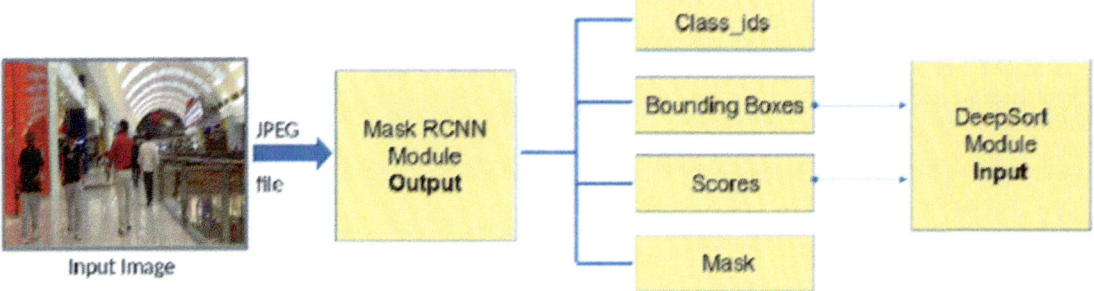

Fig. 3 Block diagram of the integration of mask R-CNN and DeepSORT

Fig. 4 **a** Ground Truth image 187 of the MOTS20-11 data set, **b** Corresponding Mask R-CNN output

(a)   (b)

After the model has been trained, the most important part is to evaluate it. The standard and commonly used technique for evaluating a Mask RCNN is Intersection over Union (IOU). The ground truth provides labels in a particular order that does not match the results predicted by the Mask RCNN model. Therefore, a matching algorithm based on the Chamfer distance principle was used to correctly match them. The mask predicted by Mask RCNN is overlapped with the corresponding ground truth data, and the mask with the highest overlap value is considered the correct label.

5 Experiment and Results

To evaluate our person detector and tracker, we follow the three evaluation techniques [6], multiple object tracking accuracy (MOTA), identification metric (IDF1), and higher order tracking accuracy (HOTA), and the results are shown in Table 2. When observing Table 2 CLEAR MOT metric carefully, despite the high number of ID switches, the accuracy of MOTA is 59.52% in the MOTS20-05 data set and

conversely 27.19% in the MOTS20-02 data set. This is because MOTA is largely based on recognition accuracy and is hardly affected by changes in IDSW. The performance of all video streams appears to be almost the same, as shown in Table 2 IDF1 metric, which focuses more on tracking.

HOTA is considered the current state of the art for evaluating multiple object tracking and is calculated by giving equal weight to detection accuracy (DetA) and association accuracy (AssA), which contributes to a balance between the two. From Table 2 HOTA metric, IOU (LocA) is more than 80%, which shows correctly detected individuals accuracy. The lowest accuracy of 42.35% was obtained for the MOTS20-02 video from HOTA, which is due to the fact that a total of 2455 individuals were missing from the MOTS20-1 video due to insufficient annotations. For the MOTS20-05 data set, the accuracy of HOTA is 46.11% lower than the other two data sets, which is due to the fact that 903 individuals were not detected in the MOTS20-05 video stream due to person reflections and human-like clothing. An example of such a false recognition can be seen in Figs. 4 and 5. The red mark in Fig. 1.4b, the MaskRCNN module recognizes a

Table 2 MOTA IDF1 and HOTA evaluation results along with their sub metrics on the MOTS20 data set

Data set	CLEAR MOT metric			IDF1 metric			HOTA metric			
	$MOTA^a$ %	$MOTP^b$ %	$IDSW^c$ no	$IDF1^d$ %	IDR^e %	IDP^f no	$HOTA^g$	$DetA^h$	$AssA^i$	$LocA^j$
MOTS20-02	27.19	77.34	52	52.61	61.83	45.78	42.35	42.66	42.40	80.21
MOTS20-05	59.52	80.67	106	54.94	51.16	59.34	46.11	53.68	40.13	82.82
MOTS20-09	52.18	82.74	33	59.76	61.0	58.57	51.21	52.34	50.43	84.65
MOTS20-11	59.27	84.89	36	58.41	57.23	59.64	53.71	56.65	51.0	86.29
Combined	49.69	81.46	227	56.14	57.62	54.74	48.35	50.77	46.44	83.96

[a] Multiple Object Tracking Accuracy
[b] Identification recall
[c] Identity Switch
[d] Identification metrics
[e] Multiple Object Tracking Precision
[f] Identification precision
[g] Higher Order Tracking Accuracy
[h] Detection Accuracy
[i] AssociationAccuracy
[j] Localization Accuracy

Fig. 5 **a** Ground Truth image 1 of the MOTS20-02 data set, **b** Corresponding Mask R-CNN output

(a) (b)

Fig. 6 **a** Ground Truth image 1 of the MOTS20-11 data set, **b** Corresponding Mask R-CNN output

(a) (b)

person reflection as a person. While in Fig. 1.5b human-like clothing is detected as a person.

Consider an example: looking closely at Fig. 6a, there are a total of 14 individuals in the ground truth, and in Fig. 6b, the Mask R-CNN module detects a total of 12 individuals. Here, a discrepancy was found between the number of individuals detected by the module and the actual labels. Looking again at Fig. 6b, the Mask R-CNN module detected 2 individuals in the right corner of the image, but in Fig. 6a, the corresponding Ground Truths are missing. This indicates poor annotations.

TrackRCNN is the method used with the MOTS20 dataset to detect and track individuals, and is considered the benchmark with MOTA as 65.1% and MOTP as 75.7%, capable of producing 2 frames per second. Our approach for the same dataset has achieved MOTA as 49.69% and MOTP as 81.46% with 5 frames per second.

Python 3.7, a user-friendly programming language, is used in this work and implemented on Google Colab Tesla K80 GPU version with TensorFlow 1.15.2 and CUDA version 11.2. Mask R-CNN generates at least 5fps and DeepSORT 16fps. The processing time of MOTS20-02 is 0.19s for one image, MOTS20-05 is 0.11s, MOTS20-09 is 0.14s, and MOTS20-11 is 0.15s. The MOTS20-02 data set is highly crowded compared to the others and therefore takes comparatively more time.

6 Conclusion

Using the TrackRCNN [8] method, the accuracy of MOTA is 65.1% at 2fps. The presented method achieves all the goals of person detection, differentiation, and tracking at 5fps. Combined accuracy of MOTA is 49.69%, HOTA is 48.35% and

IDF1 is 56.14%. Crowded scenes can lead to missing detections and id switches. In the process of evaluation of our system it became clear that some false detections are due to reflections, shadows, or clothing that resembled humans. Therefore, future research is needed to overcome these deficits. These limitations can be addressed by reflection suppression using convex optimization technique, by improving the annotations, and by training the model with more datasets.

Acknowledgements This research was funded by the Deutsche Forschungsgemeinschaft (DFG, German Research Foundation)—Project-ID 416228727—CRC 1410.

References

1. Bewley, A., Ge, Z., Ott, L., Ramos, F., & Upcroft, B. (2016). Simple online and realtime tracking. In *2016 IEEE international conference on image processing (ICIP)*, IEEE, pp. 3464–3468
2. Girshick, R. (2015). Fast r-cnn. In *Proceedings of the IEEE international conference on computer vision*, pp. 1440–1448
3. He, K., Gkioxari, G., Dollár, P., & Girshick, R. (2017). Mask r-cnn. In *Proceedings of the IEEE international conference on computer vision*, pp. 2961–2969
4. Khan, G., Tariq, Z., Khan, M. U. G., Mazzeo, P., Ramakrishnan, S., & Spagnolo, P. (2019). Multi-person tracking based on faster r-cnn and deep appearance features. In *Visual Object Tracking with Deep Neural Networks* (pp. 1–23) IntechOpen London, UK
5. Lin, T.-Y., Maire, M., Belongie, S., Hays, J., Perona, P., Ramanan, D., Dollár, P., & Zitnick, C. L. (2014). Microsoft coco: Common objects in context. In *European conference on computer vision* (pp. 740–755) Springer
6. Luiten, J., Osep, A., Dendorfer, P., Torr, P., Geiger, A., Leal-Taixé, L., & Leibe, B. (2021). Hota: A higher order metric for evaluating multi-object tracking. *International journal of computer vision*, 129(2):548–578

7. Ren, S., He, K., Girshick, R., & Sun, J. (2015). Faster r-cnn: Towards real-time object detection with region proposal networks. *Advances in neural information processing systems, 28*

8. Voigtlaender, P., Krause, M., Osep, A., Luiten, J., Sekar, B. B. G., Geiger, A., & Leibe, B. (2019). Mots: Multi-object tracking and segmentation. In *Proceedings of the IEEE/cvf conference on computer vision and pattern recognition*, pp. 7942–7951

9. Wojke, N., Bewley, A., & Paulus, D. (2017). Simple online and real-time tracking with a deep association metric. In *2017 IEEE international conference on image processing (ICIP), IEEE*, pp. 3645–3649

Modeling Intentional Complexity in Hybrid Interaction Scenarios Beyond Explicit and Implicit Communication

Martin Siefkes, Ellen Fricke, Jana Bressem, and Akira Charoensit

Abstract

The present contribution investigates the role of intentionality in human–machine interactions. Building on previous research on intentionality in human–human interactions, it is shown that different "levels of intending" (including complex nested intentions and beliefs) need to be taken into account; dichotomies such as "implicit versus explicit" are insufficient to capture the necessary distinctions. Adequately guessing the degree of intending of interaction behavior, from basic actions up to and including communication processes, requires taking into account all relevant aspects of the behavior (speech, gesture, gaze, and further aspects of movement and/or body behavior). Furthermore, an adequate "model of intentionality" is needed in order to infer the degree of intending of an interaction based on various types of cues. It will be argued that Embodied Digital Technologies (EDTs) with the capabilities necessary to adequately infer and represent intentions and beliefs in their mental models of themselves and of interactants may be able to achieve improved situational awareness and a more human-like interaction quality.

M. Siefkes (✉) · E. Fricke · J. Bressem · A. Charoensit
Faculty of Humanities, German Linguistics, Semiotics and Multimodal Communication, Technische Universität Chemnitz, Chemnitz, Germany
e-mail: martin.siefkes@phil.tu-chemnitz.de

E. Fricke
e-mail: ellen.fricke@phil.tu-chemnitz.de

J. Bressem
e-mail: jana.bressem@phil.tu-chemnitz.de

A. Charoensit
Laboratoire d'Informatique de Robotique et de Microélectronique de Mtp, LIRMM, Pôle de Recherche Mathématiques, Informatique, Physique, Systèmes, MIPS, Université de Montpellier, Montpellier, France

Keywords

Intentionality · HRI · Communication · Multimodality · Gestures · Complexity · Speech act theory

1 Introduction

All human communication is connected with intentionality. Participants intend to achieve something (at the very least to be understood), and have certain beliefs about the intentions and beliefs of the other communication partners. During communication, each participant incrementally develops a model of these intentions and beliefs for all the participants (e.g., [1–3]). In recent years, further advances in the theory of intentionality have aimed toward a description of collective states of intentionality based on shared attention [4, 5]. In the 1990s, Roland Posner developed a formal notation system for describing complex communicative intentions, including higher-level (nested) intentions and beliefs about one's own and the interaction partner's intentions and beliefs [6]. The approach is based on John Searle's concept of "intentionality" [7], on speech act theory (e.g., [8, 9]), and on discussions in linguistics and semiotics about the threshold between communication and lower-level sign processes [10, 11]. It provides a formalized approach toward communicative intentions and beliefs recognizing various "levels of intendedness," including complex nested intentions and beliefs the interlocutors develop during communication about each others' mental states [12, 13]. Posner's model explicitly considers communicative intentions in the context of human-machine interaction, building on the assumption that machines will have to recognize and adequately represent human intentions and beliefs of varying degrees of complexity.

The present contribution first introduces Posner's model, focusing on the most important aspects for the purpose of this article. It then proceeds to demonstrate, using empirical as well as constructed examples in various scenarios, the relevance of distinguishing between different degrees of intendedness, both in human-only and in human-machine interactions up to and including communication. With a specific focus on first encounter scenarios in the public space, it will be shown that a description based on Posner's semiotic model of intentionality is more adequate, at least for certain scenarios and under certain context conditions, than more simple dichotomic distinctions between "intended" and "non-indended" behavior, or between "implicit" and "explicit" cues.

The goals of this article are

- to show that it is important to distinguish between different degrees of intentionality in situations of interaction and communication,
- to demonstrate that the formalization proposed by Posner is suitable for the description of examples of varying complexity (using both empirical examples from recorded real-world interactions and constructed examples),
- to provide evidence that Embodied Digital Technologies (EDTs), if they are to interact in a human-like fashion, will have to possess a mental model of their own and their interaction partner's intentionality that allows them to adequately distinguish between different cases, to draw correct conclusions on the basis of observed clues related to intentionality, and to react in the appropriate manner.

It will be argued that in order to operationalize the concept of "levels of intentionality" for the analysis of real-world interaction situations, it is useful to distinguish between an "indicating condition" and a "communication condition." In every sign process or communication, either none of these conditions, the indicating condition, or both conditions can be fulfilled. The ability to distinguish between these three cases should lead to higher situational awareness and communicative capabilities, in both humans and EDTs.

2 Intending, Causing, and Believing: Posner's Formal Semiotic Approach

In 1993, Roland Posner published an approach that presented a description of communication processes based on a representation of their intentional foundations ([6]; an abridged German version is [14]). Posner's model formalizes the mental representations of intentional states connected with various types of sign processes. The aim of the model is twofold: "[The model] is offered as a first step toward both the reconstruction of what happens in human communication and the construction of communication behavior in artificial intelligent systems" ([6], p. 238).

Intentionality, in the sense of [7], is the assumption that both sender and receiver of a communication take a specific stance toward the message that is communicated. While intentional states can be complex, the basic building blocks are "Believing" in a proposition p, and "Intending" for a proposition p to become true. Posner (and previous models, e.g. [10]) build on these assumptions and aim to describe intentionality via nested intentions and beliefs that are connected with the production or the interpretation of a sign. The schema developed by Posner includes definitions of five different speech acts [9], which are represented in the top row of the schema (Fig. 1), but it also includes simpler sign processes below the level of communication, where intentions and beliefs already play a role. The schema includes these processes in the lower levels (cf. Fig. 1, rows 1a to 2b), while processes of communication are represented on the highest level (row 2bcom).

The formalization is simple and uses just a few concepts of intentional logic. Three predicators are used ([6], p. 221f.): $E(f)$ denotes the occurrence of an event f, $T(a, f)$ denotes that a does f, where the first argument a denotes a behavioral system and the second argument f denotes a behavior, and $Z(a)$ denotes that a is in a specific inner state.

The formalization also uses three sentence operators ([6], 222f.): "$\rightarrow$" is a two-place sentence operator denoting the cause-and-effect relation; e.g., $E(f) \rightarrow E(e)$, where f and e are event terms, denotes that the occurrence of event f causes the occurrence of event e. The second sentence operator is G, which "is a two-place operator having a behavioral system as its first argument and a proposition as its second, and denoting the relation of belief that holds between the behavioral system and the proposition." ([6], p. 221) For example, $G(a, p)$ denotes that a (which could be a human, an animal, or a robot capable of intentional behavior) believes p. Finally, 'I' is a two-place operator having a behavioral system as its first argument and a proposition as its second, and denoting the property of the behavioral system having the intention to realize the proposition." ([6], p. 223) For example, $I(a, p)$ denotes that a intends p to become true.

The lowest level of the model consists of four basic sign types, and on the cause-effect relation as the most basic type of a process that underlies any sign process:

- "[1a/I, cause:] If there are no behavioral systems involved, we are dealing with a simple causal process; f is a cause and the occurrence of f causes the occurrence of e: $E(f) \rightarrow E(e)$.
- [1a/II, signal:] If there is a reacting system a intervening between cause f and effect e, f functions as a signal for a to do r, this is because the occurrence of f causes a to do r: $E(f) \rightarrow T(a,r)$.
- [1a/III, indicator:] If the response of a to the occurrence of f is not just any event, but consists in a believing that p, then f functions as an indicator of p for a; the occurrence of f causes a to believe p: $E(f) \rightarrow G(a,p)$.

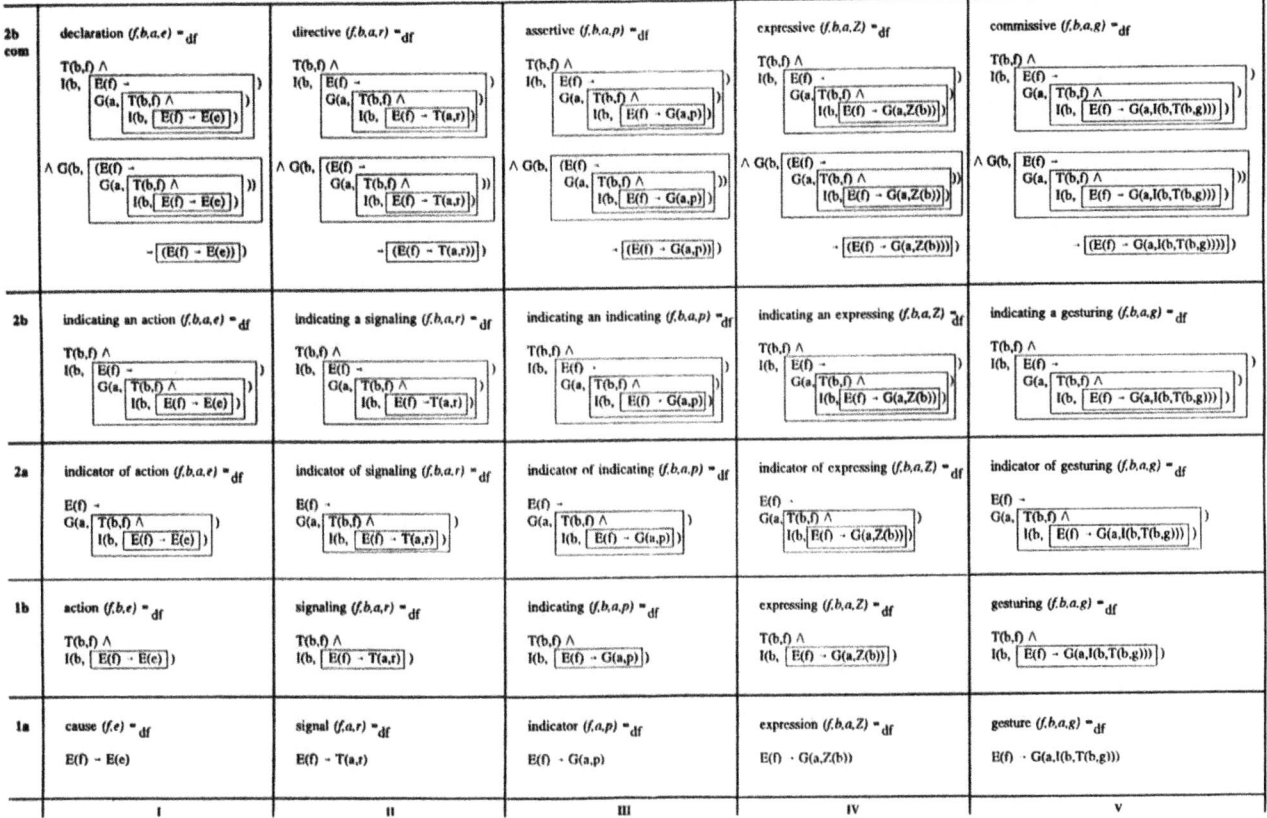

Fig. 1 Posner 's complete model, describing complex intentions and beliefs accompanying semiotic processes up to the level where it can be called "communication" (2bcom)

- [1a/IV, expression:] If what a believes upon registering f is that there is a b that produced f and that is in state Z, then f is an expression of Z for a; the occurrence of f causes a to believe in state Z of b: $E(f) \rightarrow G(a, Z(b))$.
- [1a/V, gesture = expression of intention [1]:] If the state Z of the acting system b assumed by the reacting system a is an intention to produce a further event g, then f is a gesture of b to do g for a; the occurrence of f causes a to believe in the intention I of b to do g: $E(f) \rightarrow G(a, I(b, T(b,g)))$." ([6], p. 223)

On the vertical axis, Posner's schema represents sign types with nested beliefs and intentions of increasing complexity.

Posner explains this, using the categories from column III as an example:

"When an acting system b comes into play, b can produce an event f because b intends this event to cause a to do or believe something. a can then believe that a certain event f was produced by b because b intended f to cause a to do or believe something. b can even intend this, i.e., intend a to believe that a certain event f was produced by b because b

intended it to cause a to do or believe something. And a can in turn have reason to believe in that complex intention of b, etc." ([6], p. 227).

In order to distinguish between the different levels, we can construct examples that may occur in traffic situations: (1a/III) In the most basic case, the occurrence of an event f causes a (the receiver of the sign) to believe in a specific proposition p. f acts as an *indicator* (or *index*) for a. For example, a car slows down, and an observer assumes that the driver is looking for a parking space:

$$E(f) \rightarrow G(a, p) \tag{1}$$

(1b/III) When a sender b produces the sign f with the intention that it should become an indicator for a, this is called *indicating*. For example, a driver may slow down with the intention that a pedestrian waiting on the side of the road believes she has sufficient time to cross the street.

$$T(b, f) \wedge I(b, E(f) \rightarrow G(a, p)) \tag{2}$$

(2a/III) The next level is reached if a pedestrian a sees the driver b slowing down (event f), and assumes that this is an *indicator of indicating* directed at a that it is safe to cross the street.

---

[1] More recently, Posner preferred to call this category "intention expression", in order to distinguish it from gesture in the sense of "meaningful body movement" (personal communication with M.S., ca. 2013).

$$E(f) \rightarrow G(a, T(b, f) \wedge I(b, E(f) \rightarrow G(a, p))) \qquad (3)$$

(2b/III) If the driver slows down (event f) with the intention that the pedestrian believes that the driver does f with the intention that the driver believes it is safe to cross the street, this is called *indicating an indicating*. The difference to level 1b is that b does not intend for a to believe p (that it is safe to cross the street), but rather to believe that b intends for a to believe this. In order to achieve this, b has to make this higher-level intention clear to a, for example, by breaking in a quite obvious, somewhat 'unnatural' manner, maybe by breaking stronger or earlier than necessary, thus making it clear that the breaking is intended as a sign.

$$T(b, f) \wedge I\,(b, E(f) \rightarrow G\,(a, T(b, f) \wedge I(b, E(f) \rightarrow G(a, p)))) \qquad (4)$$

(2bcom/III) In order to adequately describe communication, it is also necessary to add a further complex belief of the sender. In order for the sign process to count as a communication, the sender b has to believe that the whole formula given above actually causes the receiver a to believe in p. This condition excludes certain special cases, for example, where b is aware that a already knows p, or that b will not believe it, which Posner, building on previous discussions, does not regard as normal acts of communication.

$$T(b, f) \wedge I(b, E(f) \rightarrow G(a, T(b, f) \wedge I(b, E(f) \rightarrow G(a, p))))$$
$$\wedge\, G(b, (E(f) \rightarrow G(a, T(b, f) \wedge I(b, E(f) \rightarrow G(a, p))) \rightarrow (E(f) \rightarrow G(a, p))) \qquad (5)$$

The formal representations are intended to delineate various "levels of intending" in communication, and to distinguish the intentional underpinnings of the respective sign processes, up to and including different types of speech acts on the level 2bcom.

"The two results of this section are highly relevant for two central tenets of traditional speech act theory. Declarational, directive, assertive, expressive and commissive acts of communicating are counterparts of the five types of speech acts postulated in [9]. As Searle claims, all human speech acts belong to one of these types, but he cannot prove this claim because he does not lay open its analytical basis. If one accepts our definitions of declarations, directives, assertives, expressives, and commissives as explications of Searle's speech act types, one can use our conceptual system to deduce the universality and completeness of Searle's typology" ([6], p. 238).

One achievement of Posner's approach is that it formalizes the relationship between simple sign types and speech acts, integrating them with various signs on intermediate level into a general model giving descriptions on the intentional basis of communication. Another innovative aspect of Posner's schema is that it is not limited to *verbal* utterances. Utter-

ances possessing different materiality, as well as multimodal utterances (such as gesture-speech combinations), can function as signs of the different categories described in Posner's schema. Instead of the verbal utterance "Stop!", a gesture or a stop sign, or a combination of signs of different materiality, could be used to conduct the same utterance [15, 16].

---

3 Attributing and Indicating Intentions in Multimodal Interaction Scenarios: Interpersonal and Hybrid Settings

Imagine a future everyday scenario at Potsdamer Platz [Potsdam Square] in Berlin with humans and EDTs, such as driverless cars or delivery robots, spontaneously interacting in public space. Asking for a particular route description is a very common request. Tourists visiting Berlin for the first time and searching for the theater at Marlene-Dietrich-Platz (see Fig. 2 below) could turn to human passersby or non-human tourist guide robots and ask for advice.

Communicating a route description with the goal that the addressee will later be able to walk this route independently requires more than only indicating the guide's or adviser's intention(s) (see the upper half of the formula 2bcom/II, Fig. 1). As it is pointed out in [18], p. 1710, "it also requires the sender's belief that through indicating them he can cause the wishes to be fulfilled by the addressee" (see the lower half of the formula 2bcom/II). Thus, according to [6], p. 244ff, two conditions must be fulfilled ([18], p. 1710):

1. **indicating condition**: a sender b produces an event f with the intention that f causes the recipient a to believe that b intends a to behave in a certain way r, and

Fig. 2 Street leading to the musical theater at Potsdamer Platz in Berlin ([17], p. 366)

2. **communication condition**: b believes that his or her belief will cause a to behave in that way r.

In multimodal interaction, for example, the indicating condition is related to a repertoire of particular semiotic behavior (as part of the event f) that may indicate different degrees of intending (e.g., intensification, additional gaze directing the addressee's attention etc.) and/or levels of abstraction in Posner's schema (see Fig. 1). In principle, this fact also applies to semiotic properties of interacting EDTs and is highly relevant for their communicative design ([2, 3]).

To illustrate this, let us first elaborate the constructed example of a tourist in our Potsdamer Platz scenario, who is standing near the Potsdamer Platz subway exit and is looking for the musical theater at Marlene-Dietrich-Platz. He is looking around using head movements and other body movements, trying to orient himself. His search is an intended action (1b/I). A Berlin resident, a guide robot, or any other person will, insofar as he or she pays attention to the tourist, interpret this physical behavior as an intentional search and orientation behavior (2a/I). Nevertheless, they will not feel addressed and invited to approach the tourist in order to offer their help. The situation is different if the tourist indicates his search behavior for others with very expansive body movements and additionally modifies and reinforces it with, for example, raised or contracted eyebrows, a furrowed forehead, and other mimic behavior (2b/I). By intentionally displaying his search behavior for others, the tourist simultaneously signals his accessibility for possible offers of help by people who know the place or a guide robot in a hybrid setting. This offer of help can take the form of multimodal (e.g., verbal and gestural) directions that successfully lead the tourist to his desired destination (2bcom/II) as demonstrated by the following empirical example (1) of a pointing gesture in Fig. 3.

Example (1):

A: also der Ausgangspunkt iss am (..) U-Bahnhof Potsdamer
    Platz/ (..)
        'so the starting point is at subway station Potsdamer
        Platz'
B: an welchem rechts oder links/ (uv: xx)
        'at which right or left' (incomprehensible: xx)
A: [an **die**sem/ (..)
        'at this'
B: (uv: xx)
        (incomprehensible: xx)
A: Infobox (.) genau\]
        'Infobox, exactly'

The co-speech pointing gesture with the outstretched index finger specifies the starting point of the route that is described.

Fig. 3 Pointing at a particular subway exit at Potsdamer Platz accompanying the verbal utterance *an diesem* [at this] ([16], p. 253)

The subway station Potsdamer Platz has two different exits on the left and right. In the utterance *an diesem* [at this], the verbal deictic *diesem* [this] obligatorily requires a co-speech pointing gesture in order to direct the addressee´s attention to the exit previously chosen by the speaker as the starting point of her route description. This example illustrates both conditions (indicative and communicative) of a directive represented in 2bcom/II in Posner's schema: a sender b produces an event f (verbal and gestural deixis as a means of establishing joint attention) with the intention that f causes the recipient a to believe that b intends a to behave in a certain way r, and b believes that his belief will cause a to behave in that way r. Or paraphrased with regard to our empirical interaction scenario: The addressee b understands that the speaker a intends her to do r (looking in the direction of a particular subway exit), and this understanding makes her actually do r.

In a further step, the relevance of both conditions and the indicating condition in particular will be illustrated by a second empirical example of a route description of Potsdamer Platz in Berlin [17, 19–22] with solely human communication partners. This conversation took place in an office at TU Berlin and thus in absence of the described situation. Other route descriptions were recorded at the Potsdamer Platz itself with perceptual access to the route described (see Example (1) above).

Speaker A on the right was instructed to describe a prefixed route near Potsdamer Platz in Berlin, which she had previously walked along, to addressee B with enough precision for B to describe this route in turn to an addressee C, who should then be in a position to find the way independently ([17, 21]). Thus, the primed macro intention of this instruction is a directive according to Posner ([6], see 2bcom/II). This macro intention can be satisfied by sequences of micro intentions indicated by parts of utterances while turn-taking. Both empirical examples have been extensively analyzed from the perspective of multimodal grammar, deixis theory, and spatial cognition [17, 19, 21]. Our focus with regard to different degrees of intending in "natural" communication lies on the following basic questions:

(1) How can we recognize that hand movements or other body movements are intended?
(2) How can we recognize the more specific intention that a hand or body movement is intended to be noticed by someone else, and/or to influence them? (Level 1b of Posner's schema)
(3) And moreover, how can we recognize the sender's or speaker's intention to indicate his or her intention (indicating condition)? (Level 2b of Posner's schema).

In contrast to verbal utterances, hand movements and other speech-accompanying bodily behavior are often supposed to be not consciously executed. But the arrangement of different gesture classes in Kendon's Continuum [23–25], ranging from spontaneous co-speech gestures through convention-alized emblematic gestures like the thumbs-up gesture for 'okay' to the fully developed sign languages of the deaf, demonstrates that hand movements have the potential to be linguistic and, therefore, can also attain a particular level of semioticization (e.g., for a comprehensive overview on the field of international gesture research, see the HSK handbook "Body—Language—Communication," 2 vols, [26] and [27]). This fact makes the study of co-speech gestures and other body movements like gaze an interesting case for the question by which abstract means we indicate gradations of inten-tionality in interpersonal communication and to what extent, abstracted from this, certain principles can be transferred to human-machine communication.

If we consider the following empirical example (2), the degree of intentionality while executing hand movements can be recognized within two main areas:

1. Differences in the degree of structural and functional integration into speech ([19], p. 252).
2. Establishment of joint attention as an indicator of intended communicative relevance (e.g., [17, 28–30])
   - by eye-gaze directed toward gestures,
   - by verbal deictics (e.g., *hier* [here], *so* [like this], *da* [there]),
   - by pointing gestures,
   - by particular use of the parameter gesture space ("big gestures").

Both communication partners are sitting in an office at the TU Berlin, trying to solve the task of reconstructing a certain route at Potsdamer Platz that A has been walking along previ-ously. Speaker and addressee verbally and gesturally create a map-like representation from a bird's eyes view that stands as a complex sign for Potsdamer Platz that is not present in the utterance situation.

Example (2):

A: Right hand: [1][{ja} also wenn **hier** *so die Straße iss (.) von da Fußgängerweg und von da auch Fußgängerweg (.) und da iss McDonald's/ (xxx)*][1]

Left hand: {ja} also wenn hier so die [2][Straße iss (.) von **da** *Fußgängerweg*][2] [3][und von **da** *auch Fußgängerweg (.)*][3] [4][und **da***iss McDon*ald's/][4]

'if the street' is here like this (.) from there pedestrian path and from there pedestrian path too … and there is McDonald's'

In the utterance *wenn hier so die Straße ist* 'if the street is here like this', the speaker uses the verbal deictic expression *so* that obligatorily requires the description of a quality [17, 19, 28, 31–34], in this case instantiated by the iconic gesture of

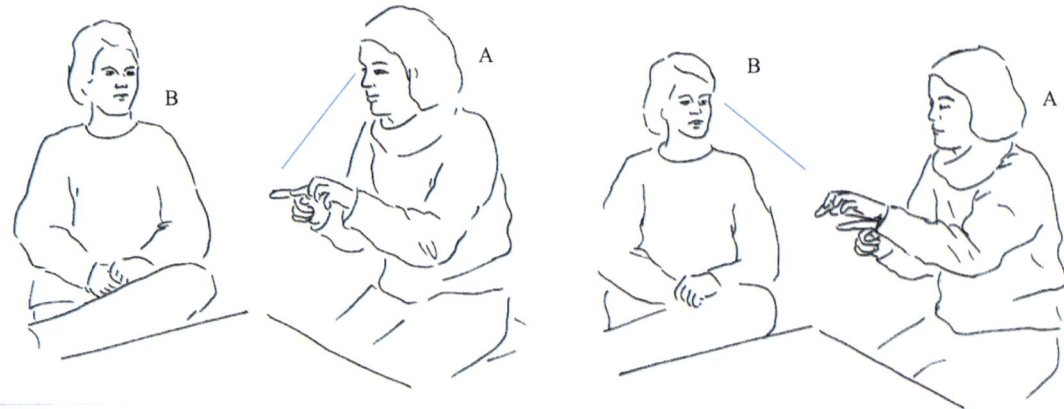

Fig. 4 Indicating intended communicative relevance of hand movements by gaze, gesture, and verbal context ([16], p. 279; [18], p. 137)

the right hand in which the index finger depicts the elongated shape of a street (see Fig. 4a on the left) as the represented object. The fact that due to its particular verbal context this iconic gesture is obligatory clearly indicates a high degree of intending. If we consider the attention-directing pointing gestures of the left hand, then the speaker, firstly, points to spatial objects such as the outstretched index finger of the right hand and, secondly, to "empty" points in space [17, 35, 36], that stand as signs for objects and spatial points at Potsdamer Platz. An additional indicator for the communicative relevance of the right-handed iconic gesture is the speaker's gaze directed to the respective hand position in the center of gesture space (e.g., [28, 34, 36]; see also Hausendorf [37] on "perceived perception" and Tomasello [29, 38, 39] on "joint attention"). To summarize: Different means of establishing joint attention between speaker and addressee as well as structural and functional integration into verbal speech indicate the speaker's intention to communicate that the respective hand movements are intended gestures. The fact that obligatory gestures are supposed to be intended leads to the hypothesis that the degree of integration correlates with the degree of intended gestural execution.

What potential relevance does this finding have for the design of human-machine interaction? First of all, relevant context information has to be considered that has the potential for turning so called "implicit" or non-intended driving cues (e.g., [40]) into "explicit" or intended ones from a recipient's perspective. Imagine two different situations in our Potsdamer Platz scenario. In the first situation, a pedestrian is waiting at a zebra crossing. A driverless car is approaching and slowing down. In this particular context the behavior of slowing down is obligatory, similar to the obligatoriness of an iconic gesture required by the verbal deictic *so* [like this] in the empirical example (2) above. The pedestrian is inclined to interpret the car's behavior of slowing down as executed with the intention to stop in front of the pedestrian and let him cross the street (2a/I). On a more abstract level, our previous hypothesis can be generalized to the effect that intentions are more likely to be attributed to participants—at least in public space scenarios—when a particular behavior is obligatory [2, 3]. A comparison to a second situation without zebra crossing seems to support this assumption: A pedestrian is standing on a sidewalk and wants to cross a street. There are neither traffic lights nor a zebra crossing. A driverless car is approaching and slowing down. In contrast to the first situation, the car is not required to stop due to certain traffic rules. Thus, slowing down might be more likely interpreted as non-intended.

What both situations have in common is the fact that the indicating condition according to Posner is not instantiated. This is where the communicative design of EDTs and robots such as driverless cars could increasingly come into play: In cars steered by humans, the actual driver

might indicate his intention to stop by bodily means while looking and addressing the respective pedestrian. Equivalents for driverless cars might be, for example, new and innovative signals of light and color or other semiotic means with the same function, namely the intention of indicating the intention to stop and let the pedestrian pass [41]. The same requirements of, first, context, and, second, the indicating condition can be generally applied to the interactive design of other traffic situations, e.g., with cyclists who want to change lanes with and without appropriate traffic signs [42].

As the examples above have shown, in natural interpersonal communication, there are certain procedures and semiotic means for indicating gradations of intending. If EDTs are to communicate successfully with people in different situations, they must be familiar with these procedures insofar as they can firstly identify them in the human sender and secondly, as non-human senders themselves, have these procedures at their disposal in order to be able to address people adequately. With regard to the indicating condition as a first step, it has been shown that Posner's formal approach has the potential to support analysis as well as design processes of complex structures of intending in hybrid interaction scenarios of humans and machines.

4 Perspectives on Complex Intending Behavior in Future Hybrid Societies

The capacity of EDTs such as autonomous vehicles or delivery robots to analyze their surroundings on the basis of image and sensor data is rapidly evolving. At least in principle, EDTs can already recognize patterns of movement such as pedestrians or cyclists changing direction or slowing down while approaching a zebra crossing. In the future, it may become possible to reliably recognize gestures in image or video data, and to react to them.

To prepare for such a future, we have to consider the role of intentionality in communication generally, and in public settings more specifically. From Posner's model, and the empirical research conducted on its basis, various consequences can be drawn. The most important results for research on hybrid societies are summarized in the following list:

i. Intentionality is complex, and cannot be reduced to the simple dichotomy between "implicit" and "explicit" cues occasionally used in technical settings or in HRI studies.
ii. An action or behavior that, regarding its purely physical side, belongs to the same category of events (e.g. "slowing down", "turning one's head," "stopping," "approaching" etc.) may, depending on the intentionality (complex nested beliefs and intentions) accompanying it, belong

to quite different categories. For example, it could be a simple action without communicative purpose, a signaling, or a communication, with different implications regarding expectations from other interactants, and expected future behavior paths.

iii. Adequately detecting these differences is therefore important for smooth and intuitive interaction between EDTs and humans. An EDT with an inadequate model of intentionality will make regular and serious errors in judging the intentions of other participants: it may, for example, mistake the simple action of slowing down for the signal "I want you to pass first", or vice versa, which imply different and, in some instances, contradictory optimal behavior responses. It will also be unable to adequately represent how other interactants understand and interpret the situation, and thus itself have an incorrect, or at best incomplete, representation of the situation. It will generally have less situational knowledge than a human would, and be inferior to humans in terms of interaction quality.

iv. The quality of the mental model of an EDT in regard to (a) the intending of all other interactants and (b) the adequate own intentions and beliefs will likely correlate with the degree to which it is understood and accepted by its interactants. In a more general perspective, if EDTs are to become an accepted and normal part of public space, they will arguably need mental models including (at least some degree of) intentionality.

5 Summary and Outlook

Posner's approach aims toward a more adequate modeling of human-machine communication. Embodied Digital Technologies (EDTs) such as robots, virtual agents could profit from representing intentions and beliefs related to the ongoing communication process. Examples show that it can make a difference for how to best react if intentions and beliefs are inferred and represented. Self-driving cars, delivery robots, even AIs in a "smart home" could profit from detailed models of intentions and beliefs. EDTs may be able to interpret human behavior (from basic signs to communicative acts) more adequately if they possess (a) an adequate model distinguishing different sign types or communication types on different "levels of intending," (b) a multimodal sensorium that is capable of detecting all relevant clues with a sufficient degree of detail; (c) a knowledge base (e.g., pre-trained model).

Posner's articles demonstrate that various sign processes exist "below" communication (level 2bcom), and that these work differently from communication, and from each other. If EDTs are only intended to "communicate" convincingly, default assumptions may suffice. Adequate reactions may be programmed on the basis of the standard assumptions

that are valid for communication (e.g., that it is intended, and that the communication partner believes that the other person knows this). However, the "lower levels" of Posner's intentionality hierarchy become highly relevant for open-world scenarios such as encounters in public space, where inferring intentions and beliefs without prior knowledge is important.

Building on these results, we intend to demonstrate in further work that EDTs with the capabilities necessary to adequately represent intentions and beliefs in their mental models of themselves and of interactants, and to continually update these representations during an interaction, may be able to achieve improved situational awareness and a more human-like interaction quality (e.g., [2, 3]). Interactions with EDTs that represent intentionality in a sufficiently detailed manner will be more intuitive, which may lead to higher acceptance especially during unsolicited encounters. In the development toward hybrid societies where interactions with EDTs will be spontaneous and non-scripted, EDTs with an adequate "model of intentionality" will have an advantage.

Acknowledgements Funded by Deutsche Forschungsgemeinschaft (DFG, German Research Foundation)—Project-ID 416228727—SFB 1410. The authors wish to thank Karin Becker for her drawings (Figs. 3, 4).

Appendix

TRANSCRIPTION CONVENTIONS USED IN THIS ARTICLE (BASED ON [17] AND [23]).

1. On the verbal level:
   - Pauses of different length: (.), (..), (3 s)
   - Intonation: rising /, falling \, constant –
   - Capital letters indicate noticable stress: [THIS one]
2. On the gestural level:
   - Square brackets indicate the beginning and end of a gesture unit. They are inserted in relation to the verbal utterance. If several gestures occur in an example sequence, they are differentiated by subscript figures: [left]$_1$ [and right]$_2$ [huge skyscrapers]$_3$
   - In the case of gestural overlaps of communication partners, the starting points of the respective gesture unit are marked by superscript figures, the end points by subscript figures: $^1[\ldots^2[\ldots]_1\ldots]_2$
   - Gestural embedding: $^1[\ldots^2[\ldots]_2\ldots]_1$
   - Bold letters indicate gestural strokes: [**left**]$_1$ [and **right**]$_2$ [**huge** skyscrapers]$_3$
   - Underlinings indicate gestural holds: [**that** <u>is the arcade</u>]

References

1. Fricke, E. (2021). Mental Spaces, Blending und komplexe Semioseprozesse in der multimodalen Interaktion: Zeichenbasierte und ontologiebasierte Mental Spaces. *Zeitschrift für Semiotik [Journal of Semiotics], 43*(1–2), 113–144.

2. Fricke, E. (2021). "Anthropomorphisierung und intentionale Komplexität. Absichten und Absichtszuschreibungen in der Interaktion von Menschen und Maschinen." In E. Fricke & M. Meiler (Eds.), *Transformations – signs and their objects in transition. Proceedings of the 16th International Congress of the German Semiotics Society (DGS) 2021* (p. 280). Chemnitz: Universitätsverlag Chemnitz.

3. Fricke, E. (2025). Anthropomorphism and intentional complexity in hybrid interaction scenarios (working title). *Zeitschrift für Semiotik [Journal of Semiotics], 47*, 1–2. in preparation.

4. Tomasello, E., & Rakoczy, H. (2009). "Was macht Erkenntnis einzigartig? Von individueller über geteilte zu kollektiver Intentionalität." In H. B. Schmid & D. P. Schweikard (Eds.), *Kollektive Intentionalität. Eine Debatte über die Grundlagen des Sozialen* (pp. 697–737). Frankfurt a.M.: Suhrkamp.

5. Tuomela, R., & Miller, K. (2009). "Wir-Absichten." In H. B. Schmid & D. P. Schweikard (Eds.), *Kollektive Intentionalität. Eine Debatte über die Grundlagen des Sozialen* (pp. 72–98). Frankfurt a.M.: Suhrkamp.

6. Posner, R. (1993). "Believing, causing, intending. The basis for a hierarchy of sign concepts in the reconstruction of communication." In R. J. Jorna, B. v. Heusden, & R. Posner (Eds.), *Signs, search and communication. Semiotic aspects of artificial intelligence* (pp. 215–270). Berlin, New York: De Gruyter.

7. Searle, J. (1983). *Intentionality: An essay in the philosophy of mind.* Cambridge, GB: Cambridge University Press.

8. Strawson, P. F. (1964). Intention and convention in speech acts. *Philosophical Review, 73*, 439–460.

9. Searle, J. (1979). *Expression and meaning: Studies in the theory of speech acts.* Cambridge University Press.

10. Meggle, G. (1981). *Grundbegriffe der Kommunikation.* De Gruyter.

11. Posner, R. (1997). "Pragmatics." In R. Posner, K. Robering, & T. A. Sebeok (Eds.) *Semiotik / Semiotics: A handbook on the sign-theoretic foundations of nature and culture, 4 vols, 1997–2004* (Vol. 1, pp. 219–246). Berlin, New York: De Gruyter.

12. Siefkes, M. (2008). Unkooperativität und andere kommunikative Randphänomene in 'Die Zeit und das Zimmer' von Botho Strauß. *Kodikas/Code. Ars Semeiotica, 31*(1–2), 91–118.

13. Siefkes, M. (2012). "Grenzfälle der Kommunikation". In E. W. B. Hess-Lüttich (Ed.), *Sign culture zeichen kultur* (pp. 503–525). Würzburg: Königshausen & Neumann.

14. Posner, R. (1996). "Sprachphilosophie und Semiotik," in Sprachphilosophie. Ein internationales Handbuch zeitgenössischer Forschung, M. Dascal, D. Gerhardus, K. Lorenz, and G. Meggle, Eds. Berlin, New York: De Gruyter, pp. 1658–1685.

15. Bressem, J. & Müller, C. (2014). "The family of AWAY-gestures." In C. Müller, A. Cienki, E. Fricke, S. H. Ladewig, D. McNeill, & J. Bressem (Eds.),. *Body – language – communication. An international handbook on multimodality in human interaction* (pp. 1592–1604) (Handbooks of Linguistics and Communication Science 38.2). Berlin, Boston: De Gruyter.

16. Fricke, E. (2024). "Negation multimodal: Geste und Rede, Text und Bild." In L. Bülow, S. Kabatnik, M.-L. Merten, & R. Mroczynski (Eds.), *Pragmatik multimodal.* Tübingen: Narr.

17. Fricke, E. (2007). *Origo, Geste und Raum: Lokaldeixis im Deutschen.* De Gruyter.

18. Lynn, U. (2014). "Levels of Abstraction." In C. Müller, A. Cienki, E. Fricke, S. H. Ladewig, D. McNeill, & J. Bressem (Eds.),. *Body – language – communication. An international handbook on multi-*

modality in human interaction (pp. 1702–1712) (Handbooks of Linguistics and Communication Science 38.2). Berlin, Boston: De Gruyter.

19. Fricke, E. (2012). *Grammatik multimodal: Wie Wörter und Gesten zusammenwirken.* De Gruyter.

20. Fricke, E. (2014). "Deixis, gesture, and embodiment from a linguistic point of view." In C. Müller, A. Cienki, E. Fricke, S. H. Ladewig, D. McNeill, & J. Bressem (Eds.),. *Body – language – communication. An international handbook on multimodality in human interaction* (pp. 1803–1823) (Handbooks of Linguistics and Communication Science 38.2). Berlin, Boston: De Gruyter.

21. Fricke, E. (2022). "The pragmatics of gesture and space." In A. H. Jucker & H. Hausendorf (Eds.), *Pragmatics of space. Handbook of pragmatics* (Vol. 14, pp. 363–397). Berlin, Boston: De Gruyter.

22. Fricke, E. (2024). "Indexicality, deixis, and space in gesture." In A. Cienki (Ed.), *The Cambridge handbook of gesture studies.* Cambridge: Cambridge University Press.

23. McNeill, D. (1992). *Hand and mind: What gestures reveal about thought.* Chicago University Press.

24. Kendon, A. (1980). "Gesticulation and speech: Two aspects of the process of utterance." In M. R. Key (Ed.), *The relationship of verbal and nonverbal communication* (pp. 207–227). The Hague: Mouton.

25. Kendon, A. (2004). *Gesture.* Cambridge University Press.

26. Müller, C., Cienki, A., Fricke, E., Ladewig, S. H., McNeill, D., & Teßendorf, S. (Eds.) (2013). *Body – language – communication. An international handbook on multimodality in human interaction* (Handbooks of Linguistics and Communication Science 38.1). Berlin, Boston: De Gruyter.

27. Müller, C., Cienki, A., Fricke, E., Ladewig, S. H., McNeill, D., & Bressem, J. (Eds.) (2014). *Body – language – communication. An international handbook on multimodality in human interaction* (Handbooks of Linguistics and Communication Science 38.2). Berlin, Boston: De Gruyter.

28. Streeck, J. (1993). Gesture as communication I: Its coordination with gaze and speech. *Communication Monographs, 60*(4), 275–299.

29. Tomasello, M. (1995). Joint attention as social cognition. In D. Moore & P. J. Dunham (Eds.), *Joint attention: Its origin and role in development* (pp. 103–130). Erlbaum.

30. Stukenbrock, A., & Dao, A. N. (2019). "Joint attention in passing: What dual mobile eye-tracking reveals about gaze in coordinating embodied activities at a market." In E. Reber & C. Gerhardt (Eds.), *Embodied activities in face-to-face and mediated settings* (pp. 177–213). Palgrave Macmillan.

31. Ehlich, K. (1987). "So – Überlegungen zum Verhältnis sprachlicher Formen und sprachlichen Handelns, allgemein und an einem widerspenstigen Beispiel." In I. Rosengren (Ed.), *Sprache und Pragmatik. Lunder Symposium 1986* (pp. 279–313). Stockholm: Almqvist & Wiksell.

32. Fricke, E. (2013). "Towards a unified grammar of gesture and speech: A multimodal approach," in [26], pp. 733–754.

33. Herbermann, C.-P. (1988). "Entwurf einer Systematik der Deixisarten," in Modi referentiae. Studien zum sprachlichen Bezug zur Wirklichkeit, Clemens- Peter Herbermann, Ed. Heidelberg: Carl Winter, pp. 47–93.

34. Stukenbrock, A. (2015). *Deixis in der Face-to-Face-Interaktion.* De Gruyter.

35. McNeill, D., Cassell, J., & Levy, E. T. (1993). Abstract deixis. *Semiotica, 95*, 5–19.

36. Stukenbrock, A. (2014). Pointing to an 'empty' space: Deixis am phantasma in face-to-face interaction. *Journal of Pragmatics, 74*, 70–93.

37. Hausendorf, H. (2003). "Deixis and speech situation revisited: The mechanism of perceived perception," in F. Lenz (Ed.), *Deictic Conceptualisation of Space, Time, and Person* (pp. 249–269). John Benjamins.

38. Tomasello, M. (2008). *Origins of Human Communication.* Cambridge, Mass.: MIT Press.
39. Tomasello, M. (2009). *The Cultural Origins of Human Cognition.* Harvard University Press.
40. Hensch, A.-C., Beggiato, M., & Krems, J. (2022). "Should I wait or should I go? – Deciphering Implict Communication Cues for Cooperative Interactions in Left-Turn Scenarios," CIVEMSA 2022, IEEE, 978–1–6654–3445–4/22.
41. Fricke, E. (2014). "Wie entstehen Gesten – und was kann das in Bezug auf die Interaktion mit fahrerlosen Autos in der Zukunft bedeuten?" Mercedes- Benz Future Talk "Robotik," Berlin 30.6–2.7.2014.
42. Odenwald, S., Fricke, E., & Einhäuser-Treyer, W. (2022). "Predicting cyclists' intentions by measuring eye-, head- and body movements," Proceedings EU-SAFETY2022, 23./24.06, Vienna, 2022.

Designing Computer-Mediated Communication with Affective Technology to Increase Feedback Acceptance

Katharina Jahn, Oliver Rehren, Bastian Kordyaka,
Sebastian Jansen, Peter Ohler, and Günter Daniel Rey

Abstract

As text-based computer-mediated communication (CMC) supported by affective technology becomes increasingly common in our daily life, new opportunities for the communication of critical information, such as negative feedback, arise. Research on affective technology has already shown that the acceptance of negative feedback can be improved by using emoticons under specific conditions. However, in which way emotion recognizing affective technology can increase the acceptance of negative feedback automatically is still unclear. We hypothesized that automatically reported stress and a low stress level increases feedback acceptance and its predictors. Addi-tionally, we hypothesized an interaction effect that could attenuate the negative effect of high stress when the stress level is automatically detected. Using a messenger that reports the feedback provider's stress level to the feedback recipient, we investigate how the automaticity of stress detection and the displayed stress level can increase negative feedback acceptance and its predictors. We conducted a 2 (stress level: low vs. high) × 2 (automaticity: automatically detected vs. self-reported) + 1 (control group) between subjects laboratory experiment, resulting in five experimental groups. Our results show that whereas an automatically detected stress level increases perspective taking, seeing a low stress level increases perceived good intention.

This publication has been partially funded by the Deutsche Forschungsgemeinschaft (DFG, German Research Foundation)—Project-ID 416228727—SFB 1410, the German Ministry of Culture and Science of the Land of North Rhine-Westphalia (reference number: 005–1706-0006), and the research initiative "Instant Teaming between Humans and Production Systems" co-financed by tax funds of the Saxony State Ministry of Science and Art (SMWK3-7304/35/3-2021/4819) on the basis of the budget passed by the deputies of the Saxony state parliament.

K. Jahn (✉) · O. Rehren · S. Jansen · P. Ohler · G. D. Rey
Chemnitz University of Technology, Chemnitz, Germany
e-mail: katharina.jahn@phil.tu-chemnitz.de

O. Rehren
e-mail: oliver.rehren@phil.tu-chemnitz.de

S. Jansen
e-mail: sebastian.jansen@phil.tu-chemnitz.de

P. Ohler
e-mail: peter.ohler@phil.tu-chemnitz.de

G. D. Rey
e-mail: guenter-daniel.rey@phil.tu-chemnitz.de

B. Kordyaka
University of Siegen, Siegen, Germany
e-mail: bastian.kordyaka@uni-siegen.de

Keywords

Affective technology · Feedback acceptance · Negative feedback · Emoticons

1 Introduction

Computer-mediated communication has been an integral part of our daily life since the emergence of e-mails, chat programs, and social media. The rise of mobile devices such as smartphones or tablets has further boosted CMC as a natural part of human interaction. Online communication is an indispensable part of today's working world, especially when work location and work time become increasingly flexible. With this development, affective technologies, as instances of affective computing, gain increasing importance. Affective technologies represent a form of technology that is able to "recognise, express, model, communicate, and respond to emotional information" [16, p. 55]. Thus, they include a diverse set of detected emotions and application areas. For individual application settings, recent studies demonstrate that affective technology has the potential to help individuals to enhance efficiency at work [1] or increase their health [2].

The increased incorporation of affect in mediated communication settings highlights the relevance of emotions, which are, in traditional face-to-face communication, not only communicated by the exchange of words but also by facial expressions and gestures. Without the possibility to see these automatic cues of the communication partner, mediated communication, and especially text-based CMC, is confronted with a range of difficulties in creating a shared understanding of each other's emotions, thoughts, and behaviors [3]. This shared understanding, also called perspective taking, is crucial when the communication is supposed to be perceived as positive, especially in ambiguous communication situations [4], such as the delivery of feedback.

Even though human feelings can be expressed in text-based CMC by simply verbalizing them or using emoticons, most people do not make use of it in work contexts because the verbal expression of feelings is resource-consuming and emoticons are ambiguous, regarded as too informal, and can even lead to being seen as less competent [5]. As a consequence, content sent by e-mail can be interpreted by the recipient more negatively than intended [4]. Current research on the integration of emotions into text-based CMC has investigated technology design mostly related to the use of self-reported emoticons. In the context of technology-supported emotion sharing, Janssen et al. [6] have investigated the effects of affective technology with respect to the number and the time emotions are shared by using emoticons in a situation in which users were watching movies. The results show that sharing self-selected emotions at user-initiated times is perceived as more intimate than sharing these emotions at computer-generated times. Additionally, a larger number of emoticons creates more intimacy.

For ambiguous communication situations, such as the delivery of feedback, research is still scarce. Although negative feedback is highly relevant for organizations and individuals because it can improve performance [7], negative feedback can be seen as threat to a positive self image, whereas positive feedback can enhance the self image [8]. Thus, the acceptance of positive feedback seems independent of surrounding circumstances, while the acceptance of negative feedback is rather dependent on situational factors [9]. One study investigating technological situational factors in form of self-selected emoticons in traditional text-messages indicates that when the feedback provider uses positive emoticons while giving specific negative feedback, the feedback recipient experiences an increased perception of good intention and a decreased perception of feedback negativity, which both increase feedback acceptance [10]. In contrast, for unspecific negative feedback, positive emoticons do not change perceptions of good intention or feedback negativity, whereas negative emoticons worsen them. This research shows that, especially in the case when the resources of the feedback provider are so limited that they cannot give specific feedback, using self-reported emoticons does not necessarily help and may even backfire.

Although these initial studies exist, the question of how increasingly available affective technologies, which record the emotions of the user automatically, influence the acceptance of negative feedback has not yet been answered. In affective technology design, the role of *automaticity* of self-reported (participants) or automatically recognized (technology) emotions in text-based CMC in the context of negative feedback acceptance is therefore still unclear. We aim to close this gap by addressing the following research question:

Research Question *How does affective technology design influence the acceptance of negative feedback?*

We use previous research on negative feedback acceptance [10] and perspective taking [11, 12] as theories to explain how an affective technology that reports the stress level of the feedback provider to the feedback recipient influences feedback acceptance.

2 Hypothesis Development

We use two design features for perspective taking: emoticons and the reported automaticity of stress detection. With regard to emoticons, Wang et al. [10] showed that the valence of the emoticon in combination with feedback specificity is crucial for influencing perceived good intention and feedback negativity. When feedback is perceived as highly specific, liking emoticons increase the perceived good intention of the feedback provider and decrease the perceived feedback negativity compared to not using any emoticons. In contrast, disliking emoticons have neither a positive nor a negative effect on the two dependent variables. The opposite pattern arises for unspecific feedback: Liking emoticons do not relate to either of the dependent variables, whereas disliking emoticons decrease perceived good intention and increase feedback acceptance.

Regarding automaticity of stress detection, we argue that reporting stress automatically facilitates perspective taking, as it provides an indicator of the stress level that cannot be influenced by the feedback provider. Thus, although detecting stress automatically by the system might provide inaccurate information on the current stress level, the information cannot be manipulated and appears to be objective. In contrast, self-reported stress levels can not only easily be manipulated by the feedback provider but can also be incorrect if the feedback provider recognizes their stress level incorrectly (e.g., when failing to recognize high stress). Consequently, a stress level automatically detected by the system can be viewed as an objective indicator of the feedback provider's stress, whereas self-reported stress might—intentionally or unintentionally—deviate from the true stress level. Therefore, we hypothesize that detecting stress automatically increases perspective taking and perceived good intention compared to self-reported stress because the feedback recipient is under the impression that they have received accurate information on the feedback provider. For perspective taking, this can be regarded as a cue on the feedback provider's emotional state, which can facilitate perspective taking. Likewise, a similar process can be expected for perceived good intention.

Hypothesis 1 *Detecting the stress level of the feedback provider automatically enhances (a) perspective taking, and (b) perceived good intention, and. (c) feedback acceptance compared to assessing the stress level through self- reporting.*

With regard to the valence of the emoticons, in contrast to using liking or disliking emoticons similar to Wang et al. [10], we use emoticons that indicate the stress level of the recipient. Thus, the stress level can vary between a lowly stressed state and a highly stressed state. To explain how these emoticons can influence feedback acceptance, we adopt the perspective of emotion contagion theory [13]. Emotion contagion theory proposes that the valence of emotional expressions can influence human behavior by conscious or unconscious processes. Specifically in the IS domain, previous research has shown that users mimic the facial expression of interaction partners, which influences subsequent attitudes and behavior [14].

In line with this previous research, we argue that the valence of emoticons can influence humans attitudes and behavior. Specifically, we draw upon emotion contagion theory [13, 15] and affect infusion theory [14, 16]. Emotion contagion theory explains how affect is transferred from one person to another during their communication. One way that has been successfully investigated in several study is that the information recipient uses mimicking behavior to experience the emotion [14, 15]. When this process has taken place, affect infusion theory proposes that decisions are influenced by emotions, especially when they are complex [16]. Therefore, we propose that displaying a negative emotional state (e.g., stressed) in the form of emoticons should decrease perspective taking and good intention because individuals mimic this negative state. As a result, these negative emotions spill over, decreasing both the motivation for perspective taking and the motivation for seeing a good intention in the feedback provider.

Hypothesis 2 *Displaying high stress decreases (a) perspective taking, (b) perceived good intention, and (c) feedback acceptance compared to displaying low stress.*

Additionally, we propose that an interaction effect for stress level and automaticity might also arise. Specifically, we expect that the negative effect reporting a high stress level will become positive if the feedback is automatically reported because it serves as a credible cue of actual stress (and thus, limited resources to give extensive feedback). On the other hand, a self-reported stress level could be interpreted as intentional decision to report stress for various reasons (e.g., reporting a high stress level to highlight the negative feedback), regardless of actual stress level.

Hypothesis 3 *There is an interaction effect between stress and automaticity for (a) perspective taking, (b) perceived good intention, and c) feedback acceptance.*

3 Method

We used a 2 (stress level: low vs. high) × 2 (automaticity: automatically detected vs. self-reported) + 1 (control group) between-subjects laboratory experiment to empirically manipulate the level of situational perspective taking. Thus, there were five experimental groups.

Overall, we collected 93 observations. The participants received 10 Euro as a reward for taking part in our experiment. Gender was quite equally distributed (46 female, 47 male) and participants had an average age of 24.14 years (SD = 2.48), ranging from 19 to 32 and most participants had finished at least the highest possible undergraduate degree in their country (96,77%).

3.1 Procedure

During the experimental task, the participants were tested in groups of two to four people. When the participants entered the laboratory, the instructor told them that they were participating in an experiment about emotions in computer-mediated communication for which they had to complete a short writing exercise. This exercise would then be evaluated by a communication partner. Afterward, they were seated at computer workplaces inside individual cubicles.

The first part of the experiment consisted of the writing task. At the start of this part, the participants were instructed in the usage of the communication program. For the participants in the automatically detected condition, it was explained that their communication partners, who provided them with feedback, would wear a stress-sensitive ring that could display the stress he or she had before reading their essay. In the self-reported condition, the participants were informed that the feedback provider selected his or her stress level before reading their essay. Additionally, the participants could see a table that displayed either high or low stress emoticons. Next, they were told that their feedback provider would be randomly selected from a group of students in another room. After they had entered their own name in the program and waited for a period of about three seconds during which their

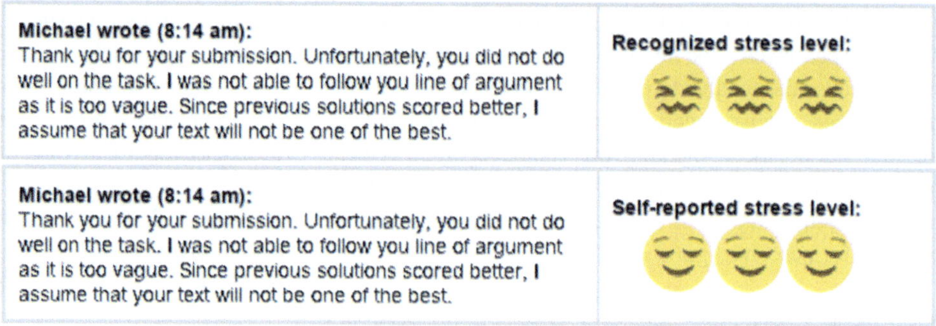

Fig. 1 Visualization of the automaticity and stress level conditions. Top: Stress-sensitive system, High Stress Condition; Bottom: Self-reported, Low Stress

communication partner was allegedly chosen, the name of their communication partner was revealed to be "Michael." Finally, they were told that they had 10 min to write an essay about the influence of smartphones on friendships.

When the time was up, the message was sent to Michael automatically, and the participants waited for his feedback. After two minutes, the program showed that Michael was typing, and a minute later the following pre-programmed feedback message was shown, combined with the respective conditions (see Fig. 1). We intentionally added a spelling mistake to create the impression that Michael was a real person.

After reading the feedback, they completed the questionnaires on the dependent and control measures. When all the participants had finished the experiment, they were debriefed about the true goals of the study.

3.2 Independent Variables and Measures

In the control condition, no emoticons were displayed.

Automaticity of stress detection. The automaticity of stress detection that the communication program displayed differed in being either automatically recognized by the system or self-reported by the feedback provider. In the automatically detected condition, the participants were told that Michael's stress level would be measured by a ring that measured electrodermal activity. In the self-reported condition, the participants were told that Michael had to select the emoticon that best represented his stress level. Additionally, the condition was displayed in the feedback message from Michael (see Fig. 1). In the control condition, the space was left blank.

Measures. Perceived good intention (e.g., "Michael was willing to support me in the creation of the essay"), feedback specificity (e.g., "When Michael gave me the feedback about my work, he provided me with specific information"), and feedback acceptance (e.g., "I agreed with the feedback that I received from Michael") were adapted from previous research on feedback acceptance [10]. Additionally, we adapted a scale to our context for situational perspective taking (e.g., "I tried to take Michael's perspectives") [17]. and used established scales for the control variables dispositional perspective taking [3]

and self-awareness [18]. All items were measured on a 7-point scale (1 = strongly disagree, 7 = strongly agree). As manipulation checks, we asked participants how stressed Michael was and how the stress level had been measured.

4 Results

4.1 Manipulation Checks

Stress level. For the stress level condition, we carried out a one-way ANOVA comparing the high stress group with the low stress and control group. The ANOVA was significant ($F(2, 81) = 55.10$, $p < 0.001$), revealing that participants in the high stress group ($M = 5.67$, $SE = .25$) reported significantly more stress than in the control ($M = 2.89$, $t(81) = 4.74$, $SE = 0.53$, $p < 0.001$) and low stress group ($M = 1.86$, $t(81) = 10.35$, $SE = 0.27$, $p < 0.001$), whereas no significant difference could be found for the control and low stress condition. Therefore, the manipulation of stress level was successful.

Automaticity of stress detection. We carried out a t-test to compare the two groups of automaticity of stress detection on the manipulation check for automaticity (we excluded participants from the control condition because the question would not have made sense for them, as no stress level was reported). The t-test showed a significant effect for automaticity ($t(73) = 2.04$, $p = 0.045$), revealing that, as expected, the scores of the automaticity manipulation check were lower in the self-reported group ($M = 2.47$, $SE = 0.96$) than in the automatically detected condition ($M = 2.53$, $SE = 1.81$). Thus, we recorded that the manipulation of the factor automaticity of stress detection was successful and showed the intended differences between the two factor levels.

4.2 Hypothesis Testing

Situational Perspective Taking. To assess if covariates influenced situational perspective taking, we first conducted a regression analysis with self-awareness and dispositional perspective taking as predictors. Results revealed a significant effect of self-awareness ($b = 0.556$, $SE = 0.155$, $p < 0.001$), whereas dispositional perspective taking showed only a marginally significant effect ($b = 0.358$, $SE = 0.185$, $p = 0.056$). Thus, we retained only self-awareness in the subsequent analysis. The results of the 2×2 ANCOVA revealed a main effect of automaticity ($F (70, 1) = 6.02$, $p = 0.017$, $\eta^2 = 0.081$), with an automatically detected stress level leading to higher perspective taking ($M = 4.72$, $SE = 0.197$) than a self-reported stress level ($M = 4.03$, $SE = 0.199$). On the other hand, the main effect for stress and the interaction did not reach significance (all $ps > 0.665$, $\eta^2 < = 0.002$). Therefore, H1a can be supported by our data while H2a and H3a cannot.

Perceived Good Intention. The regression analysis on perceived good intention revealed significant effects for situational perspective taking ($b = .170$, $SE = 0.085$, $p = 0.049$) and feedback specificity ($b = 0.65$, $SE = 0.083$, $p = 0.002$). In a subsequent 2×2 ANCOVA, there was a significant main effect of stress ($F (70, 1) = 5.35$, $p = 0.006$, $\eta^2 = 0.072$), showing that participants in the high stress condition ($M = 2.46$, $SE = 0.147$) perceived less good intention than participants in the low stress condition ($M = 2.96$, $SE = 0.157$). Additional, there was no significant main effect for automaticity ($p = 0.358$, $\eta^2 = 0.012$) or the interaction between both factors ($p = 0.636$, $\eta^2 = 0.003$). There- fore, H2b can be supported by our data, whereas H1b and H3b cannot be supported.

Feedback Acceptance. A regression on feedback acceptance revealed situational perspective taking ($b = 0.248$, $SE = 0.101$, $p = 0.017$) and perceived good intention ($b = 0.467$, $SE = 0.147$, $p = 0.002$) as significant predictors, whereas neither self-esteem nor specificity showed a significant effect (all $ps > 0.055$). The subsequent 2×2 ANCOVA revealed only a significant interaction effect ($F (69, 1) = 4.65$, $p = 0.035$, $\eta^2 = 0.063$), whereas neither the main effect for automaticity ($p = 0.729$, $\eta^2 < 0.001$) nor stress level ($p = 0.595$, $\eta^2 = 0.003$) reached significance. The interaction indicated highest feedback acceptance for an automatically detected high stress level ($M = 3.87$, $SE = 0.246$), and the self-reported low stress condition ($M = 3.81$, $SE = 0.263$), followed by the self-reported high stress condition ($M = 3.40$, $SE = 0.261$) and the automatically detected low stress condition ($M = 3.17$, $SE = 0.279$). Thus, H3c can be supported by our data, whereas H1c and H2c cannot.

5 Discussion

With regard to our research question, our results indicate that using affective technology in computer-mediated communication provides the opportunity to increase the acceptance of negative feedback. Specifically, detecting stress automatically increases perspective taking (H1a). On the other hand, we could find no support for a negative stress deteriorating perspective taking or feedback acceptance (H1b, H1c). Addi-

tionally, displaying a negative stress level through emoticons decreased perceived good intention (H2b) and we found an interaction effect between stress level and automaticity for feedback acceptance (H3c). We discuss implications of these findings for existing theories below.

First, we could show that affective technology can serve as a design feature for perspective taking (H1a), which can be used as a starting point for other research in need of design features for perspective taking [e.g., 7]. Previous research has indicated that individuals perceive a different degree of intimacy depending on whether they believe that an affective technology or a real human has decided on *the time* at which they share their emotions [6]. Our study is the first to show that the information of whether a real human or an affective technology detects the *valence* of an emotion can be used to influence the perceptions of the relation of the user and the communication partner. Additionally, the hypothesis that a low stress level leads to higher perceived good intention could be supported (H2b), whereas we could find no support for the hypothesis that the automaticity of stress detection influences perceived good intention (H1b). This could imply that theoretical approaches to explaining negative feedback acceptance have to be adapted for affective technology.

Surprisingly, the hypothesized effects that a high displayed stress level leads to lower perspective taking (H1b) and feedback acceptance (H1c) than a high stress level could not be supported by our data. This is in contrast to our predictions based on affect infusion and emotion contagion literature [14]. For perspective taking, a reason for this unexpected result might be that we measured perspective taking as a subjective measure by asking the participants whether they thought that they could understand what Michael was feeling when he provided the feedback. Other studies have investigated the success of perspective taking with more objective measures, such as allowing participants to choose between a postcard with a text in the recipient's language (high perspective taking) or in a language foreign to the recipient [1, low perspective taking]. We did not use more objective measures because perceived good intention is a subjective measures as well, and we found perceived perspective taking to be more adequate than actual perspective taking, as the mere perception of perspective taking should be sufficient to change perceived good intention. However, assessing perspective taking through more objective measures could unveil additional relationships. Regarding the interaction effect only found for feedback acceptance (H3c), which could not be found for good intention (H1b) and perspective taking (H1a), this result could hint toward additional predictors for feedback acceptance (e.g., credibility) that might be relevant in the context of affective technology.

Moreover, our study has several methodological strengths that further con- tribute to the validity of our results. Because we stay close to the methodology of Wang et al., especially regarding the use of emoticons, we contribute to the body of knowledge related to the effects of emoticons in the context of feedback acceptance. Although using emoticons as feedback design might seem rather trivial at first, choosing a "simple design" has the advantage that we can draw conclusions specifically related to a small entity (i.e., emoticons), making confounding effects within the design more unlikely. Therefore, possible methodological explanations for the different effects of valence (in our case, stress level) in our study compared to Wang et al. [10] include the use of an affective technology that is designed to express the stress level of the feedback provider and the slightly different task, but not in the concrete instantiation of emotion expression. Additionally, because we defined the meaning of the emoticons clearly by telling the participants that they reflected high or low stress, we can further draw the conclusion that our results are valid for the area of stress-related emoticons. Investigating the effect of emoticons in other affect-related areas remains a promising path for future research.

Regarding practical implications and design science, since automatically detecting the stress level of the feedback provider enhances perspective taking in CMC, it can be concluded that the implementation of automatic affect detection could be fruitful in CMC systems. When misunderstandings do not occur in ambiguous communication situations, employees do not have to invest resources into solving these misunderstandings and can instead focus on their work, possibly resulting in more efficient work processes and higher productivity.

Moreover, the inclusion of perspective taking as an additional goal of technology design can lead to innovative design solutions for improving human-to-human communication or human–computer interaction. Here, the use of emoticons as a means for expressing stress or other emotions by affective technologies could be reasonable in multiple contexts, such as in private communication settings, gaming, or the interaction with artificial intelligence.

As with every study, our study has some limitations. First, we did not assess physiological measurements. Future research could use physiological measurements related to feedback acceptance, for example, the stress level after receiving feedback (e.g., measured by heart rate variability). Second, as we focused on the context of negative feedback holding specificity constant, conclusions on specific negative feedback and differences between feedback with low and high specificity cannot be drawn from our study. Finally, regarding the low sample size used, the conclusions from our results should be interpreted with caution because the power of our study was only sufficient to detect medium to high effects. Future research should try to expand the sample size to test the stability and power of

our findings. Additionally, future research on the acceptance of negative feedback could investigate the effect of affective technologies that are not related to stress but to other emotions.

References

1. Bailey, B. P., & Konstan, J. A. (2006). On the need for attention-aware systems: Measuring effects of interruption on task performance, error rate, and affective state. *Computers in Human Behavior, 22*(4), 685–708.
2. Picard, R. W. (2002). Affective medicine: Technology with emotional intelligence. In *Future of Health Technology, Studies in Health Technology and Informatics, IOS*. Press
3. Davis, M. H. (1983). The effects of dispositional empathy on emotional re- actions and helping: A multidimensional approach. *Journal of personality, 51*(2), 167–184.
4. Byron, K. (2008). Carrying too heavy a load? The communication and miscommunication of emotion by email. *The Academy of Management Review, 33*(2)
5. Brown, S. A., Fuller, R., & Thatcher, S. M. B. (2016). Impression formation and durability in mediated communication. *Journal of the Association for Information Systems, 17*(9), 614–647.
6. Janssen, J. H., Ijsselsteijn, W. A., & Westerink, J. H. (2014). How affective technologies can influence intimate interactions and improve social connectedness. *International Journal of Human Computer Studies, 72*(1), 33–43.
7. Ang, S., Cummings, L. L., Straub, D. W., & Earley, P. C. (1993). The effects of information technology and the perceived mood of the feedback giver on feedback seeking. *Information Systems Research, 4*(3), 240–261.
8. Anseel, F., & Lievens, F. (2006). Certainty as a moderator of feedback reactions? A test of the strength of the self-verification motive. *Journal of Occupational and Organizational Psychology, 79*(4), 533–551.
9. Stockton, R. A., & Morran, D. K. (1981). Feedback exchange in personal growth groups: Receiver acceptance as a function of valence, session, and order of delivery. *Journal of Counseling Psychology, 28*(6), 490–497.
10. Wang, W., Zhao, Y., Qiu, & Zhu, Y. (2014). Effects of emoticons on the acceptance of negative feedback in computer-mediated communication. *Journal of the Association for Information Systems, 15*(8), 454–483
11. Dadgar, M., & Joshi, K. D. (2018). The role of information and communication technology in self-management of chronic diseases: An empirical investigation through value sensitive design. *Journal of the Association for Information Systems, 19*(2), 86–112.
12. Fiori, M., Krings, F., Kleinlogel, E., & Reich, T. (2016). Whose side are you on? Exploring the role of perspective taking on third-party's reactions to workplace deviance. *Basic and Applied Social Psychology, 38*(6), 318–336.
13. Hatfield, E., Cacioppo, J. T., & Rapson, R. L. (1994). *Emotional contagion*. Cambridge University Press.
14. Fehrenbacher, D. (2017). Affect infusion and detection through faces in computer-mediated knowledge-sharing decisions. *Journal of the Association for Information Systems, 18*(10), 703–726.
15. Olszanowski, M., Wróbel, M., & Hess, U. (2020). Mimicking and shar- ing emotions: a re-examination of the link between facial mimicry and emotional contagion. *Cognition and Emotion, 34*(2), 367–376
16. Forgas, J. P., Chan, N. Y. M., & Laham, S. M. (2001). Affective influences on thinking and behavior: Implications for clinical, applied

and preventive psychology. *Applied and Preventive Psychology,* *10*(4), 225–242.

17. Grant, A. M., & Berry, J. W. (2011). The necessity of others is the mother of invention: Intrinsic and prosocial motivations, perspective taking, and creativity. *Academy of Management Journal, 54*(1), 73–96.

18. Scheier, M. F., & Carver, C. S. (1985). The self-consciousness scale: A revised version for use with general populations. *Journal of Applied Social Psychology, 15*(8), 687–699.

19. Abbate, C. S., Isgr'o, A., Wicklund, R. A., & Boca, S. (2006). A field experiment on perspective-taking, helping, and self-awareness. *Basic and Applied Social Psychology, 28*(3), 283–287

Research Environments and Methods

RDMflow: Managing Research Data Workflows with Micro-Frontends

Jan Ingo Haas⊙, Christoph Göpfert⊙, and Martin Gaedke⊙

Abstract

Research data management activities are commonly visualized using the *Research Data Management Lifecycle*. This lifecycle describes the various phases that research data undergo from the planning phase of a project to the reuse of the data in a subsequent research project. To ensure the reusability of data, the FAIR principles in particular play an important role. According to the FAIR principles, researchers should adopt established best practices of their research domain and use domain-specific standards for data annotation. However, researchers are often not aware which best practices and domain-specific standards apply to their dataset. Additionally, the complexity of some domain-specific standards may overwhelm researchers. With the RDMflow approach, we aim to improve the research data publishing procedure by adequately considering the context of research data. We use the modeling language BPMN to model processes of the research data workflow. Components of the workflow are subsequently mapped to micro-frontends which implement the underlying activity. Workflows may describe arbitrary processes such as publishing a paper or research data. The flexible and modular architecture created by combining micro-frontends and BPMN models makes it easy to define workflows that help annotate research data from a wide range of fields. This supports researchers in implementing standards and established practices, improving expressiveness, machine-readability, interoperability and reusability of research data.

J. I. Haas (✉) · C. Göpfert (✉) · M. Gaedke
Distributed and Self-Organizing Systems Group, Chemnitz University of Technology, Chemnitz, Germany
e-mail: jan-ingo.haas@informatik.tu-chemnitz.de

C. Göpfert
e-mail: christoph.goepfert@informatik.tu-chemnitz.de

M. Gaedke
e-mail: martin.gaedke@informatik.tu-chemnitz.de

Keywords

Research data management · BPMN · Micro-frontends · Data publishing · Scientific data management

1 Introduction

An invaluable part of advancing research is building on prior results. Many research projects generate research data, which are essential for the reproducibility of results and subsequent use [1].

Researchers are increasingly encouraged or even required to make their research data available by publishing them in a research data repository, such as Zenodo [1] [2, 3]. The data registration processes of these repositories share similarities in their flow. Researchers proceed through a web form in which various types of information (metadata) about the research data are gathered. Currently, repositories rely predominantly on free-text input for data entry. While this is convenient for users, it can be a hindrance to subsequent use of the data. (1) Free text entries can lead to ambiguities, (2) they do not allow for any input validation and (3) are thus inherently more susceptible to erroneous input, (4) they are difficult to relate to semantically similar terms (broader/narrower terms, synonyms, etc.), and (5) they cannot be processed by machines without intermediate steps.

In order for data to be published, a registration procedure is necessary to obtain further information required for data management and citation. The registration procedure is a fixed component of repositories and can therefore either not be integrated into external systems at all, or only by communicating with an API provided by the repository system [4]. However, even by using an API, registration cannot be integrated into existing workflows without additional effort. Most repositories release the data directly for publication after the

---

[1] https://zenodo.org/.

registration process has been completed; leaving no room for an intermediate step such as for a review. Rather, they rely on self-monitoring and self-correction of their users.

The RDMflow approach enables researchers to create data publishing workflows that can be tailored to the respective research context of data. Thus, our approach represents an improvement of metadata descriptions in terms of reusability. Data registration workflows created with our approach are easy to integrate into existing workflows of a researcher or a research group and allow the definition of arbitrary intermediate steps, such as a review step in advance of publishing data, thereby seeking to improve collaboration among researchers.

The remainder of this paper is structured as follows: in Sect. 2, we present current issues in data publishing workflows, taking a fictitious scenario as example, and derive criteria for an enhanced, prospective solution. In Sect. 3, we describe an approach for modeling extensible research data workflows, which are rendered using micro-frontends. Section 4 provides a discussion of our approach and Sect. 5 offers a comparison to related work. Section 6 summarizes our work and gives an outlook for future work.

2 Problem Analysis

In the following, we will use a fictitious example scenario in which research data are to be published to emphasize issues in common data publishing procedures.

Hannah is an early career researcher in Professor Jane's research group. Most recently, Hannah authored a paper to which she intends to publish associated survey data. She found a suitable research data repository and is going through its data registration procedure. It requires her to enter general metadata using simple text fields. Unfortunately, there is no way for Hannah to provide more specific information about her survey data, such as demographic information. Therefore, she is left with two unsatisfactory options: either she does not provide any specific information at all, or she utilizes a more general input field (e.g., "description") to add specific details instead. After submitting the form, Hannah uses the export function to e-mail Professor Jane the data description draft and asks for feedback.

From the scenario, two fundamental problems are apparent—the lack of an option to specify domain-specific information and the lack of an option to request feedback from a third party.

Most repositories use static web forms for their data registration procedure, disregarding the research data's context. Although this has the advantage that users are confronted with the same form repeatedly and thus become familiar with it more quickly, it comes at the expense of the quality of the data's metadata description. Missing or incomplete domain-specific information may result in subsequent users being unable to reproduce findings and may even render subsequent use impossible [5, 6]. For this reason, established domain-specific standards should be used to describe research data. This goes in accordance with the FAIR (Findable, Accessible, Interoperable, Reusable) principles, calling for metadata descriptions to meet *domain-relevant community standards* [6].

Implementing the FAIR principles' reusability criteria is challenging from several points of view. One obstacle is in finding a standard that is appropriate for describing the research data at hand. This issue is further complicated in the case of results from interdisciplinary research. Since the weighting of the importance of the data properties is inherently subjective and the viewpoint may differ even more depending on the field of research.

Another hurdle is the metaphorical maze of existing standards [7] which makes the selection of appropriate standards complicated. Depending on the type of research data (measurement data, survey data, etc.), several established standards may already exist, and their purpose and function may overlap partially. This makes the decision-making process increasingly time-consuming, as it is necessary to study the available candidate standards in depth to familiarize oneself with them before being able to make an informed decision. Once a suitable standard has been selected, the final hurdle is in its application. The reasons for this can be found in the complexity of the standards. For example in the nesting of information which is dependent on each other through conditions, insufficient or missing documentation, or required expert knowledge, for instance through implicit instructions.

Besides reusability, interoperability is another key component of the FAIR principles. The FAIR principles do not only call for interoperability w.r.t. data, but also of "tools from non-cooperating resources" [6]. However, few solutions focus on integrating Linked Data technologies in the data publishing workflow by means of providing endpoints for metadata export in an RDF format such as JSON-LD. We observed that the data publishing workflow itself is seldomly integrable into other workflows. A common approach of most research data repositories is to provide their users with simple registration forms inquiring general metadata, such as information required for citation, often followed by a form to upload related datasets, or linking to an external location where the resource is already available at. While this is enough to satisfy the interoperability principle of the FAIR principles, it does not fulfill the interoperability problem as stated in their original publication [6].

Our *Overall Objective* is to improve the research data publishing procedure in a way that the data's research context is adequately considered, as recommended by the FAIR principles. The *Purpose* for this is to improve the research data's expressiveness, machine-readability, interoperability and reusability, ultimately improving interdisciplinary knowledge-exchange for subsequent users. In order

Fig. 1 Examples of micro-frontends

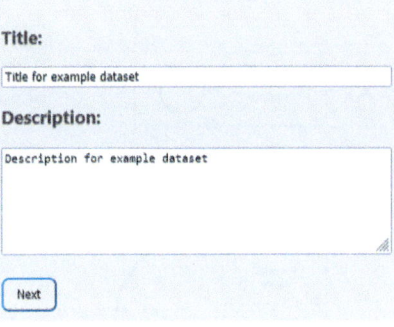

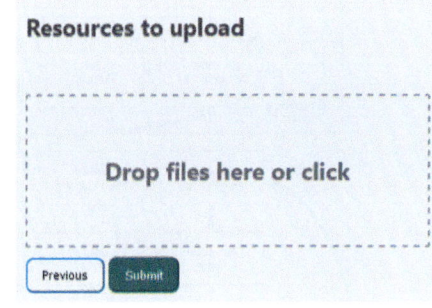

(a) Annotating research data.　　　　　　(b) Uploading data resources.

to realize our objective, the following goals need to be achieved:

1. Improving reusability of research data by tailoring data publishing workflows to the context of research data.
2. Extending the data publishing workflow to be customizable. For instance in terms of feedback loops, user roles and task lists.
3. Enabling researchers to integrate the data publishing workflow into their own workflows without disruptions.

3　RDMflow Approach

Our approach combines the Business Process Model and Notation (BPMN) [16] with the concept of micro-frontends. We use BPMN to describe workflows, such as a research data publishing procedure. Workflows consist of *Events* and *Activities*. *Events* may be the start of the publishing procedure or submission of data. *Activities* are sub-procedures of a workflow. *Activities* and *Events* are connected through *Connections*. *Connections* may be forked by gateways under certain conditions. Conditions may be influenced by variables which can be altered and processed as needed across several *Activities*.

In our approach, *Activities* of a workflow are mapped to micro-frontends. For our prototypical proof-of-concept implementation,[2] we realized micro-frontends in form of Web Components [8]. These components contain data for presentation (HTML, CSS) and logic/functionality (Javascript). As such, they can be used as building blocks in web pages (cf. Fig. 1). A micro-frontend may contain an entire registration form—or only parts of it. This decision is up to the developer of the micro-frontend. Thus, a micro-frontend focuses on one single functionality, like registration, uploading data or annotating data. If needed, user input captured by one micro-frontend can be passed onto another micro-frontend of a subsequent *Activity*, or affect the selection of a downstream activity in the workflow.

Web Components can be distributed using Content Delivery Networks (CDN) to make them available for other web developers. Other web developers are then enabled to easily embed the web components into their own websites, without further complicated development effort. For security reasons, reuse of the components can and should be restricted to certain domains.

3.1　Initial Setup

For the realization of our approach, a BPMN engine is required for execution of workflows, as well as a component for storing the research data and their respective metadata description. For the latter, a generic database or a data repository platform suffices.

Prior to deployment, an initial setup step for configuration is required. In this step, basic workflows must be specified. These may still be modified at a later stage or even entirely new workflows may be added after deployment. Moreover, the required micro-frontends need to be developed, which serve as user interfaces for processing individual *Activities* of a workflow. Worker processes to process data acquired in the micro-frontends must be developed. After data processing, uploaded data and their metadata descriptions must be transferred by the worker to a repository or database and be stored permanently.

Micro-frontends must be developed in advance because they realize *Activities* of workflows. They need to be structured as adaptively as possible. That is, especially in the context of data registration forms, the structure should be sufficiently flexible so that it may be extended by any desired number of input fields, and are able to handle different data types like text, numbers, date, etc. and provide grouping options for input fields. Moreover, they should provide researchers further assistance when publishing data, e.g., by means of instructions, help texts or example input data. This information needs to be considered when developing the micro-frontends.

As was stated previously, finding suitable standards may pose an obstacle, either because they are not known to

---

[2] https://gitlab.hrz.tu-chemnitz.de/vsr/rdmflow/microfrontends.

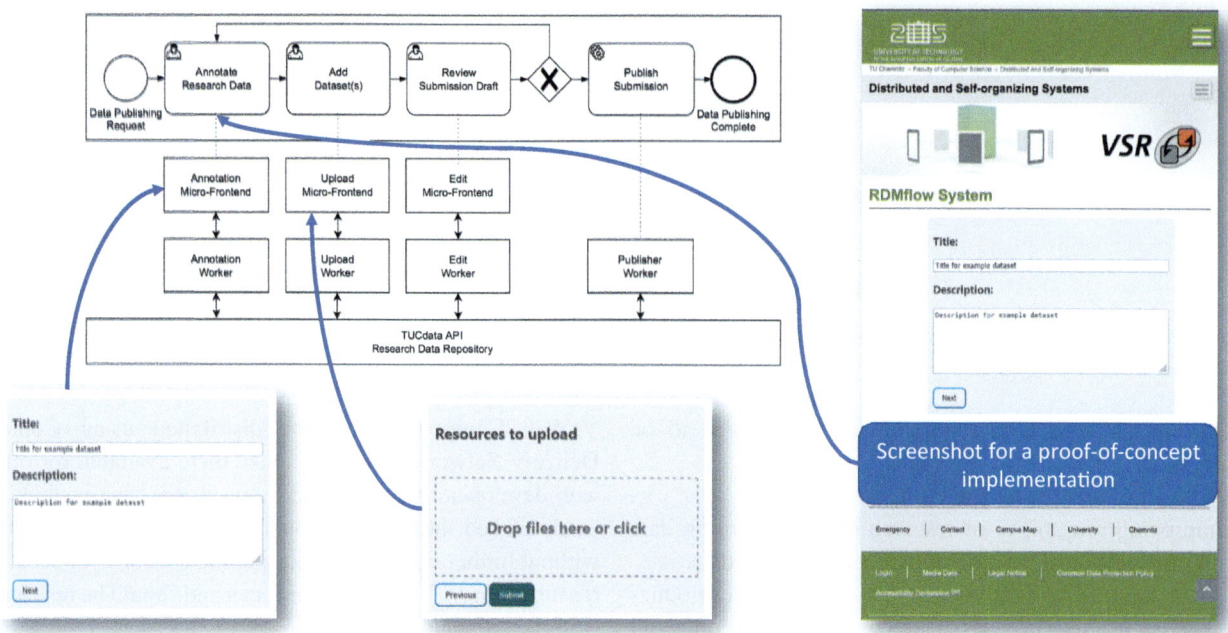

Fig. 2 Conceptual overview of the RDMflow approach

researchers or because expert knowledge is required to apply them. With the help of our approach, it can already be determined during the creation of the workflow under which conditions a specific standard should be recommended to the researcher. Accordingly, workflows should be created by an expert of the research domain, in the initial scenario this would be Professor Jane.

Users interact with the system initially via the interface of a BPMN engine such as Camunda's BPMN Workflow Engine.[3] In such a system, open tasks assigned to the user are listed in a dashboard-like user interface. The user interfaces for the actual processing of the tasks is represented by micro-frontends. The micro-frontends forward received user input to corresponding worker jobs, where the data are further processed. Data processing is not limited and may include validation of input data, processes for suggesting input or auto-completion, retrieving suitable domain-specific standards for describing the data in more detail, or sending notifications to colleagues.

3.2 Data Publishing Workflow with RDMflow

In the following, we consider a registration procedure as an example. We use the scenario described in Sect. 2 as basis. The following *Activities* can be derived from the scenario:

- Annotate Research Data
- Add Dataset(s)
- Review Submission Draft
- Publish Submission.

To realize more complex workflows such as the feedback loop described in the scenario, user roles may be used. These can be related to specific *Activities* in a workflow so that users of a certain role are exclusively assigned to certain tasks depending on their role. A solution approach for the workflow described in the scenario may involve the roles *Group Member* and *Group Manager*. The *Activities Annotate Research Data* and *Add Dataset(s)* are processed by users of the role *Group Member*, the activity *Review Submission Draft* by the role *Group Manager*. Figure 2 serves as an illustration. The workflow is processed as follows:

1. Hannah logs into her personal user account at the BPMN platform and starts a new process instance of the workflow *Data Publishing*. The system assigns a new task instance of the activity *Annotate Research Data* to Hannah. Hannah starts working on the task. She is redirected to a common web form, which allows her to annotate her research data. It is worth noting at this stage that the workflow can easily be customized to retrieve context-specific information as well. To do so, the *Annotate Research Data* activity must be followed by a second activity, so that the annotation process is carried out as a two-step procedure.

[3] https://camunda.com/.

2. Once Hannah completed annotation, the BPMN engine creates a new task *Add Dataset(s)*. Hannah is redirected to a web form which is represented by an *Upload Micro-Frontend* component. In the displayed form, Hannah has the option either to upload or to specify an existing location of her research data. The provided data sets are then either uploaded or linked through the *Upload Worker* in the research data repository TUCdata.

3. In the next step of the workflow, Professor Jane is assigned to a new task *Review Submission Draft*. A summary of the currently obtained meta-data is shown, which Professor Jane may modify as desired. If multiple changes are necessary, she can initiate a revision process. This will repeat the workflow starting with the *Annotate Research Data* activity until no more revisions are necessary.

4. Finally, the service activity *Publish Submission* follows, for which no user action is required. The BPMN engine directly calls the *Publisher Worker* to publish the research data together with their metadata description. Depending on the configuration set by Hannah, associated data sets are published publicly or with access restrictions.

Finally, the provided information is used to generate a metadata description of the research data which is made available by the data repository in both a human-readable format (e.g., PDF) and a machine-readable format (e.g., JSON-LD).

4 Discussion

We see various advantages in the RDMflow approach over traditional static registration forms:

Firstly, the approach can be quickly learned and applied by researchers. Due to the fact that workflows are described using BPMN, the learning curve is steep. Users are thus rapidly capable of modifying existing workflows or creating new workflows, customized to their needs. For users who merely wish to use existing workflows, there is no need to familiarize themselves with the approach. All they need to do is to log in to the BPMN platform which will redirect them to a web page containing web forms, as they are already used to. Thus, users only see the micro-frontends that affect themselves. A participant with the upload user role, for example, only has to take care of the upload of the resources, and does not see the micro-frontend for metadata annotation. This lowers the entry barriers to the whole system and the learning curve for participants of a certain user role. No complex familiarization steps are required.

Secondly, the workflows are customizable and consequently more easily integrable into other workflows. Users may add new *Activities* and new roles. Users may add intermediate steps of their own to the workflow as desired. In the previous section, we demonstrated this with a workflow that included an intermediate revision step before the data is published. Furthermore, the RDMflow approach is ideal for processes that need to be described in great detail. This applies, for example, to experiments involving humans, in which study procedures need to be precisely described, as well as to the planned processing of research data. For such scenarios, RDMflow enables the creation of tailored workflows with targeted assistance to be provided to researchers in describing their research data, implementing best practices and adhering to domain-specific standards. Furthermore, the workload of researchers can be reduced by automating repetitive tasks. Communication effort can be reduced as well by setting up automatic notifications to be sent through RDMflow after completion of an activity.

We believe that our approach can be particularly valuable for interdisciplinary collaboration, such as in a Collaborative Research Center. Interdisciplinary research generates a lot of relevant data that should be shared. The DFG's *Guidelines for Safeguarding Good Research Practice* recommend to make research data available to the public for at least 10 years [2]. RDMflow is a suitable approach to implement this guideline in a reusable manner.

Furthermore, a CRC brings together researchers with different levels of experience in scientific practice. As a result, data management is handled differently. Early career researchers often have not yet internalized best practices in handling research data. In order to support them in their research, it is a good idea to create workflows in RDMflow that are based on best practices. Researchers then only need to follow the workflow to ensure best practices are followed. Workflows can be customized to meet the needs of the collaborative research center. This will digitize best procedures and help researchers adhere to them, especially early career researchers.

Finally, we believe that the greatest potential of our approach lies in improving the quality of research data descriptions. On the one hand, this can be achieved through additional activities as shown with the control step of revision to catch erroneous or incomplete data. On the other hand, through the additional options of providing domain-specific information which are essential for the findability and reusability of research data.

5 Related Work

The authors of [9] describe a Capability Maturity Model (CMM) for Scientific Data Management. The CMM gives an indication of the existing level of maturity in the research data management process of an institution. It also shows a clear path from immature, mostly intuitive and not clearly defined management processes to mature processes with clear objectives and best practices. Four key areas for Scientific Data Management are further identified by the authors, for which best practices have been defined that achieve a high level of maturity. The method presented in our work allows to define Research Data Management processes that have clear objectives and adhere to best practices and thus have a high degree of maturity according to the CMM.

Qin et al., [10] emphasize the importance of metadata in finding records. Since datasets differ completely from domain to domain, it is very difficult to define standards for metadata. Based on this finding, two questions are formed and answered regarding the function of metadata and the model of a possible standard for metadata. From this work, we can infer that standardized metadata sets are important for dataset discovery (interoperability). In this respect, our approach offers the possibility to populate discipline-independent metadata records, but may also provide a way to dynamically annotate datasets.

Rucio [11] is an open-source framework for research data management. Developed as a platform for managing research data generated by Large Hadron Collider experiments, Rucio is based on a holistic approach to managing large volumes of data. Furthermore, there is the Dataverse project [12], which is also in the field of data management and has been classified as FAIR by [6] because it fulfills all FAIR principles. Other popular data management platforms are, for example, CKAN,[4] EUDAT [13] or Zenodo. In contrast to the frameworks mentioned above, our approach is not a data management platform, but uses the infrastructure of such a platform. Therefore, the RDMflow approach is not bound to a specific platform either. Another difference is that our proposed approach offers the user greater flexibility in specifying data management workflows using BPMN workflows. The presented frameworks have this functionality only to a very limited extent.

The Research Data Management Organizer [15] takes care of planning, managing and executing data management processes by summarizing information related to the necessary steps in a data management plan. Other tools for creating data management plans exist, such as ARGOS,[5] DataWiz,[6] or Data Stewardship Wizard [14]. Tools for creating data management plans differ from our proposal in that such tools do not cover all areas of the research data lifecycle. Our tool can assist the researcher at each stage by creating appropriate workflows.

Acknowledgements Funded by the Deutsche Forschungsgemeinschaft (DFG, German Research Foundation)—Project-ID 416228727—SFB 1410

6 Conclusion and Future Work

In this paper, we presented the RDMflow approach, a flexible and customizable approach for publishing research data. The approach uses the modeling language BPMN to model workflows. In these workflows, user roles, conditional connections and variables may be used. Workflow *Activities* are mapped to micro-frontends which are building blocks that can be easily embedded in web pages. In our prototype, these building blocks were implemented as Web Components.

The workflows created with the RDMflow approach enable the context of research data to be taken into account during the registration *Activity* and thus attempt to improve the reusability of research data. In addition, the publishing procedure can be easily integrated into already existing workflows. For future work, we plan to extend the approach by means of a taxonomy service. The taxonomy service will provide researchers with a broad range of curated Linked Data concepts for describing their research data. Furthermore, we intend to evaluate the RDMflow approach by means of user tests.

References

1. Sayogo, D. S., & Pardo, T. A. (2013). Exploring the determinants of scientific data sharing: Understanding the motivation to publish research data. *Government Information Quarterly, 30,* S19–S31
2. DFG. (2015). Leitlinien zum Umgang mit Forschungsdaten
3. European Commission. (2017). H2020 Programme—Guidelines to the Rules on Open Access to Scientific Publications and Open Access to Research Data in Horizon 2020
4. Borycz, J. (2021). Implementing data management workflows in research groups through integrated library consultancy. *Data Science Journal.* https://doi.org/10.5334/DSJ-2021-009
5. Hruby, G. W., McKiernan, J., Bakken, S., & Weng, C. (2013). A centralized research data repository enhances retrospective outcomes research capacity: A case report. *Journal of the American Medical Informatics Association.* https://doi.org/10.1136/amiajnl-2012-001302
6. Wilkinson, M., Dumontier, M., Aalbersberg, I. J., Appleton, G., Axton, M., Baak, A., Blomberg, N., Boiten, J.-W., Bonino da Silva Santos, L. O., Bourne, P., Bouwman, J., Brookes, A., Clark, T., Crosas, M., Dillo, I., Dumon, O., Edmunds, S., Evelo, C., Finkers, R., & Mons, B. (2016). The FAIR guiding principles for scientific data management and stewardship. Scientific Data 3. https://doi.org/10.1038/sdata.2016.18
7. Riley, J. (2018). Seeing standards: A visualization of the metadata universe. https://doi.org/10.5683/SP2/UOHPVH
8. World Wide Web Consortium. (2014). Introduction to Web Components
9. Crowston, K., & Qin, J. (2011). A capability maturity model for scientific data management: Evidence from the literature. *Proceedings of the American Society for Information Science and Technology, 48,* 1–9. https://doi.org/10.1002/meet.2011.14504801036

---
[4]https://ckan.org/.

[5]https://argos.openaire.eu/splash/.

[6]https://datawiz.leibniz-psychology.org/DataWiz/.

10. Qin, J., Ball, A., & Greenberg, J. (2012). Functional and architectural requirements for metadata: Supporting discovery and management of scientific data. In *Proceedings of the 2012 International Conference on Dublin Core and Metadata Applications, Dublin Core Metadata Initiative, DCMI'12*, pp. 62–71

11. Barisits, M., Beermann, T., Berghaus, F., Bockelman, B., Bogado, J., Cameron, D., Christidis, D., Ciangottini, D., Dimitrov, G., Elsing, M., Garonne, V., di Girolamo, A., Goossens, L., Guan, W., Guenther, J., Javurek, T., Kuhn, D., Lassnig, M., Lopez, F., Magini, N., Molfetas, A., Nairz, A., Ould-Saada, F., Prenner, S., Serfon, C., Stewart, G., Vaan- dering, E., Vasileva, P., Vigne, R., & Wegner, T. (2019). Rucio: Scientific data management. *Computing and Software for Big Science 3*(1), 11

12. King, G. (2007). An introduction to the dataverse network as an infrastructure for data sharing. *Sociological Methods & Research, 36*(2), 173–199. https://doi.org/10.1177/0049124107306660

13. Lecarpentier, D., Wittenburg, P., Elbers, W., Michelini, A., Kanso, R., Coveney, P., & Baxter, R. (2013). EUDAT: A new cross-disciplinary data infrastructure for science. *International Journal of Digital Curation 8*. https://doi.org/10.2218/ijdc.v8i1.260

14. Pergl, R., Hooft, R., Suchánek, M., Knaisl, V., & Slifka, J. (2019) Data stewardship wizard: A tool bringing together researchers, data stewards, and data experts around data management planning. *Data Science Journal 18*. https://doi.org/10.5334/dsj-2019-059

15. Neuroth, H., & Engelhardt, C. (2018). Aktives Forschungsdaten-management—das DFG-Projekt Research Data Management Organiser (RDMO). https://opus4.kobv.de/opus4-bib-info/frontd oor/index/index/docId/3688

16. Object Management Group. (2011). Business Process Model and Notation (BPMN), Version 2.0

Facilitating Ethics Application and Review for Interdisciplinary Human-Participant Research via Software-Based Guidance and Standardization

Alexandra Bendixen⬛, Thomas G. G. Wegner⬛, and Wolfgang Einhäuser⬛

Abstract

Research on human-technology interaction has become highly interdisciplinary. As a consequence, many fields are now performing research on human participants. From an ethical point of view, such research requires careful consideration of the well-being of all involved persons. Review boards for assessing ethical aspects of human-participant research are therefore confronted with steadily rising application volumes. Here we present a software-based approach ("Ethiktool") that facilitates ethics application and review procedures via a user-oriented dialogue system. Based on the user's responses in the guided dialogue, the ethics application as well as the information and consent forms to be used for the study participants are automatically created. This ensures consistency and standardization of all documents, which benefits all involved parties alike: the individual researchers applying for ethics approval and the review boards (who both can focus on ethical content) as well as the study participants (who can rely on transparent and standardized information forms). The software-based guidance helps to raise awareness of ethical issues in research, which is highly relevant in view of the ever- increasing possibilities of gathering data from human participants. Our approach will thus foster ethical responsibility of academic institutions toward a broad range of scientific disciplines and society as a whole.

Keywords

Ethics approval · Human participants · Software-based guidance · Ethiktool · User-oriented dialogue · Review board

1 Introduction

Research is becoming more and more interdisciplinary, reflecting the complexity of modern society's challenges. This is particularly evident in the field of human-technology interaction, which requires the interplay of many disciplines. Due to the digital transformation and the ever-increasing development of novel technology, our society is experiencing huge structural shifts. It is being discussed that we are moving toward a "hybrid society" in which humans and (embodied) digital technologies will co-exist and coordinate their actions with one another within the same shared space. Shaping such developments needs informed contributions by psychology, cognitive science, robotics, computer and data science, mechanical and electrical engineering, and many more disciplines. Interdisciplinary collaboration ensures that human expectations, attitudes, and abilities are taken into account when designing new technology, and vice versa, that technological possibilities are considered when asking humans about wishes and concerns they might have. Therefore, studies involving human participants have become the rule rather than the exception in all fields contributing to this interdisciplinary area. Such studies are ethically acceptable only if the well-being of all persons directly or indirectly involved in the research is given highest priority. Regulations ensuring ethical treatment have existed for many decades in medical research and in research on non-human animals. For non-medical research on human participants, which is in the focus of this work, such ethical considerations were initially left to the individual researchers. Individual considerations were informed by general guidelines as laid down,

A. Bendixen (✉)
Cognitive Systems Lab, Chemnitz University of Technology, Chemnitz, Germany
e-mail: alexandra.bendixen@physik.tu-chemnitz.de

T. G. G. Wegner · W. Einhäuser
Physics of Cognition Group, Chemnitz University of Technology, Chemnitz, Germany
e-mail: thomas.wegner@physik.tu-chemnitz.de

W. Einhäuser
e-mail: wolfgang.einhaeuser-treyer@physik.tu-chemnitz.de

© The Author(s) 2026
B. Meyer et al. (eds.), *Hybrid Societies*, Advances in Science, Technology & Innovation,
https://doi.org/10.1007/978-3-032-03488-5_37

for instance, in the Declaration of Helsinki [1] for the case of medical research. In recent years, psychology has taken up the notion that researchers should be assisted in the risk assessment of their human-participant research by independent ethics boards, who carry out formalized review procedures. Awareness has risen—within individuals, institutions, publishers and funding agencies alike—that an independent evaluation of planned research on human participants is not only ethically desirable, but also increases overall standards and quality of research.

While it seems natural to extend formalized ethics review to any research project in human-technology interaction, this poses severe practical challenges. In the German academic system, independent ethics review is typically carried out by local departmental or institutional review boards. These are often run as part of academic self-governance (i.e., based on voluntary time investment of the board members). Therefore, the application volume they can handle is limited, and boards often restrict evaluations to members of their own department. With research on human participants extending beyond its traditional fields, this puts institutions and individual researchers to a dilemma. Even if ethics review boards would like to extend their reviewing capacities, they do not want to take the risk of sacrificing their standards in terms of review scrutiny given their limited time. Researchers, in turn, want to have their research evaluated independently, but they may lack access to ethics review boards at their departments or institutions [2]. Even if they have access, they often lack experience and academic training with regard to ethics review. As a consequence, they might perceive the ethics-review procedures (e.g., which information is required for evaluation, how should information and consent forms be structured, etc.) as a substantial challenge. In turn, the fear that inexperienced applicants will cause a disproportionate burden may prevent departmental boards from extending access to other departments or further researchers.

Here we introduce a solution that has the potential to remove both these obstacles at once—that is, it lowers the threshold to seek and obtain independent ethics review and it reduces the burden on review boards without compromising reviewing standards. The key idea of our solution is to provide automated dialogue-oriented guidance for the preparation of research projects for ethics review. This is provided by a software solution called *Ethiktool*. The guidance offered by the software-based approach automatically standardizes the ethics applications, simplifies their compilation, and thereby supports both the individual researcher applying for ethics approval and the review board evaluating the applications. Researchers at any level of academic training will benefit from such guidance and standardization, as templates and information about review procedures are the means of support most frequently sought from ethics boards even by experienced

researchers [2]. In the following, we introduce the *Ethiktool* approach and discuss its potential for interdisciplinary research involving human participants.

2 General Approach

A. Starting Points and Challenges
Considering ethics in interdisciplinary research comes with a number of challenges, most of which are solvable once they are recognized and made explicit, as outlined in the following.

(1) Different Concepts of Research Ethics
Bringing up ethics as a topic in an interdisciplinary collaboration usually triggers very different concepts. These can broadly be sorted into three facets: First, ethics is often associated with good scientific practice (behaving responsibly as and within a scientific community, avoiding and reporting scientific misconduct, improving working conditions for scientists, appropriately representing individual contributions in authorship order, and the like [3]). Second, science has ethical responsibilities toward society in the sense of which questions may be asked and which technologies may be developed. This facet of ethics requires careful consideration of potential societal, economic or military use of the research results beyond the original purpose of the research project ("dual use"). The ethical dilemma of such dual use is currently quite prominent, for example, in the life sciences and biotechnology [3], and—closer to the human-technology interaction aspect—in the context of artificial intelligence (AI): not only has AI the potential to be misused to develop weapons or toxins [4], but there are also seemingly less dramatic— yet still ethically debatable—developments such as basing consequential decisions on self-learning AI algorithms [5]. Third, having settled upon a research question, it is important to carry out the research itself in an ethical manner toward all involved humans—that is, the experimenters, potentially third parties (such as bystanders) and of course the participants (e.g., informing them beforehand about the nature of the research and ensuring their free choice whether to participate or not throughout the study, avoiding unnecessary risks or harm of any kind [1]).

Undisputedly, all three facets of research ethics deserve attention, and responsible research can only be conducted after in-depth consideration of all of them. The first facet is vividly being discussed nowadays [6], and institution-wide committees for dealing with cases of scientific misconduct are established in many places. The second facet requires strong disciplinary knowledge in order to imagine dual use cases and the harm they might inflict. Notwithstanding its importance, it can hardly be standardized across disciplines (more specifically, the procedures of how such ethical questions are addressed can be standardized, but not the content as to which information must be sought to address ethical pitfalls). The third facet is specific to human-participant research (in whatever discipline it is conducted), and thus has a chance to be handled in a standardized manner across all disciplines. Yet the third facet is also the one that is most often overlooked when thinking about research ethics. In fact, it may be regarded as less important by some even within psychology and related disciplines that have been conducting human-participant research for decades. Especially in cases where research is seemingly harmless ("just questionnaires") and seemingly non-vulnerable groups are investigated ("just university students"), academic culture is only slowly proceeding toward an appreciation of ethics review as a necessary part of the workflow of any research project. In those disciplines that have only been starting to work more systematically with human participants, this offers a chance to "catch up" and set the standards for ethics review as an integral component right from the start.

(B) Responsibility for Human-Participant Ethics

Due to the stronger formalization of ethics review procedures in psychology compared to other disciplines, interdisciplinary projects face the danger that the responsibility for the third facet of research ethics (how to work ethically with human participants) is delegated to the psychology-related contributors or to social scientists in a wider sense. This is aggravated in many places by lack of access to an ethics review board for researchers outside psychology, who may in addition lack the experience in how to write ethics proposals for such a board. While this seems a reasonable distribution of labor, leaving the ethics responsibility with the psychologists or social scientists in the project alone misses significant opportunities. We argue that all members of an interdisciplinary research team should understand why these ethical aspects are important to guide their current and future research decisions, and in turn to shape best practices in the field of human-technology interaction.

(C) Terminological and Conceptual Challenges

Aiming toward shared responsibility for human-participant ethics across the whole interdisciplinary team requires easy access to the terminology and consensual guidelines typically used in this domain. Principles like prior information, voluntary consent, and right to abstain should be broadly clear to (former) students of psychology, but they might put significant obstacles for team members who have no academic training in human-participant research. Just like terminological misunderstandings make interdisciplinary collaboration challenging in many other respects [7], this is also true for filling out an ethics application form that has been designed by researchers with a strong psychology mindset. For example, answering a question like "Is participation in the study volun-

tary?" requires a thorough understanding of the conditions under which true voluntariness is challenging to ensure (such as in classroom or among subordinates like within the research team itself). Depending on the concept of voluntariness a researcher entertains, the answer given to this question may be more or less informative to the ethics review board. Mutual misunderstandings at this level can impede the ethics review process in various ways: it can either lead to a suboptimal decision because the different understandings of voluntariness never become explicit (i.e., the researcher checks "yes" and gives no further explanation, and the review board does not catch the possible misunderstanding of the concept of voluntariness), or it can lead to a second round of reviews as the board asks for additional explanation on how true voluntariness is ensured. The latter option delays the process, increases workload for applicant and board, and may cause frustration on both sides. As denoted in Section I, it may even make review boards hesitant to open their review process to researchers without ethical training in human-participant research. Vice versa, it may make the individual researcher hesitant to seek ethics approval for another study in the future, as it leaves them with the impression that ethics review is a complicated and long-winded procedure.

Based on experience from various ethics boards, misunderstandings of integral parts of the ethics review process happen on even much more elementary levels, such as failing to submit all relevant documents. Most ethics review boards require the researchers to submit a pre-structured description of their research project along with the documents that will eventually be used for the participating humans (i.e., information and consent forms). Apart from the elementary error that these documents are not submitted (which can be easily caught by pre-screening), another surprisingly prevalent observation is that the documents for the participants are inconsistent with the description for the review board. Such inconsistencies can, of course, be intentional under special circumstances (such as when there is good scientific reason not to disclose the full purpose of a study), but in many cases the inconsistencies reflect basic errors (e.g., caused by last-minute changes, or by copy-pasting from previous documents). Likewise, it is often observed that the written instructions for ethics applications by the review board are not comprehensively followed, or that template texts are not used in the intended way. This may reflect that applicants are not as experienced or do not share a common scientific ground with the ethics board members, possibly due to lack of training in ethical aspects of human-participant research in their research field. Consequently, a substantial fraction of reviewing is concerned with uncovering inconsistencies and reporting back how to fix them or asking for more information, all of which detracts from the actual purpose—that is, from evaluating the ethical aspects of the planned research.

In order to improve experience and outcome for all involved parties, it is desirable to develop solutions to these challenges.

2 Guided Standardization as a Solution

Many of the issues mentioned above can be mitigated by guided standardization. Using the benefits of digitalization, one can conceptualize a system that provides interactive content-oriented user guidance to prompt researchers to consider all aspects relevant for a well-founded ethics assessment. Ideally, this not only ensures that all relevant aspects are covered, but also raises the researcher's awareness for ethical issues already in early stages of research planning. Having a software tool identify formal errors as well as inconsistencies in content takes away the burden of initial screening from the review board, which then can focus on ethically relevant aspects. Besides guiding the researcher through the application process and validating all entries and their dependencies, the software can also use the information gathered in the interactive process to generate all relevant documents, including the ethics application form itself as well as participant information and consent forms. This leads to a standardization of the documents not only for the ethics review board but also for future research participants. Adherence to local regulations (e.g., are the requirements for an abbreviated review process met?) can be checked automatically, and thus using wrong submission tracks (e.g., asking for an abbreviated review process with a vulnerable participant group) can be avoided. As another beneficial side effect, no documentation and further instruction about the review process is needed as all relevant information can be covered within the software, directly at the step where it is required. Key terms and concepts can be clarified when they first occur, thereby avoiding potential misunderstandings and reducing the disciplinary barriers described above. Substantial parts of documentation can be omitted altogether by making questions dependent on the information gathered beforehand (e.g., explanations such as, "If you answer *yes* to question 24, continue with questions 25–29, else continue with question 30" can be replaced by responsive software elements that will include questions 25–29 or not depending on the answer to question 24). Such principles are easy to implement in any modern software system once all content dependencies can be algorithmically described. In this sense, a software-based ethics review system also aids formalization and standardization of the review process by making the rules explicit and thereby transparent.

An interactive software will reduce the burden on the researcher applying for ethics review, as many documents are generated in a standardized form, with formal requirements and content consistency across documents automatically checked. Such a solution will also reduce the burden on ethics review boards, as all applications and documents have a consistent format (making information easy to retrieve) and

basic content consistency is automatically ensured. This gives both the applicant and the board more resources for focusing on the ethical core of the application. In fact, this focus on the question of what is or is not ethically permissible should be at the heart of ethics review. Streamlining the process up to this point can free the capacity of the board members to get into in-depth discussion about the actual ethical issues, and into communication with the individual researchers about these aspects of their project, which increases the quality and transparency of the evaluation process. Thus, while the software can obviously not solve ethical questions by itself, it can give room to all involved parties for thorough deliberation and constructive communication.

As we will outline in Section III, we have developed and implemented a first version of a software offering the described functionality, called *Ethiktool*, which is being used by the ethics review board of Chemnitz University of Technology since June 2021.

3 Software Implementation

We conceptualized and implemented a software system called *Ethiktool* that provides interactive user guidance through the compilation of the ethics review documents, including participant information and consent forms. The *Ethiktool* software validates all entries and identifies formal errors as well as inconsistencies in content (e.g., combining fully anonymous data processing with post-experiment remuneration, declaring minimal risk for an invasive procedure, asking 4-year-old participants for their written informed consent). Based on this automatic screening, the *Ethiktool* provides constructive feedback that helps the researcher to identify potential issues and allows them to reconsider ethically relevant aspects of their research protocol before submitting it for ethics review. Moreover, it provides template texts for recurring issues, thereby increasing the consistency of wording across research projects both in the ethics application form and in the participant information and consent forms. This will be beneficial not only for ethics review boards but also for the eventual study participants, as they can understand the study procedures and risks in a consistent and transparent manner across different studies.

Specifically, in its version at the time of writing (version 2.1 as of August 2022), the *Ethiktool* queries—besides basic information about the researchers and their role in the project—a series of distinct topics about a study, such as inclusion criteria, planned interventions, expected risks, way of consent, and subject compensation. The dialogues typically first consist of simple menu items, which expand following certain selections. For example, users are asked whether participants are fully informed about the study prior to participation. If users reply "yes", they will be asked (among other items) about the form (oral/in writing/both). If prior information is presented in written form, relevant texts for the information sheet are queried. If users respond "no" to prior information, a justification for ethics review will be requested and it will be queried whether there is a debriefing in which participants are informed about their participation afterwards. In addition, there is a check for consistency across topics. For example, if a user selects that participants are informed about participation only after the study and at the same time states that they are able to withdraw at any time, this inconsistency will be flagged. Through this direct feedback, the *Ethiktool* also offers hands-on teaching and learning opportunities: Researchers can see by example of their own research projects how ethical principles interact and what needs to be considered for implementing them in a proper way.

The *Ethiktool* asks its users to answer closed questions by simple choices and to enter free text to open questions (Fig. 1). All entered information is then combined to generate the ethics review document itself as well as the information and consent forms for the participants. The users can generate these documents in pdf format at any moment during the process to double-check how their answers and free texts are integrated in the overall documents. However, they can prepare the documents for submission to the ethics review board only once they have replied to all items and no inconsistencies are found anymore.

The *Ethiktool* graphical user interface (GUI) deliberately has a plain style (Fig. 1) as the focus of this first software development was on implementing the general principle and

the underlying dependencies. Once the algorithmizing of the review process and documents is fully in place, the GUI can easily be updated to comply with current expectations for human–computer interaction. A short demonstration video showing the 2022 version of the tool in operation can be found at https://mytuc.org/pcys (please note that this shows the 2022 prototype; the current *Ethiktool* version is available at https://ethiktool.org).

The *Ethiktool* already allows for several complex designs (e.g., studies consisting of several parts or measurements). It also provides easy-to-use options for replicating the core information of a proposal. This facilitates the submission and review of studies similar to earlier applications with small changes and amendments. Again, a straightforward routine for implementing and for flagging changes is beneficial to both involved parties—for the applicant, there is little incremental effort of asking for ethics approval of another study that is similar to one already approved. For the ethics review board, reviewing similar research projects becomes much less time-consuming as only the changes need to be assessed. Since all the information is provided in standardized digital format, identifying changes is much easier than with the traditional procedures.

4 Evaluation

The digitalization and standardization approach of ethics review as offered by the *Ethiktool* must be systematically

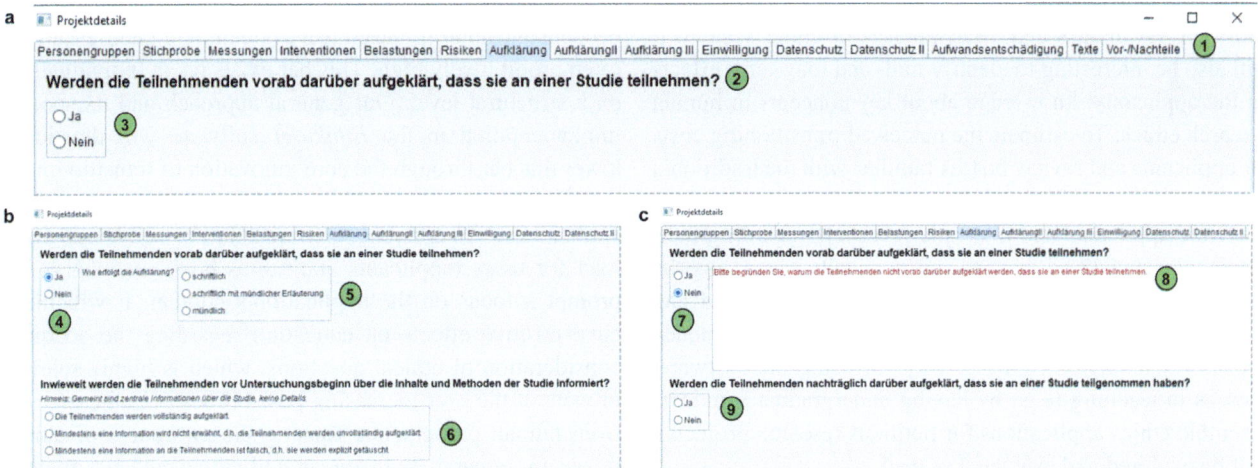

Fig. 1 Example screen from the *Ethiktool*. **a** Users are guided through the application process on a sequence of tabs (1). The example shows the question about whether participants are told in advance that they participate in a study (2), a response (yes/no) is required, but no default is given (3). **b** If the user selects "yes" (4), they will be asked about the form (in writing, in writing with additional oral information, oral) (5), and to what extent the users are informed prior to the study (fully or partly, with or without deception) (6). **c** If they respond "no" (7), a free-text justification

is required (8) and users are required to respond as to whether the participants are informed after the study (9). Depending on the responses in (6) and (9) further queries may follow. It should be noted that the graphical interface has a deliberately plain style as the initial focus was on implementing the general principle and the underlying dependencies. For a video demonstration of the *Ethiktool* in action, please visit https://mytuc.org/pcys (please note that Figure 1 shows the 2022 prototype; the current *Ethiktool* version is available at https://ethiktool.org)

evaluated for identifying the benefits and drawbacks of this approach. This is a critical step in deciding about its wide-range deployment and sustainability. So far, the anecdotal responses from users at Chemnitz University of Technology are highly encouraging. In particular, first-time ethics applicants report that the dialogue-oriented guidance has massively helped them to consider all relevant aspects and put them in a structured form. A next critical step is a comprehensive evaluation to quantify the effects of software-based standardization on the review process. Such an evaluation could not be conducted in parallel to the roll-out of the *Ethiktool* at Chemnitz University of Technology, as it coincided with the general restructuring of the reviewing processes, in particular the transition from several departmental review boards to a single university-wide review board. This would have led to a severe confound in any comparison of the review process with the *Ethiktool* as opposed to the traditional procedure. Yet, a systematic comparison is planned at other sites where the traditional procedures and the *Ethiktool* will run in parallel for an extensive evaluation period.

A continuous unbiased evaluation should comprise several outcome variables, both in terms of time (for compiling an application on the side of the applicant, and for reviewing an application on the side of the board) and in terms of quality (full coverage of necessary information, absence of errors and conceptual misunderstandings, intelligibility of the resulting documents for review board members and for human participants, consistency of the reviews). The software-based approach of the *Ethiktool* will be compared in these regards with the template-based processes that are currently in place in many review boards. This will allow for an in-depth evaluation of advantages and disadvantages of either system. It will also be interesting to identify mid- and long-term effects on the applicants' knowledge about key concepts in human research ethics. To estimate the perceived transitioning costs for applicants and review boards familiar with the traditional procedures, these experienced groups will be compared with first-time applicants and first-time reviewers regarding their *Ethiktool* usage. In addition to systematically comparing the general approaches, the software itself will be tested and improved in terms of its usability (effectiveness, efficiency and user satisfaction). Further ways of using the software, such as in teaching (e.g., by having undergraduate students assemble ethics applications for fictitious research projects), will be explored and evaluated as well.

As laid out in Section I, it is highly likely that the demand for independent ethics reviews will quickly rise across many disciplines. The *Ethiktool* offers the chance of fostering this development: If automatic consistency checks and standardization reduce the load per ethics applications for the review boards, the boards can increase their efficiency and open their processes to other departments or institutions, and they can spend more time on in-depth discussion and communication with the involved researchers on those projects where ethical acceptability is not straightforward to decide upon. The number of review boards eventually using the *Ethiktool* will be an informative evaluation criterion. At the same time, individual researchers without access to independent review can still use the tool to support them in considering ethical implications and providing participants with information and consent documents that meet the standards of research on human participants. This cannot replace independent ethics review, but will alleviate potential issues, and thus support ethical principles across disciplines and institutions.

5 Conclusion

In this paper, we present an approach toward facilitating ethics application and review procedures for research on human participants. This field of research is becoming increasingly interdisciplinary. Ensuring that such research is conducted in an ethically impeccable manner is pivotal to sustainable innovation. We have started from the observation that for researchers from fields without a longstanding tradition in studies involving human participants, the bar to obtain independent and appropriate ethical evaluation of their research is often prohibitively high. This bar exists on an individual and on a structural level. Our general approach and its specific implementation in the *Ethiktool* software will drastically lower this bar through the core innovation of transitioning to a user-oriented dialogue-based process of developing ethics applications. An efficient software solution will lower workload for users (applicants and ethics board members) and prompt a focus on the application's content. It will likely have positive effects on education regarding the adequate consideration of ethical questions, which is highly relevant in view of the ever-increasing possibilities of gathering data from human participants. Raising the awareness of ethical issues in research both on an individual and a structural level will have a lasting impact on research practices for the good of human research participants, thereby fostering ethical responsibility of academic institutions toward society as a whole.

Acknowledgements The work was funded in part by the Deutsche Forschungsgemeinschaft (DFG, German Research Foundation) project ID 416228727—SFB 1410.

References

1. World Medical Association (WMA) Declaration of Helsinki. https://www.wma.net/policies-post/wma-declaration-of-helsinki-ethical-principles-for-medical-research-involving-human-subjects/
2. Strobel, A., Zeiler, A., & Schaar, K. (2022). Der Stand der Dinge zur forschungsethischen Begutachtung in den Geistes- und Sozialwissenschaften–Ergebnisse einer Umfrage unter Forschenden [The state of affairs in ethics evaluation in the humanities and social sciences—Results of a survey among researchers]. *RatSWD Working Paper 278/2022.* Berlin: Rat für Sozial-und Wirtschaftsdaten (RatSWD). https://doi.org/10.17620/02671.67
3. Deutsche Forschungsgemeinschaft (2022). Guidelines for safeguarding good research practice. *Code of Conduct.* https://doi.org/10.5281/zenodo.6472827
4. Selgelid, M. J. (2009). Governance of dual-use research: An ethical dilemma. *Bulletin of the World Health Organization, 87,* 720–723. https://doi.org/10.2471/blt.08.051383
5. Shankar, S., & Zare, R. N. (2022). The perils of machine learning in designing new chemicals and materials. *Nature Machine Intelligence, 4,* 314–315. https://doi.org/10.1038/s42256-022-00481-9
6. Einhäuser, W., & Bendixen, A. (2019). Das Vorurteil der Maschine [Machine prejudice]. https://www.wissenschaftsjahr.de/2019/neues-aus-der-wissenschaft/das-sagt-die-wissenschaft/das-vorurteil-der-maschine/
7. Leising, D., Dshemuchadse, M., Schönbrodt, F., & Scherbaum, S. (2021). Genug ist genug: Unethisches Verhalten in der Wissenschaft muss endlich systematisch angegangen werden! [Enough already: unethical behavior in science must finally be systematically tackled!]. https://doi.org/10.5281/zenodo.5126637
8. Wudarczyk, O. A., Kirtay, M., Kuhlen, A. K., Abdel Rahman, R., Haynes, J.-D., Hafner, V. V., & Pischedda, D. (2021). Bringing together robotics, neuroscience, and psychology: Lessons learned from an interdisciplinary project. *Frontiers in Human Neuroscience, 15,* Article 630789. https://doi.org/10.3389/fnhum.2021.630789

Gyro Gearloose or Little Helper?—Two Perspectives of AI and Patent Law

Dagmar Gesmann-Nuissl and Stefanie Meyer

AI is probably the best or worst thing that can happen to humanity.

–Stephen Hawking, Physicist

Abstract

Recently, patent offices around the world were concerned with the patent application for a blinking light and a fractal food container, and extensively discussed the requirements for granting a patent. This unprecedented case sparked extensive discussions as "Dabus AI" was filed as an inventor—an autonomously acting software. This case raised a number of issues concerning patent law, in particular whether an invention in the field of artificial intelligence (AI) is patentable, and moreover, whether the invention made by an AI is patentable. In this paper, we address these two perspectives by discussing the legal requirements for filing a patent application and applying the two perspectives described to these requirements. We conclude that, as of now, only the human inventor Gyro Gearloose can be a patentable inventor. The time when his little helper will occupy a similar position is still in the future—but it is worth looking beyond the horizon into the future.

A statement made by physicist Stephen Hawking at the opening of the Leverhulme Center for the Future of Intelligence (CFI) in Cambridge in October 2016: https://www.cam.ac.uk/research/news/the-best-or-worst-thing-to-happen-to-humanity-stephen-hawking-launches-centre-for-the-future-of (accessed on 25.04.2024).

D. Gesmann-Nuissl (✉) · S. Meyer
Private Law and Intellectual Property Rights, Faculty of Economics and Business Administration, Chemnitz University of Technology, Chemnitz, Germany
e-mail: dagmar.gesmann@wiwi.tu-chemnitz.de

S. Meyer
e-mail: stefanie.meyer@wiwi.tu-chemnitz.de

Keywords

Artificial intelligence · Patent law · Inventor · AI inventions · AI-generated inventions

1 Introduction

Artificial intelligence (AI) and increasingly complex algorithms are influencing our daily lives more than ever before. The application areas are becoming more diverse and the possibilities more far-reaching. While AI applications already exceed the performance of many human experts, it is expected that AI performance will be further optimized by the algorithms used and reach superhuman levels [1]. In addition to the discussion of potential application areas, the discussion of responsibility for AI is also increasingly coming to the fore, involving many disciplines such as technology, law, psychology, and sociology in equal measure.

When legal experts talk about responsibility, their thoughts revolve around guilt, liability and compensation. But responsibility means much more than the question of liability. Responsibility also includes implementing social values and promoting innovations for the common good. Thus, it can also be asked who is responsible for the benefits of AI [2] and the innovations it generates in a positive sense. Is it—to stay with Carl Barks—Gyro Gearloose as the master of every idea, who should alone share in its benefits, even if he uses AI, or can the "little helper", the AI, be the creator and beneficiary?

To answer this question, it is necessary to take a closer look at patent law and its treatment of AI, in particular (1) the patentability of an invention that (partially) involves AI, and (2) the patentability of an invention created by AI.

B. Meyer et al. (eds.), *Hybrid Societies*, Advances in Science, Technology & Innovation, https://doi.org/10.1007/978-3-032-03488-5_38

2 Purpose of Patent Law and Its Dimensions in Relation to AI

In general, two questions can be asked about patent law in the context of AI. What is the purpose of patent law in the first place, and what role can an AI play in this overall construct of the patent? The purpose of patent law is as independent of the subject matter claiming protection as it is of the person or persons claiming to be the inventor. In general, patent law is intended to give the inventor a legal status for their creative achievement that entitles them to exclusively exploit the invention for a certain period of time. It thus protects against imitations. The more detailed purpose has been split into four patent law theories since Machlup [3], who was arguably the first to formulate these theories to critically examine the patent law system, but they interact sufficiently: the property theory, the reward theory, the incentive theory, and the disclosure theory. The property theory assumes that every intellectual creation is inherently the property of the individual who produced it—a technical invention therefore belongs to its inventor by virtue of natural human rights. The reward theory states that justice demands that every service rendered to the general public be rewarded according to its usefulness; since the inventor increases the accessible technical knowledge, the patent is their share. The incentive theory is based on technical progress and states that inventions are only made and used if there is a prospect of a return. The disclosure theory is based on the fact that the inventor receives a kind of consideration by making their invention available to the general public—without this consideration, secrecy would be more desirable for the inventor [4–6]. However, these four theories are not necessarily excluding each other, but are interrelated and complementary.

In what dimensions can patent law and AI be related? Considering a topic that is subject to extensive research, discussion and debate, it is surprising that there is currently no universally accepted definition of AI—especially in the field of patent law. In general, AI can be defined as the field of computer science that investigates the properties of intelligence by synthesizing intelligence [7]. AI is based on the execution of mathematical methods or algorithms by a computer implementation. These methods or algorithms are able to learn from data and process the data in a way that demonstrates intelligence [8]. This was also described in a 2017 study by the European Patent Office (EPO): The field of AI is described as providing machines with a kind of understanding [9]. By "understanding" is presumably meant behaviors that humans perceive as intelligent, such as learning, understanding, inferring, and deciding, where in the narrow sense, of course, it is not the machine that learns, but a process running on it, such as an algorithm [10]. The EPO also concretizes this in its examination guidelines by requiring that

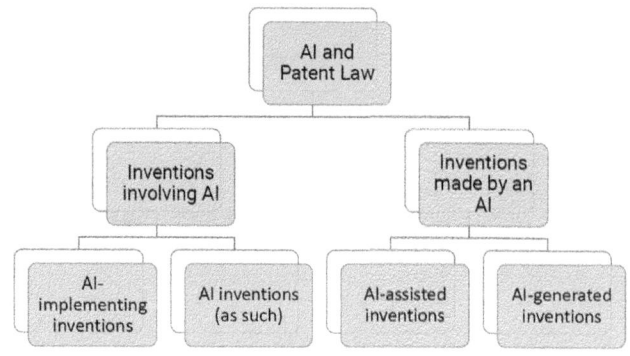

Fig. 1 Dimensions of patent law in relation to AI

AI and machine learning be discussed in the context of mathematical models rather than computer programs, since they are "inherently abstract mathematical in nature, regardless of whether they are trained with training data" [11].

Two distinct perspectives are regularly associated with AI and patent law: (1) inventions whose doctrine relies on AI, and

(2) inventions that are developed by or with the assistance of AI, but do not necessarily use AI as part of the doctrine of the invention as such [10, 12]. In the first category, a distinction can be made between processes that use an AI method as a partial feature of a particular technical application or adapt it for a particular technical implementation (AI-implementing inventions) or inventions that relate to an AI method as such (core AI) [13, 14]. In the second category, a distinction can be made between AI-assisted inventions and AI-generated inventions [15], (Fig. 1).

3 Inventions Whose Doctrine Makes Use of AI

For granting patents in the field of AI-implementing or fully AI-based inventions, the same requirements apply as otherwise for the general requirements for granting protection in patent law: According to Sec. 1 (1) of the German Patent Act (PatG) or according to Art. 52 (1) of the European Patent Convention (EPC), patents are granted for inventions in all fields of technology, provided that they are new, involve an inventive step and are susceptible of industrial application.

Since there are only a few application scenarios for patenting requirements of AI, it is useful to have a look at the case law on computer programs and computer-implemented inventions (cf. Sec. 1 (3) No. 3 PatG and Art. 52 (2) (c) EPC) due to their implementation on computers. However, the distinction between the invention AI as such and AI-implementing inventions becomes apparent as follows. The latter are presumably to be subsumed under the category of computer-implementing inventions [16], whereas AI as such is more likely to belong to the field of mathematical methods due to its basis in computational models and algorithms (cf. Sec. 1 (3) No. 1 PatG and Art. 52 (2) (a) EPC) [10]. The wording of the legislation also distinguishes between the mathematical method as such and the application of the same.

1. "New"

According to Sec. 3 (1) cl. 1 PatG and Art. 54 EPC, an invention has to fulfill the requirement of novelty. In this context, an invention is new if it does not belong to the state of the art, Art. 54 (1) EPC. According to Art. 54 (2) EPC, the state of the art includes all prior art which has been made available to the public prior to the priority date by means of a description, by use or in any other way [17]. This applies both to inventions incorporating AI and to inventions that involve no AI.

The examination of this mandatory requirement of novelty is carried out by the Patent Office. For this purpose, there are several steps it has to perform. In a first step, it has to identify the main claim, which is usually described in the first sentence of the set of claims. If there are several main claims, they all have to be determined. In a second step, the prior art has to be determined, i.e. it has to be examined which technical doctrines of the patent claim to be examined have already been demonstrably disclosed by third parties, i.e. made available to the public in any manner. This may have occurred by means of disclosures or by means of acts of use. The perspective that the Patent Office has to take is that of the person skilled in the art. What the latter takes from a pre-published material with their technical expertise and wealth of experience is disclosed in a manner detrimental to novelty. Only if, from this point of view, the invention to be applied for is to be considered "new" since it does not belong to the state of the art, the Patent Office may pursue the application (at least as far as the other requirements are fulfilled).

2. "Inventions […] that Involve an Inventive Step"

The prerequisite for patentability according to Sec. 1 (1) PatG and Art. 52 (2) EPC is the existence of a (technical) invention and the non-existence of one of the conditions set out in Sec. 1 (3) and (4) PatG and Art. 52 (2) and (3) EPC. When examining this requirement for patentability, the BGH and the EPO proceed differently, although both institutions come to the same conclusion.

According to the converging presumption, an invention in the field of technology is given if the claimed doctrine involves the use of technical devices, in particular computers [14]. With regard to AI, both for claiming the AI method as such and especially for claiming AI-implementing inventions (Fig. 1), the application of AI to a computer has to establish existence in the field of technology in a first step [19]. In order to exceed the exclusion criteria (Sec. 1 (3) and (4) PatG and Art. 52 (2) and (3) EPC), it is necessary in a second step

that the claimed invention contains instructions which serve to solve a specific technical problem by technical means or at least to affect this problem [21–23], so-called technicity.

Here, the difference between patenting the AI system as such and the AI-implementing system as shown in Fig. 1 is evident. To prevent any of the exclusion criteria (Sec. 1 (3) and (4) PatG and Art. 52 (2) and (3) EPC) from being applicable, it is decisive that the claimed doctrine contain instructions which serve to solve a specific technical problem by technical means or at least to affect such a problem [14]. If these requirements are met, patentability is to be affirmed—regardless of the degree of autonomy. However, the application of these criteria also leads to the fact that an AI system as such, i.e. a software model of a neural network as a mere software system or as a mathematical model, will not be patentable due to the exclusion of programs for data processing and/or mathematical models, even if it makes use of a computer [14]. The achievements of this AI system as such are usually entirely exhausted in the collection, processing, storage, evaluation and/or transmission of data and/or the provision of information, so that no technical problem is solved [24–27]. Nevertheless, the implementation of an AI system in the invention to solve a specific technical problem by technical means does not lack the exclusion criterion [20, 28]. A claimed sequence of methods using AI to solve the problem sufficiently addresses the findings of the claimed doctrine based on technical considerations.

3. "Susceptible of Industrial Application"

Section 5 PatG and Art. 57 EPC state that an invention is considered to be susceptible of industrial application if its subject matter can be made or used in any field of commerce, including agriculture. Regarding this requirement, it is not relevant whether the invention comprises or constitutes artificial intelligence or not.

4. Interim Conclusion

In contrast to AI-implementing systems, which should in principle be patentable, a different approach seems to apply to AI systems as such. An AI system whose performance is exhausted in the evaluation of information by a decision process by means of processing input data in an AI structure or in the comprehensive description of objects is not a technical solution of a technical problem. The patentability does not comply with Sec. 1 (3) and (4) PatG and Art. 52 (2) and (3) EPC [14]. A different situation is likely to apply if the AI system is also capable of controlling "how" the underlying technical data is determined [29] and thus goes beyond the decision-making and evaluation process. However, the learning structure of an AI system alone cannot justify patentability. An algorithm is neither patentable nor a program for a data processing system. A specific technical problem is required that the algorithm solves [29].

In addition, the use of AI leads to another consideration regarding the standard of assessment underlying patent law. In particular, the involvement of AI in the assessment process of the invention described above can have fundamental effects with regard to the principle of novelty and with regard to the inventive step: (1) prior art objections that are detrimental to novelty can be identified more quickly due to the improved availability and retrievability of knowledge [18], and (2) the capabilities of the "average" person skilled in the art, who has previously evaluated the prior art, change. The requirements for novelty and inventive step are based on the prior art that is available to the public on the filing or priority date. An invention is considered to involve an inventive step if it is not evident to an "average" person skilled in the art according to the state of the art. Conversely, a disclosed solution is to be denied protection as an invention if the solution appears as an obvious approach derived from what is already known [17].

The solution disclosed in the patent specification is thus assessed from the point of view of a fictitious person, the "average" person skilled in the art [30]. Their knowledge is considered when determining whether the problem can be solved by the method described in the patent specification. In this context, even obvious (human) misconceptions are tolerated, since the fictitious person skilled in the art may have a lot of experience in the relevant field, but is certainly not an outstanding expert. Furthermore, regarding the question whether a solution is to be classified as evident, the professional experience resulting from daily practice is to be included in the consideration of what is evident [31–35]. These two assessment criteria used by case law no longer fit if AI knowledge is used as the standard of examination. On the one hand, the *"average"* person skilled in the art would mutate into an expert, since the AI has considerably larger knowledge resources at its disposal than a human being. The latter would inevitably lead to the fact that the "excess" of the evident would reduce the area for the novelty,

i.e. the inventions. On the other hand, the wealth of experience that derives from handling and practicing technical applications cannot currently be reflected by an AI.

4 Application of Patent Law to AI

Regarding the patentability of AI-generated inventions, the question arises as to the requirement of a human inventor and an inventive step that can be ascribed to a human being. The latter consideration is accompanied in particular by the question of the extent to which the human inventive contribution, which is exhausted in the design, the weighting of the adaptations and the selection and structuring of the training data, is still reflected in the invention as such [14]. Alternatively,

it is worth considering whether an AI is capable of creating patentable outputs on its own and therefore can be an inventor or creator [36] this issue will be addressed in depth below.

The decisive step why patent offices and patent attorneys had to deal with the issue was that two patents were filed in 2019 at the European Patent Office, the Intellectual Property Office in the United Kingdom and the United States Patent and Trademark Office, namely for a fractal food container and a flashing light [37–39]. However, it was not the company Imagination Engines that was named as the inventor, but "Dabus AI", i.e. autonomously acting software. To illustrate that the inventions in the patent specification do not differ, at least in appearance, from other inventions (which, however, were created by humans through creative effort), it is recommended to take a look at the technical illustrations of the patentable inventions on the WIPO website [40], as listed under the heading "drawing" [41].

As described in Fig. 1, a distinction can be made in this scenario between AI-assisted inventions and AI-generated inventions. Since the issue of the necessity of a human inventor is not a problem in the first case, the explanations focus on AI- generated inventions.

5 "New" and "Susceptible of Industrial Application"

With regard to the requirements of novelty (Sec. 3 (1) cl. 1 PatG or Art. 54 EPC) and susceptibility for industrial application (Sec. 5 PatG or Art. 57 EPC) already defined above, there are no reservations that an invention of an AI can fulfill these.

6 "Inventions […] that Involve an Inventive Step"

Since filing the patent application with "Dabus AI" as the inventor of the fractal food container and the blinking light, there has been much discussion about how to deal with AI-generated inventions. Does the invention require an inventive step by a human being as a prerequisite for patentability [14]? One objection to the requirement of a human inventor is that patent law does not inquire into the process of development of the invention and also measures its protectability by objective criteria (such as, in particular, the average professional, novelty and lack of obviousness compared to the prior art) and not subjective achievements of the inventor [42]. In contrast however, an objection can be raised that the concept of "invention" is virtually immanent to a human inventive achievement [14]. German and European patent law, pursuant to Sec. 6 PatG and Art. 60 and 62 EPC, refers to the concept of the inventor as the person from whose creative activity the invention originated [43], who developed the knowledge of how a certain technical problem can be solved by certain technical methods [14, 37]. Thus, if an invention is made with assistance of a computer program (i.e., an AI), the inventor is not the computer program as such, but the natural person who designs the program and finds the solution to the technical problem by evaluating the results provided by the computer [44]. This is an expression of the inventor principle.

Studies and comparisons with other legal systems [7] show that this is not an exclusively German or European phenomenon. With regard to the patent applications of "Dabus AI", the patent offices of Great Britain and the USA as well as the EPO denied patentability. They all basically introduced the absence of a natural person—albeit in different ways, namely on a formal level (Art. 81 and Rule 19 EPC) or by way of substantive examination of the legal principles (35 U.S.C. § 115 (a)) and their interpretation (Sec. 7 (3) and Sec. 13 (2) (a) UK Patent Act) [31–33]. The investigation of some Asian countries [7]—specifically Japan and Korea—showed that it is at least not excluded that an AI system can also fulfill the requirements for inventorship. In a similar direction, the Australian Federal Court, in contrast to the Australian Patent Office, decided that the concept of inventor in the Patent Act 1990 does not require human creation [45]. According to the court, only the applicant has to be a human being, but not the inventor. These different interests cannot (yet) necessarily be reconciled at the moment. Currently, at least, there is always (also) a discernible human (co-)inventor who can claim the patent for themselves. However, this may change in the foreseeable future and at the latest then legal experts in general, and patent offices in particular, will have to deal with this question.

7. Interim Conclusion

In some respects, of course, such questions are still a future scenario. For now, AI-generated inventions still seem to be a distant scenario, even if there are corresponding applications. But it is precisely these applications that require timely consideration. The capabilities of AI are constantly evolving; the solutions created by an AI will therefore increasingly resemble or perhaps even surpass inventive human development [36]. It remains to be seen whether the patent authorities will adhere to this straightforward approach in the future or whether a change in the legislation will perhaps lead to an adjustment of the current patent system. When AI is not only used as a tool, but can independently set its own new tasks and solve them autonomously, the question starts to be urgent. A decision will have to be made at this point as to whether to open up patent protection for autonomous AI inventions and allow the AI itself to become the inventor. It is therefore necessary to ask whether the patent law concept of inventor should not also be opened up to non-human creators.

Can an AI be an inventor? The answer depends on the purpose of the patent system [36, 46]: The property and reward theories, already mentioned above, aim at presenting patent protection as an imperative of justice and the individual inter-

ests of the inventor as worthy of protection. If, according to these theories, the purpose of the patent system is seen to be a monetary incentive effect for the inventor, no such incentive can exist for an AI. Inventors are primarily interested in gaining personal reputation and prestige, which can give them personal satisfaction. However, an AI is not in a position to feel justice or to seek high personal prestige. However, if one sees the purpose of the patent system as promoting technical progress and making technical developments useful to the general public, as the incentive and revelation theories do, then the allocation of an inventor's property is conceivable. For the progress and the usefulness of the broad masses it is irrelevant whether a human or an AI makes an invention, if it only facilitates the everyday life and the social life. However, the incentive or disclosure theory, which focuses on the benefit to the general public and making technical progress accessible, also focuses on the individuals behind it, i.e. the inventors. These theories outlining the purpose of patent law also take into account the fact that without the consideration "patent" no inventor will be willing to disclose his invention. So, again, the patent serves the individual interests of the inventor—interests that an AI can hardly develop. Therefore, not only the lack of humanity of the AI speaks in favor of denying it an independent patent claim, but also the actual purpose of patent law, which repeatedly refers to the inventor as a human individual with corresponding characteristics.

5 Conclusion

As a result of our investigations, it can be stated that—related to the various dimensions of AI and the perspective of patent law—AI systems as such are difficult to be patented. This is particularly due to the fact that algorithms often do not meet the novelty criteria of Sec. 3 (1) cl. 1 PatG and Art. 54 EPC and that the required technicality is lacking, Sec. 1 (1) PatG and Art. 52 (2) EPC. In particular, this characteristic is rejected because the AI system's activity is limited to performing a decision- making process and evaluating it, and is hardly capable of solving a technical problem.

As a second result of our investigations, it can be stated that the invention resulting from the inventive process of an AI cannot be an invention of the AI itself. On the one hand, this is due to the fact that it has no legal capacity to assume the rights and obligations associated with being an inventor. On the other hand, all legal systems, including the EPO and the Federal Court of Justice—as of now—assume that there is always a natural person who can claim the patent for themselves, and therefore also require a natural person to be the inventor.

Those observations of the current legal situation are consistent and entirely logical. Considering that the patentability of an AI invention as such is rejected on the basis that it is not new or does not sufficiently solve a technical problem, it seems at least dubious to attribute the sufficient inventive step and creative activity to such an AI that only performs its algorithms.

To this effect, AI will initially remain the small, prudent "helper", only clever, helpful and without any claim to the legal status of "inventor".

* The original manuscript was submitted in August 2022; it was revised in December 2022. The conference and the corresponding talk took place in March 2023. The paper has not been revised in terms of content since then.

Acknowledgements Funded by the Deutsche Forschungsgemeinschaft (DFG, German Research Foundation)—Project-ID 416228727—SFB 1410.

Author Contributions Conceptualization, formal analysis, and writing, D.G.-N. and S.M. in equal measure; supervision, D.G.-N. All authors have read and agreed to the revised version of the manuscript.

Conflicts of Interest The authors declare no conflict of interest.

References

1. Mannino, A., Althaus, D., Erhardt, J., Gloor, L., Hutter, A., & Metzinger, T. (2015). Künstliche Intelligenz: Chancen und Risiken. *Diskussionspapiere der Stiftung für Effektiven Altruismus, 2*, 1–17.
2. Turner, J. (2019). Robot rules: Regulating artificial intelligence. Cham: Palgrave Macmillan. https://doi.org/10.1007/978-3-319-96235-1
3. Machlup, F. (1958). *An Economic Review of the Patent System*. US Government Printing Office.
4. Götting, H.P., des Patentrechts, G. (2003). In Schulte, M. (Eds.), Handbuch des Technikrechts. Enzyklopädie der Rechts- und Staatswissenschaft Abteilung Rechtswissenschaft, Berlin, Heidelberg: Springer, pp. 209–260. https://doi.org/10.1007/978-3-662-07707-8_7
5. Ann, C. (2022). Rechts- und wirtschaftspolitische Bewertung des Patentschutzes. In Patentrecht, Ann, C. (Eds.), München: C.H. Beck, 8th Edn, Rn. 7–12
6. Melullis, K.-J. (2019). EPÜ Art. 52 Patentierbare Erfindungen. In Ehlers, J., & Kinkeldey, U. (Eds.), Benkard: Europäisches Patentübereinkommen, Band 4a, München: C.H.Beck, 3rd Edn
7. Shemtov, N. (2019). A Study on Inventorship in Inventions Involving AI Activity. http://documents.epo.org/projects/babylon/eponet.nsf/0/3918F57B010A3540C125841900280653/$File/Concept_of_Inventorship_in_Inventions_involving_AI_Activity_en.pdf. Last accessed 18 Aug 2022
8. Stone, P., Brooks, R., Brynjolfsson, E., Calo, R., Etzioni, O., Hager, G., Hirschberg, J., Kalyanakrishnan, S., Kamar, E., Kraus, S., Leyton-Brown, K., Parkes, D., Press, W., Saxenian, A., Shah, J., Tambe, M., & Teller, A. (2016). Artificial Intelligence and Life in 2030, one Hundred Year Study on Artificial Intelligence: Report of the 2015–2016 Study Panel, Stanford University, Stanford, CA, September 2016. https://ai100.stanford.edu/sites/g/files/sbiybj9861/f/ai100report10032016fnl_singles.pdf. Last accessed 18 Aug 2022
9. European Patent Office—EPO. (2017). Patents and the fourth industrial revolution. The inventions behind digital transformation
10. Ménière, Y., & Pihlajamaa, H. (2019). Künstliche Intelligenz in der Praxis des EPA. *GRUR—Gewerblicher Rechtsschutz und Urheberrecht, 4*, 332–336
11. European Patent Office—EPO. Guidelines for Examination in the European Patent Office, Part G Chap. II 3.3.1
12. United States Patent and Trademark Office—USPTO. Request for Comments on Patenting Artificial intelligence Inventions, 84 Fed. 44889
13. European Patent Office—EPO. Guidelines for Examination in the European Patent Office, Part G Chap. II 3.6
14. Heinze, C., & Engel, A. (2020). § 10 KI und Patentrecht. In M. Ebers, C. Heinze, T. Krügel, & B. Steinrötter (Eds.), *Künstliche Intelligenz und Robotik* (pp. 355–411). C.H.Beck.
15. Comment 6 in the German Response to the WIPO Public Consultation on AI and IP, dated 13.12.2019. https://www.wipo.int/about-ip/en/artificial_intelligence/submissions-search.jsp
16. European Patent Office—EPO. (2010). Decision of 12.5.2010—G 3/08, *GRUR—Gewerblicher Rechtsschutz und Urheberrecht, 7*, 608–619
17. Melullis, K.-J. (2019). EPÜ Art. 54 Neuheit. In Ehlers, J., & Kinkeldey, U. (Eds.), Benkard: Europäisches Patentübereinkommen, Band 4a. München: C.H.Beck, 3rd Edn
18. Firth-Butterfield, K., & Chae, Y. (2018). Artificial Intelligence Collides with Patent Law. https://www3.weforum.org/docs/WEF_48540_WP_End_of_Innovation_Protecting_Patent_Law.pdf.
19. Nägerl, J., Neuburger, B., Steinbach, F. (2019). Künstliche Intelligenz: Paradigmenwechsel im Patentsystem. *GRUR—Gewerblicher Rechtsschutz und Urheberrecht, 4*, 336–341
20. Vertinsky, L. (2018). Thinking machines and patent law. In W. Barfield & U. Pagallo (Eds.), *Research handbook on the Law of Artificial Intelligence* (pp. 489–510). Edward Elgar Publishing.
21. German Federal Court of Justice—BGH. (2004). Decision of 24.05.2004—X ZB 20/03, *GRUR—Gewerblicher Rechtsschutz und Urheberrecht, 8*, 667–669
22. German Federal Court of Justice—BGH. (2010). Decision of 22.04.2010—Xa ZB 20/08, *GRUR—Gewerblicher Rechtsschutz und Urheberrecht, 7*, 613–616
23. German Federal Court of Justice—BGH. (2011). Decision of 24.02.2011—X ZR 121/09, *GRUR—Gewerblicher Rechtsschutz und Urheberrecht, 7*, 610–613
24. German Federal Patent Court—BPatG. (2015). Resolution of 09.06.2015—17 W (pat) 37/12, BeckRS 2015, 13810
25. German Federal Court of Justice—BGH. (2009). Decision of 20.01.2009—X ZB 22/07. *GRUR—Gewerblicher Rechtsschutz und Urheberrecht, 5*, 479–480
26. German Federal Patent Court—BPatG. (2011). Resolution of 03.03.2011—17 W (pat) 151/05, BeckRS 2011, 11728
27. German Federal Court of Justice—BGH. (2001). Decision of 17.10.2001—X ZB 16/00. *GRUR—Gewerblicher Rechtsschutz und Urheberrecht, 2*, 143–146
28. Hartmann, F., & Prinz, M. (2018). Immaterialgüterrechtlicher Schutz von Systemen Künstlicher Intelligenz. Tagungsband der Herbstakademie 2018 der Deutschen Stiftung für Recht und Informatik (DSRITB), pp. 769–790
29. German Federal Court of Justice—BGH. (2004). Resolution of 19.10.2004—X ZB 33/03. *GRUR—Gewerblicher Rechtsschutz und Urheberrecht, 2*, 141–143
30. Scharen, U. (2019). EPÜ Art. 69 Schutzbereich. In Ehlers, J., & Kinkeldey, U. (Eds.), Benkard: Europäisches Patentübereinkommen, Band 4a, München: C.H.Beck, 3rd Edn
31. German Federal Court of Justice—BGH. (2004). Decision of 7.9.2004—X ZR 255/01. *GRUR—Gewerblicher Rechtsschutz und Urheberrecht, 12*, 1023–1025
32. German Federal Court of Justice—BGH. (2018). Decision of 9.1.2018—X ZR 14/16. *GRUR—Gewerblicher Rechtsschutz und Urheberrecht, 4*, 390–395

33. German Federal Court of Justice—BGH. (1992). Decision of 5.5.1992—X ZR 9/91. *GRUR—Gewerblicher Rechtsschutz und Urheberrecht, 9*, 594–597

34. European Patent Office—EPO. Decision of 21.10.1987—T 378/86, BeckRS 1987 30574270

35. European Patent Office—EPO. Decision of 14.2.1996—T 39/93, BeckRS 1996 30502809

36. Schröler, P. R., & Kuß, C. (2022). 2 Rechtliche Grundsatzfragen, E. Urheber-, Patent- und Markenrecht. In Künstliche Intelligenz, K.J. Chibanguza, C. Kuß, & H. Steege, (Eds.), Baden-Baden: Nomos, pp. 171–192

37. European Patent Office—EPO. Decisions of 27.01.2020—18 275 174.3 and 18 275 163.3

38. UK Intellectual Property Office—UK IPO. Decision of 04.12.2019—BL O/741/19

39. United States Patent and Trademark Office—USPTO. Decision of 22.04.2020—16/254,350

40. WIPO. (2024). WO/2020/079499, Food container and devices and methods for attracting enhanced attention. https://patentscope.wipo.int/search/en/detail.jsf?docId=WO2020079499. Last accessed 25 Apr 2024

41. WIPO. (2024). WO/2020/079499, Food container and devices and methods for attracting enhanced attention. https://patentscope.wipo.int/search/docs2/pct/WO2020079499/pic/o8X05iEVg-54boBWMRfQV5GutESREeJ2xajNJIoP1NYkOMrnVq9UO2_w18QdWqve and https://patentscope.wipo.int/search/docs2/pct/WO2020079499/pic/AzXZbi41FxeyaDZtX6dHgIVLKpQ9WZzjBNzy_Zs6PafrSUx7-v5R8rJB6ldzhaLx. Last accessed 25 Apr 2024

42. Hetmank, S., & Lauber-Rönsberg, A. (2018). Künstliche Intelligenz—Herausforderungen für das Immaterialgüterrecht. *GRUR –Gewerblicher Rechtsschutz und Urheberrecht, 6*, 574–582

43. WIPO. (2019). Standing Committee on the Law of Patents, Background Document on Patents and Emerging Technologies

44. Tochtermann, L. (2020). Kapitel 7.3 Immaterialgüterreechtlicher Schutz von KI de lege ferenda, IV. KI als Erfinder oder Schöpfer? In Kaulartz, M., & Braegelmann, T. (Eds.), Rechtshandbuch Artificial Intelligence und Machine Learning, München: C.H.Beck, pp. 329–333

45. Federal Court of Australia—FCA. (2021). Decision of 30.07.2021—Thaler vs. Commissioner of Patents [2021] FCA 879

46. Engel, A. (2020). Can a patent be granted for an AI-generated invention? *GRUR—International Journal of European and International IP Law, 11*, 1123–1129

Advanced Methodological Approach for Designing Innovative Business Models for Hybrid Mobility Technologies

Marco Rehme, Stephan Odenwald⑩, and Uwe Götze⑩

Abstract

Successful interactions of autonomous agents and humans in traffic scenarios will require diverse novel technologies as well as corresponding products and services (e.g., smart objects, data analysis tools, and communication services). However, the exploitation of the vast technology-induced innovation potentials is a challenging task for different reasons in the form of typical innovation barriers. The contribution aims to present an advanced business model development approach that is tailored for innovations based on ground-breaking hybrid mobility technologies, such as for enhanced understanding of human perception and behavior or for human-likeness fostering smooth interactions of autonomous agents and humans. This focus implies the need to consider the specific risks and challenges of hybrid technologies (technological, social and legal aspects) and considerably higher uncertainties in the business model development and evaluation methods. The proposed approach integrates technology and business model development to enable an early consideration of customer needs and economic targets in innovation processes. Furthermore, it extends business model development from the company level to the level of interlinked value chains—comprising all interacting players that are necessary for a successful innovation. Additionally, ecological and social perspectives are included besides the economic dimension to ensure the develop-ment of sustainable business models. The contribution will demonstrate how a testbed can address challenges arising in the early development phase, such as the collection of real-world data, a timely participation of relevant stakeholders and elaborating still vague opportunities and threats. By allowing experiments and analyses of the functionality or performance of smart objects as well as associated products and services in real traffic situations such a planned "Testbed Hybrid Traffic Chemnitz" will effectively support the development of viable business models.

Keywords

Sustainable business model development · Innovation barriers · Embodied technologies · Mobility · Value creation networks

1 Introduction

Converting technological inventions and research-based business ideas into enduringly economically successful products and services faces various obstacles and challenges of different kinds. They result, among others, in the need of an integrated technology and business model development as well as the inclusion of different stakeholders involved in the supply chains of the innovative product or service. In recent years, the sustainability of innovations and business models must increasingly be considered. Concerning hybrid mobility technologies, major challenges arise from the currently nearly unpredictable user acceptance of autonomous agents in many aspects of daily living [1], the rapid technological progress in the field and supply chains still evolving and not yet established.

Against the background of these challenges, the objective of the paper is to outline an advanced methodological approach for translating mobility-related ideas based on hybrid technologies into successful innovation and busi-

M. Rehme (✉) · U. Götze
Chair of Management Accounting and Control, University of Technology Chemnitz, Chemnitz, Germany
e-mail: marco.rehme@wirtschaft.tu-chemnitz.de

U. Götze
e-mail: uwe.goetze@wirtschaft.tu-chemnitz.de

S. Odenwald
Department Sports Equipment and Technology, University of Technology Chemnitz, Chemnitz, Germany
e-mail: stephan.odenwald@mb.tu-chemnitz.de

B. Meyer et al. (eds.), *Hybrid Societies*, Advances in Science, Technology & Innovation,
https://doi.org/10.1007/978-3-032-03488-5_39

nesses. On the one hand, this approach is closely interlocking technology and business model development and furthermore referring to the supply chain level as well as to the sustainability of products and services going beyond merely economic success. On the other hand, it is tailored for the specifics of hybrid mobility technologies. A central role in this methodological approach plays the Testbed Hybrid Traffic Chemnitz: It allows to investigate human behavior in scenarios with autonomous agents as well as their acceptance and so enable to gather data being relevant for technology and business model development.

2   State of the Art

The intended development of a novel methodological approach can be based on a complex body of knowledge concerning innovation barriers and technology and business model development methods in general as well as in the field of current mobility technologies.

Innovation barriers for business models in the generic field of mobility and also specifically related to new forms of mobility (e.g., autonomous, connected, electric, and shared mobility) have already been systematized by [2]. Different categories of barriers to open innovation in the specific case of mobility as a service (MaaS), for example, are also examined in [3] by using empirical research design in the form of interview analyses. As categories were identified here: legislation, inter-organizational collaboration (incl. contract negotiations), organizational innovation cultures and decision-making schemes, corporate strategies, dependence on (dominant) key players, technical integration as well as economic risk. In [4] a "barrier management process" based on risk management theory has been suggested for the purpose of mastering such innovation barriers successfully.

At very early stages of their design process, technologies should already be developed closely interlocked with their anticipated business models. Such an iterative interplay is reflected in the concept of *integrated business model and technology development* [5]. The alignment between designing technological systems and the associated business model elements should mitigate risks and enable a successful economic exploitation after the market launch. Early managerial steering impulses from insights into the user and market perspective (e.g., customer needs and willingness to pay) can be used to determine development or adjustment requirements and to select from alternative technology variants.

Another important prerequisite of establishing well-functioning and lasting business models is the early consideration of the economic rationales of major actors and stakeholders in *entire value chains*, especially with regard to relevant collaborative business models. In [6] management challenges of multimodal mobility systems have been worked out. These challenges arise from the characteristics of business ecosystems, such as a high level of interdependencies, co-opetition and co-evolutionary development processes between value creating partners. It is to be expected that business models based on hybrid mobility technologies will also have such distinct characteristics of business ecosystems and that special attention must therefore be paid to the functionality of the complete value chain.

The increasing importance of *sustainability* with its three pillars economic viability, environmental protection and social responsibility, leads to take a concept for a respective pre-evaluation and subsequent verification into account while designing and evaluating new business models. In [7] the business model canvas for sustainability (BMCS), depicted in Fig. 1, is proposed. It is an adaptation of the nine elements of the broadly known canvas according to [8] and requires the consideration of all sustainability dimensions in the analysis and design decisions of each of the canvas elements. The BMCS builds on a life cycle perspective centered around the value proposition of a product or service, to which the required input stakeholders, activities and resources are assigned on the left and the output stakeholders, from the client to the end-of-life actors, as well as output-related relationships and channels, are mapped on the right. Supplementary [9] suggest a method for sustainability-related strategic evaluation of business models as well as design ideas and options (and their roughly estimated burdens and benefits) in the early phase of business model development.

Referring to open mobility innovations, [10] investigates MaaS and proposes an own business model canvas for a shared mobility market. In [11], van den Berg et al. differentiate MaaS systems according to three different business models and considers the effects of mobility on welfare and behavior. Likewise, [12] reflects on MaaS business models by developing a framework that merges the business model canvas with morphological analysis, with morphological boxes for value proposition, infrastructure category, customer category and revenue structure each.

Due to its early state of development and the lack of broadly market-ready products and services, analyses of business models based on hybrid mobility technologies are largely in their infancy. Existing work seems to focus on autonomous driving and new mobility concepts, such as the MaaS, without recognizing the possibilities of human-machine interactions in a broader sense. Interactions of artificial agents or smart objects with humans, for example, are only considered to a limited extent. Remane et al. [13] identify 14 new digital business model types. These include, e.g., the manufacturer of autonomous products/robots, IT-enabled self-service provider, seller of sensor information, and sensor-enabled service innovator.

Noteworthy preliminary work of participating researchers at University of Technology Chemnitz stems from interdis-

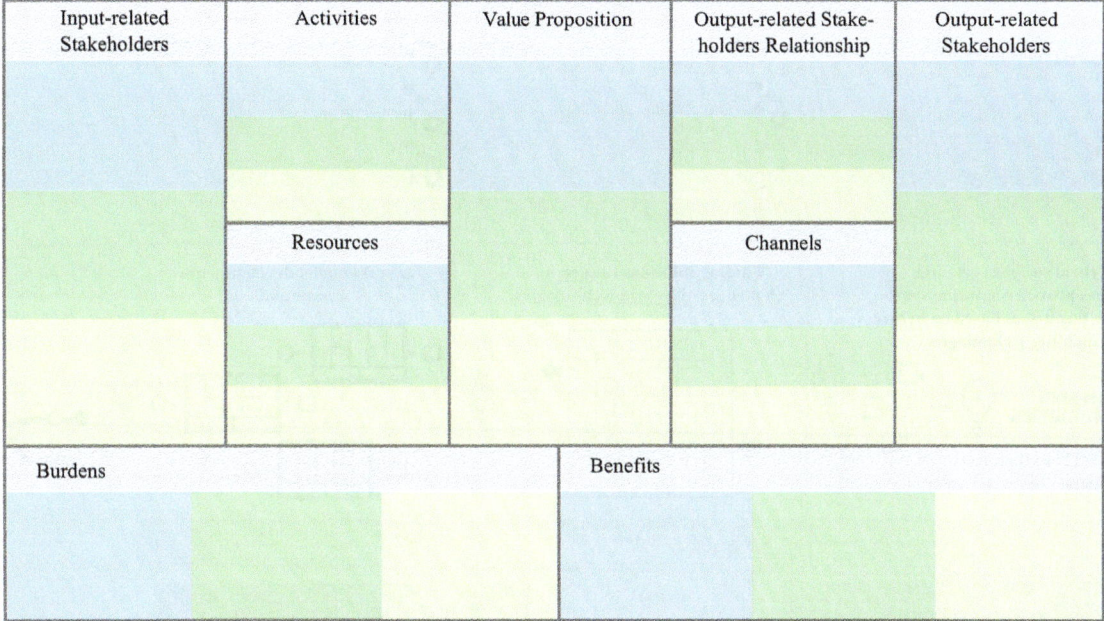

Input-related Stakeholders	Activities	Value Proposition	Output-related Stake-holders Relationship	Output-related Stakeholders
	Resources		Channels	
Burdens			Benefits	

Fig. 1 Business model canvas for sustainability (blue—economic aspects; green—environmental aspects; yellow—social aspects) [6]

ciplinary projects with a focus on novel business models and value networks. This notably includes results from the ESF junior research group EcoMobility [14], the electro-mobility showcase projects NeMoS and VEReMO [15] and from a techno-economic analysis of 5G networks within the fast2020 (fast actuators sensors & transceivers) research cluster [16].

Concluding, a couple of approaches exists that can serve as a basis for designing innovative business models for hybrid mobility technologies. However, not all the existing challenges outlined above are addressed in a coherent way. This motivates the intended development of an advanced methodological approach for designing such business models.

3 Methods

The aforementioned own preliminary work has proven its worth by adapting and validating the respective state-of-the-art methods used in various interdisciplinary research activities at the interface between engineering, natural, human and economic sciences, and in particular in the field of mobility. In this context approaches for an integrated technology and business model development as well as life-cycle oriented models, methods and concepts for economic, ecological and sustainability assessment and management have been created and applied to many innovations close to the market. The proposed advanced methodological approach for designing innovative business models will follow these proven exam-

ples and integrate identified parameters and methodology modules.

The path to the development and application of the envisaged approach is summarized and displayed together with the methodological key elements and four working tasks in Fig. 2.

Starting point are the scientific and technical insights generated in the Collorative Research Center (CRC) Hybrid Societies projects with mobility-related applications. Together with reflections and assessments from different perspectives in the CRC, these findings serve as ideas provider and initial spark for potential new businesses based on hybrid mobility technologies.

In a first step, the *specific challenges and risks* of innovations based on hybrid mobility technologies are to be worked out, analyzed in detail and effective *solutions for their overcoming* are to be found here as an important sub-goal. These solutions will be converted and combined into a coherent approach for the development of business models in this technology and application field. This step will mainly comprise research tasks of knowledge generation and method refinement, but also the conceptualization and preparation of demonstrators and test environments.

A second step, mainly comprising of transfer tasks, will deal with the actual *development and initiation of promising business models*, based on marketable business ideas from thematically relevant Hybrid Societies projects. In doing so, the previously elaborated coherent approach for the development of business models is applied to specific applications or demonstrators and validated. For this purpose, the individual methodological modules will be utilized, e.g., the BMCS, which may be adapted or extended for the research object, is

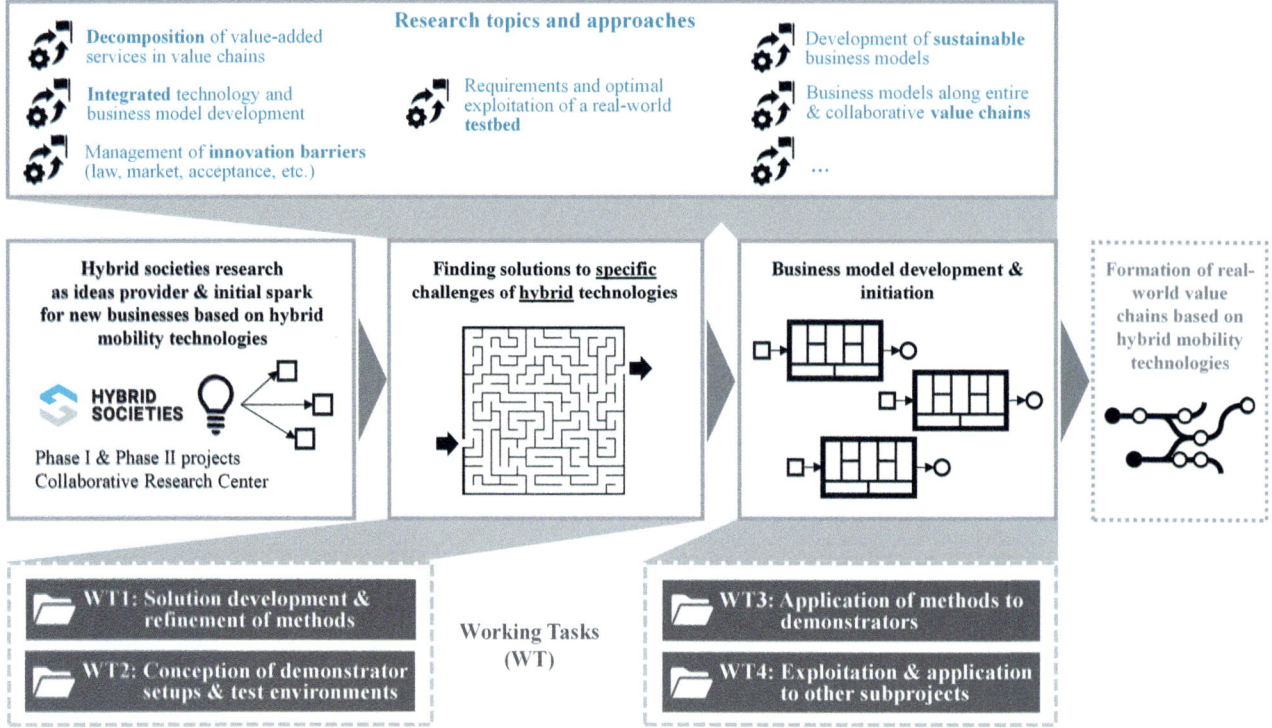

Fig. 2 Framework of research topics, approaches and working tasks

then applied to the various business ideas under consideration. The envisaged innovative testbed will play an important role in the empirical testing. If necessary, reasonable modifications and extensions of the approach will be implemented as a result of the practical findings and experiences gained. In addition, regarding exploitation and application, the transferability of the concept to other Hybrid Societies subprojects, which have not yet provided any business ideas or demonstrators in the preparatory work (including such that are somewhat further away from the scope of mobility), and toward business ideas for hybrid technologies in traffic scenarios in general is to be examined by theoretical-conceptional analyses. Transfer efforts will strive for the broadest possible impact of the created methods across and beyond the CRC. Limits of suitability and applicability of the elaborated methodology are to be identified. The tasks of the second step lead to the sub-goal of initiated and well-prepared new business models that can assert themselves in the real market (in the form of start-ups, spin-offs or new product fields of industry partners). In a possible subsequent third step, the transfer successes of these initiation activities are then evaluated. Such transfer successes would manifest themselves in the fruitful *integration into new value chains* that are being formed around hybrid mobility technologies in the following years.

4   Results

In connection with the challenges mentioned earlier, a variety of relevant research topics and approaches should be taken into account for the two steps of solution finding and business model development. These include, without claim to be exhaustive, the decomposition of value chains, the meshing of technology and business model development, the dealing with innovation barriers, ensuring sustainability, the holistic view of value chains, and the utilization of real-world testbeds.

In the following, initial considerations are presented for three of the methodological building blocks, namely a generic value creation network model for products and services based on hybrid mobility technologies, a description of the role, features and benefits of a real-world testbed and a systematization of innovation barriers of hybrid mobility technologies.

The *generic value creation network model* for products and services based on hybrid mobility technologies depicted in Fig. 3 can serve as a starting point when considering value net aspects in business model generation. It was derived based on previous work within the innovation network "Smart Rail Connectivity Campus" [17] and indicates the breadth and complexity of the associated value creation processes, starting with the development and provisioning of hardware and software components through to possible end-of-life activities. The potential usages of hybrid mobility

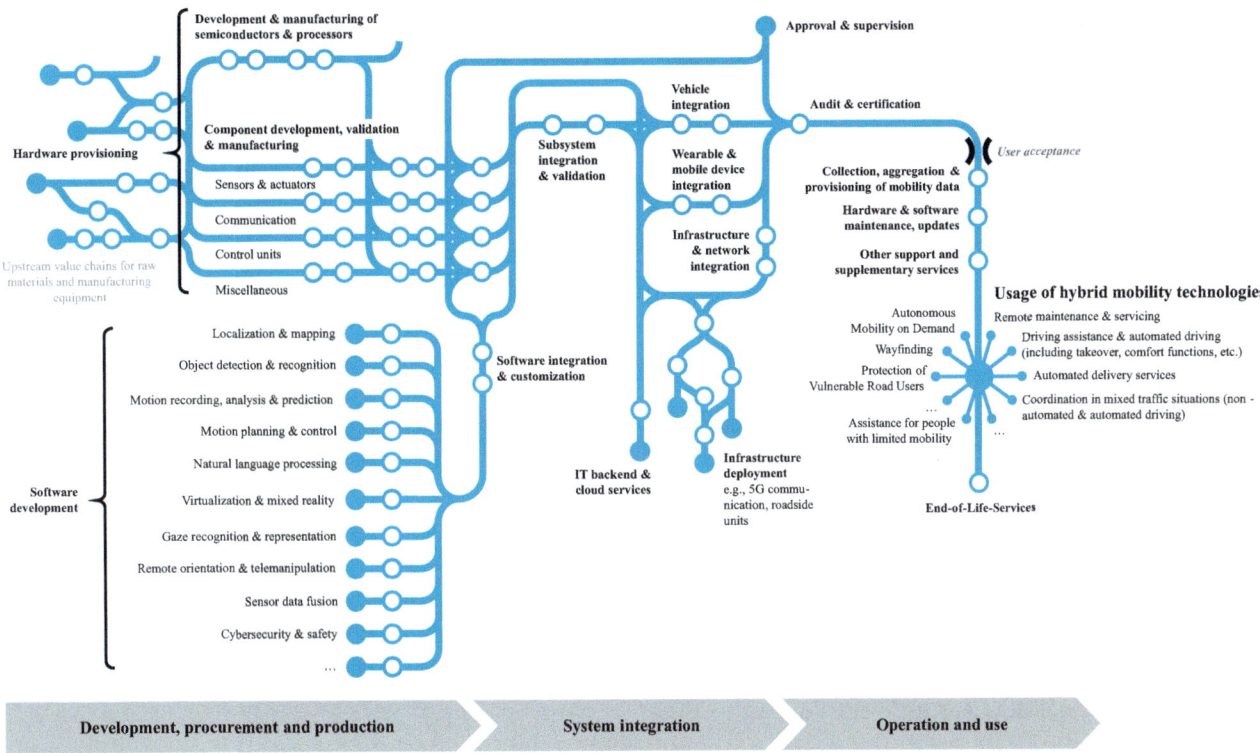

Fig. 3 Value creation network of products and services based on hybrid mobility technologies

technologies themselves have an extraordinarily wide range. The field of mobility here includes not only automotive-related applications, but also those that affect all users of traffic systems. Coordination between non-automated and automated traffic participants, control transitions in partially or conditionally automated vehicles, tele-operated maintenance and servicing, Autonomous Mobility on Demand as well as delivery services, assistance and protection for individual travelers or specifically for Vulnerable Road Users (VRU) and people with limited mobility are listed here only as examples.

In this context, the *"Testbed Hybrid Traffic Chemnitz"* is to play a special role. The testbed is designed to analyze behavior and interaction of road users with special emphasis on traffic situations that involve VRU, namely pedestrians and cyclists. Mobile road-side units equipped with innovative technologies are to be tailored to a flexible operation at changing locations (points of interest). Latest sensor technology and in-build data processing is paired with broadband communication for easy set-up and quick removal. In addition to the observation of human-human encounters the testbed is equipped with several tele-operated and autonomous vehicles and devices for planned interventions.

It will allow (a) experiments and analyses of the functionality and performance of smart objects as well as other products and services in real traffic situations, (b) the inclusion of possible members of future value chains in projects, and (c)

the reflection of technology and business model development in a realistic environment. The testbed can not only be used for cognitive research but also for preparation for commercialization (e.g., for assessing user acceptance, organizational feasibility, and legal compliance). The activities revolve around the optimal selection of implementation variants of demonstrators, the optimal design of their technical and business model-relevant properties, as well as their actual realization and preparation. The experiments thus made possible will allow a systematic data collection that can be used to support design decisions that affect both the applied technologies and the associated business model(s). The data obtained could, e.g., be used to carry out economic and sustainability assessments and to make adjustments to obtain a marketable product.

Thereby, the testbed can help to cope with existing *innovation barriers* as well. Such barriers threaten the feasibility and success of innovative business models. Therefore, the advanced approach for designing business models includes their systematic identification and consideration. Building on the barrier systematizations mentioned in the state of the art and the challenges identified so far in the CRC, seven relevant categories of innovation barriers were recognized. While the innovations barriers depicted in Fig. 4 can already be quite clearly identified based on insights from the first phase of the CRC, the barriers mentioned in Fig. 5 have to be specified and validated in the second phase. To illustrate the relation of this work with the activities of the CRC, subprojects concerned

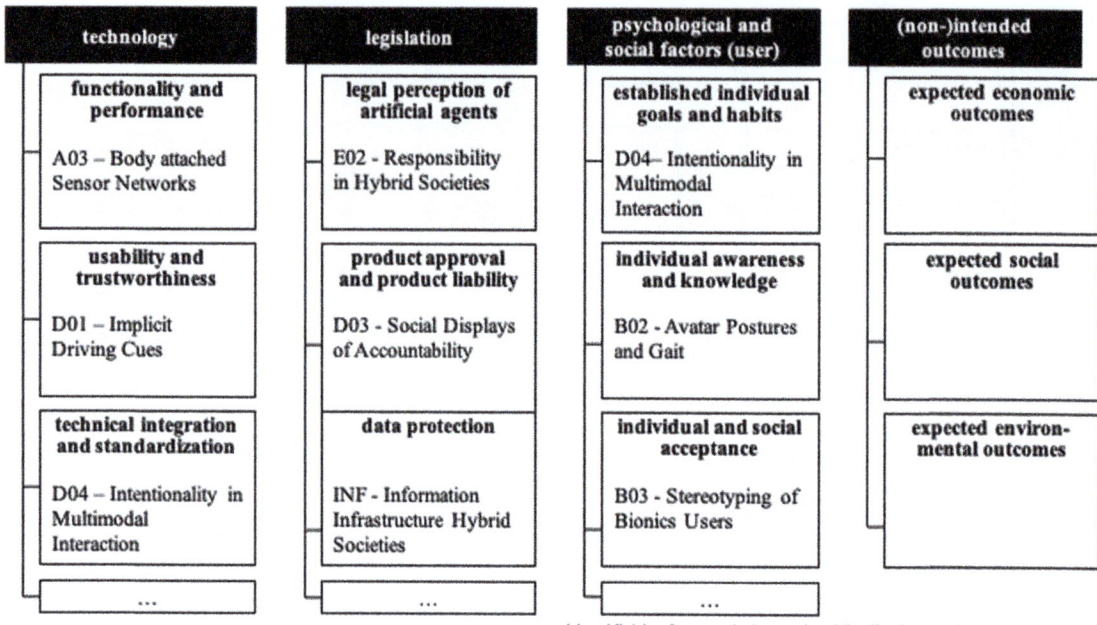

identifiable from existing scientific findings of the CRC phase 1

Fig. 4 Systematization of innovation barriers of hybrid mobility technologies. First group: barriers that can already clearly identified based on insights from the current phase of the CRC

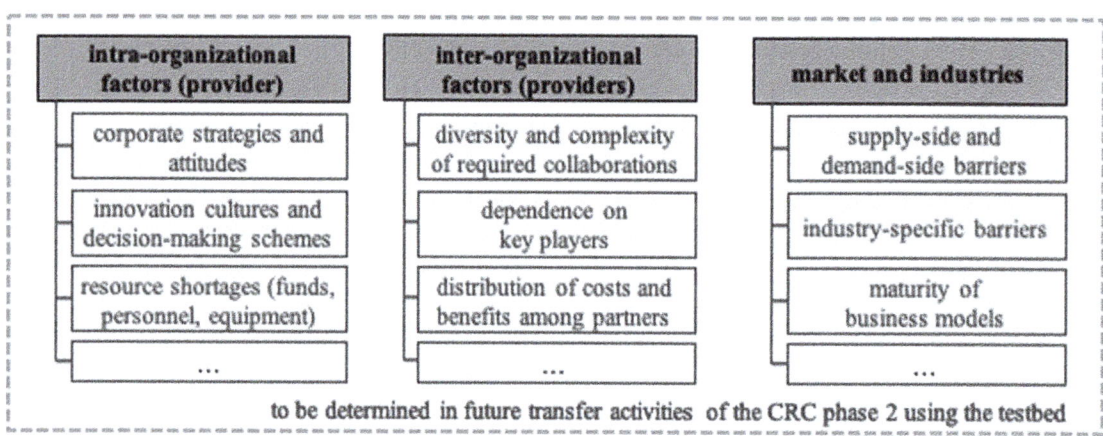

to be determined in future transfer activities of the CRC phase 2 using the testbed

Fig. 5 Systematization of innovation barriers of hybrid mobility technologies. Second Group: The three areas represent sources of innovation barriers to be elaborated and specified with experts inside an outside

with at least one of the barriers are mentioned. The selection is not complete and may reflect the currently ongoing research only in part.

The first category of major innovation barriers is about *technology*-related issues. The *functionality and performance* of technologies is to be benchmarked to human capabilities for ensuring smooth interaction. A probable barrier in the subproject "Detecting gestures by using Body Attached Sensor Networks" (A03) is the not yet happened step to integrate the algorithms into a wearable device. This lack of functionality is an innovation barrier, which is expected to be overcome soon. The barrier tackled in the subproject "Implicit Driving Cues" (D01) is the potential lack of trust,

e.g., of passengers in autonomous shuttles (at least initially). The *usability and trustworthiness* of technologies in traffic situations is an important factor in general. *Technical integration and standardization* might be inhibited in disciplines such as "multimodal communication" (D04). The gesture-code varies amongst social and regional groups. This builds up an innovation barrier for technologies, aiming to understand the meaning of, e.g., human gestures.

A second category of innovation barriers are imposed by the *legislation*. While the *legal perception of artificial agents* is investigated in E02 "Responsibility in Hybrid Societies", the project D03—regarding *product approval and product liability*—might serve as a valuable source on how Embodied

Digital Technologies (EDTs) design can support or damage user acceptance, as a main factor for future technologies. Barriers for innovation are raised also by the regulations on *data protection* that constrain business model development in every domain, mostly in research and marketing.

The third category—*psychological and social factors*—is central and refers to the user perspective on the topic. *Established individual goals and habits* are basis for "Multimodal Interaction" (D04) which requires the matching set of communication tools and skills on both sides. The human's *individual awareness and knowledge* requires certain qualities while interacting with human and also digital actors. The work in B02—"Avatar Postures and Gait"—aims to understand the triggering parameters in mutual perception. An example how to amplify *individual and social acceptance* and to overcome such innovation barriers is established in B03—"Stereotyping of Bionics Users"—by immersing non-amputees in the situation of wearing a bionic arm prosthesis.

The last category of innovation barriers remains without the notion of specific subproject, as the *outcome*-related innovation barriers are an overall subject of the CRC Hybrid Societies. "Outcome-related" innovation barriers mainly come from unexpected results and impact in *economic, social, or environmental aspects*. Each project in the CRC aims to show the opportunities and also the risks of emerging technologies. And this is the kind of information needed for designing innovative business models.

The second group of innovation barriers can be roughly clustered into intra-organizational factors, inter-organizational factors and markets and industries according to the involved stakeholders. *Intra-organizational factors* comprise the existing technology and product strategies which might not be focused on (or aligned with) hybrid mobility technologies yet, the innovation culture and decision-making procedures (e.g., risk aversion, underestimation of innovation potentials, hierarchies hostile to innovation) as well as restricted resources (required competencies or funds). *Inter-organizational factors* refer to the interplay of the various actors involved in realizing the intended innovations: the number and diversity of partners, the necessity to include dominant partners that might not be willing to cooperate or abuse their position as well as the creation of fair procedures for distributing cost and benefits. Finally, barriers at the level of the *markets and industries* may exist, e.g., insufficient access to suppliers or customers, industry-specific barriers to entry (like high regulatory hurdles or switching costs) and a lack of mature and stable business models for novel services as well as ways to provide new kinds of value to customers. The presented systematization should be useful to identify more of the relevant innovation barriers. Once identified, these barriers should be systematically "managed" by removing them or by adapting the technology and/or business model respectively

their implementation processes. This can be done following the "barrier management process" already mentioned [4].

5 Discussion and Conclusion

Converting technological inventions and research-based business ideas into enduringly successful products and services has to face different challenges. This holds true for technology-based innovations in the field of hybrid mobility technologies automated mobility as well. This motivates to suggest an advanced approach for designing business models. As described above, this approach considers different viewpoints or fields of management research: innovation management, business modeling, supply chain management, sustainability management as well as the interface to engineering and technology research.

The suggested approach is still in its early life cycle phases. Some of its elements have not been elaborated until now. This will follow according to the framework depicted in Fig. 2. Thereby, the close collaboration with the projects of the CRC will play a central role: They are invited to provide ideas for innovations as well as data and insights that are relevant for the successful design of business models. Furthermore, the planned testbed serves as an important enabler of the application as well as the validation and refinement of the suggested approach.

References

1. Mandl, S., Bretschneider, M., Meyer, S., Gesmann-Nuissl, D., Asbrock, F., Meyer, B., & Strobel, A. (2022). Embodied digital technologies: first insights in the social and legal perception of robots and users of prostheses. In *Frontiers in Robotics and AI, Sec. Ethics in Robotics and Artificial Intelligence, 9.* https://doi.org/10.3389/frobt.2022.787970
2. Rehme, M., Lindner, R., & Götze, U. (2016). Barrieren bei Geschäftsmodell-Innovationen der Neuen Mobilität—Implikationen für Innovationsmanagement und Geschäftsmodellentwicklung. In Proff, H., & Fojcik, T. M. (Eds.), *Nationale und internationale Trends in der Mobilität, Technische und betriebswirtschaftliche Aspekte*, Wiesbaden: Springer, pp. 63–81
3. Smith, G., Sochor, J., & Karlsson, M. (2019). Public–private innovation: Barriers in the case of mobility as a service in West Sweden. *Public Management Review, 21*(1), 116–137. https://doi.org/10.1080/14719037.2018.1462399
4. Rehme, M., Oehme, S., Götze, U., & Claus, S. (2020). Smart Rail—Bewertung von Innovationsideen und Management von Innovationsbarrieren am Beispiel integrierter Mobilitätsketten für ländliche Räume. In Proff, H. (Ed.), *Neue Dimensionen der Mobilität. Technische und betriebswirtschaftliche Aspekte*, Wiesbaden:Springer, pp. 107–125
5. Götze, U., Jacobsen, B., Finke, H., Rother, S., & von Unwerth, T. (2019). Integrierte Geschäftsmodell-und Technologieentwicklung für Brennstoffzellen-systeme. In *Konferenzband der ersten FC³ Fuel Cell Conference Chemnitz 26. und 27. November 2019*, Chemnitz: Universitätsverlag Chemnitz

6. Rehme, M., Wehner, U., Rother, S., & Götze, U. (2021). Management von Wertschöpfungsnetzwerken multimodaler Mobilität. In Proff, H. (Ed.), *Making Connected Mobility Work. Technische und betriebswirtschaftliche Aspekte*, Wiesbaden: Springer, pp. 39–59

7. Cardeal, G., Höse, K., Ribeiro, I., & Götze, U. (2020). Sustainable business models–canvas for sustainability, evaluation method, and their application to additive manufacturing in aircraft maintenance. *Sustainability, 12*, 9130. https://doi.org/10.3390/su12219130

8. Osterwalder, A., & Pigneur, Y. (2010). Business model generation: A handbook for visionaries, game changers, and challengers. *Hoboken*, NJ: John Wiley & Sons. ISBN 9780470876411

9. Höse, K., Süß, A., & Götze, U. (2022). Sustainability-related strategic evaluation of business models. *Sustainability, 14*, 12:7285. https://doi.org/10.3390/su14127285

10. Turón, K. Open innovation business model as an opportunity to enhance the development of sustainable shared mobility industry. *Journal of Open Innovation: Technology, Market, and Complexity, 8*, 37. https://doi.org/10.3390/joitmc8010037

11. van den Berg, V., Meurs, H., & Verhoef, E. (2022). Business models for Mobility as an Service (MaaS). *In Transportation Research Part B, 157*, 203–229. https://doi.org/10.1016/j.trb.2022.02.004

12. Krauss, K., Moll, C., Köhler, J., & Axhausen, K. (2022). Designing mobility-as-a-service business models using morphological analysis. *Research in Transportation Business & Management*. https://doi.org/10.1016/j.rtbm.2022.100857

13. Remane, G., Hildebrandt, B., Hanelt, A., & Kolbe, L. (2016). Discovering new digital business model types - a study of technology startups from the mobility sector. In *Pacific Asia Conference on Information Systems. PACIS 2016—Proceedings*, ISBN 978-986049102-9

14. Born, B., Günther, M., Jacobsen, B., Jähn, B., Müller-Blumhagen, S., Temmler, A., & Teuscher, J. (2019). ECoMobility—Connected E-Mobility. *Vernetzte Elektromobilität am Beispiel der Technischen Universität Chemnitz", final report on the research project*, Chemnitz: Universitätsverlag der Technischen Universität Chemnitz, ISBN 978–3–96100–082–1. http://nbn-resolving.de/urn:nbn:de:bsz:ch1-qucosa2-327884

15. Apel, M., Dod, M., Leonhardt, V., Lindner, P., Müller, M., Ramm, S., Rehme, M., Richter, S., & Talatzko, P. (2016). Abschlussbericht Perspektiven der Vernetzten eMobilität (VEReMO) im Rahmen des Schaufensters. ELEKTROMOBILITÄT VERBINDET. Oberlungwitz: IVM Institut für Vernetzte Mobilität gGmbH, unpublished

16. Schulz, P., Wolf, A., Fettweis, G., Waswa, A., Soleymani, D., Mitschele-Thiel, A., Dudda, T., Dod, M., Rehme, M., Voigt, J., Riedel, I., Wankhede, T., Nitzold, W., & Almeroth, B. (2019). Network architectures for demanding 5G performance requirements: Tailored toward specific needs of efficiency and flexibility. *IEEE Vehicular Technology Magazine, 14*(2), 33–43.

17. Götze, U., & Proksch, T. (2022). Smart rail connectivity campus. WIR!-Zwischenevaluation. Chemnitz

The manufacturer's authorised representative in the EU is Springer
Nature Customer Service Centre GmbH, Europaplatz 3, 69115 Heidelberg,
Germany. If you have any concerns regarding our products, please
contact ProductSafety@springernature.com

Printed and bound by CPI Group (UK) Ltd, Croydon, CR0 4YY

23/04/2026

02095599-0001